ROUTLEDGE HANDBOOK OF RELIGION AND ECOLOGY

The moral values and interpretive systems of religions are crucially involved in how people imagine the challenges of sustainability and how societies mobilize to enhance ecosystem resilience and human well-being.

The *Routledge Handbook of Religion and Ecology* provides the most comprehensive and authoritative overview of the field. It encourages both appreciative and critical angles regarding religious traditions, communities, attitudes, and practices. It presents contrasting ways of thinking about "religion" and about "ecology" and how to connect the two terms. Written by a team of leading international experts, the *Handbook* discusses dynamics of change within religious traditions as well as their roles in responding to global challenges, such as climate change, water, conservation, food, and population. It explores interpretations of indigenous traditions regarding modern environmental problems, drawing on concepts such as lifeway and indigenous ecological knowledge. This volumes uniquely connects the field of religion and ecology with new directions in the environmental humanities and the environmental sciences.

This interdisciplinary volume is an essential reference for scholars and students across the social sciences and humanities and for all those looking to understand the significance of religion in environmental studies and policy.

Willis Jenkins is Associate Professor of Religion, Ethics, and Environment, University of Virginia, USA.

Mary Evelyn Tucker and **John Grim** are Directors and Founders of the Forum on Religion and Ecology 1998 to present, as well as Senior Lecturers and Research Scholars at the School of Forestry and Environmental Studies and at the Divinity School, Yale University, USA.

ROUTLEDGE HANDBOOK OF RELIGION AND ECOLOGY

Edited by Willis Jenkins, Mary Evelyn Tucker
and John Grim

LONDON AND NEW YORK

First published 2017 by Routledge

2 Park Square, Milton Park, Abingdon, Oxfordshire OX14 4RN
711 Third Avenue, New York, NY 10017

Routledge is an imprint of the Taylor & Francis Group, an informa business

First issued in paperback 2018

British Library Cataloguing-in-Publication Data
A catalogue record for this book is available from the British Library

Library of Congress Cataloging-in-Publication Data
A catalog record for this book has been requested

ISBN: 978-1-138-78957-9 (hbk)
ISBN: 978-1-138-31593-8 (pbk)

Typeset in Bembo
by FiSH Books Ltd, Enfield

CONTENTS

Contents

Contents

ILLUSTRATIONS

Figures

Table

CONTRIBUTORS

Peter Adriance, US Baha'i Office of Public Affairs
Peter Adriance has served as Representative for Sustainable Development in the U.S. Baha'i Office of Public Affairs since 1990. He has represented the Baha'i International Community and the Baha'is of the U.S. at numerous national and international fora on sustainable development issues. In 2009, he received the Interfaith Bridge Builder's Award from the Interfaith Conference of Metropolitan Washington DC "for his passionate commitment to inter-religious care for the earth." He helped found and serves on the governing board of the International Environment Forum—a Baha'i-inspired organization addressing the environment and sustainable development.

Miguel Angel Astor-Aguilera, Arizona State
Miguel Astor-Aguilera is an Arizona State University Associate Professor of Religious Studies whose scholarship concentrates on ethnography, material culture, archaeology, and socio-religious theory. He specializes in Mesoamerican cosmovisions and their historical traditions, that is, pre-Columbian, colonial, and contemporary. His work focuses on Maya ritual specialists in the Yucatán peninsula and their ecological environment. He conducts research focused on ecologically focused ritual production amongst the Maya peoples. Astor-Aguilera currently conducts ethnographic and archaeological investigations of Maya cenote-sinkholes in the Yucatán peninsula.

Zainal Bagir, Gadjah Mada University, Indonesia
Zainal Abidin Bagir is the Director of the Center for Religious and Cross-cultural Studies (CRCS), a Master's program at the Graduate School of Gadjah Mada University (GMU), Yogyakarta, Indonesia. At CRCS he teaches Religion, Science and Ecology, and Religion and Politics. His recent articles are published in *Zygon* and *Review of Faith and International Affairs*.

Subhankar Banerjee, Port Townsend, WA
Subhankar Banerjee is an artist, writer and environmental humanities scholar. He is Land Arts of the American West Endowed Chair and Professor of Art and Ecology in the Department of Art and Art History at the University of New Mexico. He is author of *Arctic National Wildlife Refuge: Seasons of Life and Land*, and editor of *Arctic Voices: Resistance at the Tipping Point*. His

photographs have been exhibited in more than fifty museums and galleries around the world; and essays in several anthologies and journals. Banerjee received a Greenleaf Artist Award from the UNEP and a Cultural Freedom Award from Lannan Foundation.

Sigurd Bergmann, Norwegian University of Science and Technology
Sigurd Bergmann is professor of Religious Studies at the Norwegian University of Science and Technology, Trondheim. Previous studies have investigated images of God and nature in late antiquity, contextual theology, indigenous arts in the Arctic, as well as visual arts, architecture, and religion. Ongoing projects investigate the relation of space/place and religion and "religion in climatic change." Bergmann initiated and chaired the *European Forum on the Study of Religion and Environment*. He is a member of the Royal Norwegian Society of Letters and Sciences and fellow at the Rachel Carson Center, München. He edits *Studies in Religion and the Environment* (LIT, Berlin).

J. Baird Callicott, University of North Texas, emeritus
J. Baird Callicott retired as University Distinguished Research Professor and formerly Regents Professor of Philosophy at the University of North Texas. He is co-Editor-in-Chief of the *Encyclopedia of Environmental Ethics and Philosophy* and author or editor of a score of books and author of dozens of journal articles, encyclopedia articles, and book chapters in environmental philosophy and ethics. Callicott is perhaps best known as the leading contemporary exponent of Aldo Leopold's land ethic and has elaborated an Aldo Leopold Earth ethic, *Thinking Like a Planet* (OUP, 2013), in response to global climate change. He taught the world's first course in environmental ethics in 1971 at the University of Wisconsin-Stevens Point.

Christopher Key Chapple, Loyola Marymount University
Christopher Key Chapple is Doshi Professor of Indic and Comparative Theology and Director of the Master of Arts in Yoga Studies at Loyola Marymount University. A specialist in the religions of India, he has published several books in the area of religion and ecology, including *Nonviolence to Animals, Earth, and Self in Asian Traditions* (SUNY, 1993), and edited volumes including *Hinduism and Ecology* (with Mary Evelyn Tucker, Harvard, 2000) and *Jainism and Ecology* (Harvard, 2002); *Yoga and Ecology* (Deepak Heritage, 2008). He edits the journal *Worldviews: Global Religions, Culture, and Ecology* (Brill).

Douglas E. Christie, Loyola Marymount University
Douglas E. Christie, Ph.D., is Professor of Theological Studies at Loyola Marymount University. He is author of *The Word in the Desert: Scripture and the Quest for Early Christian Monasticism* (Oxford) and *The Blue Sapphire of the Mind: Notes for a Contemplative Ecology* (Oxford). He is founding editor of the journal *Spiritus* and recently served as co-director of the Casa de la Mateada Program in Córdoba, Argentina.

Ernst M. Conradie, University of the West Cape, South Africa
Ernst Conradie is Senior Professor in the Department of Religion and Theology at the University of the Western Cape where he teaches systematic theology and ethics. He specializes in the field of Christian ecotheology and is the author of recent monographs such as *The Earth in God's Economy: Creation, Salvation and Consummation in Ecological Perspective* (LIT Verlag 2015). He is the leading editor (with Sigurd Bergmann, Celia Deane-Drummond and Denis Edwards) of *Christian Faith and the Earth: Current Paths and Emerging Horizons in Ecotheology*

(T&T Clark, 2014). He has won the Andrew Murray prize for theological publications in Afrikaans twice (in 2007 and 2013).

Arthur Dahl, International Environment Forum

Arthur Dahl is President of the International Environment Forum and a retired Deputy Assistant Executive Director of the United Nations Environment Programme (UNEP). He is an international consultant on indicators of sustainability, environmental assessment and observing strategies, coral reefs, biodiversity, islands, environmental education, and social and economic development. He organized the Pacific Regional Environment Programme and represented the Baha'i International Community at the UN Conference on the Human Environment in Stockholm in 1972, the Earth Summit in Rio in 1992, the World Summit on Sustainable Development (Johannesburg, 2002), and Rio+20 in 2012. He teaches in various university programs, lectures widely, and has published many scientific papers and books including *Unless and Until: A Baha'i Focus on the Environment.*

Heather Eaton, St Paul's University

Heather Eaton holds an interdisciplinary doctorate in theology, feminism, and ecology from Saint Michael's College at the University of Toronto and is a professor in Conflict Studies at Saint Paul University in Ottawa, Canada. She works in engaging religions on ecological, social, and ethical issues. She has published extensively on ecofeminism, ecospirituality, cosmology and ecojustice, as well as the intersection of science, evolution, and religion. Her many publications include: *The Intellectual Journey of Thomas Berry: Imagining the Earth Community,* ed. (2014); *Ecological Awareness: Exploring Religion, Ethics and Aesthetics,* (with Sigurd Bergmann, 2011); and *Introducing Ecofeminist Theologies* (2005). Heather works as a socially engaged academic with various national and international groups on religion, ecology, social issues, nonviolence, and peace.

Jessica J. Goddard, University of California Berkeley

Jessica J. Goddard teaches in the Energy and Resources Group, an interdisciplinary graduate program at University of California, Berkeley. She currently researches urban stormwater, drawing on science and technology studies and urban geography. Her background is in English literature and environmental sciences & engineering.

John Grim, Yale University

John Grim is a Senior Lecturer and Research Scholar at Yale University. He teaches courses in Native American and Indigenous religions and World religions and ecology. His published works include: *The Shaman: Patterns of Religious Healing Among the Ojibway Indians* (University of Oklahoma Press, 1983) and, with Mary Evelyn Tucker, a co-edited volume, *Worldviews and Ecology* (Orbis, 1994). With Mary Evelyn Tucker, he directed a 10 conference series and book project at Harvard on "*World Religions and Ecology,*" He edited *Indigenous Traditions and Ecology: The InterBeing of Cosmology and Community* (Harvard, 2001). He is co-founder and co-director of the Forum on Religion and Ecology at Yale. With Mary Evelyn Tucker he authored *Ecology and Religion* (Island Press, 2014).

Pengfei Guo, Minzu University of China and University of Minnesota

Pengfei Guo is a doctoral candidate in Sociology at the Minzu University of China and a visiting scholar in the Department of Sociology at the University of Minnesota. His main interests are the environmental justice studies, environmental sociology, and natural resources sociology.

David L. Haberman, Indiana University
David Haberman, Professor of Religious Studies at Indiana University, investigates the role that religion plays in the looming environmental crisis. His research identifies the contributions that religion makes to the causes as well as potential solutions to the environmental crisis by examining the manner in which religious worldviews shape human attitudes and behaviors toward the nonhuman world. While his work includes an interest in the relationship between ecology and religion worldwide, the great majority of his textual and ethnographic work has taken place within the Hindu traditions of India. He is working on a book entitled *Loving Stones* that is about the human conceptions and interactions with mountains and stones in Hindu India.

George B. Handley, Brigham Young University
George Handley teaches Interdisciplinary Humanities at Brigham Young University where he serves as Associate Dean in the College of Humanities. He has published widely in the areas of comparative literature of the Americas and environmental humanities, including a number of seminal essays on Mormonism and the environment, a co-edited book, *Stewardship and the Creation: LDS Perspectives on the Environment*, and an environmental memoir, *Home Waters: A Year of Recompenses on the Provo River*, which explores The Church of Jesus Christ of Latter-day Saint (LDS) theology, Mormon pioneer history, and contemporary environmental awareness in the American West. He serves on the boards of Utah Interfaith Power and Light and LDS Earth Stewardship and is involved in various other environmental fronts.

Melanie L. Harris, Texas Christian University
Dr. Melanie L. Harris is Associate Professor of Religion and Ethics and AddRan College of Liberal Arts Administration Fellow at Texas Christian University. A writer and scholar Dr. Harris is the author of *Gifts of Virtue: Alice Walker and Womanist Ethics and co-editor with Kate M. Ott of Faith, Feminism and Scholarship*. Her forthcoming book *Ecowomanism* is being published by Orbis Press. Dr. Harris research areas include African American Environmental History, Religious Social Ethics, Womanist Religious Thought, African American Literature, and Media and Religion. A former broadcast journalist, she has worked for ABC, CBS and NBC news affiliates.

Laura M. Hartman, University of Wisconsin-Oshkosh
Laura M. Hartman teaches environmental studies at the University of Wisconsin-Oshkosh. Her research interests exist at the intersection of Christian virtue ethics and environmental concerns. She has written about food, Sabbath keeping, environmental modesty, geoengineering, and transportation. Her first book, *The Christian Consumer: Living Faithfully in a Fragile World*, was published by Oxford University Press in 2011. She is also the editor of *Flourishing: Comparative Religious Environmental Ethics* (Oxford, forthcoming).

Graham Harvey, The Open University
Graham is Head of the Department of Religious Studies at the Open University. His publications include *Animism: Respecting the Living World* (2007) and *Contemporary Paganism* (2011). His interest in religion, location, and ecology and the cultures of indigenous peoples has led him to spend time with various generous and interesting hosts, including in Aotearoa, Australia, Hawaii, Newfoundland, Nigeria, the Ojibwe traditional territories, and Sápmi. Most of his research about contemporary indigenous religious traditions has been about "animism": the varied ways in which people engage with the larger than human world. His monograph, *Animism: Respecting the Living World* (2005), was followed by the edited *Handbook of*

Contemporary Animism (2013), a multidisciplinary engagement at the cutting edge of this new field.

Mānuka Hēnare, Auckland University

Mānuka is Associate Professor in Māori Business Development in the Department of Management and International Business at the University of Auckland. Mānuka is also the foundation Director of the Mira Szászy Research Centre for Māori and Pacific Economic Development. He teaches Māori business and economic history, strategy, and management of tribal enterprises. Mānuka has just completed a seven-year term as government appointee to the Council of Te Wānanga o Aotearoa, NZ's largest tertiary institution. He has advised government departments, local authorities, and other institutions on ambicultural or bicultural governance and management policies and also served on government advisory committees on development assistance, peace and disarmament, archives, history, social policy, andenvironmental risk.

Yong Huang, Chinese University of Hong Kong

Yong Huang teaches at the Chinese University of Hong Kong. His research focus has been on moral (both ethical and political) issues from an interdisciplinary and comparative perspective. Dr. Huang is co-chair of the Confucian Tradition Group of American Academy of Religion. He has also served as the President of Association of Chinese Philosophers in America (1999–2001). During this tenure, among other things, he inaugurated a book series and a journal, *Dao: A Journal of Comparative Philosophy*. Recently he has initiated a new book series, *Dao Companions to Chinese Philosophy*. Three collections of his essays have been published in Chinese by the National Taiwan University: *Ethics in a Global Age*; *Politics in a Global Age*; and *Religion in a Global Age*.

Mike Hulme, King's College, London

Mike Hulme is professor of climate and culture at King's College London. His works include: *Weathered: Cultures of Climate* (SAGE, 2016) and *Why We Disagree About Climate Change* (CUP, 2009), chosen by *The Economist* magazine as one of its science and technology books of the year. Since 2007 he has been the founding Editor-in-Chief of the review journal *Wiley Interdisciplinary Reviews (WIREs) Climate Change*. He served on the United Nations' Intergovernmental Panel on Climate Change (IPCC) and has advised the EU Commission, the UK Government, international agencies, and many private and third sector organizations about climate change.

Maria Ivanova, John McCormack School of Policy & Global Studies, University of Massachusetts—Boston

Maria Ivanova is an international relations and environmental policy scholar specializing in governance and sustainability. Her research focuses on international environmental institutions and their performance, the implementation of global environmental conventions, and the policy processes around the Sustainable Development Goals. She directs the Center for Governance and Sustainability, is a member of the Scientific Advisory Board of the UN Secretary-General, and a Board member of the UN University Institute for the Advanced Study of Sustainability (UNU-IAS) and of the Ecologic Institute in Berlin. In 2015, she was named an Andrew Carnegie Fellow. She edits the Governance and Sustainability Issue Briefs and the Global Leadership Dialogues series at UMass Boston.

Christopher Ives, Stonehill College
Christopher Ives teaches in the area of Asian Religions with a focus on modern Zen ethics. In 2009 he published *Imperial-Way Zen*, a book on Buddhist social ethics in light of Zen nationalism, especially as treated by Buddhist ethicist Ichikawa Hakugen (1902–86). Currently he is engaged in research on Zen approaches to nature and Buddhist environmental ethics. His other publications include *Zen Awakening and Society* (1992); *The Emptying God* (co-edited with John B. Cobb, Jr., 1990); and *Divine Emptiness and Historical Fullness* (edited volume, 1995). He serves on the editorial board of *The Journal of Buddhist Ethics*.

Willis Jenkins, University of Virginia
Willis Jenkins is Associate Professor of Religion, Ethics, and Environment at The University of Virginia and is author of two award-winning books: *Ecologies of Grace: Christian Theology and Environmental Ethics* (2008) and *The Future of Ethics: Sustainability, Social Justice, and Religious Creativity* (2013).

Frederic Laugrand, Laval University
Frederic Laugrand teaches in the department of Anthropology at Université Laval, Quebec, Canada. He has conducted research in many areas of the Canadian Arctic with a special interest in the reception of Christianity by the Inuit and shamanic traditions. His current research projects are conducted in partnership with several Aboriginal groups (Inuit and Innu in Quebec; Mangyan and Blaan in the Philippines) and relate to human/animal relationships, the intergenerational transmission of knowledge and practices, orality, space and territory, nomadism and forced relocations, residential schools, and governance and education models in Aboriginal communities. He is the former director of the Interuniversity Centre for Aboriginal Studies and Research (CIÉRA) and is director of the journal *Anthropologie et Sociétés*.

Thomas E. Lovejoy, George Mason University
Thomas E. Lovejoy is a professor at George Mason University. He works on the interface of science and environmental policy. In the 1970s he helped bring attention to the issue of tropical deforestation and in 1980 published the first estimate of global extinction rates (Global 2000 Report to the President). He coined the term "biological diversity" and founded the PBS series *Nature*. He has served in advisory and executive roles for the United Nations Foundation, World Bank, Smithsonian Institution, and World Wildlife Fund-US. He has served on advisory councils in the Reagan, George H.W. Bush, and Clinton administrations.

Najiyah Martiam, Gadjah Mada University, Indonesia
Najiyah Martiam is a member of staff of public education at the Center for Religious and Cross-cultural Studies (CRCS). She teaches sufism and mysticism at Islamic College of Sunan Pandanaran, Yogyakarta, Indonesia. Currently she is conducting research on Islam and Ecology in Indonesia for CRCS.

James Miller, Queens University
James Miller is professor of Chinese Studies at Queen's University in Kingston, Ontario, Canada, and Director of the Interdisciplinary Graduate Program in Cultural Studies. His research has focused mainly on traditional Chinese views of nature and environment, and he has published five books on this topic, with a sixth in preparation. His current research concentrates on the contemporary period and investigates: how the process of modernization and the ideology of modernity has transformed Chinese cultural views of both nature and religion;

how Chinese religions are changing as a result of climate change and the widespread sense of a global ecological crisis; and how Daoism can be understood as China's green religion.

Jesse N. K. Mugambi, University of Nairobi
Jesse N. K. Mugambi is Professor of Philosophy and Religious Studies, University of Nairobi. He has been a Resource Person in the Commission for Higher Education (now Commission for University Education) since 1988, focusing on Philosophy, Theology, Religious Studies and Applied Ethics. As a member of the Editorial Board for the *Cambridge Dictionary of Christianity* (2010) he contributed thirteen articles therein. His academic specializations include Philosophy, Theology and Religious Studies. For the World Council of Churches (WCC) he has served as Member of the Commission on Faith and Order; Working Group on Church and Society; Working Group on Climate Change; Working Group on Inter-religious Dialogue and Cooperation.

Nalini Nadkarni, University of Utah
Dr. Nadkarni's research interests include the ecological roles that canopy-dwelling plants play in forests at multiple spatial and temporal scales and the effects of forest fragmentation on community function. She is also interested in the development of database tools for canopy researchers; dissemination of research results to non-scientific audiences; and partnering of scientists and artists to enhance conservation of forests. She carries out research in Washington State and in Monteverde, Costa Rica, and is supported by the National Science Foundation, the National Geographic Society, the Mellon Foundation, and the Whitehall Foundation.

John Copeland Nagle, Notre Dame Law School
John Copeland Nagle is the John N. Matthews Professor at the University of Notre Dame. Nagle is the co-author of casebooks on "The Practice and Policy of Environmental Law" and "The Law of Biodiversity and Ecosystem Management. His book *Law's Environment: How the Law Shapes the Places We Live*, was published by Yale University Press. In 2002, he received a Distinguished Lectureship award from the J. William Fulbright Foreign Scholarship Board to teach at the Tsinghua University Law School in Beijing and another Fulbright award to teach at the University of Hong Kong in 2008. He also worked in the United States Department of Justice (attorney in the Office of Legal Counsel and a trial attorney conducting environmental litigation).

Melissa K. Nelson, San Francisco State
Melissa K. Nelson is an ecologist, writer, media-maker, and indigenous scholar-activist. She is an associate professor of American Indian Studies at San Francisco State University and president of The Cultural Conservancy, an indigenous rights organization. Melissa is Anishinaabe/Métis/Norwegian and an enrolled member of the Turtle Mountain Band of Chippewa Indians. Her work is dedicated to indigenous revitalization, environmental restoration, intercultural understanding, and the renewal and celebration of community health and cultural arts. Her first edited anthology, *Original Instructions: Indigenous Teachings For a Sustainable Future* (Inner Traditions, 2008), focuses on the persistence of traditional ecological knowledge by contemporary Native communities. Melissa was the co-producer of the award-winning documentary film, *The Salt Song Trail: Bringing Creation Back Together*.

Richard B. Norgaard, University of California, Berkeley, emeritus
Richard B. Norgaard is Professor Emeritus of Energy and Resources at University of

California, Berkeley. He holds a Ph.D. in economics from the University of Chicago. Among the founders of ecological economics, his research addresses how complexity challenges scientific understanding and policy; how ecologists and economists understand systems differently; and how globalization affects environmental governance. Norgaard was a lead author on the 5th Assessment of the Intergovernmental Committee on Climate Change and is currently a member of the Delta Independent Science Board addressing the adequacy of the science being used to resolve California's ongoing water crises.

Patricia Paladines, Stony Brook University.
Patricia Paladines works in the Institute for the Conservation of Tropical Environments at Stony Brook University.

David N. Pellow, University of California, Santa Barbara
David N. Pellow is the Dehlsen Professor of Environmental Studies and Director of the Global Environmental Justice Project at the University of California, Santa Barbara. He is coauthor of *The Slums of Aspen: Immigrants vs. the Environment in America's Eden* and author of *Resisting Global Toxics: Transnational Movements for Environmental Justice* and *Garbage Wars: The Struggle for Environmental Justice in Chicago.*

Christiana Z. Peppard, Fordham University
Christiana Z. Peppard is Assistant Professor of Theology, Science and Ethics in the Department of Theology at Fordham University, where she is affiliated faculty in Environmental Studies and American Studies. Professor Peppard is the author *of Just Water: Theology, Ethics, and the Global Water Crisis* (Orbis, 2014), and articles in *Journal of Environmental Studies and Sciences*; *Journal of Feminist Studies in Religion*; *Journal of Catholic Social Thought*; and *Journal of the Society for Christian Ethics*. Her research engages religious environmental ethics at the intersection of Catholic social teaching, ecological anthropology, natural law theory, and the scientific fields of hydrology and geology. Professor Peppard holds a Ph.D. in Ethics from Yale University, Department of Religious Studies.

Anna Peterson, University of Florida
Anna Peterson is professor in the Department of Religion at the University of Florida. She received her Ph.D. from the University of Chicago Divinity School and her AB from the University of California at Berkeley. Her research interests are in religion and social change, especially in Latin America; environmental and social ethics; and animal studies. She has published a number of articles, chapters, and books in these areas. She is working on two new book projects: one examining the role of practice in ethical theory; the other exploring the politics and ethics of companion animal rescue.

Carl Safina, State University of New York, Stony Brook
Carl Safina, an ecologist, is founder of The Safina Center at Stony Brook University, where he is the inaugural holder of the Endowed Chair for Nature and Humanity and steering committee co-chair of the Alan Alda Center for Communicating Science.

Jalel Sager, University of California, Berkeley
Jalel Sager is a lecturer and post-doctoral scholar at the Energy and Resources Group at the University of California-Berkeley. He teaches ecological economics and economic history; his research focuses on energy price fluctuations and long economic cycles. Jalel does field work

on renewable microgrid electricity systems, including their design, financing, and social/environmental impact.

Hava Tirosh-Samuelson, Arizona State University
Hava Tirosh-Samuelson is Irving and Miriam Lowe Professor of Modern Judaism and Director of Jewish Studies at Arizona State University. An intellectual historian, she writes on Jewish philosophy and mysticism, Judaism and ecology, and science, religion, and technology. She is the author of award- winning *Between Worlds: The Life and Thought of Rabbi David ben Judah Messer Leon* (SUNY Press, 1991) and *Happiness in Premodern Judaism: Virtue Knowledge and Well-Being* ((Monographs of the Hebrew Union College, 2003) and is the editor of several volumes including *Perfecting Human Futures: Technology, Secularization, and Eschatology* (2015). She is the editor-in-chief of the Library of Contemporary Jewish Philosophers (2013–2016), a series of 20 volumes that features influential living Jewish philosophers.

A. Whitney Sanford, University of Florida
Whitney Sanford teaches and researches in the areas of Religion and Nature and Religions of Asia. Her current work lies at the intersection of religion, food (and agriculture), and social equity, focusing on South Asia and the Americas. Her current book, *Being the Change: Food, Community and Sustainability,* explores contemporary intentional communities in the United States to discern how they translate values such as non-violence and voluntary simplicity into practice (University of Kentucky Press, 2017). *Growing Stories from India: Religion and the Fate of Agriculture* (University Press of Kentucky, 2012) uses Hindu agricultural narratives to consider how we can provide food in a sustainable and just manner. Her first book, *Singing Krishna: Sound Becomes Sight in Paramanand's Poetry* (SUNY, 2008), focuses on Braj devotional traditions and explores the role of devotional poetry in ritual practice.

Scott Slovic, University of Idaho, Moscow
Scott Slovic is professor of literature and environment and chair of the English Department at the University of Idaho. He served as founding president of the Association for the Study of Literature and Environment (ASLE), and since 1995 has edited the journal *ISLE: Interdisciplinary Studies in Literature and Environment*. His 23 books include *Seeking Awareness in American Nature Writing* (1992) and *Going Away to Think: Engagement, Retreat, and Ecocritical Responsibility* (2008).

Leslie E. Sponsel, University of Hawaii, emeritus
Leslie E. Sponsel has taught at seven universities in four countries, two as a Fulbright Fellow. He joined the Anthropology faculty at the University of Hawai'i in 1981 to develop and direct the Ecological Anthropology Program. Although retired as a Professor Emeritus in 2010, he still teaches one course a semester, including on Sacred Places, Spiritual Ecology, and Anthropology of Buddhism. The rest of his time is devoted to research and publications. Sponsel has published numerous journal articles, book chapters, and encyclopedia entries as well as four edited books. His recent monograph, *Spiritual Ecology: A Quiet Revolution*, won the science category of the Green Book Award in San Francisco in 2014. The companion website is: http://spiritualecology.info.

Mary Evelyn Tucker, Yale University
Mary Evelyn Tucker is co-director of the Forum on Religion and Ecology at Yale, where she also teaches. With John Grim she organized 10 conferences on World Religions and Ecology

at Harvard. They are the series editors for the 10 resulting volumes. She co-edited *Confucianism and Ecology, Buddhism and Ecology*, and *Hinduism and Ecology*. She authored, with John Grim, *Ecology and Religion* (Island Press, 2014) and edited Thomas Berry's books including *Selected Writings* (Orbis, 2014). With Brian Swimme she wrote *Journey of the Universe* (Yale, 2011), and is the executive director of the Emmy award-winning *Journey* film that aired on PBS. She served on the International Earth Charter Drafting Committee and was a member of the Earth Charter International Council.

Gretel van Wieren, Michigan State University

Van Wieren's courses and research focus on religion, ethics, and the environment. She is author of *Restored to Earth: Christianity, Environmental Ethics, and Ecological Restoration* (Georgetown University Press, 2013). Van Wieren is a participant in the Values Roundtable of the New Academy for Nature and Culture and the Blue River Quorum of the Spring Creek Project for Ideas, Nature, and the Written Word. She is a contributor to the Center for Humans and Nature City Creatures Blog and Planet Experts. Van Wieren is 2015 recipient of the spring Writers-in-Residency program at the H.J. Andrews Experimental Forest of Oregon State University.

Paul Waldau, Canisius College

Paul Waldau is an educator, scholar, and activist working at the intersection of animal studies, law, ethics, religion, and cultural studies. A Professor at Canisius College in Buffalo, New York, Paul is the Director of the Master of Science graduate program in Anthrozoology. Paul also taught Animal Law at Harvard Law School from 2002 to 2014. The former Director of the Center for Animals and Public Policy, Paul taught veterinary ethics and public policy at Tufts University School of Veterinary Medicine for more than a decade. His most recent volume is *Animal Studies – An Introduction* (Oxford University Press, 2013). He is also co-editor of *A Communion of Subjects: Animals in Religion, Science, and Ethics* (Columbia University Press, 2006).

Donald Worster, University of Kansas, emeritus

Donald Worster is currently a visiting foreign expert at Renmin University of China. Before retirement he taught history at the University of Kansas. He has held fellowships from the Guggenheim Foundation, the Australian National University, the National Endowment for the Humanities, the Mellon Foundation, and the American Council of Learned Societies. His book, *A Passion for Nature: The Life of John Muir*, was published by Oxford in 2008 and was named the best work of non-fiction by the Scottish Arts Council and won the Ambassador Award for Biography from the English Speaking Union. His earlier books have won more than dozen prizes. He is former president of the American Society for Environmental History.

Robert J. Wyman, Yale University

Robert J. Wyman is Professor of Molecular, Cellular, and Developmental Biology at Yale. His central research has been in neuro-genetics. In the field of religion, he has traced the idea of original sin as a determinant of behavior into the current debates on Nature and Nurture. Wyman believes that the earth's exploding population is the most important issue facing humanity as a major driver of both human and environmental misery. He has conducted research at the Nobel Institute, Stockholm Sweden, The Medical Research Council, Cambridge, England and the Biozentrum, Basel, Switzerland as well as the Scripps Oceanographic, Kerkhoff and Woods Hole Marine Laboratories. His publications have appeared widely in both biological and demographic journals.

Guigui Yao, Jianghan University

Guigui Yao is Professor and Director of Center for American Culture Studies, School of Foreign Languages, Jianghan University, Wuhan, China. She is currently on-leave at Yale University as an Associate Research Scholar in American Studies. She was previously a Senior Visiting Fellow at Yale under the sponsorship of the US–China Education Trust, where she studied the activities of the American Christian Right in attacking the Chinese One-Child Policy and the UN Fund for Population Activities, which was claimed to be complicit in Chinese human rights abuses. Her forthcoming volume is *American Agrarianism: Dream and Reality*.

Dan Smyer Yü, Yunnan Minzu University

Dan Smyer Yü, Professor and Founding Director of the Center for Trans-Himalayan Studies at Yunnan Minzu University, received his Ph.D. in anthropology from the University of California, Davis. He works in the areas of modern Tibetan Buddhism, ecology of sacred sites, emotionality of landscape, transboundary governance of natural heritages, religion and peace-building, and comparative studies of Eurasian secularisms. He is the author of *The Spread of Tibetan Buddhism in China: Charisma, Money, Enlightenment* (2011) and *Mindscaping the Landscape of Tibet: Place, Memorability, Eco-aesthetics* (2015). He is also an award-winning documentary film-maker.

ACKNOWLEDGEMENTS

The editors are indebted to Luke Kreider, who served as an exceptional Assistant Editor for this project, offering careful proofreading along with substantive feedback to authors. Many of the contributors to this volume also helped review other chapters. In addition, we are grateful to Whitney Bauman, Sam Mickey, Lisa Sideris, and Julia Watts-Belser for very helpful reviews of chapters. Tara Trapani prepared the contributor biographies.

PART I

Introducing religion and ecology

1

THE MOVEMENT OF RELIGION AND ECOLOGY

Emerging field and dynamic force

Mary Evelyn Tucker and John Grim

As many United Nations (UN) reports attest, we humans are destroying the life-support systems of the Earth at an alarming rate. Ecosystems are being degraded by rapid industrialization and relentless development. The data keeps pouring in that we are altering the climate and toxifying the air, water, and soil of the planet so that the health of humans and other species is at risk. Indeed, the Swedish scientist, Johan Rockstrom, and his colleagues, are examining which planetary boundaries are being exceeded (Rockstrom and Klum, 2015).

The explosion of population from 3 billion in 1960 to more then 7 billion currently and the subsequent demands on the natural world seem to be on an unsustainable course. The demands include meeting basic human needs of a majority of the world's people, but also feeding the insatiable desire for goods and comfort spread by the allure of materialism. The first is often called sustainable development; the second is unsustainable consumption. The challenge of rapid economic growth and consumption has brought on destabilizing climate change. This is coming into full focus in alarming ways including increased floods and hurricanes, droughts and famine, rising seas and warming oceans.

Can we turn our course to avert disaster? There are several indications that this may still be possible. On September 25, 2015 after the Pope addressed the UN General Assembly, 195 member states adopted the Sustainable Development Goals (SDGs). On December 12, 2015 these same members states endorsed the Paris Agreement on Climate Change. Both of these are important indications of potential reversal. The Climate Agreement emerged from the dedicated work of governments and civil society along with business partners. The leadership of UN Secretary General Ban Ki Moon and the Executive Secretary of the UN Framework Convention on Climate Change, Christiana Figueres, and many others was indispensable.

One of the inspirations for the Climate Agreement and for the adoption of the SDGs was the release of the Papal Encyclical, *Laudato Si'*, in June 2015. The encyclical encouraged the moral forces of concern for both the environment and people to be joined in "integral ecology". "The cry of the Earth and the cry of the poor" are now linked as was not fully visible before (Boff, 1997 and in the encyclical). Many religious and environmental communities are embracing this integrated perspective and will, no doubt, foster it going forward. The question is how can the world religions contribute more effectively to this renewed ethical momentum for change. For example, what will be their long-term response to population growth? As this is addressed in the article by Robert J. Wyman and Guigui Yao, we will not take it up here.

Instead, we will consider some of the challenges and possibilities amid the dream of progress and the lure of consumption.

Challenges: the dream of progress and the religion of consumption

Consumption appears to have become an ideology or quasi-religion, not only in the West but also around the world. Faith in economic growth drives both producers and consumers. The dream of progress is becoming a distorted one. This convergence of our unlimited demands with an unquestioned faith in economic progress raises questions about the roles of religions in encouraging, discouraging, or ignoring our dominant drive toward appropriately satisfying material needs or inappropriately indulging material desires. Integral ecology supports the former and critiques the latter.

Moreover, a consumerist ideology depends upon and simultaneously contributes to a world-view based on the instrumental rationality of the human. That is, the assumption for decision-making is that all choices are equally clear and measurable. Market-based metrics such as price, utility, or efficiency are dominant. This can result in utilitarian views of a forest as so much board feet or simply as a mechanistic complex of ecosystems that provide services to the human.

One long-term effect of this is that the individual human decision-maker is further distanced from nature because nature is reduced to measurable entities for profit or use. From this perspective we humans may be isolated in our perceived uniqueness as something apart from the biological web of life. In this context, humans do not seek identity and meaning in the numinous beauty of the world, nor do they experience themselves as dependent on a complex of life-supporting interactions of air, water, and soil. Rather, this logic sees humans as independent, rational decision-makers who find their meaning and identity in systems of management that now attempt to co-opt the language of conservation and environmental concern. Happiness is derived from simply creating and having more material goods. This perspective reflects a reading of our current geological period as human induced by our growth as a species that is now controlling the planet. This current era is being called the "Anthropocene" because of our effect on the planet in contrast to the prior 12,000 year epoch known as the Holocene.

This human capacity to imagine and implement a utilitarian-based worldview regarding nature has undermined many of the ancient insights of the world's religious and spiritual traditions. For example, some religions, attracted by the individualistic orientations of market rationalism and short-term benefits of social improvement, seized upon material accumulation as containing divine sanction. Thus, Max Weber identified the rise of Protestantism with an ethos of inspirited work and accumulated capital.

Weber also identified the growing disenchantment from the world of nature with the rise of global capitalism. Karl Marx recognized the "metabolic rift" in which human labor and nature become alienated from cycles of renewal. The earlier mystique of creation was lost. Wonder, beauty, and imagination as ways of knowing were gradually superseded by the analytical reductionism of modernity, such that technological and economic entrancement have become major inspirations of progress.

Challenges: religions fostering anthropocentrism

This modern, instrumental view of matter as primarily for human use arises in part from a dualistic Western philosophical view of mind and matter. Adapted into Jewish, Christian, and

Islamic religious perspectives, this dualism associates mind with the soul as a transcendent spiritual entity given sovereignty and dominion over matter. Mind is often valued primarily for its rationality in contrast to a lifeless world. At the same time we ensure our radical discontinuity from it.

Interestingly, views of the uniqueness of the human bring many traditional religious perspectives into sync with modern instrumental rationalism. In Western religious traditions, for example, the human is seen as an exclusively gifted creature with a transcendent soul that manifests the divine image and likeness. Consequently, this soul should be liberated from the material world. In many contemporary reductionist perspectives (philosophical and scientific) the human with rational mind and technical prowess stands as the pinnacle of evolution. Ironically, religions emphasizing the uniqueness of the human as the image of God meet market-driven applied science and technology precisely at this point of the special nature of the human to justify exploitation of the natural world. Anthropocentrism in various forms, religious, philosophical, scientific, and economic, has led, perhaps inadvertently, to the dominance of humans in this modern period, now called the Anthropocene. (It can be said that certain strands of the South Asian religions have emphasized the importance of humans escaping from nature into transcendent liberation. However, such forms of radical dualism are not central to the East Asian traditions or indigenous traditions.)

From the standpoint of rational analysis, many values embedded in religions, such as a sense of the sacred, the intrinsic value of place, the spiritual dimension of the human, moral concern for nature, and care for future generations, are incommensurate with an objectified monetized worldview as they not quantifiable. Thus, they are often ignored as externalities, or overridden by more pragmatic profit-driven considerations. Contemporary nation-states in league with transnational corporations have seized upon this individualistic, property-based, use-analysis to promote national sovereignty, security, and development exclusively for humans.

Possibilities: systems science

Yet, even within the realm of so-called scientific, rational thought, there is not a uniform approach. Resistance to the easy marriage of reductionist science and instrumental rationality comes from what is called systems science and new ecology. By this we refer to a movement within empirical, experimental science of exploring the interaction of nature and society as complex dynamic systems. This approach stresses both analysis and synthesis – the empirical act of observation – as well as placement of the focus of study within the context of a larger whole. Systems science resists the temptation to take the micro, empirical, reductive act as the complete description of a thing, but opens analysis to the large interactive web of life to which we belong, from ecosystems to the biosphere. There are numerous examples of this holistic perspective in various branches of ecology. And this includes overcoming the nature–human divide (Schmitz 2016). Aldo Leopold understood this holistic interconnection well when he wrote: "We abuse land because we see it as a commodity belonging to us. When we see land as a community to which we belong, we may begin to use it with love and respect." (Leopold, 1966)

Collaboration of science and religion

Within this inclusive framework, scientists have been moving for some time beyond simply distanced observations to engaged concern. The Pope's encyclical, *Laudato Si'*, has elevated the level of visibility and efficacy of this conversation between science and religion as perhaps never before on a global level. Similarly, many other statements from the world religions are linking the wellbeing of people and the planet for a flourishing future. For example, the World Council of Churches has been working for four decades to join humans and nature in their program on Justice, Peace, and the Integrity of Creation.

Many scientists such as Thomas Lovejoy, E. O. Wilson, Jane Lubchenco, Peter Raven, and Ursula Goodenough recognize the importance of religious and cultural values when discussing solutions to environmental challenges. Other scientists such as Paul Ehrlich and Donald Kennedy have called for major studies of human behavior and values in relation to environmental matters (Ehrlich and Kennedy, 2005). This has morphed into the Millennium Alliance for Humanity and the Biosphere (mahb.standford.edu). Since 2009 the Ecological Society of America has established an Earth Stewardship Initiative with yearly panels and publications. Many environmental studies programs are now seeking to incorporate these broader ethical and behavioral approaches into the curriculum.

Possibilities: extinction and religious response

The stakes are high, however, and the path toward limiting ourselves within planetary boundaries is not smooth. Scientists are now reporting that because of the population explosion, our consuming habits, and our market drive for resources, we are living in the midst of a mass extinction period. This period represents the largest loss of species since the extinction of the dinosaurs 65 million years ago when the Cenozoic period began. In other words, we are shutting down life systems on the planet and causing the end of this large-scale geological era with little awareness of what we are doing or its consequences.

As the cultural historian Thomas Berry observed some years ago, we are making macrophase changes on the planet with microphase wisdom. Indeed, some people worry that these rapid changes have outstripped the capacity of our religions, ethics, and spiritualities to meet the complex challenges we are facing.

The question arises whether the wisdom traditions of the human community, embedded in institutional religions and beyond, can embrace integral ecology at the level needed? Can the religions provide leadership into a synergistic era of human–Earth relations characterized by empathy, regeneration, and resilience? Or are religions themselves the wellspring of those exclusivist perspectives in which human societies disconnect themselves from other groups and from the natural world? Are religions caught in their own meditative promises of transcendent peace and redemptive bliss in paradisal abandon? Or does their drive for exclusive salvation or truth claims cause them to try to overcome or convert the Other?

Authors in this volume are exploring these topics within religious and spiritual communities regarding the appropriate responses of the human to our multiple environmental and social challenges. What forms of symbolic visioning and ethical imagining can call forth a transformation of consciousness and conscience for our Earth community? Can religions and spiritualities provide vision and inspiration for grounding and guiding mutually enhancing human–Earth relations? Have we arrived at a point where we realize that more scientific statistics on environmental problems, more legislation, policy or regulation, and more economic analysis, while necessary, are no longer sufficient for the large-scale social transformations needed? This is where the world religions, despite their limitations, surely have something to contribute.

Such a perspective includes ethics, practices, and spiritualities from the world's cultures that may or may not be connected with institutional forms of religion. Thus spiritual ecology and nature religions are an important part of the discussions and are represented in this volume. Our own efforts have focused on the world religions and indigenous traditions. Our decade-long training in graduate school and our years of living and traveling throughout Asia and the West gave us an early appreciation for religions as dynamic, diverse, living traditions. We are keenly aware of the multiple forms of syncretism and hybridization in the world religions and spiritualities. We have witnessed how they are far from monolithic or impervious to change in our travels to more than 60 countries.

Problems and promise of religions

Several qualifications regarding the various roles of religion should thus be noted. First, we do not wish to suggest here that any one religious tradition has a privileged ecological perspective. Rather, multiple interreligious perspectives may be the most helpful in identifying the contributions of the world religions to the flourishing of life.

We also acknowledge that there is frequently a disjunction between principles and practices: ecologically sensitive ideas in religions are not always evident in environmental practices in particular civilizations. Many civilizations have overused their environments, with or without religious sanction.

Finally, we are keenly aware that religions have all too frequently contributed to tensions and conflict among various groups, both historically and at present. Dogmatic rigidity, inflexible claims of truth, and misuse of institutional and communal power by religions have led to tragic consequences in many parts of the globe.

Nonetheless, while religions have often preserved traditional ways, they have also provoked social change. They can be limiting but also liberating in their outlooks. In the twentieth century, for example, religious leaders and theologians helped to give birth to progressive movements such as civil rights for minorities, social justice for the poor, and liberation for women. Although the world religions have been slow to respond to our current environmental crises, their moral authority and their institutional power may help effect a change in attitudes, practices, and public policies. Now the challenge is a broadening of their ethical perspectives.

Traditionally the religions developed ethics for homicide, suicide, and genocide. Currently they need to respond to biocide, ecocide, and geocide. (Berry, 2009)

Retrieval, reevaluation, reconstruction

There is an inevitable disjunction between the examination of historical religious traditions in all of their diversity and complexity and the application of teachings, ethics, or practices to contemporary situations. While religions have always been involved in meeting contemporary challenges over the centuries, it is clear that the global environmental crisis is larger and more complex than anything in recorded human history. Thus, a simple application of traditional ideas to contemporary problems is unlikely to be either possible or adequate. In order to address ecological problems properly, religious and spiritual leaders, laypersons, and academics have to be in dialogue with scientists, environmentalists, economists, businesspeople, politicians, and educators. Hence the articles in this volume are from various key sectors.

With these qualifications in mind we can then identify three methodological approaches that appear in the still emerging study of religion and ecology. These are retrieval, reevaluation, and reconstruction. Retrieval involves the scholarly investigation of scriptural and commentarial sources in order to clarify religious perspectives regarding human–Earth relations. This requires that historical and textual studies uncover resources latent within the tradition. In addition, retrieval can identify ethical codes and ritual customs of the tradition in order to discover how these teachings were put into practice. Traditional environmental knowledge (TEK) is an important part of this for all the world religions, especially indigenous traditions.

With reevaluation, traditional teachings are evaluated with regard to their relevance to contemporary circumstances. Are the ideas, teachings, or ethics present in these traditions appropriate for shaping more ecologically sensitive attitudes and sustainable practices? Reevaluation also questions ideas that may lead to inappropriate environmental practices. For example, are certain religious tendencies reflective of otherworldly or world-denying orientations that are not helpful in relation to pressing ecological issues? It asks as well whether the material world of nature has been devalued by a particular religion and whether a model of ethics focusing solely on human interactions is adequate to address environmental problems.

Finally, reconstruction suggests ways that religious traditions might adapt their teachings to current circumstances in new and creative ways. These may result in new syntheses or in creative modifications of traditional ideas and practices to suit modern modes of expression. This is the most challenging aspect of the emerging field of religion and ecology and requires sensitivity to who is speaking about a tradition in the process of reevaluation and reconstruction. Postcolonial critics have appropriately highlighted the complex issues surrounding the problem of who is representing or interpreting a religious tradition or even what constitutes that tradition. Nonetheless, practitioners and leaders of particular religions are finding grounds for creative dialogue with scholars of religions in these various phases of interpretation.

Religious ecologies and religious cosmologies

As part of the retrieval, reevaluation, and reconstruction of religions we would identify "religious ecologies" and "religious cosmologies" as ways that religions have functioned in the past and can still function at present. Religious ecologies are ways of orienting and grounding whereby humans undertake specific practices of nurturing and transforming self and community in a particular cosmological context that regards nature as inherently valuable. Through cosmological stories, humans narrate and experience the larger matrix of mystery in which life arises, unfolds, and flourishes. These are what we call religious cosmologies. These two, namely religious ecologies and religious cosmologies, can be distinguished but not separated. Together they provide a context for navigating life's challenges and affirming the rich spiritual value of human–Earth relations.

Human communities until the modern period sensed themselves as grounded in and dependent on the natural world. Thus, even when the forces of nature were overwhelming, the regenerative capacity of the natural world opened a way forward. Humans experienced the processes of the natural world as interrelated, both practically and symbolically. These understandings were expressed in TEK, namely, in hunting and agricultural practices, such as the appropriate use of plants, animals, and land. Such knowledge was integrated in symbolic language and practical norms, such as prohibitions, taboos, and limitations on ecosystems' usage. All this was based in an understanding of nature as the source of nurturance and kinship. The Lakota people still speak of "all my relations" as an expression of this kinship. Such perspectives will need to be incorporated into strategies to solve environmental problems. Humans are part of nature and their cultural and religious values are critical dimensions of the discussion.

Multidisciplinary approaches: environmental humanities

We are recognizing, then, that the environmental crisis is multifaceted and requires multidisciplinary approaches. As this book indicates, the insights of scientific modes of analytical and synthetic knowing are indispensable for understanding and responding to our contemporary environmental crisis. So also, we need new technologies, such as industrial ecology, green chemistry, and renewable energy. Clearly ecological economics is critical along with green governance and legal policies as articles in this volume illustrate.

In this context it is important to recognize different ways of knowing that are manifest in the humanities, such as artistic expressions, historical perspectives, philosophical inquiry, and religious understandings. These honor emotional intelligence, affective insight, ethical valuing, and spiritual awakening.

Environmental humanities is a growing and diverse area of study within humanistic disciplines. In the last several decades, new academic courses and programs, research journals, and monographs, have blossomed. This broad-based inquiry has sparked creative investigation into multiple ways, historically and at present, of understanding and interacting with nature, constructing cultures, developing communities, raising food, and exchanging goods.

It is helpful to see the field of religion and ecology as part of this larger emergence of environmental humanities. While it can be said that environmental history, literature, and philosophy are some four decades old, the field of religion and ecology began some two decades ago. It was preceded, however, by work among various scholars, particularly Christian theologians. Some ecofeminists theologians, such as Rosemary Ruether and Sallie McFague, Mary Daly, and Ivone Gebara led the way.

The emerging field of religion and ecology

An effort to identify and to map religiously diverse attitudes and practices toward nature was the focus of a three-year international conference series on world religions and ecology. Organized by us with other religious scholars, ten conferences were held at the Harvard Center for the Study of World Religions from 1996 to 1998 that resulted in a ten volume book series (1997–2004). Over 800 scholars of religion and environmentalists participated. The Director of the Center, Larry Sullivan, gave space and staff for the conferences. He chose to limit their scope to the world religions and indigenous religions rather than "nature religions", such as wicca or paganism, which the organizers had hoped to include.

Culminating conferences were held in Fall 1998 at Harvard and in New York at the UN and the American Museum of Natural History where 1,000 people attended and Bill Moyers presided. At the UN conference we founded the Forum on Religion and Ecology, which is now located at Yale. We organized a dozen more conferences and created an electronic newsletter that is now sent to over 12,000 people around the world. In addition, we developed a major website for research, education, and outreach in this area (fore.yale.edu). The conferences, books, website, and newsletter have assisted in the emergence of a new field of study in religion and ecology. Many people have helped in this process including Whitney Bauman and Sam Mickey who are now moving the field toward discussing the need for planetary ethics. A Canadian Forum on Religion and Ecology was established in 2002, a European Forum for the Study of Religion and the Environment was formed in 2005, and a Forum on Religion and Ecology @ Monash in Australia in 2011.

Courses on this topic are now offered in numerous colleges and universities across North America and in other parts of the world. A Green Seminary Initiative has arisen to help educate seminarians. Within the American Academy of Religion there is a vibrant group focused on scholarship and teaching in this area. A peer-reviewed journal, *Worldviews: Global Religions, Culture, and Ecology*, is celebrating its twenty-fifth year of publication. Another journal has been publishing since 2007, the *Journal for the Study of Religion, Nature, and Culture*. A two volume *Encyclopedia of Religion and Nature*, edited by Bron Taylor, has helped shape the discussions, as has the International Society for the Study of Religion, Nature and Culture he founded. Clearly this broad field of study will continue to expand as the environmental crisis grows in complexity and requires increasingly creative interdisciplinary responses.

The work in religion and ecology rests in an intersection between the academic field within education and the dynamic force within society. This is why we see our work not so much as activist, but rather as "engaged scholarship" for the flourishing of our shared planetary life. This is part of a broader integration taking place to link concerns for both people and the planet. This has been fostered in part by the twenty-volume Ecology and Justice Series from Orbis Books and with the work of John Cobb, Larry Rasmussen, Dieter Hessel, Heather Eaton, Cynthia Moe-Loebeda, and others. The Papal Encyclical is now highlighting this linkage of ecojustice as indispensable for an integral ecology.

The dynamic force of religious environmentalism

All of these religious traditions, then, are groping to find the languages, symbols, rituals, and ethics for sustaining both ecosystems and humans. Clearly there are obstacles to religions moving into their ecological, ecojustice, and planetary phases. The religions are themselves challenged by their own bilingual languages, namely, their languages of transcendence, enlightenment, and salvation; and their languages of immanence, sacredness of Earth, and respect for nature. Yet, as the field of religion and ecology has developed within academia, so has the force of religious environmentalism emerged around the planet. Roger Gottlieb documents this in his book *A Greener Faith* (Gottlieb 2006). The Ecumenical Patriarch Bartholomew held international symposia on "Religion, Science and the Environment" focused on water issues (1995–2009) that we attended. He has made influential statements on this subject for 20 years. The Parliament of World Religions has included panels on this topic since 1998 and most expansively in 2015. Since 1995 the UK-based Alliance of Religion and Conservation (ARC), led by Martin Palmer, has been doing significant work with religious communities under the patronage of Prince Philip.

These efforts are recovering a sense of place, which is especially clear in the environmental resilience and regeneration practices of indigenous peoples. It is also evident in valuing the sacred pilgrimage places in the Abrahamic traditions (Jerusalem, Rome, and Mecca) both historically and now ecologically. So also the attention to sacred mountains, caves, and other pilgrimage sites stands in marked contrast to massive pollution in East Asia and South Asia.

In many settings around the world, religious practitioners are drawing together religious ways of respecting place, land, and life with understanding of environmental science and the needs of local communities. There have been official letters by Catholic Bishops in the Philippines and in Alberta, Canada alarmed by the oppressive social conditions and ecological disasters caused by extractive industries. Catholic nuns and laity in North America, Australia, England, and Ireland sponsor educational programs and conservation plans drawing on the ecospiritual vision of Thomas Berry and Brian Swimme. Also inspired by Berry and Swimme, Paul Winter's Solstice celebrations and Earth Mass at the Cathedral of St. John the Divine in New York city have been taking place for three decades.

Even in the industrial growth that grips China, there are calls from many in politics, academia, and NGOs to draw on Confucian, Daoist, and Buddhist perspectives for environmental change. In 2008 we met with Pan Yue, the Deputy Minister of the Environment, who has studied these traditions and sees them as critical to Chinese environmental ethics. In India, Hinduism is faced with the challenge of clean up of sacred rivers, such as the Ganges and the

Yamuna. To this end in 2010 with Hindu scholars, David Haberman and Christopher Chapple, we organized a conference of scientists and religious leaders in Delhi and Vrindavan to address the pollution of the Yamuna.

Many religious groups are focused on climate change and energy issues. For example, InterFaith Power and Light and GreenFaith are encouraging religious communities to reduce their carbon footprint. Earth Ministry in Seattle is leading protests against oil pipelines and terminals. The Evangelical Environmental Network and other denominations are emphasizing climate change as a moral issue that is disproportionately affecting the poor. In Canada and the US the Indigenous Environmental Network is speaking out regarding damage caused by resource extraction, pipelines, and dumping on First Peoples' Reserves and beyond. All of the religions now have statements on climate change as a moral issue and they were strongly represented in the People's Climate March in September 2014. *Daedalus,* the journal of the American Academy of Arts and Sciences, published the first collection of articles on religion and climate change from two conferences we organized there (Tucker and Grim, 2001).

Striking examples of religion and ecology have occurred in the Islamic world. In June 2001, May 2005, and April 2016 the Islamic Republic of Iran led by Presidents Khatami and Rouhani and the UN Environment Programme sponsored conferences in Tehran that we attended. They were focused on Islamic principles and practices for environmental protection. The Iranian Constitution identifies Islamic values for ecology and threatens legal sanctions. One of the earliest spokespersons for religion and ecology is the Iranian scholar, Seyyed Hossein Nasr. Fazlun Khalid in the UK founded the Islamic Foundation for Ecology and Environmental Science. In Indonesia in 2014 a *fatwa* was issued declaring that killing an endangered species is prohibited.

These examples illustrate ways in which an emerging alliance of religion and ecology is occurring around the planet. These traditional values within the religions now cause them to awaken to environmental crises in ways that are strikingly different from science or policy. But they may find interdisciplinary ground for dialogue in concerns for ecojustice, sustainability, and cultural motivations for transformation. The difficulty, of course, is that the religions are often preoccupied with narrow sectarian interests. However, many people, including the Pope, are calling on the religions to go beyond these interests and become a moral leaven for change.

Renewal through *Laudato Si'*

Pope Francis is highlighting an integral ecology that brings together concern for humans and the Earth. He makes it clear that the environment can no longer be seen as only an issue for scientific experts, or environmental groups, or government agencies alone. Rather, he invites all people, programs and institutions to realize these are complicated environmental and social problems that require integrated solutions beyond a "technocratic paradigm" that values an easy fix. Within this integrated framework, he urges bold new solutions.

In this context Francis suggests that ecology, economics, and equity are intertwined. Healthy ecosystems depend on a just economy that results in equity. Endangering ecosystems with an exploitative economic system is causing immense human suffering and inequity. In particular, the poor and most vulnerable are threatened by climate change, although they are not the major cause of the climate problem. He acknowledges the need for believers and nonbelievers alike to help renew the vitality of Earth's ecosystems and expand systemic efforts for equity.

In short, he is calling for "ecological conversion" from within all the world religions. He is making visible an emerging worldwide phenomenon of the force of religious environmentalism on the ground, as well as the field of religion and ecology in academia developing new ecotheologies and ecojustice ethics. This diverse movement is evoking a change of mind and heart, consciousness, and conscience. Its expression will be seen more fully in the years to come.

Conclusion

The challenge of the contemporary call for ecological renewal cannot be ignored by the religions. Nor can it be answered simply from out of doctrine, dogma, scripture, devotion, ritual, belief, or prayer. It cannot be addressed by any of these well-trodden paths of religious expression alone. Yet, like so much of our human cultures and institutions the religions are necessary for our way forward yet not sufficient in themselves for the transformation needed. The roles of the religions cannot be exported from outside their horizons. Thus, the individual religions must explain and transform themselves if they are willing to enter into this period of environmental engagement that is upon us. If the religions can participate in this creativity they may again empower humans to embrace values that sustain life and contribute to a vibrant Earth community.

References

Berry, Thomas. 2009. *The Sacred Universe: Earth Spirituality and Religion in the 21st Century.* (New York: Columbia University Press).

Boff, Leonardo. 1997. *Cry of the Earth, Cry of the Poor.* (Maryknoll, NY: Orbis Books).

Ehrlich, Paul and Donald Kennedy. 2005. "Millennium Assessment of Human Behavior," *Science,* 22 July, 562–3

Gottlieb, Roger. 2006. *A Greener Faith: Religious Environmentalism and Our Planetary Future.* (Oxford: Oxford University Press).

Grim, John and Mary Evelyn Tucker. eds. 2014. *Ecology and Religion.* (Washington, DC: Island Press).

Leopold, Aldo. 1966. *A Sand County Almanac.* (Oxford: Oxford University Press).

Rockstrom, Johan and Mattias Klum. 2015. *Big World, Small Planet: Abundance Within Planetary Boundaries.* (New Haven: Yale University Press).

Schmitz, Oswald. 2016. *The New Ecology: Science for a Sustainable World.* (Princeton: Princeton University Press).

Taylor, Bron. ed. 2008. *Encyclopedia of Religion, Nature, and Culture.* (London: Bloomsbury).

Tucker, Mary Evelyn. 2004. *Worldly Wonder: Religions Enter their Ecological Phase.* (Chicago: Open Court).

Tucker, Mary Evelyn and John Grim. eds. 2001. "Religion and Ecology: Can the Climate Change?" *Daedalus,* Vol. 130, No.4.

2

DEVELOPMENTS IN RELIGION AND ECOLOGY

Sigurd Bergmann

The research field of "religion and ecology," even termed as "religion and the environment" or "religion, nature and culture," behaves, due to its short but dynamic history, like a child still finding its feet. It can take the hands of its parents, theology and religious studies, and find support among older siblings such as philosophy, history, anthropology, biology, and others. Asymmetries, unbalances, and tumblings are natural, as are the joys of moving, seeing with different eyes and harvesting first fruits. Nevertheless, spreading one's wings requires balance: between employing established theories and methods and forging new unproved ones in other lands. Given this open and fresh context, this chapter will not map the whole but focus on some selected creative developments in what emerges as a new and flourishing research landscape.

As the notion of "nature" is essential for the self-understanding of the whole Western civilization, also religions have in their long history contributed to the development of the concept of nature. "Nature" in the three Abrahamic religions is interpreted as "creation," which exists out of its relation to God. "Nature" is less central in African and Asian cultures, where "Life" and "Earth" play more important roles. "Land" is the analogous category in indigenous traditions and other spiritualities that grow out of and within relations to specific bioregional spaces.

Beliefs in Creation, Life, and Land are changing as "nature" turns into "the environment" – that is, as nature is affected radically by human social and technical activities. The distribution of Ernst Haeckel's concept of ecology, the emergence of environmental science, and the world-views and values of environmentalism within social movements catalyze this process even more. Religions have faced and responded in many ways to the environmental challenge. The emergence and rapid dynamic development of our field mirrors this change in the interconnected concepts of life/nature/land and the Sacred. Therefore one can ask if the change from nature to environment offers such a deep and common challenge to all world inhabitants and believers of all kinds that we at present are moving over a threshold towards a common planetary, global, though still locally differentiated, world religion, where the differences between life, nature, and land-based belief systems merge into one common colorful earth religion. In my view it is still too early to formulate such a statement but the idea of a terrapolitan belief system (Deudney 1998) might nevertheless serve as a useful working hypothesis. At the same time religion might be analyzed as a human construct that functions not only constructively but also destructively. This deep ambiguity impacts the more-than-human life worlds in our "Mit-Welt" (co-world). Both sides – the pathology of religion, which attracts many younger scholars

13

today, as well as its liberative force – must be indissolubly connected to each other. This is important not least because our theories and methods are not simply reflections *on* but implicit parts *of* life taking place.

Religions offer substantial cultural skills.[1] Beside the skills of meaning making, ritualizing, mapping, and tracing, religion enables the human activity of "making-oneself-at-home." It locates believers in a world and at a place that is inhabited by the Divine (cf. Tuan 2009, 70). Humans do not land on Earth as travelling strangers; our history is fully entangled with the evolution of material, bodily life on Earth. Humans, including believers and scholars, are earth-lings. Religious practices therefore certainly "reflect the natural environments and ways of life in which they emerged" (Buttimer 2006, 200). Natural environments embed, carry and nurture human life and thereby also faith. Faith, religion, belief, and spirituality appear in such a view as deeply natural forces. Even "thinking is a process of nature" (Picht 1989, 12). Analyses of religion, therefore, must respect not only the subjective, sociocultural, and historical dimensions of religious traditions, but also the ecological functions of faith.

What follows is a preliminary discussion of how three phenomena are driving novel devel-opments in the field: climate change, technology, and space/place. All have in common the capacity to crisscross established academic discourses. Expressions of faith appear in new terri-tories and symbolic systems, and the strong transgressing capacity of religion becomes manifest.

Climate change

Global climate change represents one of the most demanding challenges facing humanity in this century and it has provoked different responses. Climatology represents one of the most successful transdisciplinary developments of recent decades. Nevertheless, current discussions about mitigation and adaptation to climate change are dominated by propositions for techno-logical and economic solutions. However ecologically informed, they are largely shaped by the limits of mechanistic and economy-oriented worldviews. Instead, we need a deeper under-standing of the cultural dimensions of anthropogenic environmental change (Hulme et al. 2009) – to which the study of religion and ecology makes substantial contributions.

Several Christian theologians have in the last years offered exciting reflections about how climate change affects faith, ethics, and the image of God (Northcott 2007, Conradie 2008, McFague 2008, Bloomquist 2009, Primavesi 2009, Northcott and Scott 2014). In particular, Michael Northcott's plea for a new political theology in this context should be taken seriously (Northcott 2013). Hydroclimatologist Dieter Gerten and I have initiated a process of research-ing religion *in* climate change (Bergmann and Gerten 2010, Gerten and Bergmann 2011), and recently Robin Globus Veldman et al. have published a collection of essays on religious responses to climate change (2014). An increasing number of studies in social anthropology also offer needed insights into the human, spiritual and cultural dimensions of anthropogenic climate change (see Crate and Nuttall 2009).

Globus Veldman et al. consider four factors that "help religions in general engage with climate change" (Globus Veldman et al. 2014, 309–313). First, religions are exerting an influ-ence on believers' *worldviews*, which can be in harmony and even in conflict with other cultural and political influences. The influence by worldviews remains ambivalent as it either can strongly motivate climate activism or encourage quietism and denial of on-going change. Second, the social scientists emphasize that many people are reached and affected by the *moral authority* wielded by religions. Religious arguments for climate justice are clearly growing and also interfaith collaborations are fertilized by the global change challenge. Arguably, in Europe and North America, the strong exchange of religious and secular environmental social movements raises the question to what degree new global ecological morality is emerging within a *global citizenship*. Third, religions' *institutional and economic resources* are important. Education, access to transnational networks, leadership, and also ownership of land and capital represent important resources that should not be underestimated as roughly 90% of the world population identifies as belonging to religious traditions. Furthermore, religions provide *social connectivity* and *collective action*. Common faith can be an important form of social cohesion, and the overlappings of religious communities and the civil society are many. Religion, for me, works as a sociocultural driving force that enhances and deepens communicative and communitarian skills and processes. Sometimes religions can mobilize transformative countervailing power with regard to existing power constellations. While social science and also economistic worldviews tend to reduce and fragmentize the human and spiritual dimension, the scholar of religion and the environment needs to constantly struggle to keep the perspective open.

I would highlight four further dimensions, drawing attention not only to how climate change impacts religion, but also to the ways religion can make a difference in the Anthropocene (Bergmann 2009b). For one, religions' responses to the production of suffering and violence offer *passiological* skills highly relevant in times where anthropogenic impacts on global and local life worlds produce radically new modes of suffering while reinforcing conventional ones. For example, the historical event of God's own crucifixion in Christianity now also functions as global metaphor. Who is crucified? What in fact is the Cross? Where does the anthropogenic *Via Dolorosa* lead? Religions also provide a diversity of non-verbal cultural skills such as built environments, rituals, topographies, memorials, images, music, and drama, as well as gardening, weather belief, arts of cooking, and social care. This "*aesth/ethical*" dimension allows a synergy of empathic forces where our bodily lives are connected to a new constructive imagination of our place and role in the world. Furthermore, the culture of *money*, as Georg Simmel has called it, represents a crucial force and driver of climatic change. In many ways, the invented abstraction of money underpins the colonization of life worlds in all scales of life in the Anthropocene. Religions have in their long history of human culture continuously developed antidotes to the misuse of money. Finally, a general insight in the context of the Anthropocene is that religions seem to accelerate their *spatial turn* (Bergmann 2007). Not only "our common future" as the Brundtland process has demanded but also "Earth, our common home" has become a central theme for the religious construction of meaning about the world amidst dangerous climatic and environmental change. For me, one of the central analytic questions herein is how religions provide support for *making oneself at home* in such a world (Bergmann 2014).

Technofutures

Although the highly advanced technical skills of human beings impact the environment, the field of technology has all too long been left only to engineers and specialists. Fortunately that is changing, and today scholars in many disciplines ask questions about the ethics of technology and its cultural, social, political, economic, and also ecological significance. Nevertheless, the entanglement between religion, nature and technology appears still to be a lacuna where one might expect a more intense research activity in the near future.

When the absence of thinking about technology's sociocultural and ecological implications is characteristic for the initial phase of technological development, ethical problems that arise in its application can only be discussed in a very limited way afterwards. Democratic principles of participatory decision-making are regularly set aside in the sphere of engineering.

An interesting exception is taking place at present in the field of climate change where geoengineering has been forwarded as a central solution. Leading scholars have already in the initial phase asked critical questions about their own activity and coined the notion of the Anthropocene (Steffen et al. 2007) in order to establish an open public discourse about how humans could and should impact all spheres of the world. Interestingly enough, they also encourage faith communities to be involved as central interlocutors as the challenge of global engineering the climate and more raises unprecedented questions about the scale of humanity's technically aided impact on the planet (Lawrence 2015).

Other scholars have approached the challenge by understanding our late modern state of being as postnatural – that is, a state where prior ancient assumptions about the natural world are no longer secure. At the same time, the postnatural does not negate our dependence on the natural world. Still others have shown how implicit religious driving forces impact on technological practices (Deane-Drummond, Bergmann and Szerszynski 2015) and how an understanding of the Sacred is always at work implicitly in concepts and practices of technology (Szerszynski 2005, 159ff). Technology furthermore often includes strong claims about salvation, sometimes even identifying itself as a tool of salvation for humanity, an optimistic view that is hard to agree with today.

The false notion that technology is value-neutral is often used to obscure the deeply problematical implications of so-called technical "innovation," and to safely quarter engineers and economic interests within a supposedly subpolitical sphere. Instead, many technical artifacts are a physical outcome of complex social processes involving the production and sharing of power among humans as well as between human and non-human life forms. The invention and production of technical artifacts takes place in the triangle of natural/environmental, sociocultural, and human-subjective dimensions, where each impacts the others.

From a different angle, one can regard machines and high-tech systems as animated artifacts for human survival. Dead things are made alive. Machines can impact other life forms. The discourse about animism and neo-animism, where all beings, things, and places are regarded as inspirited entities, is therefore significant for the discourse about technology and the environment. Furthermore the notion of fetishization allows a fresh and exciting new understanding of technology. Technical artifacts and also the abstract construction of money value offer examples of how a fetish is, in a deeply spiritual process, loaded with meaning and power to affect other beings. For Karl Marx, fetishism was "the religion of sensuous appetites," (Marx and Engels 1982, 22) and the Marxist analysis of technology as driver of alienation and commodification offers insights not only on the spiritual, social, and ecological power executed by the machines but it also explains how religious belief systems are affected and threatened by the colonization of life worlds through intricate technical systems. Expanding the concept of fetishism, one can identify processes of technocratization in many spheres of social life, where everything is believed to be possible.

Faith communities face challenges in such a context because religions perceive and receive life – including human capacities for technology – as a gift. The ability to do everything and the "technological imperative" where one has to do whatever one is able to do seems to be radically opposed to the foundational attitude towards life in religion. The commodification of life through fetishized artifacts and money appears contradictory to the belief in life as a gift. There are many religious traditions that allow an alternative understanding of life and also environmental engineering. Christian believers might remind themselves about the life-giving Holy Spirit who penetrates all life forms from within. Indigenous people might regard the land itself as the power of life, from where also the power of artifacts springs so that the human use of artifacts must take place in harmony with, and not against, the spirits of the land. New green spiritualities might depart from a general understanding of all life forms and places as inspirited and instead ask for alternative ecological practices characterized by respect and dignity. Undoubtedly the practices of engineering and the intrinsic power of technical artifacts represent one of the central conditions for modern life and also one of the most central threats to its sustainability. One might wonder if the future will offer a dramatic change of our understanding and practice with regard to technology development and what role religious believers might play in this.

The spatiality of faith

One of the driving forces behind civilization is the development of city space that began more than 10,000 years ago (Soja 2000, 35). The accelerating process of urbanization is now turning the whole planet into one single "postmetropolis." A majority of the world population now lives in urban areas. This affects the reshaping of landscapes and regions all over and it deeply affects the development of religious processes.

Following philosopher Henri Lefebvre, Soja has made a theoretically useful distinction between types of space: physical, imagined, and lived space (Soja 1996). The concept of "lived religion," as it has been developed for the phenomenological study of religious practices, cooperates well with this concept and it offers an exciting tool to analyze the spatiality of faith (Bergmann 2008). How is religion at work within built environments, and how are natural and built environments impacting belief systems?

A general insight in the context of the Anthropocene that has been catalyzed through the accelerating dynamics of dangerous anthropogenic change of climatic, water, and land systems is that faith communities need to accelerate their *spatial turn*. While Western thinkers in the twentieth century have mainly depicted their culture in terms of time and history the environmental challenge clearly turns our focus to the spatiality of life.

The all-embracing space for human life represents for all religions one common gift of life. It reveals its glory in the complexity and interconnectedness of life systems in one single planetary space for all (Primavesi 2009). "Earth is our home," the Earth Charter succinctly states (Earth Charter Commission 2000). The statement sounds simple but it summarizes a deep wisdom that has been guarded by religions for many ages. "The Earth is the Lord's," it sounds in the Hebrew Bible (Exodus 9:29, Psalm 24:1), and Paul is very clear in his letter to the Romans that the earth and humans as God's icons are interconnected in a communion of suffering and "hope for liberation" (Rom. 8.21) In Islam, Allah created humans as guardians of nature, within the concept of trusteeship (Khalifa), and the unity of Allah (Tawheed) and of humanity and nature is emphasized as a central force. Buddhism rejects the illusion of separateness and emphasizes the interconnectedness of all beings. Many more references can be approached in the rich literature about religions and ecology and also through the Alliance of Religions and Conservation (ARC), which has been discussed in a previous chapter.

Scholars of religion and theology have developed multiple approaches to connect the discourse about space and place and the discourse about religion in general and religion and the environment in particular. Stanley D. Brunn has in an extensive publication recently shown how the world religion map is globally stated in a process of change (Brunn 2014). For Kim Knott, the notion of locality offers an important tool (Knott 2005), while others prefer to begin with the understanding of place (Inge 2003). Some have started to depict "geographies of religion" (Ivakhiv 2006, Kong 2010), while others prefer to talk about "sacred lands" (Park 1994). Furthermore themes of mobility have become interesting, where for example pilgrimage and tourism offers creative spatial expressions of faith (Bergmann and Sager 2008, Stausberg 2011).

In Christian theology the "spatial turn" will undoubtedly accelerate due to the demanding

17

experiences of change in a common planetary space. The challenge to renew faith traditions hereby is to explore how the life-giving and all-embracing space of the Creator is a gift to his/her creatures. At the same time, believers inhabit a global space where risks and damages are socioeconomically distributed in a violent and unjust way. Will God's good all-embracing space turn into a catastrophic space where some are victimized for the survival of others? How does God's love to the poor relate to situations where the most vulnerable become the most victimized? What does climate justice imply and how can it turn into a global and local spatial justice?

My own work has employed the notion of "Raum" (space/place) (Bergmann 2014). Here religion is understood as a skill of *Beheimatung* (making-oneself-at-home). Beliefs help people to root and inhabit as well as to move, transform, and creatively adapt to a world "in turmoil," to use Rilke's striking expression. Religion not only interacts with spatial processes but serves as *Raum* itself.

It should be self-evident that a spatial and platial understanding of religion is significant for the study of religion and the environment. Terms such as nature, ecology, and the environment offer spatial root metaphors for interpreting human bodily being alive, and one can only wonder why scholars have so often marginalized the spatiality of existence in favor of its temporality.

Heimat (home) offers another central metaphor for this spatial turn (cf Scott and Rodwell 2016). While the skill of inhabitation is common for all organisms that have to interact with their specific surroundings in order to survive, it also serves as a metaphor to describe the Sacred within the world. In Christianity, God's acting in and for the world is depicted as the Holy Spirit's indwelling, and animism, both in its traditional and modern ecospiritual forms, works with a view of life where all beings carry the spiritual source of their existence both within and around themselves. Furthermore, contemporary human migrations and other forms of uprootedness driven by the technology and economics of globalization seem to catalyze an increasing sense of homelessness, which challenges religious modes of making-oneself-at-home. This geopolitical situation produces an increasing existential homelessness among the rich and a violent dislocation among the poor with a growing number of people and peoples who cannot stay in their traditional environments due to dramatic environmental change. Migration flows are connected to regional patterns of global warming, and increasing global economic injustice draws new maps of so-called developed and developing countries, where the latter could often be described as de-developing. Will we ever come home? How does this changing world religion map affect religious faith? And what change might religion itself bring for a new "topophilia" (Tuan 1990)? Can it foster love toward the earth?

To live and believe in the times of the Anthropocene is to be continuously aware of being a receiving as well as an acting part of nature, or what Alexander von Humboldt had entitled as *Naturgemälde* (the painting of the world) (Humboldt 1845). Human beings are both painted by the world and painters of the world. Rituals and prayers, artworks and technologies, doctrines and values, as well as cosmologies and images of faith and earth are simply human brushstrokes in an ever-evolving process of iconography (cf Bergmann 2009a). The role of religions remains crucial in the Anthropocene (cf Deane-Drummond, Bergmann and Vogt 2017).

The diversity of methods

The theme of nature allows a broad range of approaches, making it necessary to decide about one's preferences. In general one can begin with a hypothesis that images of nature and images of God/the spirits/the Sacred are deeply interconnected, so that any change to one has an impact on the other. Images and practices are also entangled, so that the scholar can approach these interconnections by analyzing practices as well as ideas. Also, space/place-studies of religion and the environment add original insights.

While ecotheologians have cultivated the field with methods from historic–systematic theology, biblical studies, ethics, and practical theology, perspectives from church history are unfortunately still lacking (although the field of environmental history would offer excellent potential for cooperation). Scholars in religious studies often apply methods from cultural studies and cooperate closely with environmental anthropologists who usually operate with more or less diffuse concepts of religion. In particular, Roy R. Rappaport's influential work has inspired many scholars (Rappaport 1968, 1999).

Ethical perspectives can of course be made relevant in many fields, such as climate justice, landscape preservation, species extinction, and much more. Here mostly theologians and philosophers have been at the forefront while many (younger) scholars in religious studies prefer to explore the pathology of religion rather than its emancipatory power. Phenomena such as environmentalism and ecospirituality in other spheres than explicit religion have although been investigated intensely.

Many creative new interdisciplinary adventures have taken place in the fields of climate science, where a diversity of approaches have been offered, although a comparative worldwide research agenda on religion in climatic change has not yet been established where all religious traditions are compared in a balanced way. Nevertheless, the conditions for such a project are growing all the time. Other promising transdisciplinary projects have mined deeper into the aesthetic dimension, where, for example, environmental arts, architecture, and the art history of climate have been explored together with scholars of religion. Ritual studies offer furthermore a unique toolbox of methods with creative significance for to combine the practical and ideational analysis of religion. Biology has so far mainly transferred its insights directly into politics for preservation, and hereby not satisfyingly included local cultures and inhabitants. Yet environmentally committed biologists have also established cooperations with scholars in the environmental humanities in order to find new forms of environmental care for landscapes. In this context scholars established in 2011 the Sacred Natural Site network, where a manifold of different habitats that are administered by religious communities as sacred sites or landscapes are monitored in a highly ambitious way. Might knowledge about how religion is at work as a skill with regard to a habitat help to construct new modes of social support and new modes of care for specific environments?

This short, in no way exhaustive, methodological survey intends only to underline the exciting creativity and diversity in the field. Can one regard this creative diversity in itself as an expression of evolutionary power, not only in nature but also in academic culture?

My understanding of religion as a skill of both perception and action intends to enrichen this diversity even more, and to encourage scholars to explore the richness of this skill: gladly in adjectival forms such as passiological, aesthetical, localizing, critical, imaginative, conflictory, dividing, empathic, collaborative, caring, and others. Skills of faith might then appear as a complex deep driving force in the interplay of nature and culture, not only with regard to climate change, technofutures, and spatiality but as a human skill of central significance for encountering the demands of Earth our common home.

Note

1   For Tim Ingold skills are "not … techniques of the body, but the capabilities of action and perception of the whole organic being (indissolubly mind and body) situated in a richly structured environment." (Ingold 2000, 5).

Sigurd Bergmann

References

Bergmann S. (2007) "Theology in its spatial turn: Space, place and built environments challenging and changing the images of God," *Religion Compass* 1:3, 353–379.
Bergmann S. (2008) "Lived religion in lived space," in Streib H., Dinter A. and Söderblom K. eds., *Lived Religion: Conceptual, empirical and theological approaches; Essays in honor of Hans-Günter Heimbrock*, Brill, Leiden and Boston 197–209.
Bergmann S. (2009a) *In the Beginning is the Icon: A liberative theology of images, visual arts, and culture,* Equinox/Routledge, London.
Bergmann S. (2009b) "Climate change changes religion: Space, spirit, ritual, technology – through a theological lens," *Studia Theologica – Nordic Journal of Theology* 63:2, 98–118.
Bergmann S. (2014) *Religion, Space and the Environment,* Transaction Publishers, New Brunswick and London.
Bergmann S. and Gerten D. eds. (2010) *Religion and Dangerous Environmental Change,* LIT, Berlin.
Bergmann S. and Sager T. eds. (2008) *The Ethics of Mobilities: Rethinking place, exclusion, freedom and environment,* Ashgate, Farnham.
Bloomquist K. ed. (2009) *God, Creation and Climate Change,* Lutheran World Federation, Geneva.
Brunn S. D. ed. (2014) *The changing World Religion Map: Sacred places, identities, practices and politics,* Springer, New York.
Buttimer A. (2006) "Afterword: Reflections on geography, religion, and belief systems," *Annals of the Association of American Geographers* 96:1, 197–202.
Conradie E. (2008) *The Church and Climate Change,* Cluster Publications, Pietermaritzburg.
Crate S. A. and Nuttall M. eds. (2009) *Anthropology & Climate Change: From encounters to actions,* Left Coast Press, Walnut Creek.
Deane-Drummond C., Bergmann S. and Szerszynski B. eds. (2015) *Technofutures, Nature and the Sacred: Transdisciplinary perspectives,* Ashgate, Farnham.
Deane-Drummond C., Bergmann S. and Vogt M. eds. (forthcoming 2017) *Religion in the Anthropocene: Challenges, Idolatries, Transformations,* Wipf & Stock/Cascade, Eugene.
Deudney D. (1998) "Global Village Sovereignty: Intergenerational Sovereign Publics, Federal-Republican Earth Constitutions, and Planetary Identities," in Litfin K. T. ed., *The Greening of Sovereignty in World Politics,* MIT, Cambridge MA, 299–325.
Earth Charter Commission 2000 The Earth Charter (www.earthcharterinaction.org). Accessed 25 May 2015.
Gerten D. and Bergmann S. eds. (2011) *Religion in Global Environmental and Climate Change: Suffering, values, lifestyles,* Continuum, London and New York.
Globus Veldman R., Szasz A. and Haluza-DeLay R. eds. (2014) *How the World's Religions are Responding to Climate Change,* Routledge, London and New York.
Hulme M., Boykoff M., Gupta J., Heyd T., Jaeger J., Jamieson D., Lemos M.C., O'Brien K., Roberts T., Rockström J. and Vogel C. (2009) "Conference covered climate from all angles," *Science* 324, 881–882.
Humboldt A. von (1845) *Kosmos: Entwurf einer physischen Weltbeschreibung,* vol. 1, Cotta, Stuttgart and Tübingen.
Inge J. (2003) *A Christian Theology of Place,* Aldershot, Ashgate.
Ingold T. (2000) *The Perception of the Environment: Essays on livelihood, dwelling and skill,* Routledge, London.
Ivakhiv A. (2006) "Toward a Geography of 'Religion': Mapping the Distribution of an unstable signifier", *Annals of the Association of American Geographers* 96:1, 169–175.
Knott K. (2005) *The Location of Religion: A spatial analysis,* Equinox, London and Oakville.
Kong L. (2010) "Global shifts, theoretical shifts: Changing geographies of religion," *Progress in Human Geography* 34:6, 755–776.
Lawrence M. (2015) Geoengineering: Hope or hubris? Keynote at the fifth conference of the European Forum for the Study of Religion and the Environment on "Religion in the Anthropocene: Challenges, idolatries, transformations," 14–17 May, Munich.
Marx K. and Engels F. (1982) *On Religion,* Scholars Press, Atlanta.
McFague S. (2008) *A New Climate for Theology: God, the world, and global warming,* Fortress, Minneapolis.
Northcott M. S. (2007) *A Moral Climate: The ethics of global warming,* Darton, Longman & Todd, London.
Northcott M. S. (2013) *A Political Theology of Climate Change,* Eerdmans, Grand Rapids.
Northcott M. S. and Scott P. eds. (2014) *Systematic Theology and Climate Change: Ecumenical perspectives,* Routledge, Abingdon and New York.

20

Park C. C. (1994) *Sacred Worlds: Introduction to geography and religion*, Routledge, London.

Picht G. (1989) *Der Begriff der Natur und seine Geschichte*, Klett Cotta, Stuttgart.

Primavesi A. (2009) *Gaia and Climate Change*, Routledge, London.

Rappaport R. R. (1968) *Pigs for the Ancestors: Ritual in the Ecology of a New Guinea People*, Yale University Press, New Haven.

Rappaport R. R. (1999) *Ritual and Religion in the Making of Humanity*, Cambridge University Press, Cambridge.

Sacred Natural Sites (2011) (http://sacrednaturalsites.org). Accessed 25 May 2015.

Scott P. M. and Rodwell J. eds. (2016) *At Home in the Future: Place & belonging in a changing Europe*, LIT, Berlin.

Soja E. W. (1996) *Thirdspace: Journeys to Los Angeles and other real-and-imagined places*, Blackwell, Malden and Oxford.

Soja E. W. (2000) *Postmetropolis: Critical studies of cities and regions*, Blackwell, Oxford.

Stausberg M. (2011) *Religion and Tourism: Crossroads, destinations and encounters*, Routledge, London and New York.

Steffen W., Crutzen P. J. and McNeill J. R. (2007) "The Anthropocene: Are humans now overwhelming the great forces of nature?" *Ambio* 36:8 614–21.

Szerszynski B. (2005) *Nature, Technology and the Sacred*, Blackwell, Oxford.

Tuan Y-F. (1990) *Topophilia: A study of environmental perception, attitudes, and values*, Columbia University Press, New York.

Tuan Y-F. and Strawn M. A. (2009) *Religion: From place to placelessness*, The Center for American Places at Columbia College Chicago, Chicago.

3

WHOSE RELIGION? WHICH ECOLOGY?

Religious studies in the environmental humanities

Willis Jenkins

The field of "religion and ecology" investigates religious dimensions of ecological relations. What makes a dimension "religious" or a relation "ecological"? Lively debates contest the field's two organizing concepts and ways of connecting them. Scholars argue over whose conceptions of religion illuminate the relevant phenomena, sometimes suggesting that "culture" or "spirituality" would be more fitting concepts. They also argue over which senses of ecology create the most important intersections, sometimes arguing that "nature" or "environment" would be more inclusive or more accurate.

Debates over method are a sign of intellectual vitality. Robust academic fields typically foster arguments about what deserves recognition as signal work, how much diversity a coherent field can accommodate, and what terms of inquiry should inform new scholarship. This essay sketches the field's major methodological debates as a way towards interpreting the diversity in this volume and beyond it, as well as broader arguments over the intersection it represents.

Yet argument is made possible by implied agreement. Among the minimal assumptions that make debate possible across diverse research programs is at least this: humanity's ecological relations have religious and cultural dimensions. Failure to recognize and interpret those dimensions impoverishes environmental understanding, whereas engaging them has the potential to connect environmental questions with fundamental human questions of meaning, value, and purpose.

Religious studies in the environmental humanities

The field of religion and ecology is part of a broader intellectual collaboration of culture-focused approaches to environmental topics, many of which gather under the rubric of "environmental humanities." Inquiry in the environmental humanities takes many forms but is generally distinguished by humanistic practices; it employs historicizing methods, cultural memory, interpretive tools, evaluative skills, or critical dispositions in order to interpret environments and engage ecological questions.

The field of religion and ecology often (but not always) employs the methods and tools of religious studies. As it does, it extends central arguments in religious studies into new domains.

This chapter shows how perennial disciplinary arguments over conceptions of religion and over appropriate roles for constructive and normative scholarship appear within ecological domains, and then are changed by that extension.

Extending religious studies to environmental questions has sometimes encountered confusion from other religionists, who have wondered: is ecology really a proper arena for religious inquiry? Something like that confusion is encountered by all the environmental humanities. Humanistic practices were developed to interpret the human, so skeptics wonder whether they can legitimately be employed to interpret the natural world, which seems the province of the sciences. To answer skepticism, the various fields of the environmental humanities seem to devote their first generation of scholarship to establishing the coherence of their research program within their own discipline.

While each establishes itself differently, fields in the environmental humanities share a common tactic: they typically critique a key assumption of the skepticism by illustrating the social construction of the idea of nature and its supposed distinction from culture. If ever it made sense to bifurcate knowledge into humanities and sciences, pervasive anthropogenic ecological change renders defunct a separation of spheres. Environmental issues are "inescapably entangled with human ways of being in the world;" understanding them must include ways of studying the human (Rose et al. 2012, 1).

For religious studies, that thought carries two basic implications. First, however a scholar imagines religion, it is inescapably entangled with environments. Religion is ecological, at least in the sense that it is a significant part of humanity's evolutionary history (Bellah 2011). Particular religious inheritances, traditions, communities, or practices cannot be fully understood apart from the environmental history from which they emerged and whose ecological relations they in turn influenced. Second, because environmental issues are entangled with human ways of being in the world, they are entangled with religion. Insofar as religion is involved in how people inhabit and interpret their world, it is involved in ecologies.

It is now standard to find tools of religious scholarship deployed to engage environmental questions. In addition to the ethical projects where one expects engagement with contemporary issues, environmental themes have begun to inform textual, historical, theological, and ethnographic studies of religion. Where once an anthology like this one would have had to argue for the very idea of connecting religion and ecology, this handbook can instead follow the many different research directions emerging from widely recognized entanglement. This volume therefore highlights transdisciplinary directions facilitated by the emergence of environmental humanities and underscores distinctive contributions of religious studies to environmental understanding.

Nonetheless, accounts of the environmental humanities sometimes omit religious inquiry. Even when it includes "underlying cultural and philosophical frameworks that are entangled with the ways in which diverse human cultures have made themselves at home in a more than human world," an overview of environmental humanities may fail to recognize roles for religious studies (Rose et al. 2012, 2). Perhaps making a case for humanistic inquiry into environmental matters already seems tenuous without evoking religion (too far a trespass of modernity's intellectual etiquette). Yet if ecological dynamics are "entangled with human ways of being in the world," they are also entangled with all the ways that religion haunts, animates, influences, and interprets those ways of being.

In the first issue of the journal *Environmental Humanities,* the editors write that work in the area "engages with fundamental questions of meaning, value, responsibility and purpose in a time of rapid, and escalating, change" (Rose et al. 2012, 1). From its beginning, religion and ecology has attracted scholars from other disciplines, including the sciences, seeking a place to think about the fundamental questions of meaning involved in environmental change. So religion and ecology has been not only an arena in which scholars extend the disciplinary practices of religious studies, but also a transdisciplinary intersection where researchers, artists, and activists from many fields take up questions of meaning and purpose amidst changing ecological relationships.

Situating religion and ecology within the environmental humanities allies the field with a broad intellectual challenge to divisions of knowledge that continue to impede public understanding of complex ecological questions. In a context of pervasive anthropogenic change, everything that shapes human action – from imaginations and ideas to technology and economic policies – reverberates throughout ecological systems. Disciplines interpreting the human condition therefore bear renewed importance for understanding the changing human role in nature (Nye et al. 2012). How *could* one understand water or climate change without deep cultural analyses? The chapters on those two problems in this volume demonstrate not only what lacks in investigations that fail to include religious inquiries, but also how to begin assembling knowledges in ways that let researchers and students think about meanings, values, and purpose.

When research fields organize as environmental humanities, they underscore a shared epistemological claim: there can be no adequate ecological knowledge without understanding the culturally embedded humans who are changing ecological systems. Without environmental humanities, there can be no adequate understanding of environments. To that claim the field of religion and ecology adds the corollary: if there is no avoiding the human in attempts to understand Earth, there is no avoiding religious dimensions of the human experience.

Whose religion?

Any project in religion and ecology faces a version of the basic methodological question for religious studies: how to investigate religious dimensions of ecological relations when "religion" is a category that encompasses so much difference? Not only are there many self-identifying religious traditions with conflicts about how to interpret their differences from one another and internal arguments over boundary and meaning, there are also fields of personal experience, cultural creativity, and political life that involve dimensions apt for religious analysis. Attempts to gather all eligible phenomena into a theory of religion invite debates between theories, and those debates sometimes include suspicion of the very idea of religion. Religion is an irreducibly pluralist concept; any field that would anchor itself to it as foundation will find itself dropping into unfathomable waters.

So why stick with "religion"? The available alternatives seem less satisfying, as they do for the broader field of religious studies. This volume employs sections named "traditions" (a very short list of world religions and two emerging traditions with global aspirations); "cosmovisions" (indigenous lifeways for which conventional categories of religion and tradition may not fit); "spiritualities" (including experiences and practices sometimes excluded from religion); "regions" (a geographical approach bringing into view the hybrid and overlapping religiosities in a particular area); and "challenges" (where ecological problems present sites of religion-involved cultural stress and creativity). None of those categories would on their own allow attention to the range of scholarship that shapes the field and that is needed to understand ecological relations. "Religion" is usefully unfathomable, supporting many different research programs that recognize one another, at least as participants in the debate over the meaning of religion.

Can the field of religion and ecology have a shared conversation if it includes so many methods and phenomena? While environment-related work was pursued in various domains of religious studies as early as 1960, a shared field of exchange could not emerge until connective terms of inquiry were forged. An important event for making that conversation possible was a conference series on "Religions of the World and Ecology" held at Harvard University from 1996 to 1998, which helped organize many dispersed projects into a field with shared research interests. Led by Mary Evelyn Tucker and John Grim, the conference series gave rise to the Forum on Religion and Ecology (now at Yale). The Harvard conference participants developed three basic theses that made it possible for scholars of many traditions and methods to recognize a shared conversation: (i) that religious worldviews are significant for environmental behavior, (ii) that scholars should critically engage religious traditions with the ecological values and ideas needed for humanity to find "new and sustaining relationships to the Earth amidst an environmental crisis," and (iii) that environmental crises are also cultural crises of a religious depth (Tucker and Grim 1998, xv).

The first thesis redirected debates over whether and which religious traditions are culpable of exploitative environmental attitudes by focusing on a minimal assumption of argument: that religious worldviews matter for environmental behavior. Whether or not Lynn White was correct that modern instrumentalism could be traced to medieval Christianity, the notion that current behavior can be explained by worldviews motivates a line of inquiry into the "ecological attitudes" of all the traditions (White 1967). The second thesis then combined that notion with a commitment to critically re-examine traditions. "The Harvard project identified seven common values that the world religions hold in relation to the natural world," write Tucker and Grim, which can orient academic work of retrieving, re-evaluating, and reconstructing religious traditions for modern circumstances (Grim and 2014, 8). The third thesis then underscores the cultural significance of that academic work by suggesting that modern environmental problems put into question fundamental human–Earth relations, about which "the shared symbol-making capacity that has endured in world religions can be a source of wisdom" (Grim and Tucker 2014, 170). Those three ideas oriented shared inquiry across religious worlds and connected diverse projects in a coherent cultural endeavor.

All three theses have been subject to criticism by subsequent scholarship. Understanding criticism of the foundational stage of research in the field is important for understanding the range of subsequent work. The arguments made possible by those initial terms of inquiry have extended the boundaries of the field and given rise to different terms.

Why, some scholars have asked of the first thesis, should we assume that the religious worldviews of the major traditions are important drivers of environmental behavior? Scholars with a more materialist view of history investigate connections that run the other way: how changing environmental conditions may drive changes in how people think about their world, including religious thought. If worldviews determine how scholars think about religion and its relation to society, it can lead to organizing work by a list of the world religions and a comparative assessment of their ecological ideas.

Whose religion is included on a list like that? The ideas of a "world religion" and of a "global tradition" were born in colonial projects seeking to sort primitive superstition from civilized belief, in order to safely exclude the former from the rights and dignities of the latter. The categories of "religion" or "tradition" exert both constructive and reductive influence (Masuzawa 2005; Chidester 2014). Comparative fields have strong incentives to construct simple and discrete religious units that can be made subject to common terms of inquiry ("ecology"). In order to make comparison possible, scholars may reduce complex, pluralist, and contested inheritances into an essentialized picture of ecological thought.

As a result, serious scholars have sometimes found themselves grappling with hopeless questions like: "Is Buddhism an ecological religion?" (See Christopher Ives in this volume on the career of that question.) One would need to know whose enactment of which Buddhism in what contexts, before even beginning to wonder what it means for the phenomena to count as "ecological." The result can also be repressive and exploitative, as when applying the category of "religion" to indigenous cultures and traditional ecological knowledge imposes on them a typically western separation of religion from ordinary life. As Grim notes in his introduction to the section on Indigenous Cosmovisions in this volume, some indigenous communities therefore reject the category of religion. If their bioregional practices are seen as religious, it could allow modern secular polities (for which religion appears as private and subrational) to dismiss indigenous ecological sovereignty as mere religion, thereby keeping their lands open for resource exploitation.

Moreover, some scholars in religious studies have criticized conceptions of religion that focus on beliefs and institutions, calling for more attention to embodied experiences, hybrid cultural flows, and everyday practices. What about forms of ecospirituality that are not tied to conventional memberships? Catherine Albanese's *Nature Religions in America* helped draw attention to a stream of thought and spirituality that may not show up as religious because it cuts across the conventional taxonomies; nature functions as symbolic center for some within major traditions, for dissenters from those traditions, and for those who do not think of themselves as religious at all (Albanese 1991). Following implications of those critiques, Bron Taylor, founding figure of the Society for the Study of Religion, Nature, and Culture and editor of *The Encyclopedia of Religion and Nature*, writes that initial methods in the field of religion and ecology "left much nature-related religiosity out of sight" (Taylor 2005, 1375–6). Marginal communities and hybridizing bricolage may not appear visibly religious, but they are important for interpreting the proliferation of various "Earth-based spiritualities" that are influential within streams of environmental culture (Taylor 2001). Consider Gary Snyder's bioregionalist borrowing from indigenous, pagan, and Buddhist traditions, or the way that practices like surfing or fly-fishing might be important spiritual experiences for individuals, or how nature-themed religiosity appears in popular culture.

Graham Harvey's chapter on paganism and animism in this volume focuses on forms of nature-based religiosity that have sometimes been overlooked because regarded as primitive vis-à-vis the "world religions." In fact, as David Haberman shows, the concept of a world religion was developed in contrast to notions of animism as childish and uncivilized. In the case of Hinduism, the construct encouraged scholarly excision of nature-focused pieties from the devotional practices that made up the "tradition" (Haberman 2013). So, in that case, the very idea of a tradition was hostile toward nature-based spiritualities. Harvey thinks religious studies still bears some of that hostility, and calls for the discipline to attend more broadly to "the relationships that constitute, form, and enliven people in everyday activities in the material world" (Harvey 2014, 2).

Now, rejoinders could be made to this first line of criticism that it is not really true; work in the Harvard series included attention to indigenous cosmovisions, hybridizing practices, and Earthen spiritualities – even an essay by Gary Snyder. Some of the tension here might be explained by Evan Berry's distinction between substantive and functionalist conceptions of religion operative in the field. The former focuses on institutional memberships and the environmental implications of their beliefs and practices; the latter looks for implicit religious characteristics of a wide cultural range of environmental behaviors (Berry 2013). Taylor's criticism, however, targets another zone of methodological controversy in the study of religion: the role of confessional and normative commitments. When scholars understand themselves to be contributing to ecological reform of the traditions they study, and imagine such revisions contributing to an overarching civilizational movement – as the second and third theses of the

Harvard project hold – then, Taylor thinks, they have become "confessional" and "activist" in a way that is inappropriate to the academic discipline of religious studies. Taylor describes the *Encyclopedia* as placing more emphasis on historical and social-scientific description of religious phenomena and less emphasis on reforming traditions. Provocatively mapping that distinction onto a broader disciplinary debate sometimes cast as "theology" (confessional) versus "religious studies" (descriptive), Taylor criticizes the received terms of the field as being, effectively, too theological (Taylor 2005, vii–xxii, 1374–6).

However unreliable that distinction and however accurate Taylor's application of it, the question of normative commitments is critical for the field of religion and ecology – perhaps even more acutely than for the wider discipline of religious studies. The question comes up in two major issues: the role of theological work and the field's relation to environmental activism. I begin with theology and take up activism in the next section.

Christian ecotheology proliferated early, catalyzed by responses to ecological critiques of Christianity, and therefore played a relatively larger role than other traditions of belief and modes of study in giving initial shape to the field of religion and ecology. Some have therefore wondered whether Christian theological debates have distortively influenced environmental inquiries in other traditions and even environmental thought in other disciplines (Haberman 2013; Jenkins 2005; Callicott in this volume). That suspicion raises a broader question: how should a pluralist field receive confessional scholarship? Does it have shared terms by which to critically assess work developed with the particular logics internal to some tradition of faith?

Without some shared terms to guide comparative inquiry, the field would risk becoming either too inclusive (an uncritical clearing house for any expression of religious environmentalism) or too exclusive (excising major forms of scholarly production because modally improper). Either response to tradition-internal work would impoverish the field's pluralism. When constructive religious production happens because some religious world responds to environmental stress with its interior logic of reflection – for example, reading scriptures with ecocritical tools – the issue for the field is not whether it belongs, but what should be asked about it.

"Theological" does not always mean "confessional" or even "god-related." As a modal qualifier, theology can stand for a mode of constructive work with a tradition, often but not always understood by the author to be accountable to key lines of historical argument over the meaning of that tradition. Constructive work may bear analytic or illuminative significance beyond the tradition and beyond religious studies. For example, Stephanie Kaza's eco-Buddhism is "theological" in this modal sense because she goes beyond explication of Buddhism to make proposals for its practice and interpretation (Kaza 2010). Scholars of religious studies are sometimes wary of constructive modes of work in a tradition because they seem linked to confessional identities or seem overly deferential toward received authorities, but Kaza is primarily interested in showing how interpretations of Buddhism make illuminative contributions to environmental thought and connect with challenges faced by activists.

For some religionists, theology stands for a pluralist, pragmatist commitment to work with the cultural inheritances that inform how particular communities of discourse interpret the world around them. Theology allows constructive conversation in the moral vernaculars already in use in some community and may be a form of respecting the local differences of an interpretive world. In that mode, theology can function as a pluralist mode of discourse that nurtures creativity, helps realize cultural possibilities, or supports movements for adaptive ecological change (Kearns and Keller 2007). Movements for change take particular priority in liberative forms of constructive religious thought. In different ways, ecofeminist, indigenous, agrarian, black, latin@, queer, and other environmental justice thinkers have deployed theological analyses to think from solidarity with resistance movements in order to critically illuminate ecologies of violence.

The austere attitude of uncommitted description in some forms of religious studies thus faces especially acute pressure in the domain of religion and ecology. For while the cultural dimensions of ecological relations can certainly be investigated without evaluative concern over contemporary problems, doing so amidst such stark ecologies of violence and fragilization of planetary systems seems irresponsible, if not complicit. Constructive religious thought (a.k.a. "theology") may represent a particularist commitment to the sources and logics that shape cultural conversations, and a liberative intent to help important cultural conversations better interpret and relate to their ecology. Yet that begs a question about the second major term of the field: what is ecology?

Which ecology?

"Ecology" is variously used to refer to (a) the scientific study of organisms in relation to their environment, (b) an ethical worldview about appropriate human relations to their environment, (c) a political movement for adaptive social change, (d) a metaphor of interconnectedness, or (e) a materialist research frame for interpreting religious phenomena. That range of meanings construes work in the field so differently that some scholars would prefer the field's title instead join religion with "nature" or "environment."

The primary meaning of ecology in a university setting is (a) the scientific discipline that studies organisms in relation to their environment. Like other natural sciences, ecological research attempts to bracket issues of cultural meaning and moral value in order to focus on describing its objects of research. However, ecological science is uniquely entangled with ethics and culture, in part due to its history and in part due to the character of its research problems.

Ernst Haeckel, a founding figure of the new science, thought that knowledge about the relations among organisms would yield knowledge about the order and harmony of things, after which societies could pattern themselves. So from its beginnings, the discipline of ecology understood itself both as science and as worldview. Whether or not ecologists now think that their science can lead civilizations to better understand themselves as parts of a living whole, that is exactly how other realms of culture have appropriated it (Bauman et al. 2011).

In popular culture and in many of the fields that now make up the environmental humanities, "ecology" names (b) a worldview shaped by appreciation for sustaining goods that emerge from complex interactions of living creatures. It values those creatures individually, the relations among them, and especially the qualities that emerge from those relations. This worldview stands counter to the basic cosmology of modern industrialism, which – viewed in critical contrast – values things only as instrumentally useful to humans, who are imagined as more separate from than participant in the living Earth.

For many of its scholars, the field of religion and ecology orients to that second sense of ecology – as an ethical worldview grounded in natural science and critically different from the worldviews in which industrial capitalism emerged. Ecological science seems to support an ecological cosmology whose principles can guide resolution of ecological problems. Aldo Leopold, famous ecologist and cultural critic, often serves as intellectual icon of this meaning of ecology: "A thing is right when it tends to preserve the integrity, stability, and beauty of the biotic community" (Leopold 1949, 224). The field of religion and ecology then investigates how religion variously supports or obstructs what is right, in Leopold's basic sense of what is normatively ecological.

In that case the field has a foundational relation with environmental activism, which raises anew the question of normative commitments. Insofar as the ecological worldview differs from the values and politics that have led to environmental degradation, the field's sense of ecology seems to imply (c) a commitment to social change. Roger Gottlieb puts that commitment at the center of the field's research; in his construal, "ecology" stands for a commitment to polit-ical and cultural transformation, and "religion" for the sort of cultural resources adequate to the depth of transformation required (Gottlieb 2010). For Gottlieb, the central phenomenon for the field is a broad movement of religious environmentalism, in which he identifies any form of activism that "roots the general environmental message in a spiritual framework" (Gottlieb 2006, 231).

What exactly is that message? Establishing the ecological worldview, let alone its precise political message, by appeal to ecological science proves elusive. It is not clear that the science can supply a natural foundation of social order. Certainly the science cannot, on its own, deliver objective accounts of integrity, stability, and beauty. Leopold's qualifiers beg evaluative ques-tions, which introduce some ambiguity into the message of social change. (What sort of beauty?) Moreover, disruption, degeneration, and chaos also seem important to understanding ecological systems. Lisa Sideris argues that ecotheologians have chronically downplayed those elements in order to champion a picture of nature as harmonious, balanced, and even peaceful. The ironic result is that "ecological" work in religion sometimes appeals to pictures of nature at odds with the science of ecology (Sideris 2003). The point here is not that religionists should ground their values in a more accurate picture, but rather that activist notions of ecology require interpretive commitments beyond those warranted by the science of ecology.

Some scholars have turned about the question by analyzing environmentalist movements in religious terms (Dunlap 2015). To the extent that environmentalism resembles religious activ-ity (spreading "the message"), then assuming an activist commitment among scholars in religion and ecology might indeed look like assuming a confessional commitment. Another reason some scholars prefer the category "religion and nature" is that it can deflate some of the normative pressure of ecology as environmentalism. Adrian Ivakhiv, for example, argues that "religion and nature" permits a broad constellation of interests and interdisciplinary practices, without identifying the field with a political movement (Ivakhiv 2007). Ecology as political commitment, Ivakhiv suggests, narrows the field's pluralism.

Some scholars therefore deploy "ecology" as (d) a metaphor of interconnectedness. On this meaning, the field's title refers loosely to the study of relations among things, especially human-ity's connections with its environments, particularly as religion matters to them. Scholars investigate "ecologies of religion" in that sense when they examine some special thing ("organ-ism") inside a religious world ("ecosystem"), especially with a view to how the relations of that religious ecosystem involve connections to its physical environment (Jenkins 2008; Taylor 2007). In that case ecology is used in an analogical sense and the field's title might more accu-rately use "environments," since the research investigates how religious systems interpret or involve connections of cultures and environments (Bohannon 2014).

Included in that sense of ecology are critical accounts of how power circulates through vari-ous human arrangements of their environment – sometimes called "political ecology." Some scholars research how cultural patterns of environmental exploitation correlate with exploita-tion of vulnerable humans. Ecofeminist analyses, such as Heather Eaton's in this volume, investigate connections between human domination of Earth and male domination of females. The political ecology frame investigates how environmental conflicts are often conflicts over access to resources and exposure to risks – conflicts shaped by the relative social positions of participants. Ecologies of racism, poverty, migration, and globalization show how axes of power and vulnerability reproduce or interpret certain environmental arrangements.

There is an implicit moral anthropology to this approach: if persons are vulnerable to ecologically mediated forms of political violence that implies there are inalienable ecological dimensions of human dignity. Disproportionate flow of toxins through air and water into the bodies of disempowered populations is the most visible example of that connection. In the United States, environmental injustice shows up in a racist landscape of toxic exposure. Material flows of white power push toxins into the bodies of racial minorities. Melanie Harris's chapter in this volume thus connects police violence against black men with sick air quality in black neighborhoods, invoking the last words of Eric Garner (who was choked to death by a New York City policeman): "I can't breathe."

Tracing those lines of interconnection (ecology d) exerts critical pressure on environmental politics (ecology c). Environmental thought in the Global North has sometimes focused on the human–Earth relationship in general and has skipped injustices and dysfunctions within particular social relations. Preservationist impulses sometimes unthinkingly defend the landscape of power represented by some environment. Religion and ecology has sometimes reproduced the methodological whiteness of North American environmental thought by averting thought from social violence. However, if ecological flows are political flows, then political struggles are renegotiations of general human–environment relations and of the imaginaries that govern them. Seen that way, environmental justice struggles may be diverse sources of an adaptive cultural intelligence, sometimes drawing on religious inheritances (Jenkins 2013, Ch. 5).

Where scholars treat religious systems as more or less adaptive relationships with their environments, they attempt to rejoin cultural analysis with the primary sense of ecology as a science. In this frame, ecology denotes (e) a materialist research frame for interpreting religion as part of the physical ecology of the planet. "From this perspective," writes Gustavo Benavides, "religions can be understood as the attempt to come to terms with constraints of all kinds, a task that is accomplished by exploring through speculation and ritual the range of options open to the kind of organisms that humans happen to be," and may include maladaptive "attempts to escape limitations" (Benavides 2005, 548). Or one might interpret religion, as Stephen Kellert does, as an expression of genetic human affinity for nature (biophilia), and evaluate its variety of expressions according to adaptive fit (Kellert 2007). An implication of this view is that human–environment relations exert evolutionary pressure on religious systems (Wilson 2002). That theme remains underexplored in histories of religion, although some scholars have begun to shape inquiries into how, for example, climate change may have shaped religious experience in Europe's "Little Ice Age," perhaps fueling fear of witches in Europe and the energies leading to the Protestant Reformation (Fagan 2001).

A materialist sense of ecology will seem reductive to some religion scholars. It would be odd to think of the Reformation as merely a consequence of a climate aberration, and to consider its consequences only in terms of adaptive fit. On the other hand, consider how treating religion as an adaptive relation with an environment warrants regard for the ecological knowledge of indigenous communities and traditional ways of life. If sciences once regarded traditional ecological knowledge with suspicion because it is transmitted through religious narratives and spiritual practices, interpreting those narratives and practices in terms of their ecological function permits outsiders – especially outsiders with modernist scruples – to learn from them (Berkes 2008).

Using criteria of adaptive fit brings back the question of normative commitments in the cultural entanglement of ecological science. What constitutes adaptive success for a culture, or for a religious system? The science of ecology may not supply a moral foundation that can fully answer the question, but insofar as it researches problems that matter to human flourishing, ecology produces knowledge that matters for how cultures assess their values and understand the human condition. In conditions of pervasive anthropogenic change, ecology investigates problems that lead cultures to question their values and reassess the human condition. Researchers of global environmental change have been asking societies to take responsibility for how humans are changing Earth and to intentionally develop responsibility for its systems. How to do so well?

It would be impossible to argue well about such a basic question without understanding religious inheritances and the way that they orient many diverse environmental imaginations. In a nonfoundationalist sense, the ecological questions of an era marked by pervasive human influence are irreducibly entangled with questions of a religious depth. That is, ecological questions have become entangled with questions about what it means to be human and how to live well, about where the living world has come from and where it is going, and why. It is not obvious or given which values, virtues, practices, or narratives should answer those questions – and yet the questions cannot be avoided.

References

Albanese, C. L. (1991) *Nature Religion in America: From the Algonkian Indians to the New Age,* University of Chicago Press, Chicago.

Bauman, W., Bohannon II, R. R. and O'Brien K. J. (2011) "Ecology: what is it, who gets to decide, and why does it matter?", in W. Bauman, R. R. Bohannon II and K. J. O'Brien (eds.), *Grounding Religion: A Field Guide to the Study of Religion and Ecology,* Routledge, New York, 81–95.

Bellah, R. (2011) *Religion in Human Evolution: From the Paleolithic to the Axial Age,* Harvard University Press, Cambridge.

Benavides, G. (2005) "Ecology and Religion", in *Encyclopedia of Religion and Nature,* Bron Taylor ed., Continuum, New York.

Berkes, F. (2012) *Sacred Ecology,* Routledge, New York.

Berry, E. (2013) "Religious Environmentalism and Environmental Religion in America", *Religion Compass,* 7:10, 454–66.

Bohannon, R. ed. (2014) *Religions and Environments: A Reader in Religion, Nature, and Ecology,* Bloomsbury Academic, New York.

Chidester, D. (2014) *Empire of Religion,* University of Chicago Press, Chicago.

Dunlap, T. (2015) *Faith in Nature: Environmentalism as Religious Quest,* University of Washington Press, Seattle.

Fagan, B. (2001) *The Little Ice Age: How Climate Made History 1300–1850,* Basic Books, New York.

Gottlieb, R. (2010) "Religion and Ecology: What is the Connection and Why Does it Matter?", in Roger Gottlieb (ed.), *The Oxford Handbook of Religion and Ecology,* Oxford University Press, New York.

Gottlieb, R. (2006) *A Greener Faith: Religious Environmentalism and Our Planet's Future,* Oxford University Press, New York.

Grim, J. and Tucker, M. E. (2014) *Ecology and Religion,* Island Press, Washington, DC.

Haberman, D. L. (2013) *People Trees: Worship of Trees in Northern India,* Oxford University Press, New York.

Harvey, G. (2014) *Food, Sex and Strangers: Understanding Religion as Everyday Life,* Routledge, New York.

Ivakhiv, A. (2007) "Religion, Nature and Culture: Theorizing the Field", *JSRNC* 1.1, 47–57.

Jenkins, W. (2013) *The Future of Ethics,* Georgetown University Press, Washington, DC.

Jenkins, W. (2008) *Ecologies of Grace,* Oxford University Press, New York.

Jenkins, W. (2005) "Islamic Law and Environmental Ethics: How Jurisprudence (*usul al-fiqh*) Mobilizes Practical Reform", *Worldviews* 9:3, 338–64.

Kaza, S. (2010) "The Greening of Buddhism: Promise and Perils", in Roger Gottlieb (ed.), *The Oxford Handbook of Religion and Ecology,* Oxford University Press, New York.

Kearns, L. and Keller, C. (2007) *Ecospirit: Religions and Philosophies for the Earth,* Fordham University Press, New York.

Kellert, S. (2007) "The Biophilia Hypothesis", *Journal for the Study of Religion, Nature and Culture* 1:1, 25–37.

Leopold, A. (1949) *A Sand County Almanac,* Oxford University Press, New York.

Masuzawa, T. (2005) *The Invention of World Religions: Or, How European Universalism was Preserved in the Language of Pluralism,* University of Chicago Press, Chicago.

Nye, D., Rugg, L., Fleming, J. and Emmet, R. (2012) "The Environmental Humanities (a white paper for MISTRA)", *Swedish Foundation for Strategic Environmental Research*, Stockholm.

Rose, D. B., van Dooren, T., Chrulew, M., Cooke, S., Kearnes, M. and O'Gorman, E. (2012) "Thinking Through the Environment, Unsettling the Humanities", *Environmental Humanities* 1, 1–5.

Sideris, L. (2003) *Environmental Ethics, Ecological Theology and Natural Selection,* Columbia University Press, New York.

Taylor, B. (2005) *Encyclopedia of Religion and Nature*, 2 vols. Continuum, New York.

Taylor, B. (2001) "Earth and Nature-Based Spirituality (Part I): From Deep Ecology to Radical Environmentalism", *Religion* 31:2, 175–193.

Taylor, S. M. (2007) "What if religions had ecologies? The case for reinhabiting religious studies", *Journal for the Study of Religion, Nature and Culture* 1:1, 129–38.

Tucker, M. E. and Grim, J. (1998) "Series Forward", in *Religions of the World and Ecology*, vol. 1–10, Harvard University Press, Cambridge.

White, Jr., Lynn (1967) "The Historical Roots of Our Ecological Crisis", *Science* 155, 1203–7.

Wilson, D. (2002) *Darwin's Cathedral: Evolution, Religion, and the Nature of Society,* University of Chicago Press, Chicago.

PART II

Global traditions

Introduction: *Mary Evelyn Tucker*

The world's religious communities are increasingly involved in efforts toward fostering an integral ecology of humans and nature. These chapters show how the attitudes and beliefs that shape many people's concepts of nature around the planet are influenced in part by their religious and cultural worldviews and ethical practices. For each of these authors, the moral imperatives and value systems of religions are indispensible for mobilizing people to shape the trajectories of social-ecological change to enhance ecosystem resilience and human wellbeing. The authors here have a keen awareness of the problems of religions and encourage both appreciative and critical angles regarding religious traditions and communities regarding their attitudes and practices.

The ecological salience of the global traditions comes in part from the fact that the vast majority of the world's people are members of a religious tradition, and these affiliations directly and indirectly influence views of nature and its intelligent use or care. Examining the dynamic influences of religious ideas and communities is also important because world religions are being recognized in their great variety and hybridity as more than simply a belief in a transcendent deity or a means to an afterlife. Rather, religious traditions provide a broad orientation to the cosmos and the Earth and to redefining human roles in working within nature's boundaries.

In this context, then, religions can be understood in their largest sense as a means whereby humans, recognizing the limitations of phenomenal reality, undertake specific practices to effect self-transformation and genuine community within a cosmological context. Religions thus refer to those cosmological stories, symbol systems, ritual practices, ethical norms, historical processes, and institutional structures that transmit a view of the human as embedded in a world of meaning and responsibility, transformation and celebration. Religions connect humans with a divine presence or numinous force. They bond human communities and they assist in forging intimate relations with the broader Earth community. In summary, religions link humans to nature where life arises, unfolds, and flourishes. This link historically and at present, both contested and renewed, is explored in these articles.

In this spirit of exploration the authors here recognize that "religious traditions" is a term that has multiple meanings. These traditions are not monolithic, fixed entities, but are continually interacting with one another across time and cultures. Moreover, all of the traditions reflect complex interactions of various schools of thought within a tradition (such as Orthodox, Conservative, and Reform in Judaism) that represent contested differences and mutual influences. In addition, there is continual mingling of folk practices and intellectual aspects of all of these traditions. In the East Asian traditions there is notable syncretism between and among the religions such that the idea that "the three traditions are one" was prevalent in Ming China to describe the dynamic interaction of Confucianism, Daoism, and Buddhism. In Japan even today the norm is that one comes of age at a Shinto shrine, practices Confucian ethics at home, and is buried at a Buddhist temple. Thus, bricolage and hybridity are widespread throughout world religions. This is part of what makes them open to change over time.

In this section, then, we have included most of the major world religions with awareness of the complex syncretism and dialogical unfolding that marks their historical development over time. We have also included the Bahai and Mormon traditions, which increasingly have a global presence and aspiration. African and Latin American religions are covered in the chapters on those topics. Similarly, Daoism is discussed in detail in James Miller's article on China and Jainism is included in Christopher Chapple's article on India. In all of this, it becomes clear that the world's religious communities are increasingly involved in efforts toward fostering an integral ecology of humans and nature.

4

HINDUISM

Devotional love of the world

David L. Haberman

There is an outright assault on virtually every aspect of Earth's ecosystems these days: rivers are severely polluted, forests are razed at alarming rates, and mountains are demolished for a variety of industrial purposes. Enormous damage has already been done to meet the ever-increasing demands of a rapidly growing globalized consumer culture. We are now gearing up to inflict even greater damage as we prepare to harvest all remaining resources and to squeeze every last drop of fossil fuel from the planet. At the same time, there seems to be a new planetary awakening that seeks ways beyond our current unsustainable predicament to a healthier human presence on Earth, and religious traditions worldwide are increasingly contributing to this movement. How do people within the Hindu religious cultures of India regard and struggle with these challenges? Relatedly, how are natural entities such as rivers, trees, and mountains conceived within these cultures, and what kinds of practices are found within them that might serve to address the unprecedented environmental degradation of our day?

The term Hinduism is a complex one. Originally used by Persians to denote the religious ways of people who lived on the other side of the Indus River, today it is the accepted designation of a vast array of religious beliefs and practices of the majority of the Indian population. As in the case of every world religion, it is more accurate to speak not of a single Hinduism, but rather of a rich multiplicity of Hinduisms. Past overviews of Hinduism and ecology have tended to focus on the philosophical texts and practices of the ascetic traditions.[1] The notable work of Christopher Chapple, for example, highlights the contributions that the Hindu renouncer values of minimal consumption might make toward an environmentally friendly ethic (Chapple 1998). While these values are significant in considering ecological possibilities within Hinduism, I find myself in agreement with Vasudha Narayanan who has pressed for a shift away from an emphasis on the ascetic traditions in our understanding of Hinduism and ecology to the *bhakti* devotional texts and rituals, since "devotional (*bhakti*) exercises seem to be the greatest potential resource for ecological activists in India" (Narayanan 2001, 202). I propose to take up this recommendation with a presentation of a popular mode of Hinduism and ecology that has received little academic attention in general works on religion and ecology. This essay is not intended to be a survey, but rather a representation of a fairly widespread form of religiously informed ecological activism that I have encountered both explicitly and implicitly in my explorations of Hinduism and ecology (Haberman 2006, 2013).

Before examining some contemporary instances of Hindu ecological engagement, I briefly take up the question, "Is Hinduism ecofriendly?" We would be justified in rejecting this question altogether, for it is a simplistic and misleading formulation that both reduces a complex tradition and calls for an answer never intended by any tradition. Like all world religious traditions, Hinduism is a multifaceted cultural phenomenon that consists of many varied and sometimes contradictory voices. There is evidence for what could be identified as ecologically damaging views and practices within Hinduism, and there is evidence for what could be identified as ecologically friendly views and practices within Hinduism. Questioning the concept of the "oriental ecologist," Ole Bruun and Arne Kalland have argued that Asian philosophies have done little to prevent environmental disasters in a number of Asian societies (Bruun and Kalland 1995, 2–3). Nonetheless, while acknowledging some aspects of Hinduism have been detrimental to the environment, a number of writers have maintained that Hinduism has much to contribute to addressing the environmental crisis. Rita Dasgupta Sherma, for example, insists, "In the case of Hinduism, resources exist for the development of a vision that could promote ecological action" (Sherma 1998, 89–90).

We must keep in mind, however, that the current scope of the environmental crisis is a drastically new experience that demands new responses; no religious tradition in its present form is fully prepared to address the current problems. Poul Pedersen reminds us, "No Buddhist, Hindu, or Islamic scriptures contain concepts like 'environmental crisis,' 'ecosystems,' or 'sustainable development,' or concepts corresponding to them. To insist that they do is to deny the immense cultural distance that separates traditional religious conceptions of the environment from modern ecological knowledge" (Pedersen 1995, 226).[2] Religious traditions are always changing in the face of new historical circumstances, and one of the greatest challenges today is the environmental crisis, which is already reshaping religious traditions worldwide. With these precautions in mind, we can proceed to examine an emerging ecological development within Hindu India and those dimensions of the tradition that might serve as resources for those who employ a Hindu cultural perspective in their struggle with the environmental crisis.

Past representations of Hinduism that were heavily dependent on the ascetic philosophy of Shankaracharya's Advaita Vedanta often ignored Hinduism's most common aspect: the devotional cultures of India that focus on interaction with embodied forms of divinity and generally promote a very positive view of the world. To explore what this theistic Hinduism means for environmental thought and action, it would be useful to examine the views expressed in the Bhagavad Gita. This popular text gives representative expression to concepts that inform much theistic Hinduism. The Bhagavad Gita has also been important for many involved in early environmental movements in India, especially those influenced by Gandhi who used it for daily meditations.[3] Well-known environmental activists in India, such as the Himalayan forest defenders, have organized readings of the Bhagavad Gita as part of their strategy for environmental protection, and some have used it to articulate a specifically Hindu ecological philosophy.[4] One can even find examples of such use of the Bhagavad Gita in science-based environmental publications such as *Down to Earth*, a periodical published by the Centre for Science and Environment in New Delhi.[5] Most importantly for our considerations, the Bhagavad Gita provides a theological framework for understanding the religious thought that informs much devotionally based environmentalism within Hinduism.[6]

Like Narayanan, I too have found a preeminence of *bhakti* rituals and devotional texts in my studies of Hinduism, nature, and ecology; my comments, therefore, focus on certain devotional beliefs and practices that relate to environmental conceptions, concerns, and practices. One of the major debates between the ascetic and devotional traditions relates to the status of the world we experience with our senses. While many of the ascetic traditions teach that the phenomenal world is ultimately an illusion to be transcended, the devotional traditions have tended to affirm the reality of the world, often honored as a divine manifestation. The position of the Bhagavad Gita is relevant. Narayanan explains that, "central to the *Bhagavadgita* is the vision of the universe as the body of Krishna" (Narayanan 2001, 185).

To introduce a common way of thinking about Hinduism and environmentalism in present-day practice, I highlight four Sanskrit terms prevalent in discourse about conceptualizations of and interaction with natural entities. Since all four terms begin with an "*s*," I refer to them as the four "*s*"s. Although all of these terms are drawn from the Bhagavad Gita, they are used generally in Hindu discussions about ecology and the natural environment and come up frequently during conversations about environmental activism in India. The form of environmental activism represented by these terms is not well known outside of India; nonetheless, it is quite popular within India, as it is firmly rooted in the devotional practices that center on worshipful interaction with embodied forms of divinity. The first term, *sarvatma-bhava*, has to do with the worldview that informs much environmental activism within Hindu culture; the second, *svarupa*, relates to the devotional object of environmental activism; the third, *seva*, is increasingly used to denote environmental activism itself; and the fourth, *sambandha*, identifies the desired outcome of the action.

Among the four terms *sarvatma-bhava* is perhaps the least utilized in everyday language, but the notion it signifies is prevalent within Hinduism. It is a technical compound word that proclaims that everything is part of a unified and radically interconnected reality, called alternatively Atman or Brahman, and refers to the largely accepted viewpoint that all is sacred. In common parlance, this is often expressed theologically as God is everything and everything is God. This is a concept with deep roots in many Hindu scriptures. The highly influential Brihadaranyaka Upanishad, for example, declares, "the whole world is Brahman" (1.4). Granted most Hindus do not have detailed knowledge of Upanishadic texts, but this is an idea that is expressed repeatedly in many later texts and everyday discussions about religion. The better-known Bhagavad Gita states this notion most succinctly in the concise declaration: "Vasudeva (Krishna) is the entire world" (7.19). Without necessarily referencing texts such as these, many people articulate a similar notion while discussing the relationship between the world and Krishna. "Everything in this world is a part of Krishna and therefore worthy of reverence (*pujaniya*)," a man explained to me while discussing the natural landscape.

Early foundational texts, such as the Brihadaranyaka Upanishad, also asserted that there are two aspects of ultimate reality or Brahman: one is identified with all forms (*murta*); the other is identified with the realm of the formless (*amurta*) (2.3). Brahman as all forms is everything that is manifest and transitory, whereas Brahman as the formless is unmanifest and unchanging. These are not two separate realities, but rather different modes of the same unified reality. Although most people are not directly familiar with the Brihadaranyaka Upanishad, they are conversant with this principle. A Hindi-speaking woman expressed this concept theologically: "Some people think of God as with form (*sakar*), and others think of God as formless (*nirakar*). These are just different ways of thinking of God, but God is one (*Bhagavan ek hi hai.*)."

The Bhagavad Gita confirms these two aspects of ultimate reality, and adds a third that encompasses and surpasses both: the divine personality called Purushottama. Devotional traditions aim to establish a relationship with this supreme form. Although these three dimensions of reality are regarded in a hierarchical fashion, it is important to remember that all three are aspects of divinity, in this instance Krishna. That is, while it is assumed to be only a portion of a much vaster and unmanifest reality, the entire manifest world of multiple forms that we perceive with our senses is fully divine. While manifestly diverse, the sense of reality denoted by *sarvatma-bhava* is that everything is also simultaneously interconnected and unified; in short, the entire world is divine. The tripartite conception of reality is at the very core of many

Hindu schools of thought, and is considered to be the vital foundation for all spiritual development and productive work in the world. It is also important for understanding specific forms of devotional environmentalism in India.

Though the whole world is divine, human beings are not good at connecting with abstract universalities. We are embodied beings designed to connect with concrete particularities. Universal love is a noble sentiment, for example, but it cannot begin to compare with the passionate engagement of the intimate love of a particular person. Acknowledging this feature of human emotions and perception leads to the next "*s*": *svarupa*. The term *svarupa* has an expedient double meaning. It literally means "own-form," and in theological contexts frequently refers to the deity's own form or an essential manifestation of God, and is understood to refer to a full presence of divinity. An aspect of the highest reality, it comes to mean a specific embodied form divinity takes in the world. Divinity within Hinduism is typically understood to be infinite and all pervasive, but assumes particular concrete forms; accordingly, although unified at the unmanifest level, it manifests as a multitude of individual entities. The second meaning of the term *svarupa* is the worshiper's own form of divinity. The divine unified reality of Brahman is everywhere and everything, but one's *svarupa* is a personal and approachable concrete "handle" on the infinite; it is that distinct, intimate form of divinity to which one is especially attracted. That is, among the countless multitude of forms, this is the particular one to which a person is drawn and develops a special relationship. The particular physical forms of divinity that are *svarupa*s importantly include many natural phenomena, such as rivers, ponds, rocks, mountains, trees, and forests. Everything in the world is understood to be a potential *svarupa*, but there are natural entities that are favored through cultural selection. Specific examples would include rivers, such as the Yamuna, Ganges, and Narmada; sacred trees such as neem, pipal, and banyan; and mountains such as Govardhan and Arunachal. And all *svarupa*s comprise the three interrelated dimensions of reality. The form dimension of the Yamuna, for example, is the physical water of the river, the formless is the all-pervasive spiritual dimension, and the divine personality is the goddess Yamuna Devi. The form dimension of Mount Govardhan is the concrete rocky hill, the formless is the all-pervasive undifferentiated dimension, and the divine personality is Krishna in the form of Shri Govardhana Natha-ji.

Recognized as a special form of divine vitality, awareness of the full nature of the *svarupa* is considered by many to be a key component to beneficial environmental attitudes and actions. The physical *svarupa* of a neem tree, for example, is connected with her goddess identity. A woman who worships a particular neem tree everyday in Varanasi told me: "Ma's powerful presence is in this tree. This tree is her *svarupa*. I worship her here everyday and now have a special relationship with this tree" (Haberman 2013, 144). Accordingly, she – as well as many others who share her understanding – would never think of harming or cutting a neem tree. The *svarupa* of the Yamuna River too is associated with her divine identity. A man who lives in a town located on the shore of this river links an awareness of this identity to current environmental concerns: "The people who are not aware of the *svarupa* are polluting her. If we could get people to see the goddess in the river, they would worship her and stop polluting her. People who don't understand the *svarupa* of Yamuna-ji are polluting her. We must make them understand the real nature of Yamuna-ji, and then they will stop polluting" (Haberman 2006, 187). I observed a group of villagers living at the base of Mount Govardhan stop a man from even putting a shovel into the soil of the mountain to plant a tree, so great was their concern for the sensitive personality of the mountain.

How does one connect with a *svarupa*? This question leads to another of the four "*s*"s: *seva*, a term that means concrete acts of "loving service" or simply "acts of love." In the context of the religious culture associated with natural sacred entities a few decades ago, the word *seva* would have referred almost exclusively to ritual acts of honorific worship, such as offering flowers, hymns, and incense. However, in this age of pollution this term is increasingly being used to designate acts that would in the West be labeled "environmental activism." For example, previously it was assumed that Yamuna was a powerful and protective Mother who cared for her human children, but now there is a growing conviction that her children need to care for her. The term used for this care is *seva,* loving actions that now take a variety of forms – from picking garbage out of the river to political and legal action aimed to protect it.

In these *bhakti* traditions, love has two aspects: feelings and actions, and these two are significantly interconnected. Feelings set actions in motion, and actions engender further feelings. Since feelings are more difficult to access than actions, actions are the entryway into an ever-expanding circle of love. Moreover, the specialness of a being is revealed in the presence of love; while we shrink back into a protective shell when confronted with hostility, we come out and expose ourselves more fully in the presence of love. Likewise, the deep sacrality of the world reveals itself only in the face of love. Awareness of the true nature (*svarupa*) of something generates love for it, and that love enables one to see that true nature more clearly. Loving acts toward a being generate loving feelings toward that being – rivers, trees, and mountains included – which motivate more loving acts. This point was driven home to me one day while watching a young man perform acts of worshipful *seva* to the Yamuna River. He told me about the transformation in his own life that led him to become a daily worshiper of Yamuna. "I used to see Yamuna-ji as an ordinary polluted river. I used to wear my sandals down to her bank (He now views this as a grave insult.). But then I met my guru, and he told me to start worshipping Yamuna-ji. At first I was a little resistant, but I did what he said. Soon, I began to see her *svarupa* (true form) and realized how wonderful she really is. So now I worship her everyday with love. The main benefit of worshipping Yamuna-ji is an ever-expanding love" (Haberman 2006, 185).

Most importantly for ecological considerations, those who reflect on the environmental crisis from this viewpoint say that this deep perspective is the one most needed to restore a healthy relationship with the world. For the devotees of Yamuna immersed in this perspective, this means opening oneself to the river to the point where one can perceive the *svarupa* of Yamuna. Once this occurs, polluting the river becomes as impossible as dumping garbage on the face of one's lover. Worshipful acts, then, are the very doorway into an inner world of realization; they are concrete levers for opening up new perspectives that lead to environmental awareness and activism. While environmental degradation, I was told again and again, is the result of a very limited perspective on the world, many devotees stressed that a positive and ecologically healthy relationship with the natural world depends on a loving awareness of its true nature, or *svarupa*, which is realized through loving acts of *seva*. An awareness of the true nature of reality leads one into a world of divine love wherein destruction and pollution become unthinkable.

Love, therefore, is both a means and an end. The Hindu ecotheologian Shrivatsa Goswami maintains that, "Love is the key to all sustainability" (Haberman 2006, 157). Many environmental activists I spoke with in India articulated their actions as expressions of love. The environmental activist Sunderlal Bahuguna, who worked many years attempting to stop the Tehri Dam on a major Himalayan branch of the Ganges River, told me that his work was motivated by devotional love: "I love rivers because they are God; they are our Mother. In our philosophy we see God in all nature: mountains, rivers, springs, and other natural forms" (Haberman 2006, 71). And love for Mount Arunachal as an essential form of Shiva led Ramana Maharshi and his followers to protect the sacred hill from developmental plans and to initiate its reforestation.

The love generated in *seva*, then, leads not only to a joyful realization of the true form of the "object" of that *seva*, but also to a deep concern for it. Because of his sentiments toward Yamuna, a pilgrimage priest who resides near the Yamuna River in Mathura experienced much pain while confronting the massive pollution of the river. "When people come to Mathura and see the condition of the Yamuna," he reports, "it hurts them and they leave with a broken heart." This man's anguish spurred him into environmental action aimed at cleansing the river. "When Mother is sick," he explained, "one cannot throw her out of the house. We must help her. Therefore, I do Yamuna *seva*" (Haberman 2006, 144). *Seva*, or loving service, was a word I heard many times in conversations with environmental activists working to restore the Yamuna to health. This activist priest, who organized demonstrations to raise awareness of the plight of the Yamuna and was the primary instigator of a successful court case that imposed a ban on the release of untreated domestic sewage and industrial effluents into the Yamuna in the Mathura District, represents his environmental activism with this religious term, as do many others in India. For him, restoring the river is a deeply religious act, performed not primarily for the benefit of humans, but for the river herself.

The culminating result of this divine love affair is a firm "connection" or "relationship" with some aspect of the sacred world. This relationship is called *sambandha* – the final "s". Worshipful acts of *seva* designed to honor a particular being have the additional effect of stimulating a deeper loving connection – *sambandha* – with that being. As a recipient of loving acts of *seva*, natural entities such as rivers, trees, and stones from sacred mountains are typically personified and sometimes even adorned in an anthropomorphic manner. The Yamuna River is draped from shore to shore with a long decorative cloth made from 108 colorful saris on her birthday and other special occasions. Neem trees in Varanasi are wrapped with ornate cloth and humanlike facemasks are attached to the trunks at eye level. Faces with prominent eyes are also added to stones from Mount Govardhan, which are then adorned with clothing and jewelry. Worshipers of these natural entities report that this *seva* practice is more than a way of honoring the natural entity; they also do this to develop and enhance an intimate relationship with the *svarupa*.

A neem tree worshiper told me: "The face makes *darshan* (sight) of the goddess easier. The tree is the goddess, but it is easier to have a relationship with the goddess if a face is there. It is easier to see the goddess in the tree, or the tree as the goddess with a face on it" (Haberman 2013, 154). Many tree worshipers report that the face helps them recognize and better bond with the goddess of the neem tree: "When I look into the face of the goddess on the tree," one woman explained, "I feel a strong connection (*sambandha*) with this tree" (Haberman 2013, 154). Worshipers who add faces to stones from Mount Govardhan, which are understood to be naturally embodied forms of Krishna, express similar notions. One told me that this practice "makes it easier to perceive the *svarupa*, to see the stone as Krishna." Another said: "When you put eyes and face on a Govardhan stone you feel it is a person. It is easier to see the stone as a person with the face and clothing added. Putting eyes and other ornamentations on the stone makes its personality more perceptible. This makes a loving relationship with the *svarupa* more possible."

In this context, anthropomorphism seems to function as an intentional cultural means of connecting positively with the nonhuman world. Current research by social psychologists seems to corroborate the notion that anthropomorphism can function as a means of establishing connection with some nonhuman entity, and that this connection leads to a greater concern for the anthropomorphized agent's well-being (Epley et al. 2008). This claim has been confirmed by a group of Hong Kong-based psychological researchers who have published a study that demonstrates that anthropomorphizing enhances connectedness to natural entities, and that this results in a stronger commitment to conservation behavior (Tam 2013).

Those who see the Yamuna as a divine goddess are less prone to polluting the river and more committed to restoring it; those who see trees as divine personalities avoid harming them and oversee their protection; and devotees of Mount Govardhan do not dig into the mountain and some have worked to safeguard it from extractive exploitation. But how relevant is the concern for a single natural entity toward the larger ethical concern for all such entities? The possibility of this mode of devotional environmentalism opening out to a more universal ethic was highlighted for me during an instructive conversation. One day I visited a large pipal tree shrine in Varanasi and there met a *sadhvi*, a female practitioner who had renounced ordinary domestic life to devote herself to spiritual pursuits. At one point in our conversation she explained what she thought was the real value of worshiping a tree. "From the heartfelt worship of a single tree one can see the divinity in that tree and feel love for it. After some time, with knowledge one can then see the divinity in all trees. Really, in all life. All life is sacred because God is everywhere and in everything. This tree is a *svarupa* of Vasudeva (Krishna). As it says in the Bhagavad Gita, from devotion to a *svarupa* (one's own particular form of God) comes awareness of the *vishvarupa* (universal form of God)" (Haberman 2013, 197). In brief, this knowledgeable woman was advancing the idea that the worship of a particular has the possibility of expanding to a more reverent attitude toward the universal. Regarding trees, her point was that the worship of a particular tree could lead to the realization of the sacrality of all trees – and by extension, of everything.

With the comprehension of the universal via the particular we return full circle to the notion of *sarvatma-bhava*, the idea that everything is sacred. What first began as a proposition is now directly realized in experience. Many within Hindu religious traditions maintain that it is precisely a reawakening to this deep sacred quality of all life that is the foundation for establishing a more sustainable human presence on the planet. The notions related to the Bhagavad Gita's four "*s*'s" are deeply embedded in Hindu devotionalism. Here, then, is a potential resource that is already in place within popular Hindu culture for an emerging environmental practice and ethic that can extend loving care to all of life.

Notes

1   In addition to numerous articles and several monographs, former studies of Hinduism and ecology include two useful volumes of essays written by various scholars (Chapple and Tucker 2000, Nelson 1998).
2   More recently Emma Tomalin has argued for a distinction between what she calls biodivinity and environmental concerns. Biodivinity refers to the notion that nature is infused with divinity. This is an idea that has been current in India for a long time; the environmental crisis, however, is relatively new, as are the concerns related to it. Tomalin insists, therefore, that "there is an immense difference between the priorities and concerns of the modern environmentalist and the world-views of much earlier Hindu sages, poets, and philosophers" (Tomalin, 2004, 267). This does not mean, however, that aspects of Hinduism cannot be interpreted to support contemporary environmental thinking and action. As Tomalin recognizes, "Religious traditions constantly re-invent themselves precisely through making claims about the past in order to accommodate new ideas" (268). Sacred views of nature in India might indeed now be very useful as a resource to promote the protection and care of the environment. In fact, this is precisely what is currently taking place. Tomalin has also published a book expanding on this subject (2009).
3   Ramachandra Guha maintains that "it is probably fair to say that the life and practice of Gandhi have been the single most important influence on the Indian environmental movement" (1998, 65–66). The Norwegian philosopher and founder of deep ecology, Arne Naess, was greatly influenced by Gandhi; he took the conceptually central term "Self-Realization" from Gandhi, who in turn took it from the Bhagavad Gita (Naess 1995).
4   See Guha (1999, 162). The environmental activist and Chipko spokesperson Sunderlal Bahuguna

frequently quotes from the Bhagavad Gita to support his own ecological theology. The Chipko Movement and Bahuguna have been key sources for the Indian environmental movement. "Indeed, the origins of the Indian environmental movement can be fairly ascribed to that most celebrated of forest conflicts, the Chipko movement of the central Himalaya" (Gadgil and Guha 1995, 84).

5   "Conserve ecology or perish – this in short, is one of the messages of the *Gita*, one of the most important scriptures of the Vedic way of life now known as Hinduism" (Ahmed, Kashyap, and Sinha 2000, 28).

6   Two recent publications based on Shankaracharaya's reading of the Bhagavad Gita challenge the validity of this position of the Bhagavad Gita (Jacobsen 1996, Nelson 2000). See my own critical assessment of these articles (Haberman 2006, 29–37).

References

Ahmed K., Kashyap S., and Sinha S. (2000) "Pollution of Hinduism", *Down to Earth* Science and *Environment Fortnightly* published by Centre for Science and Environment, New Delhi (February 15th edition) 27–37 (available online at: www.downtoearth.org.in/coverage/pollution-of-hinduisim-17622).

Bruun O. and Kalland A. (1995) "Images of nature: an introduction to the study of man-environment relations in Asia", in Bruun O. and Kalland A. eds. *Asian Perceptions of Nature*, Curzon Press, Richmond, UK 1–24.

Chapple C. (1998) "Toward an indigenous Indian environmentalism", in Nelson L. ed. *Purifying the Earthly Body of God: Religion and ecology in Hindu India*, State University of New York Press, Albany 13–37.

Chapple C. and Tucker M. eds. (2000) *Hinduism and Ecology*, Harvard University Press, Cambridge.

Epley N., Waytz A., Akalis S., and Cacioppo J. (2008) "When we need a human: motivational determinants of anthropomorphism", *Social Cognition* 26:2, 143–55.

Gadgil M. and Guha R. (1995) *Ecology and Equity*, Penguin Books, New Delhi.

Guha R. (1998) "Mahatma Gandhi and the environmental movement in India", in Kalland A. and Persoon G. eds. *Environmental Movements in Asia*, Curzon Press, Richmond, UK 65–82.

Guha R. (1999) *The Unquiet Woods: Ecological change and peasant resistance in the Himalaya*, Oxford University Press, New Delhi.

Haberman D. (2006) *River of love in an Age of Pollution: The Yamuna River of northern India*, University of California Press, Berkeley.

Haberman D. (2013) *People Trees: Worship of trees in northern India*, Oxford University Press, New York.

Jacobsen K. (1996) "Bhagavadgita, Ecosophy T, and deep ecology", *Inquiry* 39:2, 219–38.

Naess A. (1995) "Self-Realization: an ecological approach to being in the world", in Drengson A. and Yuichi I. eds. *The Deep Ecology Movement: An introductory anthology*, North Atlantic Books, Berkeley 13–30.

Narayanan V. (2001) "Water, wood, and wisdom: ecological perspectives from the Hindu tradition", *Daedalus* 130:4, 179–206.

Nelson L. ed. (1998) *Purifying the Earthly Body of God: Religion and ecology in Hindu India*, State University of New York Press, Albany.

Nelson L. (2000) "Reading the Bhagavadgita from an ecological perspective", in Chapple C. and Tucker M. eds. *Hinduism and Ecology*, Harvard University Press, Cambridge, 127–64.

Pedersen P. (1995) "Nature, religion, and cultural identity: the religious environmental paradigm in Asia" in Bruun O. and Kalland A. eds. *Asian Perceptions of Nature*, Curzon Press, Richmond, UK 258–76.

Sherma R. (1998) "Sacred immanence: reflections on ecofeminism in Hindu Tantra", in Nelson L. ed. *Purifying the Earthly Body of God: Religion and ecology in Hindu India*, State University of New York Press, Albany 89–131.

Tam K., Lee S., and Chao M. (2013) "Saving Mr. Nature: Anthropomorphism enhances connectedness to and protectiveness toward nature", *Journal of Experimental Social Psychology* 49:3 514–21.

Tomalin E. (2004) "Bio-divinity and biodiversity: Perspectives on religion and environmental conservation in India", *Numen* 51:3 265–95.

Tomalin E. (2009) *Biodivinity and Biodiversity: The limits to religious environmentalism*, Ashgate, Farnham, UK.

5

BUDDHISM

A mixed Dharmic bag: Debates about Buddhism and ecology

Christopher Ives

In recent decades Buddhists have started formulating responses to the climate crisis and other environmental problems. In the months leading up to the 2015 climate conference in Paris, for example, the Dalai Lama, Thich Nhat Hanh, and other Buddhist leaders signed the "Buddhist Climate Change Statement to World Leaders." In 2009 several eco-Buddhists published an edited volume, *A Buddhist Response to the Climate Emergency*, which lead to the formulation of an organization, Ecological Buddhism, and a declaration, "The Time to Act is Now: A Buddhist Declaration on Climate Change." Another group of Buddhists, many of whom are connected to Spirit Rock Meditation Center, founded in 2013 the Dharma Teachers International Collaborative on Climate Change and issued a declaration of their own: "The Earth is My Witness." A third recently formed organization, One Earth Sangha, takes as its mission "expressing a Buddhist response to climate change and other threats to our home." A range of other Buddhist organizations and institutions have been offering additional responses to the eco-crisis, including the International Network of Engaged Buddhists (led by Thai Buddhist Sulak Sivaraksa), the Buddhist Peace Fellowship, Ordinary Dharma, Green Sangha, the Green Gulch Zen Center north of San Francisco, and the Zen Environmental Studies Institute at Zen Mountain Monastery in New York State, as well as the Boston Research Center for the twenty-first century, Wonderwell Mountain Refuge in New Hampshire, the Sarvodaya Movement in Sri Lanka, the Tesi Environmental Awareness Movement in Tibet (also known as Eco-Tibet), and the headquarters of Sōtō Zen Buddhism in Japan. Parallel to the praxis of these groups, eco-Buddhists have published monographs, anthologies, and articles in journals and popular Buddhist publications. What we are seeing in these writings is the emergence of a new theoretical dimension of the Buddhist tradition: environmental ethics.

In this "greening" of Buddhism, eco-Buddhists have tapped an array of sources: texts, doctrines, ethical values, and ritual practices. The arguments and activism of these Buddhists, however, are not without controversy. Critics have claimed, for example, that Buddhism has not been as ecological as some have made it out to be, and that eco-Buddhists are engaging in acts of eisegesis by looking selectively in Buddhist sources to support the environmental ethic they brought to their practice of Buddhism in the first place.

It is important to note that eco-Buddhists are generally focused more on continuing their activism than on responding to the skeptics. In this respect, there is no ongoing debate per se, though several eco-Buddhists have responded to the main criticisms, which concern "interdependence," identification with nature, Buddhist views of nature, the status of animals, Buddhism in relation to core constructs in Environmental Ethics, and adapted ritual practices.

Interdependence and identification

Much of the debate about Buddhism and ecology has centered on interpretations of *paṭicca-samuppāda* (Skt. *pratītya-samutpāda*), which eco-Buddhists often translate as interdependence but can be more accurately translated as "dependent origination." The Buddha reportedly expressed this doctrine as a broad principle: "When this exists, that comes to be; with the arising of this, that arises. When this does not exist that does not come to be; with the cessation of this, that ceases."[1] Eco-Buddhists frequently lift up this doctrine in support of their arguments that Buddhism, based on this notion of radical interconnectedness, is ecological and that Buddhist practice fosters a strong awareness of this interconnection as well as intimacy if not identity with nature. According to leading eco-Buddhist Joanna Macy, the egotistical self is "replaced by wider constructs of identity and self-interest—by what you might call the ecological self or the eco-self, co-extensive with other beings and the life of our planet" (Macy 1990, 53); and this shift "puts one *into* the world with a livelier, more caring sense of social engagement" (Macy 1991, 190).

Critics have questioned whether recent discourse on interdependence accurately represents the Buddhist tradition. According to David McMahan, "The monks and ascetics who developed the concept of dependent origination and its implications saw the phenomenal world as a binding chain, a web of entanglement, not a web of wonderment" (2008, 153), and early Buddhist texts advocate not engagement but "*disengagement* from all entanglement in this web" (154). Mark Blum writes that early Buddhists were motivated not "to embrace, revere, or ordain nature, but to remove any and all personal *craving for and attachment to* nature within themselves so as to become aloof or indifferent (*upekṣa*)" (2009, 215). Critics also question claims that awakening to *paṭicca-samuppāda* leads us automatically to value and care for the world. Christopher Gowans writes, "Why should the realization that we human beings are interdependent parts of the natural world give us reason to value other parts of that world? That all things are interdependent would not seem to establish, all by itself, that these things have some kind of value that we should care about, appreciate or respect" (2015, 287).

The debate about early Buddhist views of the world, however, is not settled. Some have argued that the main thing that early Buddhists were rejecting was not the world or nature per se but certain ways of viewing it, responding to it, and living in it. Gowans writes, "It may be said...that in early Buddhism suffering is not an essential feature of the natural world as such, but of our unenlightened way of experiencing the world. Moreover, enlightenment is not an escape from the natural world, but a non-attached way of living in it (as exemplified by the life of the Buddha)" (2015, 284). From this perspective, nirvana is less a separate, unconditioned realm realized after one steps back from the conditioned world of samsara than a mental state attained when one frees oneself from the "three poisons" of greed, ill-will, and ignorance. This facet of Buddhist thought becomes more pronounced in the emergence of Mahāyāna Buddhism, in which philosophers like Nāgārjuna, with their critique of the distinction between nirvana and samsara, shift the focus from "transcending samsara" to living an "awakened life in the midst of the world" (McMahan 2008, 158). (As we will see, Mahāyāna Buddhists view the conditioned world (of nature) described by the doctrine of *paṭicca-samuppāda* more positively than early Buddhists did, and it is generally out of this Mahāyāna perspective that eco-Buddhists marshal their arguments.) In short, although the monks who formulated the doctrine of *paticca-samuppāda* may have seen the world as a trap, this does not mean that the doctrine constitutes a negative view of the world.

Rendering *paticca-samuppāda* as interdependence has generated derivative statements that have prompted other criticisms. As I have outlined elsewhere (Ives 2009), eco-Buddhist discourse includes claims like "everything, including us, is dependent on everything else" (Loy 2003, 85); "in an undivided world everything miraculously supports everything else" (Batchelor 1992, 35); and "We are born into a world in which all things nurture us" (Aitken 2000, 426). Some eco-Buddhists have also derived from *paticca-samuppāda* a notion of responsibility, making claims like "in being aware of interdependence we also assume responsibility for all that occurs" (Deicke 1990, 166).

We can criticize such claims as these by noting that although things may affect each other, it is not necessarily the case that I depend on *everything* else or that *all* things support and nurture me: while I am affected by the destroyed nuclear reactors in Fukushima, they do not support or nurture me and my well-being does not depend on them but depends on my becoming physically independent of them. Nor in any intelligible ethical sense do we *all* have to assume responsibility for *everything* that happens: Jewish children in the Warsaw ghetto did not bear any responsibility for the Holocaust.

Buddhist views of nature

Some eco-Buddhists lift up passages from *suttas* to claim that from the start Buddhism has valued nature. For example, they point out that early canonical sources celebrate wild places—with their solitude, silence, and abundant examples of impermanence—as good locations for meditative practice. Critics have pointed out, however, that the Pali canon also portrays them as dangerous, for it is there that one encounters large predators like tigers, poisonous snakes and insects, bandits, and others who would do one harm. The preferred nature is a garden or groomed park, and the *Cakkavatti-sīhanāda-sutta* portrays a future utopia that is more urban than wild, as noted by Ian Harris: "In Jambudvīpa cities and towns are so close to one another that a cock can comfortably fly from one to the next. In this perfect world only urban and suburban environments are left" (Harris 1991, 108). This celebration of groomed gardens and urban utopias amounts to what Lawrence Schmithausen terms the "pro-civilization strand" of early Buddhism (1991, 14–17).

At the very least, however, early Buddhists did not see nature in stark instrumentalist terms as something to be exploited for the sake of building human cities and civilization. David Eckel writes, "one does not attempt to dominate or destroy nature (in the form of either animals or plants) in order to seek a human good" (1997, 337). "But," Eckel continues, "neither is the wild and untamed aspect of nature to be encouraged or cultivated. The natural world functions as a locus and an example of the impermanence and unsatisfactoriness of death and rebirth. The goal to be cultivated is not wildness in its own right but a state of awareness in which the practitioner can let go of the 'natural'—of all that is impermanent and unsatisfactory—and achieve the sense of peace and freedom that is represented by the state of *nirvāṇa*. One might say that nature is not to be dominated but to be relinquished in order to become free" (337).

This view of nature, however, is found mainly in early Buddhism rather than in the frameworks from which many eco-Buddhists are operating: Mahāyāna texts and East Asian Buddhism. These strands of Buddhism offer a view of nature that differs from what we have sketched thus far. The *Avataṃsaka-sūtra*, for example, formulates a notion of interconnection through the metaphor of Indra's Net and lifts up the seeker Sudhana, who has "a vision of the entire cosmos within the body of the Buddha Mahāvairocana," becomes one with that cosmic buddha, and thereby stands as the prime example of "the identification of a person with a being who is the universe itself or with the underlying reality of things" (McMahan 2008, 158). This interpretation of dependent origination, more positive view of the world, and advocacy of identification with the world helped shape Zen Buddhism and, by extension, Thich Nhat Hanh's argumentation about "interbeing" as the foundation for ecological awareness and compassionate responsiveness to suffering.

Eco-Buddhists also draw upon such East Asian resources as hermitage traditions, the celebration of nature in arts influenced by Buddhism, and discourse on the Zen-inspired love of nature ostensibly felt by the Japanese.[2] Also, as is the case with early Buddhism, many East Asian Buddhists value natural settings as good places for contemplative practice and as a bountiful source of symbols for Buddhist teachings like impermanence. Granted, this is, strictly speaking, a kind of instrumental value rather than intrinsic value, but nature is indeed being valued and the view of the natural world as dangerous, ensnaring, or unsatisfactory has dropped largely out of the picture.

The status of animals

Eco-Buddhists have lifted up the *Jātaka Tales*, with an array of virtuous animals, as granting value and dignity to non-human species. They have also cited Buddhist texts that establish a kinship between humans and animals; the *Laṅkāvatāra-sūtra*, for example, in admonishing Buddhists not to eat meat, includes the passage, "In the long course of rebirth there is not one among living beings with form who has not been mother, father, brother, sister, son, or daughter, or some other relative. Being connected with the process of taking birth, one is kin to all wild and domestic animals, birds, and beings born from the womb" (Swearer 2001, 227). Eco-Buddhists have argued that this intimate karmic connection between humans and animals provides a basis for valuing animals.

Scholars have pointed out, however, that in most Buddhist texts animals are portrayed as intellectually and morally inferior to humans and exist as one of the three "unfortunate" types of rebirth: they do not restrain their desires, they can be malevolent when they prey on other animals, and they lead an unhappy existence (Schmithausen 1991, 16). As such, Ian Harris writes, "beyond the fact that they appear to be beings destined for final enlightenment, they have no intrinsic value in their present form" (1995, 107). In response, Donald Swearer has argued that Harris's "position is founded on too narrow a construction of the Buddhist view of nature and animals based on a selective reading of particular texts and traditions" and that Harris needs to take into account the *Jātaka Tales*, which do value animals (1997, 39). Gowans and Harris point out, however, that the animals in these stories are anthropomorphized and function to motivate humans to cultivate virtues like compassion. Gowans comments that these tales use the device of "depicting various living beings as proxies for human beings," and "These are mainly morality tales about human beings…" (2015, 282). Harris claims that "the often highly anthropomorphic character of the essentially pre-Buddhist folk-tradition of the *Jātakas* may be said to empty the stories of any 'naturalistic' content, thus defeating the intention of those who bring them forward as evidence in support of an authentic Buddhist environmentalist ethic" (2000, 121). Moreover, "in the *Jātaka* context the animals are not animals at all in any accepted sense of the term, for at the end of each story the Buddha reveals that the central character was none other than himself, the *bodhisattva*, in a former life" (Harris 2000, 121).

Even so, one might respond, animals are viewed there not as mere objects but as sentient beings with at least some value, even if the tradition did not—until recently—build on this to argue in a systematic way for the protection, moral standing, or rights of animals.

Buddhism in relation to Environmental Ethics

Overlapping with the debate about the proper connotation and denotation of core Buddhist doctrines like *paṭicca-samuppāda* has been a debate about Buddhism and Environmental Ethics in the formal sense. Some critics have argued that Buddhism is ill-equipped to argue for the sorts of things that typically appear as cornerstones of philosophical and religious formulations of environmental ethics, whether rights, intrinsic value, or the sanctity of nature.

Some Buddhist writers have made claims about animal rights. Philip Kapleau, for example, has written, about the rights animals "undeniably have" (1986, 13). Critics have raised the matter of what might be a legitimate Buddhist basis for claims about the possession of rights, given the Buddhist rejection of the soul and any other sort of separate, atomistic existence apart from the web of changing relationships that constitute things. In response, eco-Buddhists have argued that intrinsic value and moral standing derive from sentience, especially the ability to feel pain and suffer in a significant sense. Others have looked to buddha-nature, but Buddhists do not agree on the connotation and scope of this construct. Some think of it as the potential to become awakened while others see it as an inherent awakening. Early Buddhists ascribed it only to (sentient) animals, not to (insentient) plants, while some in East Asia extended the scope to plants and even to inorganic things like rocks and waters. Some eco-Buddhists have celebrated this broad attribution of buddha-nature as a powerful ethical resource, but in terms of the doctrine's usefulness for environmental ethics, we must address the subject of what "the view of the presence of Buddha-Nature even in plants, mountains, and rivers entails for practical behavior" (Schmithausen 1991, 24).

Entering this debate, one can argue that rather than forcing Buddhism to fit into received categories and frameworks in environmental ethics (or Western philosophical ethics more broadly), eco-Buddhists might remain true to their tradition and still construct a viable environmental ethic by taking as their primary focus the alleviation of suffering of humans and other sentient beings, or in positive terms, the promotion of their sustained well-being, which is contingent upon certain types of ecosystems.

Of course, focusing on humans and other sentient beings lands us in the arena of the debate about the respective values of individuals and the wholes of which they are part, that is to say, the ongoing debate in Environmental Ethics between individualism and holism. In large part Buddhist ethical concern—expressed through such doctrines as non-harming, loving-kindness, compassion, and the bodhisattva ideal—is directed toward individual suffering beings, not groups, species, or wholes like ecosystems.

In general, however, while they may not agree on whether the main Buddhist ethic is a virtue ethic or a form of utilitarianism, scholars and Buddhists tend to agree that central Buddhist virtues—or to put it in a way that is more faithful to Buddhism, wholesome mental states—do offer resources for environmental ethics in several senses, especially the informal sense of "sets of beliefs, values, and guidelines that get put into practice in attempts to live in an ecological manner" (Ives 2013, 544). As I have outlined elsewhere (2013), Buddhism offers a view of flourishing that is based on the cultivation of an array of "wholesome" mental states and values with clear environmental ramifications: generosity, non-acquisitiveness, simplicity, frugality, restraint, contentment, loving-kindness, non-harming, and mindfulness.

Simply put, as humanity faces the eco-crisis, Buddhism offers a value system and way of living that not only lead to greater fulfillment than materialist and consumerist living does but also prove useful for mitigating such problems as global warming and adapting to a new world in which we will all be forced to live more simply. That being said, Stephanie Kaza has laid the groundwork for an important debate with a remark about one of the Buddhist values often lifted up in eco-Buddhist discourse:"The practice of detachment to hobble the power of desire could actually work against such environmental values as 'sense of place' and 'ecological iden- tity'" (2006, 201).

While the de facto virtue ethic of Buddhism does offer resources for ecological living, the discipline of Ethics features an ongoing debate about the limitations of virtue ethics—Buddhist or otherwise—in responding to urgent problems like the climate crisis. Though the cultiva- tion of a virtuous character over the course of a lifetime may very well lead to a more sustainable way of being, it does not readily prompt the kind of immediate response that the climate crisis calls for, nor does it offer much help in making decisions about what might be effective responses to the climate crisis and other environmental problems.

Adapted ritual practices

In addition to tapping Buddhist metaphysical constructs, texts, and values, eco-Buddhists have reformulated ritual practices, invented new practices, or simply engaged in activism in response to environmental problems,[3] and these efforts have spawned debates as well. Buddhists in Thailand have been debating the practice of ordaining trees as a way to protect them from logging and protect rural farming communities that depend on forests. This practice, originating in the 1980s, immediately caused a backlash from developers and government officials whose profits, power, and agendas were threatened by the practice. Critics among the laity and the sangha administra- tion have claimed that the environmentalist monks performing the rituals cannot ordain trees, for ordination rituals can be done only for humans (Darlington 2012), and that political and economic activism is inappropriate for monks and reduces their purity. In particular, as Sue Darlington points out, the ordinations "challenge what people consider sacred—placing trees on the same level as monks goes against the sacred and social hierarchy in place" (2012, 23).

This debate in Thailand is part of a larger debate about the appropriateness of Buddhist activism. Over the years this author has heard Zen masters and other Buddhist teachers advo- cate that their students devote their efforts to intensive meditative practice and defer social activism until after they have woken up or at least reach advanced stages on the Buddhist path. Some have even said that if one tries to save the world before extricating oneself from the self- centered ego, one will only end up making things worse. As part of a critique of broader "Engaged Buddhism," some have also argued that eco-Buddhism is a watering down of Buddhism insofar as it draws attention away from sustained wrestling with existential suffering and directs it to political agitation.

An eco-Buddhist might respond to this criticism by noting that existential suffering is not the only form of suffering that the Buddha took seriously, and working to reduce social, economic, and other forms of suffering through activism falls within the scope of the founda- tional Buddhist commitment to reduce suffering in all of its forms.

Concluding remarks

Perhaps the harshest criticism to date in the debate about Buddhism and ecology has come from Ian Harris, who once wrote that eco-Buddhism consists primarily of "exogenous elements somehow tacked on to a traditional Buddhist core which is incapable, without modi- fication, of responding to the present environmental crisis" (1995, 206). Granted, some eco-Buddhists may be misconstruing doctrines, but most are simply reinterpreting them in response to the eco-crisis, and this hermeneutic should not be dismissed out of hand. In some respects they are doing the "modification" that Harris mentions, and in most cases what we are seeing are reinterpretations of doctrines and practices in ways that at the very least do not contravene the overall Buddhist worldview and may actually be drawing out its ecological ramifications in a legitimate exegetical manner. In this respect eco-Buddhists are engaging in the sort of intellectual labor that, for example, biblical theologians have been doing for centuries as they look selectively in the Bible for passages that support the constructive argument they are making (and defending as consistent with what they take to be the *core* principles of Judaism or Christianity) in response to challenges they have faced in their particular historical situations. For example, many sections of the Bible accept—or at least do not reject—slavery, but this does not mean that anti-slavery arguments that have tapped other parts of the Bible are illegitimate. Likewise, the presence of negative views of wild nature and animals in early Buddhist texts does not in and of itself delegitimize theorizing that draws from other resources in those—or other—Buddhist texts (though it does undermine broad claims like "Buddhism is an ecological religion" or "Buddhists have always revered nature").

Like other religious traditions, then, Buddhism has continuously changed as its beliefs and practices have been reinterpreted in different cultural contexts and historical moments. So as David McMahan points out,

> Simply to dismiss the current environmental and ethical discourse of Buddhist interdependence as an inadequate representation of traditional Buddhism…would fail to take seriously the process of modernity as it manifests itself on the ground…. Like virtually all normative religious reflection, this discourse is practitioners' constructive response to an unprecedented situation, not a historiographical endeavor. Pointing out the incongruities between ancient and modern cosmologies, while crucial, is not more historically important than showing how the often radical reconstitution of doctrine in terms of present circumstances has attempted to bridge these incongruities. The history of religions is precisely the history of such reconstitutions of doctrine and practice, which are themselves reconstitutions of prior versions.
>
> *(2008, 180)*

It is also important to note that some of the most important eco-Buddhists doing this modification and reinterpretation are not convert Buddhists who might be bringing exogenous elements from their Christian, Jewish, or leftist roots to bear on Buddhism but rather renowned Asian Buddhists who were brought up in the Buddhist tradition, such as the Dalai Lama, Thich Nhat Hanh, Buddhadāsa, and Sulak Sivaraksa. Harris seems to assume that eco-Buddhists are all Western converts or simply people approaching Buddhism from typically Western perspectives, but this is clearly not the case (even allowing for some degree of Western influence on Asian eco-Buddhists). For this reason, the argument that "much that masquerades under the label of eco-Buddhism…on analysis, turns out to be an uneasy partnership between Spinozism, New Age religiosity and highly selective Buddhism" (2000, 132) does not do justice to the full scope of eco-Buddhism.

At the same time, eco-Buddhists, or at least those focused on theory more than praxis, have much intellectual labor to do. For example, work needs to be done to clarify the exact resources that the doctrine of dependent origination offers. As a metaphysical construct, it does highlight how we are all embedded in nature and our actions affect everything around us and everything affects us, but, this process of interrelating or interbeing pertains to all configurations of reality, whether a relatively pristine wilderness area or a nuclear reactor that is melting down. For this reason, if we are to avoid the naturalistic fallacy of conflating the "is" and the "ought," and if we are to make wise decisions, we need to make distinctions between various configurations (such as the pristine wilderness area, the lethal reactor, this or that economic system, this or that way of living) by considering which are desirable or optimal and which are to be mitigated or eliminated. Some eco-Buddhists have begun addressing this question (Jones 1993, 2003; Loy 2003; Kaza 2008; Ives 1992, 2011, 2013), and as their formulations become more systematic we can expect further debates.

Notes

1 This appears, for example, in the eleventh section of the *Bahudhātuka Sutta* in the *Majjhima Nikākaya*. Bhikkhu Ñāṇamoli and Bhikkhu Bodhi, trs. (2001), *The Middle Length Discourses of the Buddha: A Translation of the Majjhima Nikāya* (Boston: Wisdom Publications), 927.
2 Technically, as several of us have pointed out, in Japan the nature that is valued and loved most is a tamed, distilled, miniaturized, and stylized nature, not wild creatures, ecosystems, or the wilderness (Ives 2005, 900).
3 For examples of this, see Kaza 2000 and Swearer 1997 and 2001.

References

Aitken, Robert (2000). "Envisioning the Future," in *Dharma Rain: Sources of Buddhist Environmentalism*, edited by Stephanie Kaza and Kenneth Kraft, 423–438. Boston: Shambhala.
Batchelor, Stephen (1992). "The Sands of the Ganges: Notes towards a Buddhist Ecological Philosophy," in *Buddhism and Ecology*, edited by Martine Batchelor and Kerry Brown, 31–49. New York: Cassell Publishers.
Blum, Mark (2009). "The Transcendentalist Ghost in EcoBuddhism," in *TransBuddhism: Transmission, Translation, Transformation*, edited by Nalini Bhushan, Jay Garfield, and Abraham Zablocki, 209–238. Amherst: University of Massachusetts Press.
Darlington, Susan M. (2012). *The Ordination of a Tree: The Thai Buddhist Environmental Movement*. Albany: State University of New York Press.
Deicke, Carla (1990). "Women and Ecocentricity," in *Dharma Gaia: A Harvest of Essays in Buddhism and Ecology*, edited by Alan Hunt Badiner, 165–168. Berkeley: Parallax Press.
Eckel, Malcolm David (1997). "Is There a Buddhist Philosophy of Nature," in *Buddhism and Ecology: The Interconnection of Dharma and Deeds*, edited by Mary Evelyn Tucker and Duncan Ryūken Williams, 327–349. Cambridge, MA: Harvard University Press.
Gowans, Christopher W. (2015). *Buddhist Moral Philosophy: An Introduction*. New York: Routledge.
Harris, Ian (1991). "How Environmentalist is Buddhism?" *Religion* 21: 101–114.
Harris, Ian (1995). "Buddhist Environmental Ethics and Detraditionalisation: The Case of Eco-Buddhism," *Religion* 25, no. 3 (July 1995): 199–211.
Harris, Ian (2000). "Buddhism and Ecology," in *Contemporary Buddhist Ethics*, edited by Damien Keown, 113–135. London: Curzon Press.
Ives, Christopher (1992). *Zen Awakening and Society*. London and Honolulu: Macmillan and the University of Hawai'i Press.
Ives, Christopher (2005). "Japanese Love of Nature," in *The Encyclopedia of Religion and Nature*, vol. 1, edited by Bron R. Taylor, 899–900. New York: Thoemmes Continuum.
Ives, Christopher (2009). "In Search of a Green Dharma: Philosophical Issues in Buddhist Environmental Ethics," in *Destroying Mara Forever: Buddhist Ethics Essays in Honor of Damien Keown*, edited by Charles Prebish and John Powers, 165–185. Ithaca NY: Snow Lion Publications.
Ives, Christopher (2011). "Liberation from Economic Dukkha: A Buddhist Critique of the Gospels of Growth and Globalization in Dialogue with John Cobb," in *The World Market and Interreligious Dialogue*, edited by Catherine Cornille and Glenn Willis, 107–127. Eugene OR: Cascade Books.
Ives, Christopher (2013). "Resources for Buddhist Environmental Ethics," *Journal of Buddhist Ethics* 20: 541–571.
Jones, Ken (1993). *Beyond Optimism: A Buddhist Political Ecology*. Oxford: Jon Carpenter Publishing.
Jones, Ken (2003). *The New Social Face of Buddhism: A Call to Action*. Boston: Wisdom Publications.
Kapleau, Philip (1986). *To Cherish All Life: A Buddhist Case for Becoming Vegetarian*. Rochester: Rochester Zen Center.

Kaza, Stephanie (2006). "The Greening of Buddhism: Promise and Perils," in *The Oxford Handbook of Religion and Ecology*, edited by Roger S. Gottlieb, 184–206. New York: Oxford University Press.

Kaza, Stephanie (2008). *Mindfully Green: A Personal and Spiritual Guide to Whole Earth Thinking.* Boston: Shambhala.

Loy, David R. (2003). *The Great Awakening: A Buddhist Social Theory.* Boston: Wisdom Publications.

Macy, Joanna (1990). "The Greening of the Self," in *Dharma Gaia: A Harvest of Essays in Buddhism and Ecology*, edited by Allan Hunt Badiner, 53–63. Berkeley: Parallax Press.

Macy, Joanna (1991). *World as Lover, World as Self.* Berkeley: Parallax Press.

McMahan, David L. (2008). *The Making of Buddhist Modernism.* New York: Oxford University Press.

Schmithausen, Lambert (1991). *Buddhism and Nature.* Tokyo: International Institute for Buddhist Studies.

Swearer, Donald K. (1997). "The Hermeneutics of Buddhist Ecology in Contemporary Thailand: Buddhadāsa and Dhammapiṭaka," in *Buddhism and Ecology: The Interconnection of Dharma and Deeds*, edited by Mary Evelyn Tucker and Duncan Ryūken Williams, 21–44. Cambridge, MA: Harvard University Press.

Swearer, Donald K. (2001). "Principles and Poetry, Places and Stories: The Resources of Buddhist Ecology," *Daedalus* 130, no. 4 (Fall 2001): 225–241.

6

CONFUCIANISM

Confucian environmental virtue ethics

Yong Huang

1. Introduction

Environmental ethics, as an applied ethics, may be regarded as an application of some general ethical theory to specific environmental issues. Alternatively, as traditional ethical theories are basically limited to human beings as moral patients (recipients of action), environmental ethics can also be seen as their expansion, to include the non-human parts of the ecosystem among moral patients. In either case, environmental ethics is closely related to general moral theories. Similar to ethics in general, consequentialism, which focuses on the consequences of actions, and deontology, which focuses on moral rules or principles, dominated much of the initial development of environmental ethics. However, virtue ethics, which focuses on the characters of moral agent, has now become a powerful alternative in environmental discourses (Hill 1983, van Wensveen 2000; Thoreau 1951; Carson 1956; Bardsley 2013), partially due to its own attractiveness and partially due to the respective deficiencies of deontology (see Kant 1997: 212; O'Neill 1993: 22–24; Sandler 2007: 113) and consequentialism (see Zwolinski and Schimdtz 2013), either in these theories themselves or in their applications/expansions to environmental issues.

However, there is also a problem with this virtue ethics approach to environment issues, which shifts our attention from nature to us human beings. We need to acquire virtues, including environmental virtues, because they contribute to or are even constitutive of human flourishing, which is clearly anthropocentric (O'Neill 1993; Rolston 2005; Cafaro 2015). In this chapter, I shall develop a Confucian version of environmental virtue ethics that is not anthropocentric, by focusing on the work of the neo-Confucian philosopher in the Ming dynasty (1368–1644) of China, Wang Yangming (1472–1529). He is commonly regarded as the leader of the idealist school of the neo-Confucian movement to revive Confucianism, starting from the Song dynasty (960–1279), in response to the dominance of Buddhism in the Tang period (see Tu, Cua, and Ching 1996).

2. Being in one body with the ten thousand things

Confucian ethics is primarily a type of virtue ethics. The most fundamental virtue in Confucianism is *ren*, most frequently translated as humanity or humaneness, the virtue that characterizes a human being as a *human* being. Different Confucians define this humanity differently, and Wang Yangming's understanding of it comes more directly from Cheng Hao (1032–1085), a neo-Confucian in the Song dynasty. At the very beginning of his famous essay, "On Humanity," Cheng Hao claims that, "The first thing all learners need to understand is humanity. A person of humanity is completely in one body with all things," where there is no distinction between self and things as inner and outer (Cheng and Cheng 1988: 17). To illustrate it, Cheng Hao discusses the lack of humanity, *buren*, in the medical sense: "medical books regard the numbness of hand and foot as lack of *ren*. This is the best description" (Cheng and Cheng: 15). When one's hand is numb, one cannot feel its pain and itch, and this is the lack of *ren*. Thus when one is not numb, one can feel the pain and itch of one's whole body, and this is *ren*. Then he expands it to explain the moral sense of *ren*: "a person of *ren* takes heaven, earth, and the ten thousand things all as one body, no part of which is not oneself" (Cheng and Cheng: 15). If someone or something else in the world suffers, and I don't feel it, that means that I'm numb. In other words, I lack the virtue of humanity. By contrast, a person of humanity will feel the suffering of other things in the world, because he or she is in one body with the ten thousand things.

Wang Yangming adopts and expands on Cheng Hao's interpretation of *ren*. He states that, "a person of humanity (*ren*) regards heaven, earth and the ten thousand things as being in one body (with oneself). If there is one thing that cannot live its natural life, (he or she believes that) this must be because his or her own humanity is not fully developed" (Wang 1992: 25). When one's humanity is not fully developed, there are things that one does not feel to be in one body with. One does not regard them as part of one's body. The reason that one fails to feel being in one body with everything in the universe, for Wang Yangming, is that "one has not fully got rid of one's selfish desires" (Wang 1992: 110), which blocks one's original heart–mind, just as clouds block the sunshine. This means that one's original heart–mind naturally feels to be in one body with the ten thousand things (Wang 1992: 968).

Moreover, just as a cloud cannot completely block the sunshine, selfish desires cannot completely block one's original heart–mind, which is characterized by the virtue of "humanity." Wang explains this in the following famous passage:

> When a superior person sees an infant about to fall into a well, he or she will definitely have the heart that feels the alarm and commiseration. This is because his or her *ren* makes him or her to be in one body with the infant. You may say that this is because the infant is of the same species as the superior person. However, when he or she sees a bird or land animal making sad sound and having frightened appearance, he or she will definitely have a heart that cannot bear to let it happen. This is because his or her *ren* makes him or her to be in one body with the bird or animal. You may say that this is because the bird and animal are also sentient beings. However, when he or she sees a blade of grass or a tree breaking, he or she will definitely have the heart that feels pity. This is because his or her *ren* makes him or her to be in one body with the grass or tree. You may say that this is because they are also living beings. However, when he or she sees a tile or stone getting broken, he or she will definitely have the heart that feels regret. This is because his or her *ren* that makes him or her to be in one body with the tile or stone.
>
> *(Wang 1992: 26; 968)*

In this passage, Wang Yangming repeatedly emphasizes that the reason that a person is concerned with the wellbeing of others when something bad happens to them is not simply that they belong to the same species, but that the person has *ren*, the sensibility toward the suffering of others.

Since Wang Yangming's conception of being in one body with the ten thousand things is used to explain *ren*, the most fundamental virtue in Confucianism, his ethics is a virtue ethics. Since the object of one's concern goes beyond human beings to include heaven, earth, and the ten thousand things, it is also an environmental virtue ethics. However, the familiar version of environmental virtue ethics that we see in contemporary Western discussions is largely Aristotelian. By contrast, Wang Yangming's version of environmental virtue ethics is a sentimental one, since its central concept, being in one body with the ten thousand things, is essentially a concept of empathy. In contemporary moral psychology, empathy is understood to be a kind of emotion or an emotion-generating entity. An empathic person is one who is able to feel the pain and suffering of another person, not merely to feel about the pain and itch of that person, which is characteristic of sympathy. Thus an empathic person is naturally motivated to help the other get rid of the pain, just as anyone who feels itch in his or her back will naturally move his or her hand to scratch it. As we have seen, this is precisely how Cheng Hao and Wang Yangming describe the person who is in one body with the ten thousand things.

In appearance, Wang Yangming's neo-Confucian version of environmental virtue ethics, with empathy as its central concept, also suffers anthropocentrism and even egoism. In such a version, an environmentally virtuous or empathic person is concerned about both other human beings and non-human beings because he or she feels pain that others suffer as his or her own. In this sense, it may be claimed that this virtuous person's concern with and help for others is really also egoistic in the narrow sense and anthropocentric in the broad sense.

Defenders of this Confucian version of environmental virtue ethics may be tempted to appeal to Daniel Baston and his team, who have conducted a series of experiments to show that an empathic person's concerns with others is not due to a selfish desire to escape aversive arousal, or social disapproval, or guilt, or shame, or sadness, or to increase vicarious joy. Instead, such a concern is purely altruistic: an empathic person's concern with others is for the sake of others (see Batson 2011). However, when applied to environmental virtues, one would ask, why should a virtuous person take care of non-human beings for their own sake? Is this because they have intrinsic values and not merely instrumental values to human beings? However, a thing's having intrinsic value does not necessarily mean that it has the right for our care or that we have duty to take care of it. A virus also has its own intrinsic value, but it does not automatically have a right for our care.

Wang Yangming's environmental virtue ethics, with the central idea of being in one body with the ten thousand things, can avoid the problem of anthropocentrism or egoism in a different way. Anthropocentrism assumes the separateness between human and non-human beings or between self and other. However, since a virtuous person in Wang Yangming's sense feels to be in one body with the ten thousand things, there is no such separateness. The whole universe becomes the virtuous person's single body, and the ten thousand things in the universe become different parts of this virtuous person's own body. So, a virtuous person's taking care of the forest, for example, can be seen as both for the sake of the forest and as for the sake of himself or herself, since this forest is already part of his or her body.

3. Love with distinction

However, Wang does not think that this Confucian empathic person loves, or should love, the ten thousand things equally. Some sacrifices have to be made, especially when conflicts arise among the ten thousand things, now all different parts of a virtuous person's body.

This view of Wang Yangming's is made most clear in one of his conversations with his students. The *Great Learning*, one of the Confucian classics, calls for cultivating the self, regulating the family, governing the state, and harmonizing the world. It then states that "everyone, from the emperor to common people, should regard cultivation as the root. It is impossible to have a distorted root with ordered branches, and there has never been the case when less intense care is given when the more intense care is called for or the more intense care is given when the less intense care is called for" (Wang 1992: 108). According to Zhu Xi (1130–1200), one of the most influential neo-Confucians, while the root refers to self-cultivation, the branches refer to regulating families, governing the state, and harmonizing the world. While intense care is meant for family members, the less-intense care is meant for other people in the state and the world. Having learned about Wang's teaching, one of his students asks: "since a great person feels being in one body with (the ten thousand) things, why does the *Great Learning* still talk about the distinction between more and less intense care" (Wang 1992: 108).

Here is Wang's response in a well-known passage, directly related to our concern with environmental issues:

> There is a reason (*daoli*) for this natural distinction between the more and the less intense care. For example, all in one body, we use our hands and feet to defend our head and eyes. It is not that we are discriminating against our hands and feet. There is a reason for being so. We love both birds and animals on the one hand and grasses and trees on the other, but it is bearable for us to feed birds and animals with grasses and trees. We love both human beings and birds and animals, but it is bearable for us to kill birds and animals to feed parents, to sacrifice for rituals, and to entertain guests. We love both our parents and strangers, but it is bearable for us to save our parents and not the stranger when there is only one single dish of food, with which one can live and without which one will die. Things are so for a reason. The distinction between the more and the less intense care in the *Great Learning* is the natural vein (*tiaoli*) of one's innate moral knowledge (*liangzhi*), which is the moral rightness that cannot be transgressed.
>
> *(Wang 1992: 108)*

So while calling upon us to cultivate empathy with the ten thousand things, Wang does not go to the extreme that we should treat everything in nature, including us human beings, equally or impartially. Yet other than his view about animals (which may be killed to feed our parents, etc.), Wang's view is something even the most radical ecologist would accept. Clearly, hardly any environmental activist would object to our cutting a tree if this tree contains elements that alone can cure cancer suffered by numerous patients. The question we have is on what basis Wang develops his partialist view. Just as his view of being in one body with the ten thousand things does not assume the equality of the intrinsic values of the ten thousand things, his partialist view is not based on his conception of inequality of intrinsic values of the ten thousand things. While we may be tempted to think that Wang has a hierarchical view of intrinsic values of different beings when we see him allowing us to feed birds and animals with grasses and trees and kill animals for human purposes, thinking that humans have higher intrinsic value than animals, and animals have higher intrinsic values than plants. However, we have to abandon this assumption when we see him talking about the preferential treatment of our parents over strangers along the same line, as he certainly would not think that our parents possess higher intrinsic value than strangers. If so, precisely what does he mean by the "natural reason" (*daoli* or *tiaoli*) for such a partiality that he repeatedly talks about?

We can begin to understand it by realizing that our empathy with and empathic care for the ten thousand things is naturally a gradual process starting from the near and dear. This point is made clear in another conversation between Wang Yangming and one of his students. The student asks: "Master Cheng (Hao) says that 'a person of *ren* feels to be in one body with the ten thousand things.' How then is the Mohist idea of impartial love not *ren*?" (Wang 1992: 25). From the student's point of view, to be in one body with the ten thousand things means to have impartial love for them, and, if so, it must be right for the Mohists to advocate impartial love, which, however, all Confucians regard as problematic. In response, Wang Yangming emphasizes the gradual nature of our empathy with things, with an analogy:

> Take a tree as an example. At the beginning, there is sprout, from which the vitality of the tree originates; from the sprout grows the trunk; and from the trunk grow the branches and leaves; and then its life cycle continues ceaselessly. If there is no sprout, how can there be trunk, branches, and leaves? And there can be sprout only if there is root. Only with the root can the tree grow; without the root, the tree will die. If there is no root, how can there be sprout? The love between parents and children and among siblings is the beginning of the vitality of human heart/mind; it is like the sprout of a tree. Starting from the beginning, one can be humane to all people and love all things; it is like the tree's growing the trunk, branches, and leaves. The Mohist idea of impartial love without distinction sees one's own parents, children, and siblings as no different from strangers; as a result there is no beginning (of love).
>
> *(Wang 1992: 25–26)*

Thus, one aspect of Wang's "natural reason" for the partiality in our empathy with and empathic care of the ten thousand things is that we have to start from those near and dear to us and then gradually expand our empathy and empathic care to others. However, by itself, this does not imply any preferential treatment of those near and dear to us; it only stipulates a temporal order: we love our family members first and others later. Indeed, this is not something that Mohism really has any problem with when it argues against Confucianism. For example, the Mohist Yizi in the *Mencius* states that "we Mohists hold that there should be no distinction in love, although *our love can start with parents*" (*Mencius* 3a5; emphasis added). In other words, it is fine to first love those who are near and dear, as long as such a love is equal to our love for others, which happens later. However, this is clearly not what Wang Yangming in particular and Confucians in general have in mind with their idea of love with distinction. In the passage quoted above, Wang Yangming argues that we should have more intense love for those near and dear to us than for others, especially when our love for the former comes into conflict with our love for the latter. For Wang Yangming, there is also a "natural reason" (*daoli* or *tiaoli*) for this, although Wang may take such a reason as self-evident and thus does not fully explain it.

Indeed, to explain why we ought to have more intense love of those near and dear to us may be what contemporary philosopher Bernard Williams regards as "one thought too many". Suppose I'm in a situation in which my wife and a stranger are in equal peril and yet I can only rescue one of them. I'll naturally save my wife without any further thought. If my motivating thought, fully spelled out, is, in addition to the fact that she is my wife, in situations of this kind it is permissible to save my wife, it is "one thought too many" (Williams 1981: 18). The situation that Williams imagines is almost identical to the one conceived by Wang Yangming: when there is only one dish of food, with which a person lives and without which a person dies, one can bear to use it to save one's parent and not one's stranger. This natural tendency of partiality toward those near and dear to us has recently received empirical support from contemporary moral psychology of empathy. As Martin Hoffman, one of the most influential moral psychologists studying the phenomenon of empathy, points out, there is evidence "that most people empathize to a greater degree (their threshold for empathic distress is lower) with victims who are family members, members of their primary group, close friends, and people whose personal needs and concerns are similar to their own" (Hoffman 2000: 197).

So the second aspect of Wang Yangming's "natural reason" is that we naturally tend to have stronger empathy with those near and dear to us. However, this is still not enough. What we are naturally doing or tend to do is not necessarily what we ought to do. When Williams claims, in relation to the above-mentioned case of a husband's rescuing his wife, that "some situations lie beyond [moral] justification" (Williams 1981: 18), he considers, at least from Susan Wolf's point of view, what the husband does is a non-moral good, which is as important as a moral good and so should not be trumped by the latter (Wolf 2012). In this respect, Hoffman goes a step further, thinking that this aspect of empathy is a bias that comes into conflict with moral philosophy's criterion of impartiality. Since for him, "empathic morality, at least empathic morality alone, may not be enough" (Hoffman 2000: 206), and so it is important for empathy to be supplemented with or embedded in the moral principle of justice. Even in Williams's case, while he considers it absurd for us to require the person to provide a justification for his action of rescuing his wife instead of the stranger in addition to the fact that she is his wife, he does not say that this action itself is justified from a moral point of view.

However, when Wang Yangming says that there is a "natural reason" for us to be partial to those who are near and dear to us, this natural reason is clearly not merely descriptive but also normative. This is the third aspect that I would like to highlight. Why should we give preferential treatment to those who are near and dear to us? There are a number of reasons. Love or empathy, by its nature, is more intense to those who are near and dear to and less so who are far from and unknown to us. Assume this is a deficiency of empathy or love; and then let us imagine a world in which such an empathy or love exists and another world in which it does not exist. Other things being equal, in which world do we prefer to live? I think the answer is clear: we would like to live in a world in which such empathy or love exists, even though we know it is partialistic. This line of thinking is consistent with the contemporary moral sentimentalist philosopher Michael Slote when he says that "our high moral opinion of love is inconsistent with accepting morality as universal benevolence, and I take that to constitute a strong reason to favor caring over universal benevolence" (Slote 2001: 137).

Such a justification may still not be enough, as it may be considered that we are here forced to make a choice of the lesser evil. The ancient Chinese philosopher Mozi thus provides an alternative: imagine a world that practices Confucian love with distinction and another world that practices impartial *ai*, which literally means love, but should be more appropriately translated as care, since love as an emotion cannot be impartial by its nature. To respond to the Confucian objection that one's parents would not receive as much love in a world practicing the Mohist love as in a world practicing Confucian love, since their children now are not allowed to provide more care to them than to anyone else, Mozi says that this is a misunderstanding. While parents will indeed receive less care from their children, since their children need to care for other people as much as they care for their parents, they will receive more care from people other than their children, since other people will also care for them as much as their children care them and as much as their care for their own parents. This, however, is also a misunderstanding from the Confucian point of view. In order to care (to say nothing about love) for someone appropriately, one needs the relevant knowledge of the person, i.e., what the person needs, likes, prefers, etc. Clearly one knows better, and therefore can take better care of, those who are near and dear than those who are far and unknown, fellow human beings than animals, animals than plants, and living things than non-living things.

Moreover, on the one hand, to say that we love our parents, human beings, and animals more than other human beings, animals, and plants, respectively, does not mean that we don't love the latter. It only means that we love them less intensely, and (at least part of) the reason is that we don't know the latter as well as we know the former. On the other hand, even when we face the dilemma in which our love for the former requires that we make some sacrifice of the latter, Wang Yangming uses the term *ren* (a character different from the one that means human-ity), translated as "bear" ("bearable") above, which is quite illustrative. To bear to do something implies that enduring something unpleasant in doing it.

Thus, when he says that "we love both birds and animals on the one hand and grasses and trees on the other, but it is bearable to feed birds and animals with grasses and trees," he means that we still have empathy with grasses and trees, since we also love them; otherwise there is no reason for us to *bear* to see them being fed upon by animals. A similar thing can be said about what immediately follows in the passage, about our bearing to see animals being killed to feed parents, etc., and about bearing to see a stranger starve when the only dish of food is used to prevent the starvation of our parents. We need to make an effort to "bear" such things when they happen indicates that, even though we allow or even make them happen, they are things we would like to prevent if possible at all.

The point that Wang Yangming makes here echoes what contemporary virtue ethicist Rosalind Hursthouse calls "moral residue" or "moral remainder," or, more appropriately, the latter echoes the former. When people face dilemmas such as those mentioned by Wang Yangming, Hursthouse states, "whatever they do, they violate a moral requirement, and we expect them (especially when we think in terms of real examples) to register this in some way—by feeling distress or regret or remorse or guilt, or, in some cases, by recognizing that some apology or restitution or compensation is called for. This—the remorse or regret, or the new requirement to apologize or whatever—is called the (moral) 'remainder' or 'residue'" (Hursthouse 2001: 44).

4. Conclusion

In this chapter, by drawing on the ideas mostly developed by the neo-Confucian philosopher Wang Yangming, I have argued that Confucian environmental virtue ethics can avoid some pitfalls of deontological and consequentialist approaches to environmental issues as well as those of other versions of environmental virtue ethics, particularly the Aristotelian ones. Central to this Confucian environmental virtue ethics is the idea of being in one body with the ten thou-sand things. A virtuous person in this sense feels the pain and itch of the ten thousand things, just as he or she feels the pain and itch on his or her own back. Such an ability to feel either (both) the pain and itch of the ten thousand things or (and) to be in one body with the ten thousand things is *ren*, the cardinal Confucian virtue that characterizes a human being. It is not merely cognitive but also affective for both humans and the more-than-human world. Thus, a person who feels the pain of a bird, for example, is not merely a person who knows that the bird is in pain but also a person who is motivated to help the bird get rid of the pain. So a Confucian environmental virtuous person takes care of the ten thousand things not because of their intrinsic values but because they are part of his or her own body. Despite its appearance, such a person is not self-centered, as there is nothing outside the person, or, to put it another way, everything is part of the person, while egoism assumes the separateness of the self from others.

References

Batson, C. Daniel (2011). *Altruism in Humans*. Oxford: Oxford University Press.

Cheng, Hao 程顥 and Cheng, Yi 程頤. (1988). *Complete Collections of the Two Chengs* 二程集. Beijing: Zhonghua Shuju.

Bardsley, Karen (2013). "Mother Nature and the Mother of All Virtues: On the Rationality of Feeling Gratitude toward Nature." *Environmental Ethics* 35: 27–40.

Cafaro, Philip (2015). "Environment Virtue Ethics." In *Routledge Companion to Virtue Ethics*. Edited by Lorraine Besser-Jones and Michael Slote.

Carson, Rachel (1956). *The Sense of Wonder*. New York: Harper & Row.

Ching, Julia (1976). *To Acquire Wisdom: The Way of Wang Yang-ming*. New York/London: Columbia University Press.

Cua, AS (1982). *The Unity of Knowledge and Action: A Study in Wang Yangming's Moral Philosophy*. Honolulu: The University Press of Hawaii.

Hill, Thomas E., Jr (1983). "Ideals of Human Excellence and Preserving Natural Environments." *Environmental Ethics* 5: 211–224

Hoffman, Martin (2000). *Empathy and Moral Development: Implications for Caring and Justice*. Cambridge: Cambridge University Press.

Hursthouse, Rosalind (2001). *On Virtue Ethics*. Oxford: Oxford University Press.

Kant, Immanuel (1997). *Lectures on Ethics*. Cambridge: Cambridge University Press.

Mencius (2005). In Yang Bojun 楊伯峻, *Translation [into Modern Chinese] of and Commentaries on the Analects* 孟子譯註. Beijing: Zhonghua Shuju.

O'Neill, John Francis (1993). *Ecology, Policy and Politics: Human Well-being and the Natural World*. London: Routledge.

Regan, Tom (1983). *The Case for Animal Rights*. London: Routledge and Kegan Paul.

Sandler, Ronald (2007). *Character and Environment: A Virtue-Oriented Approach to Virtue Ethics*. New York: Columbia University Press.

Rolston, Holmes, III (2005). "Environmental Virtue Ethics: Half the Truth but Dangerous as a Whole." In Ronald Sandier and Philip Cafaro, eds., *Environmental Virtue Ethics*. Lanham. MD: Rowman and Littlefield Publishers.

Slote, Michael (2001). *Morals from Motive*. Oxford: Oxford University Press.

Thoreau, Henry David (1951). *Walden*. New York: Bramhall House.

Tu, Weiming 杜維明 (1976). *Neo-Confucian Thought in Action: Wang Yang-ming's Youth (1472–1509)*. Berkeley, Los Angles, and London: University of California Press.

Wang Yangming 王陽明 (1992). *Complete Works of Wang Yangming* 王陽明全集. Shanghai: Shanghai Guji Chubanshe.

Wensveen, Louke van (2000). *Dirty Virtues: The Emergence of Ecological Virtue Ethics*. Amherst, NY: Prometheus.

Williams, Bernard (1981). *Moral Luck*. Cambridge: Cambridge University Press.

Wolf, Susan (2012). "'One Thought Too Many': Love, Morality, and the Ordering of Commitment." In *Luck, Value, and Commitment: Themes from the Ethics of Bernard Williams*. Oxford: Oxford University Press.

Zwolinski, Matt and David Schmidtz (2013). "Environmental Virtue Ethics: What It Is and What It Needs To Be." In Daniel C. Russell, ed., *Cambridge Companion to Virtue Ethics*. Cambridge: Cambridge University Press.

7

JUDAISM

Hava Tirosh-Samuelson

Awareness of the ecological crisis has transformed all academic disciplines, including the humanities, the disciplines that inquire about values, norms, meanings, languages, and cultures. Beginning in the 1970s, but increasingly in the 1980s and 1990s, a growing number of humanities scholars have begun to argue that ecological matters are not marginal but foundational to their disciplines. Thus historians traced "connections between environmental conditions, economic modes of production and cultural ideas through time" (Glotfelty and Fromm 1996, xxi) and explained the interplay of physical nature and human culture. Philosophers and ethicists explored how humans should interact with the environment as well as the theoretical justifications of these directives, extending the scope of moral considerability to non-human nature and identifying objective and universal ground, or grounds for environmental value. The result was the emergence of distinct strands of environmental thought—Social Ecology, Deep Ecology, and Ecofeminism—each with its own analysis of the causes for the ecological crisis, the salient ethical dimension, and the proper response to the crisis. And literary scholars applied the science of ecology and ecological concepts to the study of literature, giving rise to what has been known as "literature-and-environment studies," "green cultural studies," "environmental literary criticism," or "ecocriticism" (Buell 2005, 11–12; Garrard 2012). Ecocritics tie their cultural analyses to "green" moral and political agendas, while insisting that the study of "the relationship of the human and the non-human throughout human cultural history" (Garrard, 5) be based on familiarity with the science of ecology. Problematizing and erasing the boundaries between the "human" and the "non-human," ecocriticism is closely linked to other interdisciplinary, postmodern, and critical discourses such as animal studies, trans- and posthumanism, and postcolonialism. The sub-discipline of animal studies (also known as "human-animal studies") exposes the destructive impact of the traditional assumptions about the exceptionality of humans and emphasizes the cultural significance of animals; the discourse of trans- and posthumanism endorses genetic engineering and the human—machine interface that will presumably usher a new phase in human evolution; and postcolonial ecocriticism seeks to demonstrate that the western colonial enterprise perpetuated harmful human inequalities as well as environmental abuses.

Whereas these critical discourses focus on *representations* of nature in rhetorical strategies about nature, religious studies has paid attention both to the science of ecology and to action on behalf of the environment. It was no coincidence scholars of religion were the first to pay attention to the science of ecology, since the science of ecology had deep religious overtones and its history was inseparable from utopian aspirations (Tirosh-Samuelson 2012). Emerging in the 1970s, the field of religion and ecology has brought together scholars, theologians, and practitioners of religious traditions to engage in "retrieval, reevaluation, and reconstruction" (Grim and Tucker 2014, 86) of their respective traditions in order to address the ecological crisis. More than other forms of environmental humanities, religion can mobilize humanity to address the ecological crisis because religion appeals to ultimate norms and values, religion links theory, experience, and action, and religion expresses itself through narrative, myth, and symbolism, the deepest forms of human self-expression. The discourse of religion and ecology combines ecological science, humanistic scholarship, religious faith, and social action.

Judaism and ecology: ambiguities and possibilities

As the oldest of the Western monotheistic religions, Judaism is indispensable to the discourse on religious and ecology, but Judaism also occupies an ambiguous position in this discourse. To begin, Judaism problematizes a generic definition of "religion." Although Judaism had articulated the concepts that framed the western religious vocabulary (e.g., creation, revelation, covenant, prophecy, Scripture, redemption, and messiah), Judaism differs from other traditions because it is a religion of one group of people—the Jews. Thus Jewishness consists not only of beliefs, rituals, norms, and practices that cohere into a way of life, but also a collective identity, be it ethnic or national. In the modern period, however, processes of secularization have problematized Jewish existence, giving rise to secular Jews, namely born Jews who do not live by the strictures of the Jewish religious tradition. Interestingly, secular (i.e., non-observant) Jews have been at the forefront of the environmental movement, a point that has received little recognition. For example, Barry Commoner the Jewish scientist-activist made environmentalism into a political cause. Murray Bookchin, another son of East European Jewish immigrants in America, articulated social ecology, insisting that human responsibility toward nature could be carried out only if humans first eliminate social exploitation, domination, and hierarchy by developing communitarianism. Peter Singer, the son of Jewish refugees from Austria who settled in Australia, theorized the animal liberation movement, arguing that humans have an obligation to serve the interests or at least to protect the lives of all animals who suffer or are killed, whether on the farm or in the wild. Hans Jonas, the German-Jewish philosopher and student of Heidegger, is regarded as the "father" of the European Green movement. Starhawk (aka Miriam Simos), an American Jewish feminist, created the Goddess religion, giving rise to Earth-based feminist spirituality. Finally, David Abram is an American eco-phenomenologist who coined the term "the more-than-human world" to signify the broad commonwealth of earthy life which both includes and exceeds human culture. These are all born Jews who have profoundly shaped the theory and practice of contemporary environmentalism, but without appealing to Judaism as an authoritative tradition. In some cases, their ideas reflect the secularization of traditional Jewish ideas and beliefs, and more often their environmentalist vision either substitutes for a commitment to Judaism or directly critiqued Judaism for its presumed limitations or failures.

In the context of the discourse on religion and ecology, Judaism is ambiguous for yet another reason: the Bible, the foundation document of Judaism, was accused of being the very cause of ecological crisis. Lynn White Jr., the medieval historian, was the first to charge that the Bible commanded humanity to rule the Earth (Gen. 1:28), giving human beings the license to exploit the Earth's resources for their own benefit. A lay Presbyterian, White intended the charge as prophetic self-criticism that will generate self-examination (Santmire 1984), and indeed he was exceedingly successful: his short essay (White 1967) compelled Jews and Christians to examine the Bible anew in light of the ecological crisis. Is the Bible an "inconvenient text" (Habel 2009) or is the Bible a text whose ecological wisdom has been ignored or misinterpreted? Does the Bible authorize human domination and exploitation of the Earth or rather does the Bible set clear limits on human interaction with the non-human world and commands humans to care responsibly for the Earth and all its non-human inhabitants? Since 1970 Jewish religious environmentalists have examined the Bible in light of the ecological crisis whether to defend the Bible against various (Christian) misreadings (e.g., Cohen 1989), identify a distinctive ecological sensibility (Artson 2001 [1991–92]; Bernstein 2005; Benstein 2006; Troster 2008), or articulate Jewish environmental ethics of responsibility (E. Schwartz 1997; Waskow 2000; Troster 2001 [1991–92]). For the past four decades the close study of the Bible has made clear that the Jewish sacred text espouses deep concern for the well-being of the Earth and all its inhabitants, because it asserts that "the Earth belongs to God" (Ps. 24:1) and humans are but temporary care takers, or stewards, of God's Earth; their task is "to till and protect" the Earth (Gen. 2:15) not as controlling managers but as loving gardeners.

A third source of ambiguity is the fact that in Judaism ecological wisdom is found not in the natural order itself but in divinely revealed commands that instruct humans how to treat the Earth and its inhabitants. Scripture declares that the world God had created is "very good" (Gen. 1:31) but it is neither perfect nor intrinsically holy. Only human beings, who are created "in the image of God" (Gen. 1:28), are able to perfect the world by acting in accordance with divine command. At Sinai, God revealed His Will to the Chosen People, Israel, by giving them the Torah (literally, "instruction"), which specifies how Israel is to conduct itself in all aspects of life, including conduct toward the physical environment. In the Judaic sacred myth, divine revelation establishes the eternal covenant between God and Israel, an unconditional contract whose collateral is the Land of Israel. As long as Israel observes the Will of God, the Land of Israel is fertile and fecund and Israel flourishes, but when Israel sins, the Land loses its fertility and the people suffer (Deut. 6:10–15). When the sins become egregious, God punishes Israel by exiling the people from the Land. In this manner the Bible set up the causal connection between religious morality and the well-being of the environment.

Biblical law spells how Israel is to treat the Earth, vegetation, and animals. Viewing Israel as God's tenant-farmers, Scripture commands that a portion of the land's yield be returned to its rightful owner, God (Leviticus 19:23). Since creation was an act of separation, Scripture prohibits mixing of plants, fruit trees, fish, birds, and land animals thereby protecting biodiversity (Deut. 22:9–11). The human being is indeed given *responsible authority* over other animals and is allowed to consume animals, but human consumption of animals is presented as divine concession to human craving, suggesting that vegetarianism is the ideal and radically limiting what humans are allowed to consume. Scriptural legislation is also attentive to the perpetuation of species by prohibiting the killing of the mother hen with her off-spring (Deut. 22:6) or the cutting of fruit-bearing trees in time of siege (Deut. 20:19). Compassion to domestic animals is evident in the prohibition on the yoking together of animals of uneven strength and is praised as a desired virtue. Since Scripture allows for human sacrifice of animals, the relationship between humans and animals exhibits inequality, but "this inequality is relative not absolute … because it is based on an analogy: as God is to Israel, so is Israel to its flocks and herds" (Klawans 2006, 74). Most importantly, the Bible commands rest on the Sabbath for humans and domestic animals putting "moral limits to economic exchange and commercial exploitation" (Sacks 2005, 169). Extending the Sabbath to the land, the laws of the Sabbatical year (*shemitah*) protects the socially marginal (i.e., the poor, the hungry, the widow, and the orphan) by making sure that crops that grow untended are to be left ownerless for all to share including poor people and animals. In the Bible the allocation of nature's resources is a religious issue of the highest order and social justice is eco-justice. Divinely revealed environmental legislation enables Israel to sanctify itself and the Land of Israel.

In the Second Temple period (516 BCE to 70 CE) the Bible became the canonic Scripture of the Jews, shaping the life of the People of Israel in the Land of Israel and in the diaspora. With the destruction of the Second Temple by the Romans in 70 CE, Jewish political sovereignty in the Land of Israel came to an end, but Jewish religious and legal autonomy remained intact under the leadership of small scholarly elite, the rabbis. Seeking to fathom the meaning of the divinely revealed Torah, the rabbis expanded biblical legislation through creative exegesis, giving their interpretations the status of Oral Torah which became normative Judaism. For example, from Deut. 22:19 the rabbis derived the principle "Do Not Destroy" (*bal tashchit*), which prohibits wanton destruction, a precept that defines the unique Jewish contribution to environmentalism: Judaism focuses on the duties of humans toward nature as opposed to the intrinsic or inherent rights of nature. Similarly, on the basis of Deut. 22:6, the rabbis articulated the general principle of *tza'ar ba'aley hayyim* prohibits the affliction of needless suffering of animals. Although rabbinic ethics is undoubtedly hierarchical and human centered, for example, cruelty to animals is forbidden because it leads to cruelty toward humans (R. Schwarz 2012), the rabbis often presented animals as moral exemplars and recognized special animals as "animals of the righteous," who live in perfect harmony with their Creator.

While the rabbis praised virtues that can be conducive to creation care, rabbinic Judaism also generated a certain distance between Jews and the natural world, which is the fourth source of ambiguity. As the rabbis regarded Torah study to be the ultimate commandment, equal in value to all the other commandments combined, the Torah itself (both Written and Oral) became the prism through which Jews experienced the natural world. From the second century onward rabbinic Judaism has evolved as a textual, scholastic culture that privileges the study of sacred texts at the expense of interest in the natural world for its own sake. The urbanization of Roman Palestine in the third and fourth centuries, the Jewish transition from agriculture to commerce and trade in the early centuries of Islam, and the limits on Jewish ownership of land in the Christian West exacerbated the departure of Jews from agriculture and the emergence of Jewish culture as a text-based community. This is not to say that Jews were oblivious to their physical surroundings, but that pre-modern Jews interacted with the natural world through textual exegesis. Thus rationalist Jewish philosophers in Spain and Italy sought to fathom how the laws of nature (as understood by Aristotle) reflect the inner esoteric structure of Torah; Pietists in Germany regarded nature as a secret code that could be decoded by the use of secret magical, verbal, and numerical formulas; and kabbalists in Spain saw nature as a symbolic text that mirrors the structure of the Godhead. All three intellectual strands of pre-modern Judaism treated nature as a linguistic text that has to be interpreted rather than a physical reality that can be sensually experienced by embodied humans.

The centrality of sacred texts in Jewish life was critiqued in the modern period first by the Jewish Enlightenment (Haskalah) and later by Zionism. For the proponents of the Jewish Enlightenment (*maskilim*), knowledge of physical nature was necessary for the modernization of the Jews and their entry into European society and culture. In their journals, novels, and satires, the *maskilim* presented knowledge about the natural world as conditional to the healing of Jews from excessive bookishness and called for the return of Jews to productive labor, especially agriculture. Going further, Zionism, the Jewish nationalist movement that was an offshoot of the Haskalah, preached the return of Jews to the Land of Israel as the solution to the ills of exilic life. The Zionist movement generated a fifth source of ambiguity in the Jewish relationship to nature. Zionism sought to create a new type of Jews as well as a new, Hebraic, modern culture that will be rooted in the remote agricultural past of ancient Israel, by passing rabbinic Judaism. Zionism endowed the physical environment of the Land of Israel, its topography, flora, and fauna, with spiritual (albeit secular) significance, inculcating intimate knowledge of the Land through nature hikes, field trips, and camping. Paradoxically, the resulting outdoor culture has enabled secular Israelis to understand the natural imagery and metaphors of the Bible, the document that legitimized the Zionist national project. More problematically, the successes of the Zionist project exacted a toll on the fragile environment of the Land of Israel: steep rise in population, rapid urbanization, the ongoing Arab–Israeli conflict, and initial mistakes about natural resource management have generated a long list of environmental problems (e.g., air and water pollution; soil erosion; overuse of water, etc.) requiring legislative solutions. Today the state of Israel addresses these environmental challenges through a mixture of policies, legislation, and alternative technologies, and environmentalism thrives in Israel through green political parties, numerous environmental NGOs, and creative educational and training programs (Tirosh-Samuelson 2012). Many of these environmental initiatives and organizations deal with concrete environmental problems without reference to Judaism, but some organizations draw direct inspiration from Jewish religious sources in their theoretical justification and educational programs. The degree to which Israeli environmentalism should be grounded in traditional Jewish sources is hotly debated in Israel and the movement is quite different from its American counterpart.

Eco-Judaism: practice and theology

In Israel, where Jews are the majority, environmentalism encompasses advocacy, education, public policy, legislation, sustainable architecture and agriculture, science, and technology with limited appeal to the religious sources of Judaism. By contrast, in the diaspora, where Jews are small religio-ethnic minority, Jewish environmental public discourse has to be carried explicitly in religious categories. Since its emergence in the mid-1980s, Jewish environmental activism has brought about the "greening" of Jewish institutions (e.g., synagogues, schools, communal organizations, Jewish community centers, and youth movements). Today, a variety of organizations, programs, and initiatives promote sustainable practices (e.g., energy efficiency, elimination of plastics, recycling, and waste reduction programs), reduce consumption and promote new eating habits, plant community gardens, link sustainable agriculture to urban Jewish life and education, include environmental issues in the education of youngsters and adults, organize nature walks and outdoor activities, celebrate Jewish holidays (especially Sukkot, Shavuot and Tu Bishvat) with attention to environmental agricultural themes; promote justice in food production with attention to sustainable agriculture and compassionate treatment of farm animals, and encourage Jews to live sustainably. These programs transcend congregational and denominational boundaries and are often carried out in inter-faith settings in collaboration with non-Jewish organizations. Eco-Judaism consists of environmental activism and eco-theology.

As a grass root movement, Jewish environmental activism educates Jews about environmental matters, inspires Jews to lead an environmentally correct life, implements "green" communal practices, and rallies Jews to support environmental legislation and interfaith activities. The main activities of Jewish environmental organizations and initiatives consist of nature education, environmental awareness, advocacy on environmental legislation, and community building. Thus programs of Teva Learning Alliance sensitize the participants to nature's rhythms, inspiring them to develop a meaningful relationship with nature and their own Jewish practices. Through traditional Jewish rituals (e.g., blessings, prayers, and reflections) participants become aware of nature as divine creation or learn about the vital connection between Judaism and environmental stewardship. In the Elat Chayyim Center for Jewish Spirituality, programs promote environmentally concerned Judaism as a spiritual practice and offer leadership training that teaches participants to live communally and integrate organized agricultural and sustainable living skills with Jewish learning and living. The newly reconstituted Aytzim: Ecological Judaism illustrates how environmental education, advocacy, and activism link the Jewish religion with Zionism and how the Internet is used to advance environmentalism: the website Jewcology is now managed by Aytzim. These programs and the Coalition of Jewish and Environmental Life (COEJL), which focuses on educational, legislative, and interfaith programming, illustrate eco-Judaism in practice. Attention to food is another important aspect of contemporary eco-Judaism, since food is the intersection point of humans and animals as well as of diverse social groups. Hazon: Jewish Inspiration, Sustainable Communities exemplifies the Jewish Food Movement that stresses the redemptive aspect of land cultivation and just production, distribution, and consumption of food. The Jewish Food Movement is connected with other environmental initiatives such as the Jewish Farm School, Eden Village (an eco-summer camp), Shomrei Adamah (a program in Jewish day schools that emphasizes energy flow, natural cycles, biodiversity, and interdependence), and Kayam (an educational camp) all of which are designed to bring Jews to integrate hands on knowledge about food and farming with the Jewish tradition.

The concept that gives coherence to eco-Judaism is "Eco-Kosher." It connects concerns about industrial agriculture, global warming, and fair treatment of workers with the Jewish dietary laws about food production, preparation and eating. "Eco-Kosher" means that Jews should only consume products that meet both Jewish dietary laws as well as Jewish ethical standards, and Eco-Kosher consumers should encourage food producers to care for the environment, animals, and their workers. Arthur Waskow translated Eco-Kosher into a fully fledged program of environmental justice in regard to economic and racial inequity, the unjust labor practices, and the causal connection between the exploitation of the Earth resources and unjust political policies, especially in Israel (Waskow 1996). Other rabbis have fused Eco-Kosher with kabbalistic principles as well as with non-Jewish traditions, such as the ancient Chinese art of Feng Shui, an ecologically based art of spatial arrangement that incorporates human-made objects with natural surroundings. The concept of Eco-Kosher has also inspired Jewish entrepreneurs to market Eco-Kosher meat products and the Conservative Movement to issue the Magen Tzedek Initiative, a certification program that assures consumers and retailers that "kosher food products have been produced in keeping with exemplary Jewish ethics in regard to labor, animal welfare, environmental impact, consumer issue and corporate integrity" (www.magentzedek.org).

Combining sustainable agriculture, fair labor practices, and ethical treatment of animals, Eco-Kosher generates a comprehensive life style whose goal is to bring about *Tikkun Olam* (literally, "repair of the world"). In rabbinic texts (e.g., Mishna, Gittin 4:2) "*letaken olam*" means to act in accordance to Jewish law so as to usher the Kingdom of God. This utopian notion was given an abstract, cosmic, metaphysical meaning in medieval Kabbalah, especially the sixteenth-century version of Lurianic Kabbalah where human action can restore harmony in all levels of existence, including God. In the second half of the twentieth-century *Tikkun Olam* has become the slogan of Jewish social activism, including environmentalism, although few Jews who invoke the term understand its kabbalistic connotations. In Jewish environmental organizations, the goal of *Tikkun Olam* is usually linked to two other ethical values: *responsibility and interconnectedness*. The former highlights human responsibility toward the Earth and its inhabitants and the latter insists on the relationality of all living beings. Both values are derived from biblical and rabbinic sources and are invoked in a wide variety of educational programs (Tirosh-Samuelson 2012).

Several Jewish eco-theologians deliberately build on Kabbalah and Hasidism to articulate Jewish ecological spirituality to address the ecological crisis. Zalman Schachter-Shalomi, was the first to call for a "paradigm shift" within Judaism, signifying a shift from transcendence to immanence, from monotheism to pantheism, from dualistic to non-dualistic thinking, from patriarchy to egalitarianism. He called this shift "Gaian Consciousness" and argued that Judaism has a distinctive (albeit not exclusive) role to play in the healing of the cosmos: the key ecological precept of Judaism—"Do Not Destroy"—enables Jews to act in ways that prevent what he called, the crime of "planetcide." Recasting Judaism as pantheistic monism that reframes all the major themes of traditional Judaism and gives rise to new rituals, this New-Age thinker saw his project as "trying to help the Earth rebuild her organicity and establish a healthy governing principles" (Magid 2006, 65).

Schachter-Shalomi's friend and colleague, Rabbi Arthur Green, has gone further to articulate a systematic ecological Jewish spirituality promoted as "Neo-Hasidism." Green's "mystical panentheism," fuses Kabbalah and Hasidism with the theory of evolution to depict the "bio-history of the universe" as "sacred drama" (Green 2003, 111). Green presents a holistic view of reality in which all existents are in some way an expression of God and are to some extent intrinsically related to one another (Ibid, 118–119; Green 2010). Although Green's lyrical depiction of evolution is closer to medieval Neoplatonism than to Darwinian evolution, it offers contemporary Jews "a Kabbalah for the environmental age" (Green 2002). A systematic fusion of Kabbalah, Hasidism, and environmentalism is presented in the work of Rabbi David Seidenberg who argues that by "applying the principles of Kabbalah to constructive theology, we can train ourselves to see the image of God in all of these dimensions, in a species, in an ecosystem, in the water cycles, in the entirety of this planet, and so on" (Seidenberg 2015, 312). Seidenberg is one example of Jewish spiritual teachers, artists, story tellers, and healers who find in Kabbalah, as well as other spiritual traditions (either Native American or Asian) resources for a syncretistic Jewish ecological spirituality. The syncretism of Jewish ecological spirituality brought some critics to question the Jewishness of Jewish environmentalism and to view it as an unacceptable revival of paganism.

Jewish environmentalism and environmental humanities

The ecological reinterpretation of Judaism has developed with relatively little attention to the environmental humanities. Why? First, the discourse of the environmental humanities is decidedly secular, whereas Jewish environmentalism (at least in the diaspora) is a religious endeavor that uses religious categories. Second, the environmental humanities are theoretical discourses carried out within the bounds of the academy, whereas Jewish environmentalism is a grass-root movement of non-academic activists who care about praxis rather than theory. Third, the academic discourse of environmental humanities is inherently critical, displaying skepticism, distance, and irony, whereas Jewish environmentalism calls for conviction, action, and social transformation. Finally, while some environmental humanities, especially eco-criticism, have attempted to bring the material sciences to the foreground of the humanities in order to understand the relationship between human and non-human organisms, Jewish environmentalism (and one could say Jewish public discourse in general) has been insufficiently attentive to the natural sciences. This is not to say that Jewish environmentalism cannot or should not become informed of the environmental humanities. To the contrary: familiarity with the environmental humanities (i.e., the various strands of environmental philosophy and ethics and eco-criticism) can enrich Jewish environmentalism immensely, but such dialogue could take place only if the academic interlocutors become more informed about and interested in Judaism not as the culprit of the ecological crisis, but as a tradition that could creatively address the crisis.

The dialogue between Jewish environmentalism and environmental humanities could begin in the context of postmodern environmental thought (Zimmerman 1994) and the field of eco-phenomenology (Brown and Toadview 2003), since Jewish philosophers have greatly contributed to them. Eco-phenomenology is the merger of phenomenology and contemporary environmental thought, according to which the human cognition that "nature has value, that it deserves or demands a certain proper treatment from us, must have its roots in an experience of nature" (Ibid, xi). Jewish philosophers trained in the phenomenological tradition—Martin Buber (d. 1965), Hans Jonas (d. 1993), and Emmanuel Levinas (d. 1995) and Derrida (d. 2004)—contributed to eco-phenomenology by framing the relationship between humanity to the natural world in dialogical terms, emphasizing nature as a subject to whom humans are deeply *responsible*, and by erasing rigid boundaries between humans and animals.

Martin Buber was the first to speak about nature as a subject and to call for a non-instrumental (I–Thou) relationship with nature. Although Buber was not an environmentalist, his relational, dialogical philosophy has exerted deep influence on Christian environmental ethics (McFague 1997; Santmire 2008). If Buber made nature into a moral subject with whom humans can have personal relationship, Hans Jonas endowed life itself with intrinsic moral value as he exposed the ontological basis of the ethics of responsibility, and conversely made ontology informed by ethics (Jonas 1984). Jonas' philosophy of nature highlighted the purposiveness of all life, arguing that nature commands ultimate respond, allegiance, and final moral commitment (Donnelley 2008). It is the objective goodness of things that determines not only what ought to be but also what humans ought to feel, think, and do, since humans are an integral part of organic life. For Jonas, the "imperative of responsibility" encompasses human responsibility for the continued existence of life in a planet where life is seriously endangered by modern technology. Awareness of the looming disaster generates a "heuristic of fear" that guides us to act so as to protect nature from the possibility of destruction. Humanity is responsible for its own future and must act with concern toward future generations, ensuring that they will have the conditions for life. Jonas's philosophy of nature was developed in response to the devastation of WWII in which Auschwitz and Hiroshima came about because of modern technology.

Like Jonas, Emmanuel Levinas saw responsibility as the core of the ethical, but went further than Jonas by arguing that responsibility comes first: each person is responsible for the other who faces him. If Jonas argued for *collective responsibility* of humanity, Levinas argues for *infinite individual responsibility*: every person has an obligation to his/her neighbor, expanding gradually to cover all living humanity. Levinas' ethics is decidedly human-centered since he insisted that ethics is "against-nature, against the naturality of nature (Levinas 1998, 171). However, several postmodernist environmentalists have applied Levinas' ethics to nature which is identified with the absolute Other (Edelglass, Hatley and Diehm 2012). How Levinas's ethics should be applied to nature is still a matter of debate but no one can correctly understand Levinas without acknowledging his Jewishness.

Even more influential than Buber, Jonas, and Levinas, another Jewish philosopher trained in phenomenological tradition—Jacques Derrida—has stimulated postmodernist environmental thought and the field of eco-phenomenology. Derrida's deconstruction of traditional binary dichotomies characteristic of Western philosophy (e.g., nature/culture; human/animal; transcendence/immanence) exposed the connection between phallologocentrism and "carnivorous virility" (Gross 2015, 142). Derrida criticized the "sacrificial structure of subjectivity" and exposed the links between the hatred of the Other, the hatred of animals, and the hatred of Jews that run throughout Western history and culture (Benjamin 2011). The deconstruction of human/animal boundaries (Derrida 2008) has stimulated the newly emerging field of Animal Studies (e.g., Gross and Vallely 2012) although the Jewishness of Derrida is often glossed over.

Conclusion

Any generalizations about Judaism and ecology should take into consideration the ambiguity of the term "Judaism" and the fact that the Jewish experience encompasses both religious and secular forms. Indeed, the various conceptualizations of "nature" or "environment" illustrate the complexity of the modern Jewish experience. Thus the contribution of Jews to environmentalism is more extensive and the impact of environmentalism on contemporary Judaism is more profound than is commonly acknowledged. In Israel and in the diaspora the ecological crisis has generated many Jewish responses as Jewish theologians, scholars, educators, and activists have subjected the entire Jewish tradition to rigorous reinterpretation, identified relevant literary sources, distilled the ecological insights of the tradition, articulated new ecological theologies, and spelled out policies and educational programs. Jewish environmentalism is still a small but growing strand in contemporary Judaism that is attractive to previously unaffiliated Jews, to Jews who have limited or no Jewish education, to seekers who have walked other spiritual and religious paths, and to Jews who are traditionally observant. Commitment to Jewish environmentalism means different things: for some, Jewish environmentalism means extending the ethics of responsibility to include the environment; for others environmentalism means a new, holistic, ecological consciousness that overcomes the disruptive dualism of scripture and nature. and, for still others, environmentalism signifies the return to earth-based spirituality that links Judaism to other traditions. However interpreted, a plethora of Jewish environmental organizations promote communitarianism, environmental and social justice, and a range of educational programs based on outdoors activities that inculcate respect for nature. Benefiting from the creation of the Internet, Jewish environmental activism disseminates ideas and information about activities through social media and the websites of these organizations make available relevant literary sources, commentaries, organized activities, fellowship programs, and leadership training. While the work of Jewish environmentalism rarely engages the environmental humanities, the dialogue between these discourses could enrich both: Jewish environmentalism could become more theoretically informed; the environmental humanities could openly acknowledge its debt to Jewish ethics of responsibility.

References

Artson, Bradley Shavit. 2001 [1991–92]. "Our Covenant with Stones: A Jewish Ecology of Earth." In *Judaism and Environmental Ethics: A Reader*, Martin D. Yaffe ed., 161–170. Lanham, MD: Lexington Books.
Benjamin, Andrew. 2011. *Of Jews and Animals: Frontiers of Theory*. Edinburgh: University of Edinburgh Press.
Benstein, Jeremy. 2006. *The Way into Judaism and the Environment*. Woodstock, VT: Jewish Lights.
Bernstein, Ellen. 2005. *The Splendor of Creation: A Biblical Ecology*. Cleveland: Pilgrim Press.
Buell, Lawrence. 2005. *The Future of Environmental Criticism: Environmental Crisis and Literary Imagination*. Malden, MA: Blackwell.
Cohen, Jeremy. 1989. *Be Fertile and Increase, Fill the Earth and Master It: The Ancient and Medieval Career of a Biblical Text*. Ithaca, NY: Cornell University Press.
Derrida, Jacques. 2008. *The Animal That Therefore I Am*. Translated by David Wills. New York: Fordham University Press.

Donnelley, Strachan. 2008. "Hans Jonas and Ernst Mayr: An Organic Life and Human Responsibility." In *The Legacy of Hans Jonas: Judaism and the Phenomenon of Life*, Hava Tirosh-Samuelson and Christian Wiese (eds.), 261–286. Leiden and Boston, MA: Brill.

Edelglass, William, James Hatley and Christina Diehm. 2012. *Facing Nature: Levinas and the Environment*. Pittsburgh, PA: Duquesne University Press.

Garrard, Greg. 2012. *Ecocriticism: The New Critical Idiom*. London and New York: Routledge.

Glotfelty, Cheryll and Harold Fromm. (eds.) 1996. *Ecocriticism Reader*. Athens, GA: University of Georgia Press.

Green, Arthur. 2010. *Radical Judaism: Rethinking God and Tradition*. New Haven, CT: Yale University Press.

Green, Arthur. 2003. *EHYEH: A Kabbalah for Tomorrow*. Woodstock, VT: Jewish Light Publishing.

Green, Arthur. 2002. "A Kabalah for the Environmental Age." In *Judaism and Ecology: Created World and Revealed Word*, Hava Tirosh-Samuelson (ed.) 3–16. Cambridge, MA: Harvard University Press.

Grim John and Mary Evelyn Tucker. 2014. *Ecology and Religion*. Washington, DC: Island Press.

Gross, Aaron S. 2015. *The Question of the Animal and Religion: Theoretical Stakes, Practical Implications*. New York: Columbia University Press.

Gross, Aaron and Anne Vallely. (eds.) 2012. *Animals and the Human Imagination: A Companion to Animal Studies*. New York: Columbia University Press.

Habel, Norman C. 2009. *An Inconvenient Text: Is a Green Reading of the Bible Possible?* Adelaide, Australia: ATF Press.

Jonas, Hans. 1984. *The Imperative of Responsibility: In Search of an Ethics for the Technological Age*. Chicago and London: University of Chicago Press.

Klawans, Jonathan. 2006. "Sacrifice in Ancient Israel: Pure Bodies, Domesticated Animals, and the Divine Shepherd." In *A Communion of Subjects: Animals in Religion, Science, and Ethics*, Paul Waldau and Kimberly Patton eds., 65–80. New York: Columbia University Press

Levinas, Emmanuel. 1998. *Of God Who Comes to Mind*. Translated by Bettina Bergo. Stanford, CA: Stanford University Press.

Magid, Shaul. 2006. "Jewish Renewal: American Spirituality and Post-Monotheistic Theology." *Tikkun* (May/June): 62–66.

McFague, Sally. 1997. *Super, Natural Christians: How We Should Love Nature*. Minneapolis, MN: Fortress Press.

Sacks, Jonathan. 2005. *To Heal a Fractured World: The Ethics of Responsibility*. New York: Schocken Books.

Santmire, Paul H. 1984 "The Liberation of Nature: Lynn's White Challenge Anew. *The Christian Century* 102: 18: 530–33.

Santmire, Paul H. 2008. *Nature Reborn: The Ecological and Cosmic Promise of Christian Theology*. Minneapolis, MN: Fortress Press.

Schwartz, Eilon. 2001. "Bal Tashchit: A Jewish Environmental Precept." In *Judaism and Environmental Ethics: A Reader*, Martin D. Yaffe ed., 230–49. Lanham, MD.: Lexington Books.

Schwartz, Richard. 2012. "Jewish Traditions." In *Animals and World Religions*, Lisa Kemmerer ed., 169–204. Oxford: Oxford University Press.

Seidenberg, David Mevorach. 2015. *Kabbalah and Ecology: God's Image in the More-Than-Human World*. Cambridge: Cambridge University Press.

Tirosh-Samuelson, Hava. 2012. "Jewish Environmentalism: Faith, Scholarship and Activism. In *Jewish Thought, Jewish Faith*, Daniel Lasker ed. (English section), 1–53. Beer Sheba: Ben Gurion University Press.

Troster, Lawrence. 2008. "God Must Love Beetles: A Jewish View of Biodiversity and Extinction of Species." *Conservative Judaism* 60.3 (2008): 3–21.

Troster, Lawrence. 2001 [1991–92]. "Created in the Image of God: Humanity and Divinity in an Age of Environmentalism." *Conservative Judaism* 44: 14–24. Reprinted in *Judaism and Environmental Ethics: A Source Reader*, Martin D. Yaffe ed. Lanham: MD, Lexington Books.

Waskow, Arthur. 1996. "What is Eco-Kosher.?" In *This Sacred Earth: Religion, Nature, Environment*, Roger S. Gottlieb ed., 297–300. New York and London: Routledge.

Waskow, Arthur. (ed.) 2000. *Torah of the Earth: 4000 Years of Jewish Ecological Thinking*. Woodstock, VT: Jewish Lights Publication.

White, Lynn. 1967. "The Roots of Our Ecologic Crisis." *Science* 155: 153–157.

Zimmerman, Michael. 1994. *Contesting Earth's Future: Radical Ecology and Postmodernity*. Berkeley: University of California Press.

8

CHRISTIANITY

An ecological critique of Christianity and a Christian critique of ecological destruction

Ernst M. Conradie

In the field of religion and ecology the role of the Christian tradition is deeply contested. While some criticize the *Christian* roots of current environmental destruction, others treasure the pre-modern ecological wisdom of, for example, the desert fathers, Celtic earthiness, Benedictine monasticism, Hildegard's mysticism or Franciscan spirituality. This chapter uses the deep contestation as a key to depict the current state of debate in Christian ecotheology.

Christian ecotheology entails a twofold critique: an ecological critique of Christianity and a Christian critique of environmental destruction. In the four sections that follow I first situate the ecological critique of Christianity at its current edge, within concerns about climate change. I then suggest that a far-reaching ecological reformation of the various Christian traditions is already underway. This ecological reformation has become very extensive but also highly fragmented, as old disputes between rival schools and traditions, have pushed ecotheology in many directions. In order to bring those directions into view, I offer an overview of the multiple discourses that currently constitute Christian ecotheology. Finally, I explain why the fragmentation and diversity of ecotheology resist attempts to create typologies. I argue that it is better to see doing ecotheology as an ongoing journey through an uncharted landscape with diverse companions.

An ecological critique of Christianity and climate change: five theses

Ecological criticism of Christianity is common in environmental literature. This was at first prompted by the famous essay of Lynn White (1967) on the historical roots of the ecological crisis. Focusing on climate change, I here distill one articulation of the ecological critique of Christianity – which I lay out in five theses, without further explanation and without providing detailed references to a considerable corpus of literature.

1) Despite the breakthrough of the Paris agreement at COP 21 (2015) it still seems unlikely that a rise of around 4°C in average global surface temperature above the pre-industrial era by 2100 can be averted. This will have severe implications, especially in areas where a higher rise is probable. Averting a 4°C rise is unlikely for many reasons, including that an industrialized global economy would need to refrain from using oil, coal and gas reserves already discovered. What would be required is a transformation of the energy basis of the entire global economy from fossil fuels to sustainable alternatives, within a period of 60 years

from, let us say, 1992 (the Rio Earth Summit) to 2050. However, since the first two decades were the most critical, an important window period has already been missed, while the future commitments of the various Parties at COP 21 (if implemented) would still imply a calculated rise of 2.7°C.

2) In order to prevent such a likely rise in average global surface temperature a multi-dimensional approach would be required that would need to draw not only on science and technology, economic analyses, political will, environmental education and media coverage, but would also need to draw on cultural, moral and spiritual traditions to transform the visions, goals, perceptions, hearts, minds, habits and behaviour of people across the world. Climate change poses not only moral problems around justice, but also spiritual problems. If one knows what to do and how to do it, but cannot find the moral energy to do what has to be done, this is indeed a spiritual and not only a moral problem.

3) There is a correlation between historic carbon emissions and those countries where Christianity was influential by the advent of the industrial revolution. The correlation seems connected to tacit legitimation of neo-liberal capitalism and consumerist aspirations. This correlation is continued as a result of the popularity of the prosperity gospel. An ecological reformation of the various branches of the Christian tradition is therefore necessary if a rise of around 4°C in average global surface temperature is to be averted. In short, without a radical ecological reformation of historic and current forms of Christianity, it is hard to see a radical transformation of the energy basis of the global economy.

4) In order to be persuasive in the long run, an ecological reformation of the Christian tradition cannot be approached only on functional or pragmatic grounds. It would need to be deeply rooted in the symbols of the Christian traditions, i.e. in the core message of the gospel, the heart of the Christian faith in God as Trinity, an exegesis of the biblical texts, liturgical renewal and a retrieval of the virtues of faith, hope, love and joy. However, the credibility of such an ecological transformation is undermined by deep denominational and theological divisions within the larger Christian tradition. Moreover, while Christians in some contexts contribute disproportionally to carbon emissions, Christians in other contexts are or will increasingly become the victims of climate change.

5) Although such an ecological reformation of the main branches of the Christian tradition is already underway in numerous local contexts across the globe, this is "slow work" and faces constant corruption from the now global culture of consumerism. Many cultural practices are defended for obvious reasons but are not by themselves sustainable; consider, for example, the carbon footprint of tourism, universities, mega-sports events and international conferences. At this stage it therefore seems unlikely that a rise of around 4°C in average global surface temperature above the pre-industrial era by 2100 can be averted.

This negative assessment requires further reflection on the possibility and limitations of an ecological transformation of the various Christian traditions. One may well ask: What would it really help?

An ecological reformation of various Christian traditions

The emergence of Christian ecotheology, especially since the 1960s, may be regarded as an attempt to retrieve the ecological wisdom embedded in the Christian tradition as one source for responding to ecological destruction and environmental injustice. However, it is also prompted by the widespread suspicion that the root causes of the crisis are related to the impact of Christianity. Just as feminist theology engages in a twofold critique, that is, both a Christian critique of sexism and patriarchal culture and a feminist critique of Christianity, so ecotheology offers a Christian critique of the economic and cultural patterns underlying ecological destruction, and an ecological critique of Christianity. In other words, ecotheology is not only concerned with how Christianity can respond to environmental concerns; it also offers Christianity an opportunity for renewal and reformation. Indeed, this internal reformation of particular Christian traditions may, paradoxically, be the most significant contribution that Christian ecotheology can make to addressing *global* environmental concerns. Christians also need to be involved in addressing global environmental problems but changing the hearts and minds of its own adherents and getting its own house in order (addressing its own environmental footprint) may have more impact.

This implies the need to reinvestigate, rediscover and renew Christian traditions in the light of the challenges posed by environmental destruction. Such an ecological reformation of Christianity implies that there are significant flaws in the Christian tradition – else a reformation would not be necessary. It also implies that these flaws can be corrected – else a reformation would not be possible (see Nash 1996).

A reformation cannot be prearranged: It lies beyond anyone's locus of control. It is therefore a gift to be received with gratitude but also with trembling. It is usually not welcomed, not even by its own supporters and prophets. Any radical reformation of the Christian tradition would call for continuous theological explanation. The need for explanation is born from both the polemic and the prophetic nature of such a reformation. The value of such explanations should not be underestimated, since it can sustain an ecological reformation. The value should also not be exaggerated; the environmental impact of ecotheology will be limited. New ideas do not necessarily change the world (although emerging moral visions may well do so – see Rasmussen 2013, 147).

Such a reformation of particular Christian traditions may commence anywhere (as the example of the Lutheran reformation illustrates) but will soon spread to other aspects of the Christian tradition: reading the Bible, retrieval and critique of Christian histories, revisiting Christian symbols, virtue ethics, applied ethics, ecclesial praxis, liturgical renewal, pastoral care, preaching, Christian formation and education, Christian mission, institutional projects and theological reflection on other religions. It therefore almost necessarily becomes comprehensive – but this has the danger of becoming amorphous so that the original stimulus behind the reformation may easily become corrupted through institutionalization.

It is a welcome sign that such an ecological reformation is currently taking place in all the main branches of Christianity, including the Oriental Orthodox Churches, the Eastern Orthodox Churches, the Roman Catholic Church and its various orders, the Anglican Communion, the Lutheran World Federation, the World Communion of Reformed Churches, World Methodist Council, the (ana)Baptist traditions, evangelical Christianity, Pentecostal churches and various indigenous churches. In each case the form of the reformation is shaped by the distinct spirituality of that tradition. It is impossible to give an overview of the literature emerging from within these traditions, given the sheer volume, variety and geographical spread of such contributions. Any references would be necessarily selective and will tend to privilege the better disseminated publications in English emerging from the North Atlantic cultural and academic hegemony.

The papal encyclical *Laudato si'* by Pope Francis (2015) is one highly significant recent example, but this draws on many similar statements, including the path-breaking work of the "green" Patriarch Bartholomew. One would also need to mention the role of ecumenical bodies at the national, continental and global levels. The World Council of Churches (WCC), for example, recognised the call for a "Just, Participatory and Sustainable Society" already at its fifth Assembly in Nairobi (1975), and has sustained pioneering programmes on climate justice. In evangelical and Pentecostal contexts there have emerged commitments to address environmental issues, while retrieving the indigenous wisdom of the traditional custodians of land in other geographic contexts is equally important, if typically publicized by others.

It is remarkable that environmental concerns have been mainstreamed in several traditions so that a clear sense of leadership is being exercised through resolutions, documents and programmes. It is equally noteworthy that such concerns have been raised in diverse geographical contexts all around the world. This is clearly in response to contextual considerations – either expressing solidarity with the victims of environmental destruction (see e.g. WCC 2002), or as a prophetic critique of complicity in such destruction (see e.g. Moe-Lobeda 2013). What is the actual (ecological) impact of such ecclesial resolutions? Although there have emerged some evocative local examples of carbon neutral Christian communities (including the Vatican itself), the actual environmental impact of such reforms nevertheless remains limited in terms of ecclesial praxis. Such impact would best be measured in terms of the patterns of production and consumption of the lay members of such churches wherever they live and work – and not so much by official church statements.

Distinct Christian discourses on ecology

This ongoing reformation of various Christian traditions has prompted reflection in the form of an enormous corpus of publications (by now probably in excess of 10,000 books, articles and essays – see Conradie 2006). Such literature is to be welcomed but the extensiveness of the reform movement has come at the cost of fragmentation, against its own inclinations for ecological wholeness. It is no longer possible to speak of Christian ecotheology in the singular. Instead, a conflicting variety of schools and guilds in which old methodological disputes have re-emerged have become the dominant feature of Christian ecotheology. Ecotheologians worldwide do not talk primarily to each other, because the scope of the field is so large that everyone has to focus more narrowly on specific discourses. It may be helpful here to sketch this situation by identifying a number of distinct, if overlapping, discourses. The aim is not to provide a comprehensive overview of the contributions made within each discourse, but to identify the unresolved issues at the cutting edge of each discourse and to indicate the methodological tensions that prevail between these discourses.

Biblical Hermeneutics: There are many attempts to retrieve the ecological wisdom embedded in the biblical roots of the Christian tradition. Such exegetical work naturally tends to focus on texts related to God's creation that seem congenial to ecological concerns. Although the agenda is to indicate that the Bible is "filled to the brim" with ecological wisdom, the limited choice of texts does not help Christian communities to relate ecological concerns to the core message of salvation. By contrast, others have adopted a critical hermeneutics with the assumption that the domination of otherkind by humankind, religious alienation from the earth and forms of anthropocentrism, if not dualism, are reflected in the canonical texts themselves. The texts were written by humans to serve human interests so that the voice of Earth is undermined (see Habel 2000, 2009). This has resulted in deep tensions between a hermeneutics of suspicion and one of trust in the sacred texts (see Horrell et al. 2010). While some find sources in the text for a Christian critique of ecological destruction, others use such destruction as a point of departure for a critique of the same text. For example, while some bemoan our failure to respond to the call to "stewardship" in Genesis 1:28, others trace the root causes of environmental destruction to the "successful" collective response in Abrahamic traditions to this call to "subdue" the earth and to "rule" over it.

Constructive Theology: In almost every confessional tradition and theological school there have been attempts to rethink the content and significance of the Christians faith in the light of ecological concerns. Many tend to focus on the theme of creation (e.g. following Moltmann 1985), the relationship between humanity and nature or the place and responsibility of humanity within God's creation. In this regard many emphasize the need for responsible stewardship (e.g. Berry 2006), priesthood or guardianship, while others dispute the legitimacy of these metaphors used to express such responsibility (see Conradie 2011, 81–94). Yet others recognize the need to reinterpret all the core symbols of the Christian faith, including the Trinity, faith in God as Father, Son and Sprit, God's work of providence, salvation and consummation and the formation, governance, ministries and missions of the church. As I have argued elsewhere (Conradie 2013), there remain many unresolved issues here, for example on the ecological significance of God's transcendence, on the relatedness of Christ and the Spirit, on doing justice to both God's work of creation and of salvation and on the plausibility of eschatological consummation (the hope of "going to heaven"). Although there have been some collaborative and ecumenical engagements (see Hessel and Ruether 2000, Edwards 2001, Conradie et al. 2014), there is a tendency for such discourses to become insular within particular theological traditions and regional contexts. There is a vast corpus of literature but there remain deep differences that are typically related to diverging views on nature and grace, underlying worldviews, views on faith and science and on the relationship between Christianity and other religions.

A "Just, Participatory and Sustainable Society": Since the WCC Assembly in Nairobi (1975) there has emerged a huge corpus of ecumenical literature offering theological perspectives on ethical concerns over environmental sustainability, economic injustices and violent conflicts (including debates of refugees, gender-based violence and inter-religious conflict). The multiple links between these three sets of issues are widely recognised. These are often related to the etymological links between the Greek root *oikos* (understood as the whole household of God) and the English terms ecology (the underlying logic of the household), economy (the rules for managing the household) and ecumenical fellowship (the inhabitation of the household) (see Rasmussen 1996, Kaoma 2013). However, ever since the World Convocation on Justice Peace and the Integrity of Creation (Seoul, 1990 – see Niles 1992), there has been a tendency to prioritize one of these themes over the other. In over-generalized terms one may speak of a Western emphasis on sustainability (given the impact of industrialized societies), an Eastern emphasis on enhancing life amidst conflict and a Southern emphasis on justice, including economic justice (following Boff 1997) and climate justice (see SACC 2009). This continues to pose challenges for ecumenical communication and solidarity even though the need to address all three agendas together is widely acknowledged. However, there is an even deeper tension here, namely between ecclesiology and ethics (see Best and Robra 1997), between discourse on "Faith and Order" and on "Church and Society". Some emphasize what the *church* is (in terms of the Christian faith) while others focus on what the church *does* (underplaying the distinctiveness of the church). This debate on ecclesiology and ethics remains unresolved.

Ecofeminism and Christianity: In most forms of ecofeminism there is a critique of the logic of domination in the name of differences of gender, race and class underlying modern, industrialized societies. Such domination is reinforced and ideologically justified through a system of interlocking dualisms based on the binaries of male/female, culture/nature, soul/body, transcendence/immanence and heaven/earth. The argument is that such forms of domination are extrapolated towards the subjugation of the land and other forms of life for the sake of human needs and desires. This is subjected to critique in various forms of ecofeminism and ecowomanism in very diverse contexts (see e.g. Ruether 1996, Gebara 1999, Eaton 2005, Gnanadason 2005). One may observe that this critique is related to concerns over sustainability, justice and peace but that it poses specific challenges to Christian belief systems given the fundamental distinction between Creator and creature, the male terminology employed for God (as Father and Son), patriarchal authority and dominance in church leadership even where the ordination of women is endorsed. This debate is not easily resolved given ongoing theological debates on the Trinitarian confession in relation to various forms of theism, atheism, panentheism and pantheism.

Indigenous Spirituality: Concerns over sustainability, justice and peace are often interwoven with critiques of domination in the name of differences of race and culture – and its impact on land. There is some overlap here with contributions in ecofeminism but there is also considerable contestation given the relative priority attached to issues of land vis-à-vis race and gender. This is significant given the relative paucity of contributions to ecotheology within the context of North American and southern African black theologies (see Cone 2001 though). There is a widespread retrieval of indigenous wisdom by Christians within the Aboriginal, African, Latin American, Native American and Nordic contexts – so widespread that it would be inappropriate to single out specific contributions. Where such indigenous spirituality is expressed within a Christian context (e.g. in indigenous churches) one typically finds simmering tensions with mainline Christianity. The deeper debate here has to do with how pre-Christian religious understandings of God relate with the Trinitarian confession of Nicene Christianity.

Animal Theology: Christians have traditionally affirmed not only the intrinsic worth but also the equal dignity of all human beings on the basis of the belief that humans are created in the image and likeness of God. Although such equal dignity is still rather counter-intuitive given long-standing class hierarchies, this has been widely endorsed in secular human rights discourse. In more recent times the intrinsic worth of other animals and indeed, given the evolutionary history behind the emergence of hominid species, of all forms of life, have been affirmed in many expressions of animal theology (see Clough 2012, Linzey 1995, Deane-Drummond 2014). In ecumenical discourse such intrinsic worth is expressed in terms of the notion of the "integrity of creation." However, few have endorsed the notion of biocentric equality – so that the worth of various forms of life is still graded. Such grading seems necessary in order to distinguish between the needs of children and chickens, between household pets and household pests but is contested in wider debates on biodiversity, also in non-Western societies. Where such grading is accepted, this begs the question why such graded value is not applied amongst humans. This debate on biocentric versus ecocentric approaches to environmental ethics is not unique to Christianity. However, it requires special discernment since this may be contrasted with proposals for a Christocentric (e.g. on "deep incarnation"), a Pneumatological (the Spirit as "Giver of Life") or a theocentric approach (see Gustafson 1994). In general it is striking that this debate on animal theology remains dominated by British and North American voices.

Mission and Earthkeeping: The nineteenth century witnessed widespread cross-cultural missionary activities. There has typically been an unholy alliance between such missions and the colonial conquest of land (and neo-colonial exploitation of resources). This has prompted not only debates on economic justice but also on Christianity and culture. This is expressed as a critique of the destruction of indigenous cultures but also as a critique of the consumerist culture of countries that benefited from colonial exploitation. These critiques have prompted ongoing reflection on what Christian mission entails. Some acknowledge an earthkeeping dimension to mission so that creation is regarded as being "at the heart of mission" (see Keum 2013). This is often expressed in terms of the priestly responsibility for earthkeeping in local Christian communities. Such a call for responsibility seems to appeal especially to those in positions of relative power (with a large carbon footprint), while the victims of environmental

Ernst M. Conradie

injustice, quite ironically, seek refuge in those countries where the nineteenth-century mission-aries came from. In response, the latest WCC mission document *Together towards Life* (Keum 2013) speaks of mission *from* the margins and not only *to* the margins. This debate on mission and ecology may be regarded as the storm center of diverging views and methodological conflict over the relation between missionary praxis and the spiritual sources from which such praxis springs.

Christianity and Multi-faith Dialogue: Addressing global environmental concerns cannot be done by any one group, even if there are around two billion Christians worldwide. It requires Christianity to work together with other religious groups, other organizations in civil society, science and education, business and industry and the various levels of government, not to forget the symbiosis with other forms of life. This need is widely recognized and endorsed. In the South African context this is exemplified by the work of the South African Faith Communities' Environment Institute (see http://safcei.org/). It seems perfectly possible, best expressed in the *Earth Charter*, to identify common values and proximate goals to make such cooperation possi-ble (see Rockefeller 2001). However, amongst Christians there are deep divides as to how such cooperation should be interpreted. For some earthkeeping is a form of Christian witness to the triune God, aimed at spreading the Christian gospel to the ends of the Earth to establish God's reign. For others, Christianity is one religion alongside others, which is itself one form of life emerging within evolutionary history in the earth's biosphere (see Knitter 1995). This begs further questions on how Christianity relates to other religious traditions and how the distinc-tion between the ultimate and the penultimate is interpreted.

An ongoing journey instead of a typology

Given the sheer variety of forms of ecotheology in different contexts, traditions and disciplines, with an even wider range of conversation partners (from different disciplines and with diverg-ing philosophies), and the fragmentation in diverging schools and scholarly guilds with underlying methodological disputes, it is hardly surprising that many scholars have offered typologies to classify various approaches and to map the terrain. The typologies employ diverg-ing points of demarcation in terms of forms of spirituality, religious practices, symbols, traditions, orthodoxy, methodologies, sub-disciplines, conversation partners from other discourses and so forth. It seems that there are as many typologies as there are positions! This does not imply that typologies are not helpful, only that no one typology would suffice while no overview of typologies would be comprehensive (Conradie 2011, 1–4).

An added problem is that any typology would tend to reflect geographic and language divides. There is some academic exchange and ecumenical fellowship across such divides, but any mapping of the terrain would tend to reveal underlying worldviews and linguistic assump-tions. The very term "worldview" is used with diverging connotations in different discourses. This contribution, for example, is offered from within the South African context and on the basis of international collaboration mainly through the medium of English as a former colo-nial language. However, given language and other limitations, it may well underplay Roman Catholic and Eastern Orthodox views (e.g. in Greek or Russian), African indigenous perspec-tives (see Gitau 2000, Daneel 2001, Mugambi and Vähäkangas 2001 and many others), Latin American contributions (in Spanish), a considerable corpus of literature from the Indian sub-continent (albeit from diverging perspectives) and especially East Asian Christianity, for example from South Korea (see e.g. Kim 2013 besides many publications in Korean). In each context obvious differences emerge when Christians adopt conversation partners from, for example, secularism, Judaism, Islam, Hinduism or Buddhism and other eastern religions. In

Eastern debates the term "life" is often preferred to the term "ecology" (see Kim Yong Bock 2014).

One significant attempt to map the terrain in a different way is derived from a colloquium held in San Francisco in November 2011 (see Bauman, Conradie and Eaton 2013) where the metaphor of a journey through an unchartered landscape was proposed. The choice of metaphor suggests that participating in a journey rather than the exercise of mapping the terrain may be the best way to appreciate diversity in ecotheology while also addressing the underlying challenges. In other words: the ecological reformation of Christianity may be understood as an ongoing journey.

The notion of a journey suggests a combination of a sense of place (the spatial turn – see Bergmann 2007) and the recognition of a temporal axis with an uncertain but daunting future. Various dimensions of this journey may be identified and described. It is a journey from an unacceptable present but proceeding with some sense of a destination and purpose. There are diverging sources of inspiration for such a journey, including authoritative sacred texts, tradition, reason and contextual experience, but with different weights attached to such sources. Different scholars employ distinct modes of "transport" (symbols or thematic interests) in this journey.

In traveling on this journey one typically also has to "steer through" certain tensions. These include the tensions between the use of the terms "religion" and "theology" (or philosophy) to describe one's core research interests, between an emphasis on immanence or on transcendence, between (Nicene) "orthodoxy" and "heresy," between faithfulness to one tradition and a more eclectic or pluralistic use of traditions and schools, between being "evangelical" and being "progressive," between particular (narrow group) identities and contested attempts to construct inclusive identities, between a focus on the local or the global, between biocentric, ecocentric or theocentric approaches, between a focus on wilderness or on urban ecology, between a concern for the plight of non-human animals and a concern for the human victims of environmental injustices.

Although there are different theological re-descriptions of this journey, there is also a widespread recognition of the need for diverse companions and therefore for collegiality, affinity, solidarity and cooperation with many others, e.g. environmental activists, scientists, people standing within other religious traditions and those who describe themselves as secular or post-secular.

This collective effort to map the journey of doing ecotheology cannot be an aim in itself; the focus has to remain on the journey itself. This journey demands responses to a looming catastrophe beyond imagination. One may do so with the recognition that we are embedded in a "divine milieu," journeying in the midst of an ineffable mystery in which we have received and may treasure the wondrous gift of life. An ecological reformation of the various Christian traditions will require such an ongoing journey.

References

Bauman, W., Conradie, E. M. and Eaton, H. (eds) (2013) "The journey of doing ecotheology" *Theology* 116:1, 1–49.

Bergmann, S. (2007) "Theology in its spatial turn: Space, place and built environments challenging and changing the images of God," *Religion Compass* 1:3, 353–379.

Berry, R. J. (ed.) (2006) *Environmental stewardship*, T&T Clark, London.

Best, T. F. and Robra, M. (eds) (1997) *Ecclesiology and Ethics: Ecumenical Ethical Engagement, Moral Formation and the Nature of the Church*, World Council of Churches, Geneva.

Boff, L. (1997) *Cry of the Earth, Cry of the Poor*, Orbis Books, Maryknoll.

Clough, D. (2012) *On Animals: Volume I: Systematic theology*, T&T Clark, London.

Cone, J. H. (2001) "Whose earth is it anyway?" In: Hessel, D. T. and Rasmussen, L. L. (eds) *Earth Habitat: Eco-injustices and the church's response*, Fortress Press, Minneapolis 23–32.

Conradie, E. M. (2006) *Christianity and Ecological Theology: Resources for further research* SUN Press, Stellenbosch.

Conradie, E. M. (2011) *Christianity and Earthkeeping: In search of an inspiring vision* SUN Press, Stellenbosch.

Conradie, E. M. (2013) "Contemporary challenges to Christian ecotheology: Some reflections on the state of the debate after five decades", *Journal of Theology for Southern Africa* 147, 105–22.

Conradie, E. M., Bergmann S., Deane-Drummond C. and Edwards D. (eds) (2014) *Christian Faith and the Earth: Current paths and emerging horizons in ecotheology,* T&T Clark, London.

Daneel, M. L. (2001) *African Earthkeepers: Wholistic interfaith mission,* Orbis Books, Maryknoll.

Deane-Drummond, C. E. (2014) *The Wisdom of the Liminal: Evolution and other animals in human becoming,* WB Eerdmans, Grand Rapids.

Eaton, H. (2005) *Introducing Ecofeminist Theologies,* T&T Clark International, New York.

Edwards, D. (ed.) (2001) *Earth revealing – Earth healing: Ecology and Christian Theology,* The Liturgical Press, Collegeville.

Francis (Pope) (2015) *Laudato Si: Encyclical letter of the Holy Father Francis on Care for our Common Home,* Libreria Editrice Vaticana.

Gebara, Y. (1999) *Longing for Running Water: Ecofeminism and liberation,* Fortress Press, Minneapolis.

Gitau, K. (2000) *The Environmental Crisis: A challenge for African Christians,* Acton, Nairobi.

Gnanadason, A. (2005) *Listen to the Women! Listen to the Earth!* WCC, Geneva.

Gustafson, J. M. (1994) *A sense of the Divine: The natural environment from a theocentric perspective,* T&T Clark, Edinburgh.

Habel, N. C. (ed.) (2000) *Readings from the Perspective of Earth,* Sheffield Academic Press, Sheffield.

Habel, N. C. (2009) *An Inconvenient Text: Is a green reading of the Bible possible?* ATF Press, Adelaide.

Horrell, D. G., Hunt, C., Southgate, C. and Stravakopoulou, F. (eds) (2010) *Ecological Hermeneutics: Biblical, historical and theological perspectives,* T&T Clark, London.

Hessel, D. T. and Ruether, R. R. (eds) (2000) *Christianity and Ecology: Seeking the well-being of earth and humans,* Harvard University Press, Cambridge.

Kaoma, K. J. (2013) *God's Family, God's Earth: Christian ecological ethics of ubuntu,* Kachere Series, Zomba.

Keum, J. (ed.) (2013) *Together towards Life: Mission and Evangelism in Changing Landscapes with a Practical Guide,* Geneva: World Council of Churches.

Kim, G. J-S. (2013) *Colonialism, Han, and the Transformative Spirit,* Palgrave Macmillan, New York.

Kim, Y-B. (2014) "A Christian theological discourse on integral life in the context of Asian civilization," in Conradie, E. M., Bergmann, S., Deane-Drummond, C. and Edwards, D. (eds) *Christian Faith and the Earth: Current paths and emerging horizons in ecotheology,* T&T Clark, London 219–31.

Knitter, P. F. (1995) *One Earth Many Religions: Multifaith dialogue and global responsibility,* Orbis Books, Maryknoll.

Linzey, A. (1995) *Animal Theology,* SCM Press, London.

Moe-Lobeda, C. D. (2013) *Resisting Structural Evil: Love as ecological-economic vocation,* Fortress Press, Minneapolis.

Moltmann, J. (1985) *God in Creation: An ecological doctrine of creation,* SCM Press, London.

Mugambi, J. N. K. and Vähäkangas M. (eds) (2001) *Christian Theology and Environmental Responsibility,* Acton, Nairobi.

Nash, J. A. (1996) "Towards the ecological reformation of Christianity," *Interpretation* 50:1, 5–15.

Niles, D. P. (ed.) (1992) *Between the Flood and the Rainbow,* World Council of Churches, Geneva.

Rasmussen, L. L. (1996) *Earth Community: Earth ethics,* Orbis Books, Maryknoll.

Rasmussen, L. L. (2013) *Earth Honoring Faith: Religious ethics in a new key,* Oxford University Press, Oxford.

Rockefeller, S. C. (2001) "Global interdependence, the Earth Charter, and Christian faith," in Hessel, D. T. and Rasmussen, L. L. (eds) *Earth Habitat: Eco-injustices and the church's response,* Fortress Press, Minneapolis 101–124.

Ruether, R. R. (ed.) (1996) *Women Healing Earth: Third world women on ecology, feminism and religion,* Orbis Books, Maryknoll.

South African Council of Churches Climate Change Committee (2009) *Climate Change – A challenge to the churches in South Africa,* SACC, Marshalltown.

White, L. (1967) "The Historical Roots of our Ecologic Crisis," *Science* 155:1203–1207.

World Council of Churches (2002) *Solidarity with the Victims of Climate Change: Reflections on the World Council of Churches' response to climate change,* WCC, Geneva.

9

ISLAM

Norms and practices

Zainal Abidin Bagir and Najiyah Martiam

Contemporary Muslims' concern about Islamic understandings of nature can be traced back to a series of lectures in 1966 delivered by Seyyed Hossein Nasr, linking environmental degradation to spiritual and moral crises of the modern world. Yet, four decades later Nasr said, "general indifference to the environmental crisis and apathy in seeking to find solutions to it based on Islamic principles continued until the 1980s and 1990s, when, gradually, voices began to be heard concerning this issue" (Nasr 2003, 86). This does not mean that nothing happened in the four decades. For example, in the mid 1980s Fazlun Khalid developed the Islamic Foundation For Ecology and Environmental Sciences (www.ifees.org.uk/), which is now active in a number of Muslim countries. In terms of publications, the anthology *Islam and Ecology* (2003), which is part of the Harvard Series on World Religions and Ecology, has prominently marked the new development.

Locating this discourse in the broader landscape of contemporary Islamic thought, which consists mostly of responses to modernity, it is clear that the issue of ecology does not occupy an important place yet. After 9/11, political issues such as radicalism and terrorism (with the brutality of ISIS as its most recent and vivid manifestation), democracy, human rights, and the equality of women and religious others, have exhausted the energies of contemporary Muslim thinkers.

As an illustration, it is instructive to see that in *Progressive Muslims*, a book attempting "to reflect critically on the heritage of Islamic thought and to adapt it to the modern world," which was expected to mark "a new chapter in the rethinking of Islam in the twenty-first century" (Safi 2003, 5, 6), ecological concerns were completely absent. Its two main foci were gender justice and pluralism, which reflected the intention of the authors to offer alternative Islamic voices to counter negative portrayals of Islam in a freshly post-9/11 world. Another illustration is *The Blackwell Companion to Contemporary Islamic Thought* (Abu-Rabi' 2006), which featured writings on terrorism, political movements, and women, but not on environmental problems in Muslim countries. The entry on "Theology" in *The Oxford Encyclopedia of the Islamic World* (Morewedge 2013) summarizes main theological currents in Islam since its early history, but when it comes to challenges to theology in the twenty-first century, it lists only political events: Israel–Palestine conflicts, post-9/11 terrorism and its aftermath, and the so-called "clash of civilizations." The examples can be easily multiplied.

Identifying the environmental crisis as a particular problem of the modern world, as Nasr and many other scholars do, brings them quickly to two of its main components, i.e. modern science and the dominant modern economic system, both of which can legitimately be characterized to some extent as Western. Starting from the realization that at the root of the crisis is a modern metaphysics that desacralizes nature—expressed in both modern science and the capitalist economy that has led to the exploitation of nature—the solution offered is naturally an alternative metaphysics, manifested in some alternative science and economic system.

Nasr has argued for an alternative science since his first work on this issue (1967). One of the prominent arguments in the discourse of "Islamic science" that developed in the 1980s and 1990s had to do with attempts to provide an alternative mode of relation with the natural world. The "Islamic science" movement, including a variation expressed in the term "Islamization of knowledge," was booming for some time, although it did not develop far enough toward the creation of a distinct system of science and technology as hoped for by its proponents. Furthermore, even in this more proximate, discourse of Islam and science, ecological concerns have not figured importantly.

The idea of Islamic economy developed around concerns related to the negative impacts of modern economy, especially its failure to deliver justice and its exploitation of nature. Despite the mushrooming of Islamic (or *shari'a*) financial institutions, this trend too has seldom been taken up in a way that also considers its relation with environmental problems. Instead, its focus has been promoting *shari'a*-compliance, understood in very technical terms as avoiding usury.

While *shari'a* looms large in the Islam and ecology discourse, it is hardly connected to the broader discourse about the place of *shari'a* in the modern world. All this shows that, despite the rise in publication of works on Islam and ecology in the past decade, and despite the similarity of the structure of environmental and other contemporary problems as well as their methodological challenges, this matter has not been integrated with the wider issues of Muslim responses to the modern/contemporary world.

Turning to methodology, a popular strand in Islamic responses to environmental concerns is defensive, if not apologetic. A major characteristic of the discourse consists of expositions of how Islam, as shown in its traditions (especially the two canonical sources, the Quran and *hadith*), has actually been or has the potential to be a "green religion." In communities in which religion still plays a central justificatory role, such a normative understanding should not and cannot be evaded, but this kind of overtly textual exposition does not go far enough to respond to new ecological knowledge and environmental problems.

It is anachronistic to think that a centuries-old tradition should be prepared with answers to any emergent question, especially questions that have not yet been asked, at least not in the magnitude of today's environmental crisis. Considering the possibility and plurality of interpretations within Islam (an-Na'im 2008), the main issue is what kinds of interpretation can best respond to the problems at hand. What is urgent is a broader hermeneutic insight that focuses on solving the problems. At this point it becomes urgent to focus more attention on the practices of believers and not only to (textual) normative sources as means of justification.

Attention to practices, as illustrated in this chapter with examples from Muslims in Indonesia and the U.S., serves not only as a reminder of the pluralism within Islam, which originated in different readings of Islamic textual sources, but also the possibility of a different kind of normativity. This chapter shows some new developments in that direction, that is, studies that do not only look at the primary normative, textual sources of Islam, but also at the existing practices of Muslims as an alternative source of Islamic/Muslim normativity on ecological issues. This suggestion signals some difficult epistemological challenges related to how we should understand Islamic normativity, which will be discussed later.

Challenges to the theological discourse

There has been a broad agreement in Qur'anic interpretation regarding the metaphysical framework around which the discourse of Islam and ecology has developed since the 1960s, expressed in a few central concepts. The existence of the one God (*tawhid*), the Creator, on whom the existence of the created world depends absolutely, forms the basis of Islamic belief. The natural world is understood as *muslim*, in the sense that it cannot but submit (*aslama*, from which the name of the religion, *islam*, is derived) to God's will (Ozdemir 2003, 16ff). Human beings, who received revelations from God, occupy a distinctive place in the order of nature.

The single most important concept in this framework, which almost always appears centrally in discussions of Islam and ecology, is *khalifa*, the idea that human beings are created as God's vicegerents or stewards in the world. Inasmuch as it was a human being (Adam) who was taught the names of all things, humans occupy a special position in creation because, through God's education (revelation), it is they who name all creation as objects. This unequal and asymmetrical relation between humans and the rest of creation may easily give the impression that it is human beings who make decisions about other creations, including to exploit them. There are also a number of Qur'anic verses that clearly state that the natural world exists for humans to take benefit from (Ozdemir 2003, 26). Following such an interpretation of the Qur'an, it is difficult to avoid the idea that human beings are special in the order of nature.

Some critics see such anthropocentrism as detrimental to environmental preservation. For example, such a concern was voiced strongly by Afrasiabi (1998), who pointed out the prevalence of anthropocentric images derived from Islamic sources, even by scholars engaging the environmental crisis. He argued that "an alternative Islamic theology that would be capable of integrating within its horizon the fundamental ecological precepts" is required to deconstruct such an image. For Afrasiabi, such a theology would present a non-anthropocentric conception of Islam, and a view of human beings not grounded in "the stereotypical monarchical connotation of vicegerency;" it would comprise a non-utilitarian "theology of the inorganic." In a passing remark, Foltz has also mentioned that an eco-friendly interpretation of Islam should not be hierarchical (Foltz 2003, 249).

Those suggestions raise important questions. Acknowledging plurality of interpretations in Islam, one question is whether non-anthropocentric expectations are within the scope of possible interpretations of Islamic sources, especially given the strength and centrality of the notion of humankind's role as *khalifa*. Does the concept *of khalifa* necessarily lead to anthropocentrism with all its negative implications for preservation of nature? The project of reconstructing Islamic ecotheology does not need to be an all-or-nothing affair; it is more realistic to see a variety of Islamic cosmologies presenting a spectrum in which non-human beings occupy different degrees of significance relative to humans. Taking a few chapters in *Islam and Ecology* (2003) as samples, we can already discern differences in how *khalifa* is understood despite apparent universal agreement on its centrality.

First of all, the so-called "Islamic anthropocentrism" is mitigated to the extent that it becomes a type of anthropocentrism that does not necessarily lead to unduly destructive exploitation of nature. In some interpretations, *khalifa* is effective only insofar as human beings remain the obedient servant (*'abd*) of God, to whom all human and non-human beings submit (Nasr 2003; Haq 2003). In this regard, human and non-human beings are equal, the only difference between them being that the latter are necessarily *muslims*, while the former has the choice to submit to God or not. Second, the relation between humankind as *khalifa* and non-human beings is not necessarily a straightforward relation of domination as implied by Afrasiabi. Ibrahim Ozdemir (2003) for example, affirms that indeed in Islam "human beings are at the top of the great chain of being," but immediately qualifies this by saying that they are not "the owners of nature nor is the sole aim of nature is to serve human beings and their ends." While in the Qur'an and *hadith* one may find texts with the view that non-human animals are valued mainly for the services they provide for humans, there are also texts with the understanding that animals have value of their own, apart from their usefulness to humans (Foltz 2006, 3, 4). A further position in the spectrum is given in Chisti's account of *khalifa*, which goes so far as to say that "every life-form possesses intrinsic value independent of its resource worth to humanity" (Chishti 2003, 76). Within the genre of philosophical mysticism the significance of non-human beings is articulated even more strongly. A more recurrent theme in the Qur'an is nature as the object of spiritual contemplation rather than exploitation. A further position in this regard is a non-anthropocentric Muslim worldview in which humans and non-humans are regarded as equal persons, which is discussed below.

Shari'a as law and ethics

As with responses to other contemporary problems, *shari'a* features prominently in the Islam and ecology discourse. *Shari'a* has been deployed in a wide range of ways, from the narrowest of understandings in the form of *fatwa* (a non-binding religious edict) on particular issues, to *fiqh* (religious law or jurisprudence), to a methodology of ethical decision-making (Sachedina 2009, an-Na'im 2008). The main problems discussed mostly concern how to derive laws or ethics from the Qur'an, *hadith* and other authoritative works, and how to implement them.

In the Islam and ecology literature, *shari'a* is generally conceived as a means to address environmental problems related to, among other things, the use of land, preservation of precious natural resources, and conservation. For this, scholars may draw from abundant sources in the literature of *hadith*, which includes chapters on explicitly environmental issues (Haq 2003, 141–143). Many scholars do propose making *shari'a* as the law to be enforced by the State (e.g. Mawil Izzi Dien; see Johnston 2012). For example, the tradition of *hima* (protected areas) is still kept alive in Saudi Arabia, though in recent decades these conservation areas have been dwindling rapidly (Johnston 2012, 235).

As law, only rarely is *shari'a* actually enforceable, that is, only in very rare situations when it is incorporated by a government. In most cases related to environmental issues, *shari'a* serves a different function, not as putatively enforceable laws, but as a rhetorical tool and strategy to advocate for "eco-justice" (Johnston 2012, 221). In a particular context of Indonesia, Gade argues that legal public reasoning of Islam and environmental law plays an important role as an engine of social change (Gade 2015, 162). In recent progressive discourse, the concept of the *maqasid* (the purpose of) *shari'a*, defined with public interest (*maslaha*) as one of its main principles, has come to play a central role as the legal/ethical framework (Johnston 2007; Hefner 2011). In the case of Indonesia, ecological concern has been given legal primacy that makes it the core aim of the law (*maqasid*) (Gade 2015, 164).

The enrichment of Islamic ethical discourse may be developed in a number of ways. Kecia Ali (2015) compares the domination of men over women to the domination of humans over animals in Islam, and argues that both originate from patriarchal–hierarchical cosmology. Showing the intertwined nature of the subjugation of women and of animals, Ali suggests that Muslim feminists should engage with non-religious feminist ethics because it can provide underutilized resources for Muslim thinking about food ethics in particular and ethics in general (Ali 2015, 269). "Engagement with non-Islamic (though not 'un-Islamic') ethics provides a model for productive dialogue among parties who disagree about basic presumptions but agree on desirable outcomes," (Ali 2015, 269). She also makes a case for Muslim vegetarian ethics, despite the lawfulness of meat-eating, for animal welfare and ecological concerns.

At the same time, this illustration provides an insight into how Muslim thinking about food can be expanded beyond the dominant normative discussions of dietary laws, such as defining foods as *halal*. While Ali probes deep into the underlying Muslim cosmology, Magfirah Dahlan-Taylor (2015) emphasizes the need to go beyond individual consumers' interests in consuming *halal* foods, which display Muslim religious exclusivity, and connect it to political questions of food justice that also involve consideration of labor and wealth inequality. She argues that Islamic laws and ethics are not something that can be, quoting Muslim thinker Fazlur Rahman, "deduced from the Qur'an in abstracto" (Dahlan-Taylor 2015, 14–15). Dahlan-Taylor applies this principle to the politics of dietary laws, but it is also relevant to the discourse of Islam and ecology in general. Taken more generally, these insights illustrate how Muslim discourse on environmental ethics could be expanded beyond narrow legal categories (of *halal* and *haram* as applied to particular acts) in order to include broader categories and concerns of equality and justice, which are central in Islam.

In relation to this point, it is interesting to note the emergence of the notion of "eco-*halal*", which combines the Islamic dietary principle of *halal* meat and the sustainable-food movement (Barendregt 2013; Arumugam 2009). This may be another way to expand the strictly legal discourse about what foods and methods of food processing are lawful, but, inasmuch as large-scale food industries in the U.S., Malaysia, and Europe are marketing themselves as *halal*, Dahlan-Taylor's concern for justice remains a topic of importance.

Studies of practices

The increasing number of studies of ecological practices by Muslims, as individuals and as communities, constitutes another opportunity to enrich or even sidestep the discourse on theology and ethics as discussed above, which is mostly grounded in textual studies. Such practices may also give rise to further methodological debates in Islam and ecology, as will be discussed in the final section below.

A different response to the charge of Islam's anthropocentrism may be found in Samsul Maarif's (2014) portrayal of the seemingly animistic and eco-friendly practices of a small indigenous community in Sulawesi, Indonesia. This illustration at the same time shows a different way of doing Islam and ecology. The Ammatoans profess to be Muslims—and, as such, for Maarif, they are Muslims—but practice such indigenous rituals as chanting, sacrificing animals, and giving offerings to the forest, which involves the participation of a Muslim *imam*. While the rituals may be interpreted as being animistic in the conventional sense (worshipping and believing in spirits inhabiting natural objects), there is another way of understanding animism as constituting interpersonal relationships between human persons and natural objects, in which those objects are also understood as (non-human) persons. This latter understanding, argues Maarif, parallels Quranic depictions of natural objects (mountains, birds, stars, trees) as beings that glorify God (Q. 21:79, 38:18, 13:13, 55:6), which indicates their personhood as *muslim*. The Ammatoan cosmology regards and treats non-human beings as persons equal with humans. While Maarif admits that it is unlikely that those beliefs and practices are historically derived from their understanding of the Qur'an, he defends their "animistic" practices as Qur'anic and thus opens the possibility of "being a Muslim in animistic ways."

The Ammatoan community is not the only example of Muslim "folk eco-theology." Foltz mentions Shakeel Hossain's discussion of rural Bengali Muslims' traditional river festival, which was not welcomed by some Muslims who considered it *shirk* (polytheistic) (Foltz 2003,

252–253). While it is understandable that this kind of interpretation is controversial, there have always been a spectrum of interpretations, some of which are regarded as "non-mainstream" to differing degrees.

Another interesting example of Muslim practice is the Bumi Langit (literally, Earth Heaven) Farm in a village in Java, Indonesia. Founded by Iskandar Waworuntu, a Muslim convert, the Farm manifests a cosmology and ethics that blend inspiration from the Qur'an, Western and Eastern philosophies, and local practices (Rogers 2013). In developing a permaculture, Waworuntu seeks to work with, rather than against, nature in an Islamic paradigm. In his understanding, the paradigm realizes Islam as a blessing for the whole universe (*rahmatan lil 'alamin*) by emphasizing harmonious relations between all God's creatures—human and non-human, animate and inanimate—who are all praising and chanting toward God. Bypassing the anthropocentrism debate, he upholds the true *khalifa* as one who is not simply given the title by God but who earns it by the respect given by creation. For him, being an environmentalist is the most ethical way of being a Muslim, and it forms half of faith, the other half being *tawhid* (belief in God). In this worldview, *halal* (permissible) as a legal category is not sufficient, but needs to be supplemented with the ethical category of *tayyib*, meaning goodness or whole-someness, which can be extended to the politics and ethics of food. He tries to make his farm embody Islamic teachings in all aspects.[1]

Ecological practices have also been developed by many 'eco-*pesantrens*' (Islamic boarding school/community) in Indonesia (Mangunjaya and McKay 2012), who argue that this independent grassroots initiative, which combines Islamic principles of environmental protection with traditional methods of conservation, is more effective than the top-down approach, through enactment of laws initiated by the government.

Another important example of Muslim ecological practices is shown in Eleanor Finnegan's work on three Sufi Muslim farms in the U.S. (Finnegan 2011a, Finnegan 2011b). Though not all are functional farms, all three cultivate gardens, plants for healing, herbs for cooking, and vegetables and fruit for the community. Farming has helped them to create and maintain religious identities, foster community, and nurture spirituality (Finnegan 2011a, 71–73). Here, theological affirmations—e.g. that non-human creatures are *muslim*, or the understanding of *tawhid* as creation's interconnectedness—are lived in daily experiences (Finnegan 2011b, 260ff).

Finnegan contrasts the scholars who turn to textual tradition to reconstruct ethics and the people for whom everyday life experiences on the farm shape values or practices (Finnegan 2011b, 243). In a sense, however, the Javanese and American examples show that this is not only about different ways of reconstructing theology and ethics, but also about how ecological practices help believers to understand religion and be better religious (Muslim) persons. In these examples farming has become a way not (only) to implement what they understand about Islamic theology and ethics related to nature, but also a way to understand Islam itself. The concepts invoked (such as *tawhid*, *khalifa*, nature being *muslim*) are identical with what scholars derive from the Qur'an, but here they have become lived experiences. One important difference is that while scholars seeking universal ethical concepts derive their views from Islamic textual sources, the knowledge gained from lived experiences of the communities is particular to these individuals and communities.

This last characteristic may be regarded as problematic if what is "Islamic" cannot be local and has to be universal—an issue to be taken up below. By way of introduction, it has to be admitted that the question of what or which interpretations may be considered "Islamic" is contentious and has important implications for the development of Islam and ecology discourse. This is especially true when, as in the Ammatoan case, practices are taken into account. However, drawing from her own study of the practices of American sufi Muslim farmers, Eleanor Finnegan (2011a, 2011b) shows that much is lost when the field of Islam (and for that matter, religions in general) and ecology does not pay sufficient attention to practices.

Concluding methodological notes: what is "Islamic?"

The theological and ethical discourse of Islam and ecology has produced normative arguments deduced from the canonical sources of Islam. In order to go further, it needs to be supplemented by broader considerations (e.g. on food ethics), and should pay greater attention to practices. The issues discussed in this essay—indigenous Ammatoan rituals, the ethics and politics of food, and the ecological practices of individuals and religious communities—raise questions about how texts and practices function in the formation of "Islamic views of nature" and "Islamic environmental ethics." They suggest that ecological practices not only implement the normative theories, but also play roles in the formation of "Islamic" normativity.

One way to explain this difference is by drawing a distinction between what the texts say and what Muslims do. For Richard Foltz, this distinction marks the difference between Islamic and Muslim views of ecology or between Islamic and Muslim environmentalisms. Foltz argues that "the actual practices and attitudes of Muslims have always been shaped by Islamic sources in combination with extra-Islamic cultural ones" (Foltz 2006, 8) In another place, his statement is bolder: "For an idea to achieve anything approaching universal acceptance by Muslims as 'Islamic', it must be convincingly demonstrated that it derives from the Qur'an or … the example of the prophet Muhammad". (Foltz 2003, 253) Moreover, "local or regional attitudes cannot form a basis for any kind of universal Islamic ethics." But this clear-cut distinction between Islamic environmentalism and Muslim environmentalism may be misleading because it gives the false impression that there are culturally neutral "pure" Islamic interpretations, or that there are universally accepted Islamic doctrines. As a matter of hermeneutic principle, it is impossible to claim that any interpretation of the Qur'an or *hadith* is derived purely from the canonical sources without any cultural influence. Besides, it is difficult, if not impossible, to point to any idea "approaching universal acceptance by Muslims as 'Islamic'" except on very basic or trivial issues.[2]

Of course, not all cultural practices performed by Muslims should be regarded as "Islamic," just as they do not have to be rejected as "un-Islamic." Similarly, not all views supposedly derived from the Qur'an are good or even "Islamic." Determining what is "Islamic" or "un-Islamic" involves more than justifications by reference to Qur'an and *hadith* or cultural practices. Further, as Kecia Ali argues, the defensive concern with religious authenticity and the primacy given to "Islamic" identity may constitute an obstacle for a productive ethical work. In the context of the discourse on Islam and ecology, a central criterion by which any Islamic view is to be evaluated may be the premise of the requirement of an ecologically sound way of life; such a criterion, as discussed above, may be and has been interpreted as one of the principles of *maqasid al-shari'a*.

The major argument here is that one of the important keys to furthering Islam and ecology discourse is to pay more attention to the empirical study of living traditions and practices. Such studies have long existed but are not yet widely accepted because of narrow normative criteria about what makes an idea or a practice "Islamic." What needs to be explored is not (only) the consistency and coherence of an idea with canonical sources but how Muslim communities develop, justify, and defend eco-friendly practices, and form their ideas about Islam and ecology through their practices.

Realizing that the field of Islam and ecology is not only an academic discourse but also carries practical aims of responding to environmental problems, another benefit in acknowledging local views and practices, like those of the Ammatoans and Bengalis, or the Javanese and

American Sufi Muslim farmers, is in providing recognition and support for the continuation of such eco-friendly practices. This recognition may also open ways for local communities to learn from others. Of course, this possibility is necessarily limited. Ammatoan or Bengali Muslim practices cannot simply be exported to Saudi Arabia, for example; the Ammatoan way of life may not even be replicable in other areas of Indonesia. But the local is significant precisely because it is usually very deeply and uniquely rooted in its own land. It may resist universalizing, but it continues to promise operative and sustainable practices as well as to generate fresh ideas.

Acknowledgment

The authors wish to thank the team at the Resource Center of the Center for Religious and Cross-cultural Studies (CRCS), Whitney Bauman, and Siti Sarah Muwahidah for their help in providing resources for this chapter. They also thank Gregory Vanderbilt and the editors of this book who helped to make this chapter more readable.

Notes

1   Interview with Najiyah Martiam (May 15, 2015).
2   Another note on this point is the fact that even a "universal acceptance" may not mean much. A case in point is the declarations made in Jordan by more than a hundred mainstream Muslim scholars from all around the world known as *The Amman Message* (2004) and *The Common Word* (2007). These two declarations were hailed as an unprecedented *ijma'* (consensus) in terms of the number and diversity of scholars who approved them, and as such *The Amman Message* even claims that the document is legally binding upon all Muslims. Yet they do not seem to have the effects intended: Differences remain and there are criticisms of the documents by Muslims.

References

Abu-Rabi' I. M. ed. (2006) *The Blackwell Companion to Contemporary Islamic Thought,* Blackwell, Malden MA.
Afrasiabi K. L. (2003) "Toward an Islamic ecotheology" in Foltz R. C., Denny F. M., and Baharuddin A. eds., *Islam and Ecology: A bestowed trust,* Center for the Study of World Religions, Cambridge MA, 281–296.
Ali K. (2015) "Muslims and meat-eating," *Journal of Religious Ethics* 43.2, 268–288.
an-Na'im A. A. (2008) *Islam and the Secular State: Negotiating the future of shari'a,* Harvard University Press, Cambridge MA.
Arumugam N. (2009) The eco-halal revolution (www.culinate.com/articles/features/the_eco-halal_revolution). Accessed 1 May 2015.
Barendregt B. (2013) When eco chic meets halal (www.leidenanthropologyblog.nl/articles/when-eco-chic-meets-halal). Accessed 1 May 2015.
Chishti S. K. (2003) "*Fitra*: an Islamic model for humans and the environment," in Foltz R. C., Denny F. M., and Baharuddin A. eds. *Islam and Ecology: A bestowed trust,* Center for the Study of World Religions, Cambridge MA 67–82. '
Dahlan-Taylor M. (2015) "'Good' food: Islamic food ethics beyond religious dietary laws," *Critical Research on Religion* 3.3, 250–265.
Finnegan E. (2011a) "What traditions are represented in religion and ecology? A perspective from an American scholar of Islam," in Bauman W., Bohannon R., and O'Brien K. J. eds. *Inherited Land: The changing grounds of religion and ecology,* Pickwick Publications, Eugene 64–79.
Finnegan E. (2011b) Hijra and homegrown agriculture: Farming among American Muslims, unpublished PhD thesis, University of Florida.
Foltz R. C. (2003) "Islamic environmentalism: A matter of interpretation," in Foltz R. C., Denny F. M., and Baharuddin A. eds. *Islam and Ecology: A bestowed trust,* Center for the Study of World Religions, Cambridge MA, 249–279

Foltz R. C. (2006) *Animals in Islamic Tradition and Muslim Cultures,* Oneworld, Oxford.

Gade A. M. (2015) "Islamic law and the environment in Indonesia," *Worldviews: Global Religions, Culture, and Ecology* 19.2, 161–183.

Haq Nomanul (2003) "Islam and ecology: Toward retrieval and reconstruction," in Foltz R. C., Denny F. M., and Baharuddin A. eds. *Islam and Ecology: A bestowed trust,* Center for the Study of World Religions, Cambridge MA 121–154.

Hefner R. W. ed. (2011) *Shari'a Politics: Islamic law and society in the modern world,* Indiana University Press, Bloomington and Indianapolis.

Johnston D. (2007) "*Maqasid al-Shari'a*: Epistemology and hermeneutics of Muslim theologies of human rights," *Die Welt des Islams* 47.2, 149–187.

Johnston D. L. (2012) "Intra-Muslim debates on ecology: Is Shari'a still relevant?" *Worldviews: Global Religions, Culture, and Ecology* 16.3, 218–238.

Maarif S. (2014) "Being a Muslim in animistic ways," *Al-Jami'ah: Journal of Islamic Studies* 52:1, 149–174.

Mangunjaya F. and McKay J. E. (2012) "Reviving an Islamic approach for environmental conservation in Indonesia," *Worldviews: Global Religions, Culture, and Ecology* 16.3, 286–305.

Morewedge P. (2009) "Theology," in Esposito J. L. ed. *The Oxford Encyclopedia of the Islamic World. Oxford Islamic Studies Online,* Oxford University Press, Oxford (www.oxfordislamicstudies.com/Public/book_oeiw.html). Accessed 3 September 2015.

Nasr S. H. (1967) *Man and Nature: The spiritual crisis of modern man,* Kazi Publishers, Chicago.

Nasr S. H. (2003) "Islam, the contemporary Islamic world, and the environmental crisis," in Foltz R. C., Denny F. M., and Baharuddin A. eds. *Islam and Ecology: A bestowed trust,* Center for the Study of World Religions, Cambridge MA, 85–105.

Ozdemir I. (2003) "Toward an understanding of environmental ethics from a Qur'anic perspective," in Foltz R. C., Denny F. M., and Baharuddin A. eds. *Islam and Ecology: A bestowed trust,* Center for the Study of World Religions, Cambridge MA, 3–37.

Sachedina A. A. (2009) *Islamic Biomedical Ethics: Principles and application,* Oxford University Press, Oxford.

Safi O. ed. (2003) *Progressive Muslims: On justice, gender and pluralism,* Oneworld Publications, Oxford.

Rogers C. (2013) Yogyakarta farmer pioneers Islamic environmentalism *The Jakarta Globe,* 16 October 2013 (http://thejakartaglobe.beritasatu.com/features/yogyakarta-farmer-pioneers-islamic-environmentalism/). Accessed 1 May 2015.

10

BAHÁ'Í

Peter Adriance and Arthur Dahl

Religion and ecology may seem to be two distinct domains of human experience: one with many centuries of diverse traditions and practice; the other an emerging scientific perspective on complex environmental systems and humanity's often problematic relationship with the natural world. For Bahá'ís, they are in fact complementary and closely interrelated, and have converged in recent years. The science of ecology offers a systems perspective on the relations between human society and the planetary ecosystem, including interrelated economic, social, and environmental dimensions. The concern today is sustainable development—that is, how to make the transition from a consumer society plundering the earth's resources, and political and economic institutions driven by self-interest, to a just and sustainable world civilization respecting planetary boundaries and restoring the natural productivity of the Earth and its ecosystems.

Unlike many religious traditions with centuries of practice behind them, the Bahá'í Faith, dating only from the middle of the nineteenth century, has both scriptural texts that are more explicit on environmental questions, and a body of teachings and authoritative guidance that evolves to respond to current situations. Bahá'u'lláh (1817–1892), the prophet-founder of the Bahá'í Faith, designated an authorized interpreter after his passing, 'Abdu'l-Bahá (1844–1921), and provided for an elected international council, the Universal House of Justice, to legislate on matters not specified in scripture. There is thus a continuing flow of new guidance on many matters, including the environment and sustainability, most recently in messages from the Universal House of Justice and the statements of the Bahá'í International Community (BIC) at the United Nations. Our understanding of that guidance is limited and constantly evolving, often related to the many social and cultural backgrounds of the diverse Bahá'í communities around the world, so our "interpretive debate" is in challenging our inherited assumptions and being open to new meanings. As a learning community, we are still often trapped in old ways of thinking and old frames of reference, and struggle to understand the full implications of the spiritual principles we have received. This is a time of rapid change, both in the world and in our own communities.

Since environmental problems are the result of human behavior, the response needs to be as much based on spiritual and ethical principles as scientific knowledge. This discussion of Bahá'í and ecology is placed in this larger framework.

Cross-disciplinary perspectives

The elements and lower organisms are synchronized in the great plan of life. Shall man, infinitely above them in degree, be antagonistic and a destroyer of that perfection?

(Abdu'l-Bahá)

One fundamental element of Bahá'í belief is that science and religion are two complementary knowledge systems. This inevitably leads Bahá'ís to be active in current debates about the role of religion in modern society, including in providing the ethical foundations and spiritual motivation to address many contemporary environmental challenges. Science provides the knowledge of planetary ecology and the threats represented by climate change, biodiversity loss, degradation of natural resources, and widespread pollution, as well as the necessary solutions at a technical level. However, these are only symptoms of the dysfunctions in human society, including a widespread loss of spiritual purpose and its replacement by a short-term vision and materialistic orientation. The Bahá'í teachings explicitly address the need to provide spiritual foundations for an emerging world civilization, with moderation in material civilization and an emphasis on individual lives of service (see Karlberg 2014).

One of the strengths of spiritual principles is that they provide a framework and points of reference for human behavior and social organization at a fundamental level, and thus address many problems of human society simultaneously, including those of ecology and environmental management. Principles like justice, equity and the oneness of humankind lead naturally to cross-disciplinary perspectives.

For example, climate change is an important symptom of the ecological imbalances created by an economy founded on the temporary cheap energy subsidy provided by fossil fuels, containing ancient solar energy stored in earlier geological epochs but now being exhausted in a matter of decades while threatening the well-being of future generations. From a Bahá'í perspective, we should be developing all available sources of energy on the surface of the planet to meet our energy needs on a sustainable basis. It is unjust that a wealthy minority should privilege itself while imposing the risks and costs of climate change on the poor and on future generations (BIC 2008). The Bahá'í teachings call for the elimination of extremes of wealth and poverty, and this has been an important theme for recent Bahá'í International Community (BIC) interventions at the United Nations (BIC 2010, 2011, 2012).

Recent scientific opinion is that global warming of 4°C—likely in this century if a rapid transition is not made to renewable energy sources—could wipe out half of all the species on the planet and reverse millions of years of evolution. Any future civilization would need to be founded on the renewable resources of agriculture, forestry, and fisheries. The Bahá'í writings call for a federated world government able to manage the natural resources of the planet and to distribute the benefits equitably. They give high priority to agriculture and farmers, with arrangements to support the income of farmers in bad years. Sea level rise, droughts, and other natural disasters resulting from climate change risk the permanent displacement of hundreds of millions of people. Not only will their human rights be violated, but the situation will precipitate migration crises around the world. Since most human rights violations today are against immigrants, this requires proactive educational efforts in the receiving communities, and a new international legal framework for environmentally displaced persons. The Bahá'í-inspired International Environment Forum (IEF) has raised this topic in UN forums. Religion should be part of the response, since all religions have principles about honoring guests, welcoming strangers, and assisting those in distress. Bahá'ís worldwide are engaged in learning how to create unified communities from diverse backgrounds. These are precisely the tools that will be needed to deal constructively with the massive mixing of the world population in the years ahead.

Climate change is also becoming a significant force for improvements in world governance, compelling nations to rise above national sovereignty in their common interest, just as Bahá'u'lláh called on them to do more than a century ago. While adjustments to the world economy must be made at the global level, the effects of climate change will be different in each local environment, where the Bahá'í approach to development by empowerment and capacity-building at the grassroots level will help each neighborhood and community to find its own local solutions relevant to its situation. Excessive consumption is a major driver of climate change, and the Bahá'í teachings strongly support simpler material lifestyles and a civilization focused on social and spiritual values, and those intangible things like knowledge, science, art, and culture that do not face limits to growth. Recent Bahá'í guidance has been highly critical of the consumer society, and emphasized redefining prosperity in other than material terms (Universal House of Justice [UHJ] 2005; BIC 2010).

> Take from this world only to the measure of your needs, and forego that which exceedeth them.
>
> *(Bahá'u'lláh)*

Ecology and religion intersect most intimately in our attitude towards nature. Modern urban society has largely forgotten about nature. We have conveniently replaced nature with "the environment," largely of our own making and outside of us, which we can ignore as an externality in our preoccupation with economic growth. Yet Bahá'í Scriptures describe nature as a reflection of the sacred. They teach that nature should be valued and respected, but not worshipped; rather, it should serve humanity's efforts to carry forward an ever-advancing civilization. However, in light of the interdependence of all parts of nature, and the importance of evolution and diversity to the beauty, efficiency, and perfection of the whole, every effort should be made to preserve as much as possible the earth's biodiversity and natural order (BIC 1998). There are Bahá'í warnings about the threat to nature, and that humanity should not be antagonistic to and a destroyer of the perfection in nature. Humanity has ignored that warning from a hundred years ago. Climate change, biodiversity loss, deforestation, soil degradation, over-fishing, and ocean acidification are seriously reducing the natural capacity of the planet. Nature, in its undisturbed form, will soon be a thing of the past. To make the transition to sustainability, we need a new emphasis on sustainable agriculture and renewable resources, and on preserving the ecological balance of the world.

> The country is the world of the soul, the city is the world of bodies.
>
> *(Bahá'u'lláh)*

The Bahá'í appreciation for nature goes far beyond its utilitarian value, to see it as a significant source of knowledge at the spiritual level. Many religious scriptures include exhortations about respect for nature, moderation in its use, and a prohibition on waste. Nature has spiritual significance, with the qualities of the divine being reflected in it. Contemplating nature is therefore a path to spiritual understanding. The greatness, grandeur, beauty, power, and wonders of nature can invoke in us a sense of humility. This is very healthy in our struggle with our ego, and can help to draw us out of ourselves. For those who are open to it, nature can produce a deep resonance with our spirit or soul. The great spiritual teachers (e.g. Moses,

Buddha, Jesus, Mohammed, Bahá'u'lláh) retreated into the wilderness to prepare for their mission. Many people seek mystical experiences in nature, or find their deeper self or direction in life through being in nature. For Bahá'ís, contact with nature is an important part of education.

Integrated approach

Regard ye the world as a man's body, which is afflicted with divers ailments, and the recovery of which dependeth upon the harmonizing of all its component elements.

(Bahá'u'lláh)

There is a recent recognition that one of the structural problems in present-day society is its increased specialization, reflected in government ministries, academic disciplines, international agencies, and many other components of society. The call today is to "break down the silos." Systems science, emerging in part from ecology, is providing tools for a more integrated approach to today's problems (Dahl 1996; Capra and Luisi 2014).

From a Bahá'í perspective, the problems of ecology and the environment cannot be separated from all the other problems facing the world today, and require a holistic integrated approach, both scientific and spiritual. Part of the integration needs to be in multiple levels of governance. The critical need is to acknowledge the oneness of all humanity in its diversity, and for all the nations of the world to understand and follow the admonitions of Bahá'u'lláh to whole-heartedly work together in looking after the best interests of humankind. The unity of nations is essential in the search for ways to meet the many environmental problems besetting our planet. The challenge of climate change, among others, is forcing states to recognize that only global environmental governance can address problems at a planetary scale.

The emergence of the Anthropocene, the epoch where human beings have become the major transformers of the Earth's surface, highlights the challenge, as it becomes apparent that our destruction of nature and our impact on global processes like the climate mean that our societal responsibilities now extend to the management and restoration of the whole planetary system. Our governance systems and values are totally inadequate to the challenge, and Baha'is see the rapid implementation of Bahá'u'lláh's solutions as the remedy that the world requires.

While individual statements of the BIC may address specific matters, such as climate change, sustainable consumption, globalization and transition, economic reform, gender, human rights, UN reform, prosperity, and well-being, they are all part of the Bahá'í cross-disciplinary perspective, and they come back to underlying principles of the spiritual nature and purpose of human life. They combine individual refinement of character and fulfillment, and the advancement of civilization, unity of humanity, and a new world order.

Spiritual perspectives that harmonize with that which is immanent in human nature can induce an attitude, a dynamic, a will, an aspiration, which can facilitate the discovery and implementation of practical measures. For example, at the UN Social Summit in Copenhagen in 1995, the Bahá'í International Community offered an extensive discussion of "The Prosperity of Humankind" integrating its economic, social, and environmental dimensions (BIC 1995). They returned to this theme for the debate on sustainable consumption and production at the UN Commission on Sustainable Development in 2010 with a statement on "Rethinking Prosperity: Forging Alternatives to a Culture of Consumerism" (BIC 2010).

Economic reform

> The civilization, so often vaunted by the learned exponents of arts and sciences, will, if allowed to overleap the bounds of moderation, bring great evil upon men.
>
> *(Bahá'u'lláh)*

The Bahá'í teachings have much to contribute to the discussion of economic reform, at least in principle if not in the technical details. Since the ecological challenges we now face are largely the result of an economic system founded in a materialistic interpretation of reality that has become the dominant faith in much of the world, this is highly relevant to both religion and ecology. The Bahá'í critique describes a consumer culture unembarrassed by the ephemeral nature of the goals that inspire it. For the small minority of people who can afford them, the benefits it offers are immediate, and the rationale unapologetic. Emboldened by the breakdown of traditional morality, the advance of the new creed is essentially no more than the triumph of animal impulse, as instinctive and blind as appetite, released from the restraints of religious sanctions. Selfishness becomes a prized commercial resource; falsehood reinvents itself as public information; greed, lust, indolence, pride—even violence—acquire not merely broad acceptance but also social and economic value (UHJ 2005). Endlessly rising levels of consumption are cast as indicators of progress and prosperity. This preoccupation with the accumulation of material objects and comforts (as sources of meaning, happiness, and social acceptance) has consolidated itself in the structures of power and information to the exclusion of competing voices and paradigms. The unfettered cultivation of needs and wants has led to a system fully dependent on excessive consumption for a privileged few, while reinforcing exclusion, poverty, and inequality for the majority. Each successive global crisis—be it climate, energy, food, water, disease, or financial collapse—has revealed new dimensions of the exploitation and oppression inherent in the current economic system. The narrowly materialistic worldview underpinning much of modern economic thinking has contributed to the degradation of human conduct, the disruption of families and communities, the corruption of public institutions, and the exploitation and marginalization of large segments of the population—women and girls in particular (BIC 2010).

The shift towards a more just, peaceful, and sustainable society will require attention to a harmonious dynamic between the material and non-material (or moral) dimensions of consumption and production. Bahá'ís call for a new economic model that furthers a dynamic, just, and thriving social order, will be strongly altruistic and cooperative in nature, provides meaningful employment, and will help to eradicate poverty in the world (BIC 1998). In the lead-up to the United Nations Conference on Sustainable Development (Rio+20) in 2012, it was clear that the UN had devoted much attention to the elimination of poverty, but little had been said about the elimination of extremes of wealth at a time when the gap between those extremes was widening, so a statement was prepared exploring this issue (BIC 2011) and it was one of the themes of the Bahá'í contributions in Rio, including both its statement (BIC 2012) and a side event organized at the conference centre.

Action at the community and individual levels

> We cannot segregate the human heart from the environment outside us and say that once one of these is reformed everything will be improved. Man is organic with the world. His inner life moulds the environment and is itself also deeply affected by it.

The one acts upon the other and every abiding change in the life of man is the result of these mutual reactions.

(*Bahá'í Writings*)

The Bahá'í focus is not only at the global level of economic systems and governance. Bahá'ís believe that progress in the development field depends on and is driven by initiatives at the grassroots of society rather than from an imposition of externally developed plans and programs. Different communities will often devise different approaches and solutions in response to similar needs. It is for each community to determine its goals and priorities in keeping with its capacity and resources. This encourages innovation and a variety of approaches to the environment appropriate to the rhythm of life in each community (BIC 2009).

Bahá'ís all over the world are engaged in action that promotes the spiritual development of individuals and channels their collective energies towards service to humanity, including environmental responsibility. Thousands of Bahá'ís, embracing the diversity of the entire human family, are engaged in certain core activities. These activities promote the systematic study of the Bahá'í Writings in small groups in order to build capacity for service. They respond to the innermost longing of every heart to commune with its Maker by carrying out acts of collective worship in diverse settings, uniting with others in prayer, awakening spiritual susceptibilities, and shaping a pattern of life distinguished for its devotional character. They strive to provide for the needs of children and youth, and offer them lessons that develop their spiritual faculties and lay the foundations of a noble and upright character. As Bahá'ís and their friends gain experience with these initiatives, an increasing number are able to express their faith through endeavors that address the needs of humanity in both their spiritual and material dimensions (BIC 2009b). It is natural that, as these processes mature, they will lead to consultation, action, and reflection on topics like environmental responsibility and climate change, and this is already happening in many places.

Individual Bahá'ís have been active on environmental issues for many decades, with the encouragement of Bahá'í institutions, starting with Richard St. Barbe Baker, active in reforestation and forest conservation from the 1920s to 1960s (Baker 1970), who in the early 1930s worked with all the religions in the Holy Land to begin the reforestation of Palestine. William Willoya and Vinson Brown, wrote *Warriors of the Rainbow* (Willoya and Brown 1962) collecting Native American visions of world peace and environmental harmony, which helped to inspire the founders of Greenpeace (Brown and May 1989, 12–13). Bahá'ís have published books and articles on environmental and sustainability themes, including Bahá'í perspectives on environment (Dahl 1990, 1991; Bell and Seow 1994), environmental stewardship (Karlberg 1994), ecology and economy (Dahl 1996), and the transition to sustainability (Hanley 2014; Karlberg 2014). The official record of the Bahá'í community, *The Bahá'í World*, has included essays on spirituality and ecology (White 1993) and on climate change (Dahl 2007). Dimity Podger has specifically looked at the relationship between sustainability and spirituality, using as a case study the American Bahá'í Community (Podger 2009).

Bahá'í-inspired professional organizations, such as the IEF for environment and sustainability and Ethical Business Building the Future (ebbf), encourage research and organize conferences to advance thinking on sustainability based on ethics and values. The IEF was accredited by the United Nations to the World Summit on Sustainable Development in Johannesburg in 2002, and Rio+20 in 2012, as part of the science and technology major group, and contributes substantively to international debates at the interface between science and values. It has even developed an interfaith course on climate change now being used around the world.

Research

Bahá'ís in academic and research positions often draw inspiration from the Bahá'í teachings in their choice of research areas and in the solutions they explore, including in the environmental field.

One obvious area of interest is alternative forms of governance, where the Bahá'í experience is particularly relevant. For example, Professor Michael Karlberg has studied the culture of contest so prevalent in modern society, the relative advantages of cooperation over competition, and a new environmental stewardship (Karlberg 1994, 2004, 2009). Professor Sylvia Karlsson-Vinkhuyzen has researched environmental governance, the functioning of international institutions for environment, and moral leadership (Karlsson 2007; Karlsson-Vinkhuyzen 2012; Karlsson-Vinkhuyzen et al. 2012; Vinkhuyzen and Karlsson-Vinkhuyzen 2014). Arthur Dahl has worked extensively on indicators of ecology and sustainability, including an ethical dimension inspired by Bahá'í values (Dahl 2012c, 2013, 2014a, 2014b).

Another emerging research area where Bahá'ís are active is the role of values in addressing the knowledge-action gap between environmental knowledge and responsible behavior, and making values-based education more visible and measurable. Research by Podger and others on educating for sustainability has demonstrated that inconsistency between knowledge and action reflects a conflict between morality and expediency, requiring a whole-person approach to education addressing the mind, heart, and will, and a spiritually oriented approach to service-learning (Podger et al. 2010). A research project funded by the European Union on values-based indicators of education for sustainable development was able to demonstrate that values and related behavior change could be measured in the activities of a variety of civil society organizations, both secular and faith based (Burford et al. 2013; Harder et al. 2014), and the approach is now being extended to a number of contexts including higher education (Dahl 2014b) and secondary schools through the Partnership for Education and Research about Responsible Living, which is coordinated by a Bahá'í.

Bahá'í-inspired organizations and individuals have long explored alternatives to the traditional top-down forms of development assistance and their harmonization with ecological requirements. If there is a higher human purpose that is essentially spiritual, then a focus only on material well-being is an inadequate approach to development. Research and practice have shown that education and empowerment allow local communities to find their own approaches to a more just and sustainable society, while addressing environmental concerns in ways that fit with the rhythm of life in their communities. This also means redefining the indicators used to assess sustainable development (Dahl 2014a) and exploring broader measures of human well-being.

> Every man of discernment, while walking upon the earth, feeleth indeed abashed, inasmuch as he is fully aware that the thing which is the source of his prosperity, his wealth, his might, his exaltation, his advancement and power is, as ordained by God, the very earth which is trodden beneath the feet of all men. There can be no doubt that whoever is cognizant of this truth, is cleansed and sanctified from all pride, arrogance, and vainglory.
>
> *(Bahá'u'lláh)*

We conclude with a quote from the Bahá'í International Community that summarizes the Bahá'í vision of the unity of religion and ecology:

As trustees, or stewards, of the planet's vast resources and biological diversity, humanity must learn to make use of the earth's natural resources, both renewable and non-renewable, in a manner that ensures sustainability and equity into the distant reaches of time. This attitude of stewardship will require full consideration of the potential environmental consequences of all development activities. It will compel humanity to temper its actions with moderation and humility, realizing that the true value of nature cannot be expressed in economic terms. It will also require a deep understanding of the natural world and its role in humanity's collective development – both material and spiritual. Therefore, sustainable environmental management must come to be seen not as a discretionary commitment mankind can weigh against other competing interests, but rather as a fundamental responsibility that must be shouldered—a pre-requisite for spiritual development as well as the individual's physical survival.

(BIC 1998)

References

Bahá'í International Community (1995) *The Prosperity of Humankind*, Paper distributed at the United Nations World Summit on Social Development, Copenhagen, 3 March 1995, Bahá'í International Community Office of Public Information, Haifa.

Bahá'í International Community (1998) *Valuing Spirituality in Development: Initial considerations regarding the creation of spiritually based indicators for development*, Paper written for the World Faiths and Development Dialogue, Lambeth Palace, London, 18–19 February 1998, Bahá'í Publishing Trust, London.

Bahá'í International Community (2008) *Seizing the Opportunity: Redefining the challenge of climate change*, Initial considerations of the Bahá'í International Community, Poznan, 1 December 2008, Bahá'í International Community, New York.

Bahá'í International Community (2009) *Bahá'í International Community's Seven-Year Plan of Action on Climate Change*, Bahá'í International Community, New York.

Bahá'í International Community (2010) *Rethinking Prosperity: Forging alternatives to a culture of consumerism*, Bahá'í International Community, New York.

Bahá'í International Community (2011) *Initial Considerations Regarding the Elimination of the Extremes of Poverty and Wealth*, Bahá'í International Community, New York.

Bahá'í International Community (2012) *Sustaining Societies: Towards a new 'we'*, Bahá'í International Community, New York.

Baker, R. S. B. (1970) *My Life My Trees*, Lutterworth Press, London.

Bell, R. W. and Seow, J. eds (1994) *The Environment: Our common heritage*, Association for Bahá'í Studies Australia in association with Bahá'í Publications Australia, Mona Vale.

Brown, M. and May, J. (1989) *The Greenpeace Story*, Dorling Kindersley, London.

Burford, G., Hoover, E., Velasco, I., Janouskova, S., Jimenez, A., Piggot, G., Podger, D., and Harder, M. (2013) "Bringing the 'missing pillar' into sustainable development goals: Towards intersubjective values-based indicators", *Sustainability* 5, 3035–3059.

Capra, F., and Luisi, P. L. (2014) *The Systems View of Life: A unifying vision*, Cambridge University Press, Cambridge.

Dahl, A. L. (1990) *Unless and Until: A Bahá'í focus on the environment*, Bahá'í Publishing Trust, London.

Dahl, A. L. (1991) "The world order of nature", in Lerche C. ed., *Emergence: Dimensions of a new world order*, Bahá'í Publishing Trust, London, 161–174.

Dahl, A. L. (1996) *The Eco Principle: Ecology and economics in symbiosis*, Zed Books Ltd, London, and George Ronald, Oxford.

Dahl, A. L. (2007) "Climate change and its ethical challenges", in *The Baha'i World 2005–2006*, Bahá'í World Centre, Haifa, 157–172.

Dahl, A. L. (2012) "Achievements and gaps in indicators for sustainability", *Ecological Indicators*, 17, 14–19.

Dahl, A. L. (2013) "A multi-level framework and values-based indicators to enable responsible living", in Schrader, U. et al. eds, *Enabling Responsible Living*, Springer Verlag, Berlin/Heidelberg, 63–77.

Dahl, A. L. (2014a) "Putting the individual at the centre of development: Indicators of well-being for a new social contract", in Mancebo, F. and Sachs, I. eds, *Transitions to Sustainability,* Springer, Dordrecht, 83–103.

Dahl, A. L. (2014b) "Sustainability and values assessment in higher education", in Fadeeva, F. et al. eds, *Sustainable Development and Quality Assurance in Higher Education: Transformation of learning and society*, Palgrave Macmillan, Houndsmill, 185–195.

Hanley, P. (2014) *Eleven,* Friesen Press, Victoria, BC, Canada.

Karlberg, M. (1994) "Toward a new environmental stewardship", *World Order*, 25, 21–32.

Karlberg, M. (2004) *Beyond the Culture of Contest: From adversarialism to mutualism in an age of interdependence*, George Ronald, Oxford.

Karlberg, M. (2009) "Sustainability and the Bahá'í faith", in Jenkins, W. and Bauman, W. eds, *Encyclopedia of Sustainability,* Berkshire Publishing, Great Barrington, MA, 28–32.

Karlberg, M. (2014) "Religion, meaning, and a great transition", Great Transition Initiative (December 2014) (available at: www.greattransition.org/publication/meaning-religion-and-a-great-transition).

Karlsson, S. I. (2007) "Allocating responsibilities in multi-level governance for sustainable development", *International Journal of Social Economics*, 34:1/2, 103–126.

Karlsson-Vinkhuyzen, S. I. (2012) "From Rio to Rio via Johannesburg: Integrating institutions across governance levels in sustainable development deliberations", *Natural Resources Forum*, 36, 3–15.

Karlsson-Vinkhuyzen, S. I., Jollands, N. and Staudt, L. (2012) "Global governance for sustainable energy: The contribution of a global public goods approach", *Ecological Economics*, 83, 11–18.

Podger, D. (2009) "Contributions of the American Bahá'í Community to education for sustainability", *Journal of Education for Sustainable Development*, 3:1, 67–76.

Podger, D., Mustakova-Possardt, E. and Reid, A. (2010) "A whole-person approach to educating for sustainability", *International Journal of Sustainability in Higher Education*, 11:4, 339–352.

Universal House of Justice (2005) *One Common Faith*, Prepared under the direction of the Universal House of Justice, Bahá'í World Centre, Haifa.

Vinkhuyzen, O. M., and Karlsson-Vinkhuyzen, S. I. (2014) "The role of moral leadership for sustainable consumption and production", *Journal of Cleaner Production*, 63, 102–113.

White, R. A. (1993) "Spiritual foundations for an ecologically sustainable society" (updated version), in *The Bahá'í World 1992–93*, Bahá'í World Centre, Haifa, Israel 193–227. Reprinted in *The Journal of Bahá'í Studies* 7:2 (1995), 47–74.

Willoya, W. and Brown, V. (1962) *Warriors of the Rainbow: Strange and prophetic dreams of the Indian peoples*, Naturegraph, Healdsburg, CA.

11

MORMONISM

Mormon views of environmental stewardship

George B. Handley

Mormon environmentalism is no longer an oxymoron. This is not due to new teachings but because there has been a resurgent interest in Mormon doctrines of the creation that were first published in the early-nineteenth century. Starting in the late 1990s, published research, advocacy, creative writing, and activism all began to contribute to a growing awareness of environmental responsibility among members and a deeper appreciation for Mormonism's understanding of the human relationship to nature. Moreover, The Church of Jesus Christ of Latter-Day Saints (LDS Church) has recently launched two websites devoted to "Environmental Stewardship and Conservation," which for the first time provide a focused doctrinal argument for why Latter-Day Saints should embrace their responsibilities as caretakers of the earth, a history of the church's own sustainable practices, and resources for how members might live up to their responsibilities.[1] The coupling of stewardship and conservation (along with the adjective "environmental") is an important development. Stewardship is by no means a new doctrine to Mormons, but it is safe to say that over the course of the twentieth century stewardship had lost some of the implicit environmental meanings that it had in the nineteenth century. It had instead come to be understood as pertaining mainly to stewardship of financial resources and individual talents.

This loss was due to at least a few factors, most notably that the Intermountain West, where Mormonism had its beginnings and is headquartered and where today a significant number of Mormons still live, is generally characterized by a tense relationship with the federal government. This is mainly due to conflicts over environmental policies that dictate how public lands, which make up the majority of the Intermountain states, are managed. This distrust of the federal government, including federal policies about the environment, is heightened even more by the Mormon history of persecution by the federal government in the nineteenth century over the practice of polygamy. As Mormons are predominantly politically conservative, the Republican Party's move away from environmental concerns since the 1980s also contributed to greater suspicion toward contemporary environmentalism. Moreover, as happened elsewhere in the country, as the twentieth century progressed, fewer members of the LDS Church were engaged in agricultural labor and direct individual responsibility for environmental health became increasingly easy to ignore. While there is still much work to be done to effectively root out indifference and even antagonism toward environmental stewardship among members, the trends are moving in the right direction thanks to recent efforts by members and by the LDS Church itself to slowly disaggregate environmental issues from their partisan associations.

One of the central questions within the LDS tradition upon which a strong environmental ethic depends is the understanding of the human role within the creation. LDS theology is unquestionably anthropocentric. It posits that the entire human family are children of God and sent from His presence to this earth for the purposes of obtaining a body, learning to accept and use responsibly our moral agency, and preparing ourselves to return to live with God again. According to the teachings of Joseph Smith and many leaders of the church since, this earth was created for us to develop the capacities that will enable us to become like God and to do God's work.[2] Central to this plan is the experience gained in marriage and family life. Consequently it is not uncommon for believers to assume human concerns and needs have priority over the needs of other living forms. It is fair to say that Mormonism's environmental ethic has historically envisioned a careful and efficient management of natural resources for the purposes of meeting human needs, akin to Gifford Pinchot's conservation ethic. Even in its most strident moods, this ethic offered strong criticism of the dangers of excess and greed, but never a criticism of the idea of human centrality itself.[3]

With some exceptions, recent scholarship tends to uphold this anthropocentric ethic even though it has raised more urgent questions about the dangers of human self-interest and about how to better understand and motivate human responsibility in the age of such problems as climate change and rapid species extinction. In so doing, it has offered more tempered and somewhat less anthropocentric notions of the human role. Most notably, scholars and activists alike point to a number of unique features of LDS belief that challenge or at least mitigate the anthropocentric leanings of the theology. A chief example from the Mormon account of the creation is the teaching that a spiritual creation took place prior to the physical creation, a fact that defines all plants and all animals as "living souls."[4] Scholars have argued that this teaching suggests a kind of kinship between plants, animals, and human beings that traditional paradigms of human exceptionalism do not allow. Indeed, the doctrine is so explicitly stated in Mormon scripture, it places Mormonism—its anthropocentrism notwithstanding—among the most animistic expressions of Christianity in the world. Mormonism stipulates that all living things have a right to enjoy posterity, thus adding an important caveat to the mandate to Adam and Eve to multiply and replenish. Indeed, Mormonism explicitly affirms the idea found in Genesis that God wishes the fowls and fishes to likewise multiply and replenish and fill the airs and the seas (Genesis 1:22). Thus, although humans are unquestionably created in the image of God in Mormon teaching, because we find ourselves amidst a community of subjects with whom we share both the responsibility and joy of reproduction, Mormon scholars have found sympathy with such thinkers as Aldo Leopold and Lynn White.[5]

The idea of a spiritual creation corresponds to a number of related doctrines that also emphasize the holiness and inherent value of earthly life and that reject the traditional binaries of spirit/body and heaven/earth. For example, the creation account in Mormonism posits a fortunate fall wherein the "curse" of having to struggle for our existence is deemed a blessed pathway to redemption. The body is not an obstacle but an opportunity to understand the life of God who, in Mormon theology, is a God of flesh and bone. Christ's incarnation, then, and his post-resurrection meals represent the blessedness of the earthly state that would make a sacrament of all meals and a heaven of all earthly endeavors. In this theology, we seek a return to the Garden but not by bypassing or escaping the body, sexuality, or our earthly condition, but rather in and through the very conditions that accompany working for our survival and raising families. Suffering is inevitable but so too are our chances for joy. As the Book of Mormon puts it, "Adam fell that men might be; and men are, that they might have joy."[6] Indeed, in Mormon theology, the earth itself is posited as the site of the future celestial kingdom where God and Jesus Christ will reign.[7]

Other arguments for a more tempered anthropocentrism within Mormonism include arguments about why Mormonism is or at least can be particularly friendly to the findings of science. This is in part due to a Mormon understanding of the human quest for God's knowledge and the belief in human eternal progression, but also because the account of the creation, as interpreted by Joseph Smith and other church leaders, posits a creative process that took place over the course of six creative periods, an understanding that implies more freedom for accepting the implications of the Huttonian and Darwinian revolutions and the much deeper age of the earth. Although church leadership has from time to time expressed concerns about scientistic overreach and atheistic implications, the church has never denounced evolution. Indeed, the science of evolution is widely taught at all of its own institutions of higher learning.

More to the point, the Mormon account of creation is explicitly described as emerging from pre-existent but unorganized matter, rather than *ex nihilo*. Paul Cox, former Director of the National Botanical Garden in Hawaii and a prominent scientist in the Mormon tradition, argues that both of these doctrines should give members pause before using up or otherwise harming the fabric of creation, precisely because it has emerged with great care over millions of years. His is an argument for conservation based on historical process and the unfolding of natural laws, rather than from instantaneous design. He notes: "If geophysics and astronomy were the lead sciences in the Creation, then surely restoration ecology will play a leading role in the Millennium. It seems to me that Latter-day Saints, of all people, should be conservationists—protecting the world's wild places, animals, and plants, while doing everything we can to beautify our own homes and communities" (Cox 2006). Indeed, he is one example of many scientists in the Mormon tradition who have felt inspired to pursue and act on scientific findings precisely because those findings are important to our understanding of God's sovereignty and presence in the world, even if we may lack current understanding of the terms of that compatibility.

Mormonism posits continuing revelation as the mode of communication between deity and the human family. This manifests itself simultaneously as a belief in the ultimate authority of the living prophet and his counselors and the Quorum of the Twelve Apostles and in the potentially contradictory belief that leaders are not infallible and that current understandings are contingent and subject to deeper understandings.[8] While this sometimes leads to tensions in church culture, it provides opportunity to make adjustments as new knowledge emerges and in light of new experience. An example of this might be the revelation given to Joseph Smith in 1833 known as the Word of Wisdom. For most of Mormonism's relatively brief history, it has been understood to be a dietary code that restricted use of alcohol, "hot drinks," tobacco, and illicit drugs before we had scientific understanding of the damage many of these substances can cause. However, in light of what we now know about the damage we do to the earth by eating higher on the food chain and by becoming increasingly dependent on the transportation of our food, many environmentally concerned members have pointed to the Word of Wisdom's counsel to eat meat "sparingly" and to enjoy fruits and vegetables "in the season thereof" as important principles of stewardship. Indeed, its insistence that we guard even the wild animals' right to food would seem to provide an excellent incentive for more environmentally sensitive stewardship of our food consumption (see Doctrine and Covenants 89).

One final doctrine that perhaps tempers the environmentally damaging potential of Mormonism's anthropocentrism is also what has often contributed to Mormonism's reluctance to embrace environmentalism. I am speaking of the centrality of family in Mormon belief and the concomitant importance of chastity outside of marriage and fidelity within. In the teachings of the church, our capacity to reproduce is one of our most sacred gifts. Hence, marrying and having children are central to our mortal purpose on earth, and government policies that restrict or otherwise dictate family size are deemed contrary to moral principle. When environmentalist rhetoric has focused almost exclusively on population rates and population control or otherwise seemed to justify such policies, this has tended to alienate the Mormon faithful.

The emphasis on family, however, is also an emphasis on genealogy, on knowing, honoring, and respecting the past and living for future generations. The final verses in Malachi are important to this aspect of Mormonism. They are understood to refer to the great work of gathering together the families of the earth (see Malachi 4:6). Mormons understand the description of the hearts of the children being turned to the fathers as the spirit of genealogical work. Mormon scholars and activists often remind members that Malachi's prophecy is also about the hearts of the fathers being turned to the children. They find reason in this phrase to consider the long-term impacts of our environmental choices on future generations, not unlike the Native American idea of seven generations. Indeed, it would seem logical that if the sources of human life are sacred and to be protected, so too then is the earth's capacity to nurture us and to bring forth all of life abundantly.

In sum, recent scholarship argues for an expanded understanding of this central idea of stewardship to include stewardship of one's own body and family, but also of the community at large, of the bodies of all living things, and of the future. Although the environmental implications of stewardship might have been more obvious to nineteenth-century believers who were in the majority migrants and agricultural laborers struggling with new lands and with the task of building new societies, they also did not enjoy the ecological understandings of today. It is, then, especially fortunate that in light of contemporary understandings scholars and the above-mentioned church websites have revisited these principles with particular urgency in order to explore their applicability to current environmental problems.

What is especially unusual in the tradition is how explicitly revelations speak of the importance of environmental health. An environmental ethic is most clearly articulated in what is known as the law of consecration, a series of revelations that Joseph Smith received regarding the human social order. In these revelations, poverty and inequality are identified as symptoms of sin. The Lord explicitly connects poverty and inequality to proper management and consumption of earth's resources: "But it is not given that one man should possess that which is above another, wherefore the world lieth in sin. And wo be unto man that sheddeth blood or that wasteth flesh and hath no need." (Doctrine and Covenants 49: 20–21). Another revelation promises that "there is enough and to spare" of the earth's resources but only if we learn to establish equality: "Therefore, if any man shall take of the abundance which I have made, and impart not his portion, according to the law of my gospel, unto the poor and the needy, he shall, with the wicked, lift up his eyes in hell, being in torment." (Doctrine and Covenants 104: 17–18). In order to ensure sufficient resources for all and to provide not only the basic needs of food and shelter but also the privileges of nature's ability to "please the eye and gladden the heart" and "to enliven the soul" members are explicitly enjoined to use earth's resources "with judgment, not to excess, neither by extortion" (Doctrine and Covenants 59: 18–20).

Stewardship here is a human responsibility to meet the needs of the entire human family, but, as these verses indicate, those needs are bodily, aesthetic, *and* spiritual. And as suggested earlier, human stewardship includes all living things, so it would be wrong to assume, as anti-environmental rhetoric over the past 40 years so often has, that human needs and the needs of all living things must be seen as separate or even as incompatible. In Smith's revelations, God is asking for faith as well as for commitment to make such equity a reality. For Mormons, the gospel of Jesus Christ is not merely a matter of salvation in the life to come but a means to prepare a more heavenly and more just society here on earth. Indeed, Lynn White's condemnation of Christianity's tendency to favor the next life over this, the life of the spirit over the life of the body, or even human needs over all others, suggested the need for a theology that is not so binary in its views of earthly life, a theology that scholars have argued can be found within Mormonism (White 1996). Of course, Mark Stoll's recent persuasive research on the religious foundations of environmentalism in America—based quite strongly in Puritan theology—would seem to suggest White was wrong both in his understanding of the range and subtlety of Christian theology as well as about the ways in which environmental ethics stem or do not stem from belief (Stoll 2015). Mormonism has had teachings that White presumably would have approved of, but the less-anthropocentric doctrines highlighted above have not yet penetrated the consciousness of the Mormon people sufficiently to put them at the vanguard of environmental stewardship. It is at least clear that a strong concern for just human social order—to learn to distribute natural resources so as to alleviate suffering and poverty—motivate Mormon environmentalism at present.[9] Whether or not it proves adequate to the challenges we face or to motivate large numbers of Mormons is yet to be seen.

Indeed, this concern for poverty seems to be the ethic that has motivated the LDS Church to begin its effort to give more attention and emphasis to environmental stewardship. Although the two websites the church produced, both dedicated to "Environmental Stewardship and Conservation," do not break from Mormonism's tradition of anthropocentrism, they do successfully condemn any attempt to misuse our human privileges by abusing the earth. It may indeed be the case that such teachings will be far more successful in reaching the mainstream faithful of the church precisely because they appeal to well-known doctrines, centered in family and social organization, and because they come from trusted sources of authority. For this reason, even though the websites do not advance new ideas within the Mormon tradition, they represent a genuine watershed moment.

Most notably, the websites include a talk given by a General Authority of the church, the first by any church leader exclusively devoted to environmental stewardship.[10] Elder Marcus B. Nash's talk, "Righteous Dominion and Compassion for the Earth," offered what is certainly the most definitive statement regarding earth stewardship from the LDS Church that we have to date. Again, while statements about stewardship have been made by most if not all church leaders at one time or another, until Elder Nash's speech, no one had ever devoted an entire talk exclusively to the topic, least of all in the context of the last several decades of advances in scientific understanding of our impact on the globe. What stands out about the talk is the way it directly connects the plan of salvation for the human family with environmental stewardship (Nash 2013). Although some scholars have sought to identify ways in which Mormon theology is more friendly to a biocentric view (Brown 2011), his talk instead emphasizes the anthropocentrism of Mormon theology but then embeds an environmental ethic directly into our broader and exceptional human responsibilities toward one another. Better stewardship does not require, in other words, a radical rethinking of our ethics but rather a more holistic and expanded understanding of God's gifts. The earth was created for us, and it is intended to be used for human ends, but this actually amplifies the inherent value of the long-term well-being of the planet. Elder Nash makes it unambiguously clear that all human uses of natural resources must have both long-term sustainability in mind and the needs of the poor front and center. His talk is, in other words, a call to much greater modesty in consumption, deeper reverence for all of life, and a more conscientious and compassionate approach to distributing natural resources more equitably. He sums up his argument by saying that "as stewards over the earth and all life thereon, we are to gratefully make use of that which the Lord has provided, avoid wasting life and resources, and use the bounty of the earth to care for the poor."[11]

The talk serves as a vital reference point for all future discussions and can potentially provide, along with the additional information supplied on the sites, incentives for church administrators and members to develop more effective and focused efforts in homes and in local congregations to respond to environmental problems with moral urgency and practical efficiency. The websites also make it clear, since it was not already, that the church has a long history of commitment to good stewardship practices in their design of buildings, ranches, and in other areas. These facets of institutional practice merit more publicity. They should not only be more widely known by members; the doctrinal underpinnings and justifications for such practices should be unambiguously clear to all. Indeed, the websites represent a beginning, not an end, and need to develop into a more full-blown curriculum of teachings that are more readily available, more frequently iterated, and more clearly understood by all members. Until that happens, the talk will likely remain on the margins.

If it seems paradoxical that Mormons believe in both a spiritual creation—that makes living souls of all living things—while also believing in a decidedly human purpose to the whole of creation, it is. This tension needs to be more fully explored and understood, since it is clear that Mormonism, despite having an unmistakable ethic of stewardship placed directly on our shoulders, has not produced a very even record of environmentally friendly attitudes, policies, and practices. Indeed, one of the most common perceptions of Mormonism prior to the development of this recent scholarship was that Mormonism was at best ecologically indifferent and at worst ecologically hostile. For example, in his survey of faith-based environmental initiatives, Max Oelschlaeger mistakenly concludes that the LDS church is "the only denomination that has formally stated its opposition to ecology as part of the church's mission" (Oelschlaeger 1996). While there has been historically a strong tradition of environmental teaching and practice in the tradition, this fact is usually ignored by or unknown to the church's critics. What is more tragic is that the rich environmental heritage of Mormonism continues to be met with surprise among some members. Many of the faithful are largely unaware of the church's websites and some are suspicious that they were created under political pressure. This is because much of the ideas contained in them have yet to see the full light of day in the semi-annual meetings of church leaders, known as General Conference, or in any print media or lesson manuals produced by the Church.

Mormons have made exceptional efforts in recent decades to build bridges to people of other faiths and while many of those efforts have focused on humanitarian service and efforts to protect the family, recently environmental interfaith dialogue involving individual Mormons has been noteworthy, even if somewhat unusual.[12] Given the quality of Mormonism's insights, the rapid growth of its membership, and the stakes we face today on the planet, such developments in Mormon teaching and practice are especially urgent. Indeed, the membership tends to look for signals from church leadership about where their priorities lie, and as long as environmental concerns are treated infrequently or are overshadowed by other social concerns, many members will assume that environmental problems are discrete and separate from their day-to-day living, from the plight of the poor, or even from a strong ethic of stewardship of the body and of one's financial resources. This is a very unfortunate conclusion. Given the factors mentioned above—the Intermountain West's distrust of the federal government, Mormonism's history of persecution, and the predominantly anti-environmental rhetoric of the Republican Party— it is unlikely that recent scholarship, activism, or even the church's websites will be enough to change the tide of apathy about environmental concerns and the still persistent distrust of science and activists.

That said, the church's approach, it seems to me, is the right one. Rather than appealing to political rhetoric or appealing first to the science that diagnoses a particular problem, we find an appeal to fundamental values within the Mormon tradition: concern for family, for community, for future generations, and for the poor. Indeed, it is certainly true that for many members no amount of additional scientific information, political pressure from the left, or shaming of our human-centered tendencies will motivate the change that many environmental activists desire. I dare say this is true of many religious communities in the United States. What is needed is a weaving of environmental concerns into the very fabric of the stories believers tell to make sense of the world and their responsibilities in it. Indeed, in a recent study of Mormon environmental attitudes (based on a sample drawn from Logan, Utah), it was noted that while concerns about environmental problems run high among Mormons, the willingness to make sacrifices, use taxation, or advocate strongly for policies to redress environmental problems remains low (Hunter and Toney 2005).[13] Unless we can see and identify environmental concerns deeply woven into the very heart of the narrative Mormons accept about the purpose of this life and the purpose of this earth and unless that weaving is done by the General Authorities of the church, men who alone are believed to hold the sacred responsibility of receiving revelation on behalf of the entire church, it is unlikely that those concerns will get the attention of the general membership.[14] Waiting for such things to happen institutionally should not stop individual members from taking upon themselves the responsibility to act, as people of faith, on behalf of the earth, but it will help clear the air of any suspicion that such concerns are not important. Otherwise, environmental concerns will remain a minor note in the chorus of values that Mormons hold dear.

If it is anthropocentric to tie the fate of the earth to humankind, it is also biocentric to tie humankind to the fate of the earth. Again, thanks to the research of Stoll and others, it seems that environmentalism owes a great deal to Christian anthropocentrism. This is nothing to be ashamed of. Individuals and communities might have alternative preferences, but the goal is not necessarily to define the correct worldview but to find the right motivations that will speak to individuals within their respective communities and in their language of understanding. Rather than arguing over semantics, it might be enough for believers in anthropocentric religions to fully assume their responsibilities. If Mormons are serious about ending inequality and they are serious about restoring dignity to the lives of the most poor and the most vulnerable in society, as the law of consecration demands, then they should be equally serious about respecting the creation as God's, understanding property as contingent, avoiding overconsumption in all of its forms, and honoring and protecting the sacred sources of physical life. This kind of ethic will go a long way in tempering the otherwise destructive tendencies of human self-centeredness, which might be the more accurate term for the anthropocentrism of Christianity that Lynn White criticized. An anthropocentric commitment to service assumes that responsibilities to our fellow man cannot be separated from our responsibilities to the earth. As a result, environmental stewardship becomes central to what it means to bear the name of Christ.

Notes

1   A short and well-produced video, entitled "Our Earth, Our Home" and included on these links, summarizes briefly the ethos of Mormon environmental stewardship. The video and the webpage can be found at www.mormonnewsroom.org/article/environmental-stewardship-conservation

2   In 1844 Joseph Smith said: "God himself was once as we are now, and is an exalted man, and sits enthroned in yonder heavens! ... It is the first principle of the Gospel to know for a certainty the Character of God, and to know that we may converse with him as one man converses with another, and that he was once a man like us; yea, that God himself, the Father of us all, dwelt on an earth, the same as Jesus Christ himself did" (Smith 1938, 345–6).

3   Scholar Hugh Nibley, for example, wrote the most trenchant and important critique of the idea of

subduing the earth, explaining to church members in 1981 that "Man's dominion is a call to service, not a license to exterminate." A number of his essays, published in the 1970s and 1980s, were the only such works of scholarship on environmental ethics until the late 1990s.

4   For example, in the *Book of Moses*, in what is believed to be a restored account of the vision of the creation originally given to Moses, we read: "And out of the ground made I, the Lord God, to grow every tree, naturally, that is pleasant to the sight of man; and man could behold it. And it became also a living soul. For it was spiritual in the day that I created it" (Moses 3:9).

5   See, for example, Handley (2000) and Gowans and Cafaro (2003). For a more decidedly biocentric, vitalist view of Mormon theology, see especially Brown (2011). For a perspective based in the social sciences, see Bryner (2010). Two important collections of essays on Mormon environmental ideas are Williams, Smart, and Smith (1999) and Handley, Peck, and Ball (2006).

6   See 2 Nephi 2:25.

7   It is for this reason that both Mormonism's founder, Joseph Smith, and his immediate successor, Brigham Young were particularly keen on careful planning of cities. Such ideas, considered revelations for their time, have been explored recently by Craig Galli (2006) in order to test their viability and usefulness in today's context of climate change and pollution.

8   Although this has often led to tensions in the church over particular policies or positions assumed by the leadership, it has also served to balance the centrifugal forces of experimentation with the centripetal forces of dogmatism. Bryan Wallis (2011) sees Mormon theology as particularly suitable for embracing ongoing scientific understanding and environmental stewardship because it allows for what has been described as a "flexible" theology. The particularly radical implications of ongoing revelation in Mormonism are tempered by a respect for ecclesiastical authority to receive those revelations on behalf of the whole church.

9   The similarity between the principles outlined here and the recent encyclical *Laudato Si'* by Pope Francis is striking and is apparently not lost on the LDS Church, as is evident by a recent article in the LDS-owned paper, *Deseret News*, which chose to emphasize the similarity despite the reluctance of the LDS Church to mention climate change specifically. See www.deseretnews.com/article/865631022/LDS-environmental-stewardship-statement-recent-talk-share-similarities-with-Pope-Francis.html?pg=all

10   The talk was initially given at a symposium sponsored by the University of Utah Law School. The talk and the entirety of the symposium are available online at www.law.utah.edu/event/12233/

11   Elder Nash was not the first to make the link between poverty and environmental degradation in the Mormon tradition, but he is certainly the highest-ranking official of the church to do so. One member, James Mayfield, made this connection in 1999 when he insisted on the inextricability of poverty and environmental degradation within Mormon theology. He is also one of the only scholars to have directly addressed the question of population levels. The LDS church does not prohibit contraception. Instead, it strongly favors policies that respect the right of a husband and a wife to make joint decisions about their own family planning. In his essay, Mayfield advocates not only for greater concern for the poor and more modest consumption of resources in the developed world, but also for the education and empowerment of women so that family planning in marriages is more likely to be done in an equal partnership.

12   Mormons, for example, have been involved in the activities of Utah Interfaith Power and Light (UIPL), twice assuming leadership of the board and strongly advocating for clean and renewable energies. Former Board Chair, Lincoln Davies, an LDS law professor at the University of Utah also spearheaded the efforts to create the symposium, with some assistance of UIPL, at which Elder Nash spoke. UIPL's efforts have been notable on the national scale for the degree of its impact. More recently, Alisha Anderson, a student of Terry Tempest Williams at the University of Utah and a graduate of Brigham Young University, the LDS-owned university, was asked by Williams to accompany her as a keynote speaker before the Patriarch Bartholomew at Patriarchate's Halki Summit II, devoted to exploring the role of the arts in assisting religion in the fight against climate change. Anderson spoke and shared an art film she produced about the destruction of the Oquirrh mountains. The present author was also invited as a participant. You can find her film on her website and a description of the summit at the Halki Summit website. Mormons have been part of conversations about environmental work sponsored by the Nature Conservancy, Yale University, Claremont College, the national Interfaith Power and Light conference, and the most recent Parliament of Religions held in Salt Lake City in 2015, among others.

13   The church is increasingly global, multilingual, and multiracial, so such regional studies may be less

and less useful to understanding broader trends. Nevertheless, it is interesting to note that the study found that while Mormons tended to reflect greater concern for the environment than that found nationally, there was one difference: "while LDS respondents appear environmentally concerned, they also appear to believe that environmentally benign economic growth is feasible" (Hunter and Toney 2005, 30).

14 The LDS church does not have a professional clergy. Its lay members volunteer to organize on a local level. They do so according to careful instructions from the church leadership in Salt Lake City. So while its lay clergy means broad participation at all levels of church activity among its members and in many ways a more democratic and shared sense of responsibility for the church, it is also highly hierarchical. Local leaders defer to church headquarters on questions of doctrine, management of buildings, and church policy. The aim of such an organization is to diminish local differences and thereby to provide the greatest amount of unity and equality for its worldwide membership. While this has many advantages, this particular structure of the church makes it very difficult for members to know when they can speak out, or how they might work to influence cultural change when they see it is needed. Faithful members of the church do not want to be seen as dictating or criticizing what the church leaders do or say. As members of the church have become increasingly active on the environmental front, they have faced this kind of challenge. Often interfaith groups, environmental organizations, or academic conferences will seek an authoritative LDS voice and none can be found. Individual members can speak, but they generally understand that they are not to speak on behalf of the church in any official capacity. This leads to a muted quality to these voices speaking out on behalf of the creation even though those voices are motivated precisely by the very environmentally friendly beliefs in the LDS faith. At least the two aforementioned church websites now make it clear that such speaking would not be deemed out of bounds.

References

Anderson A. (2015) Mountains that never rest. Available at: www.oquirrh.org

Brown J. (2011) "Whither Mormon theology?" *Dialogue: A Journal of Mormon Thought* 44:2, 67–86.

Bryner G. (2010) "Theology and ecology: Religious belief and environmental stewardship", *BYU Studies* 49:3, 21–45.

The Church of Jesus Christ of Latter-Day Saints (2012) Scriptures and study. Available at: www.lds.org/scriptures?lang=eng

The Church of Jesus Christ of Latter-Day Saints (2013) Environmental stewardship and conservation. Available at: www.mormonnewsroom.org/article/environmental-stewardship-conservation

The Church of Jesus Christ of Latter-Day Saints (2014) Environmental stewardship and conservation. Available at: www.lds.org/topics/environmental-stewardship-and-conservation?lang=eng

Cox P. (2006) Paley's stone, creationism, and conservation. Available at: http://rsc.byu.edu/archived/stewardship-and-creation/3-paleys-stone-creationism-and-conservation

Halki Summit II (2015). Available at: www.halkisummit.com

Handley G. (2000) "The environmental ethics of Mormon belief", *BYU Studies* 40:2, 187–211.

Handley G., Peck S., and Ball T. eds. (2006) *Stewardship and the creation: LDS perspectives on nature,* Religious Studies Center, Provo. Available at: https://rsc.byu.edu/%5bfield_status-raw%5d/stewardship-and-creation-lds-perspectives-environment

Hunter L. M. and Toney M. B. (2005) "Religion and attitudes toward the environment: a comparison of Mormons and the general U.S. population" *Social Science Journal* 42.1, 25–38.

Galli C. (2006) "Stewardship, sustainability, and cities". Available at: http://rsc.byu.edu/archived/stewardship-and-creation/3-paleys-stone-creationism-and-conservation

Gowans M. and Cafaro P. (2003) "A Latter-Day Saint environmental ethic", *Environmental Ethics* 25:4, 375–94.

Mayfield J. (1999) "Poverty, population, and environmental ruin", in Smart W., Williams T. T. and Smith G., eds. *New Genesis: A Mormon reader on land and community,* Gibbs Smith, Salt Lake City, 55–65.

Nash M. (2013) Dominion and compassion for the earth. Available at: www.mormonnewsroom.org/article/elder-nash-stegner-symposium

Nibley H. (2012) Man's dominion, or subduing the earth Available at: http://publications.maxwellinstitute.byu.edu/fullscreen/?pub=1094&index=3

Oelschlaeger M. (1994) *Caring for creation: An ecumenical approach to the environmental crisis,* Yale University Press, New Haven.

Smith J. F. sel. (1938) *Teachings of the Prophet Joseph Smith,* Deseret Book, Salt Lake City.

Stoll M. (2015) *Inherit the Holy Mountain: Religion and the rise of American environmentalism,* Oxford University Press, New York.

Walch T. (2015) "LDS environmental stewardship statement, recent talk share similarities with Pope Francis' encyclical. Available at: www.deseretnews.com/article/865631022/LDS-environmental-stewardship-statement-recent-talk-share-similarities-with-Pope-Francis.html?pg=all

Wallis B. (2011) "Flexibility in the ecology of ideas: Revelatory religion and the environment", *Dialogue: A Journal of Mormon Thought* 44:2, 57–66.

White L. (1996) "The historical roots of our ecologic crisis", in Gottlieb R. ed. *This Sacred Earth: Religion, nature, environment* Routledge, New York 173–181.

Williams T. T., Smart W., and Smith G. eds. (1999) *New Genesis: A Mormon reader on land and community,* Gibbs Smith, Salt Lake City.

PART III

Indigenous cosmovisions

Introduction: *John Grim*

Cosmovision is used here as a term for differing Indigenous perspectives of a world that speaks to them. Cosmovision refers to narratives in which beings in the world tell their own stories, and in the telling make manifest the deeper meanings of the world and humans woven into it. These are not metanarratives that impose a fixed truth to which individuals and communities harmonize themselves. Cosmovision indicates a path that unfolds in the telling – and that telling arises from the lips of the world. Cosmovision lives primarily in Indigenous oral forms that both describe a world and prescribe how to live in it.

The term, Indigenous, is used to designate such societies having a shared language, kinship system, and relationships with local lands and biodiversity. Many Indigenous societies were fragmented by the political reality of the colonialist period. Since colonization began in the sixteenth century, narratives and practices of European settlers implementing a capitalist economics of acquisition and extraction dominated Indigenous societies.

During this colonialist period and into the present, Indigenous cosmovisions have often been marginalized, erased, exploited, and coopted by dominant state actors. However, regenerating Indigenous voices, often associated with *indigeneity*, are evident in UN forums and documents such as the 2007 *United Nations Declaration of the Rights of Indigenous Peoples*.

The terms, indigenous and indigeneity, are not unilaterally used by native scholars. Some African, South, and Southeast Asian scholars consider "indigenous" as a colonialist carry-over. That is, this term is seen as degrading peoples as frozen in some "traditional" past; or, as confusing historical encounters in which local peoples were displaced by settlers, as in the Americas, with other regions where no such widespread displacement occurred, as in Africa, India, or Southeast Asia. Some therefore object to the term Indigenous as an outsider perspective separating certain peoples as authentic or original to a place and excluding others who have been in the country for extended periods of time.

Acknowledging these differences, Indigenous is used in this section to describe diverse small-scale societies around the world. Cosmovision enriches these discussions by framing Indigenous views as ecologically informed and informing. Indigenous ways of being-in-the-world are also signaled by the term "lifeway." The integral character of cosmovision as something woven into cultural life is what is meant by lifeway.

Moreover, phrases such as Indigenous knowledge (IK) and traditional environmental knowledge (TEK) also open pathways into understanding lifeway as ways of knowing. As one Indigenous scholar has said:

> Perhaps the closest one can get to describing unity in Indigenous knowledge is that knowledge is the expression of the vibrant relationships between people, their ecosystems, and other living beings and spirits that share their lands....All aspects of knowledge are interrelated and cannot be separated from the traditional territories of the people concerned.... The purpose of these ways of knowing is to reunify the world or at least to reconcile the world to itself. Indigenous knowledge is *the way of living* within contexts of flux, paradox, and tension, respecting the pull of dualism and reconciling opposing forces.
>
> *(Battiste 2000)*

What emerges in the chapter of this section is a greater understanding of humans embedded in land, which is described in oral narratives. Cosmovision, then, is the way that Indigenous communities locate themselves in respectful sustenance relations with their homelands and animals and plants. Indigenous cosmovisions bring forward ancient ecological wisdom, historical memory of community of life, and awareness of Indigenous humanism into the dialogue of contemporary environmental challenges. Cosmovisions suggest grounding in a world that affirms humans along with biotic and abiotic life as having a shared future. The chapters in this section ask: can humans hear what these cosmovisions are saying?

Reference

Battiste, Marie and James Henderson. 2000. *Protecting Indigenous Knowledge and Heritage* Saskatoon, SK: Purich Publishing, 35.

12

AFRICA

African heritage and ecological stewardship

Jesse N. K. Mugambi

This chapter focuses on stewardship of Africa's ecology – past, present and future. Its contextual setting is the ecological crisis climate scientists have researched since the 1970s, particularly with regard to the impacts of rapid climate change on human habitats and livelihoods. The impacts are experienced especially in those places where the natural environment has been destroyed through settlement and pollution. Comparatively, the continent of Africa as an ecological region is most adversely affected although its inhabitants are least responsible for the industrial pollution of which they are the most afflicted victims.

First, a note on terminology. Instead of "indigenous", the adjectives "rural" and "local" are used. The reason is that the word "indigenous" refers to "native" or "original" to a place, contrasted with "exotic" or "foreign". In Africa few nations have "exotic" populations except as diplomats, missionaries, expatriates, tourists, investors and so on. In the context of Tropical Africa the most striking demographic distinction is between *rural* and *urban* habitation. Rural inhabitants are the overwhelming majority, most adversely affected by ecological degradation.

The current global ecological crisis is a cumulative result of emissions of greenhouse gases into the atmosphere, especially since the 1850s. These emissions have been mainly produced from factories, motorized travel, mechanized farms and energy requirements for cities especially in the most industrialized countries. Africa's per capita contribution of GHG emissions is less than 3%, half of which comes from Egypt and South Africa. The remaining 52 African nations contribute less than 1.5% of all greenhouse gas emissions.

Africa's urban population is concentrated in a few cities, such as Cairo, Ibadan, Johannesburg, Nairobi, etc. The rest of Africa's population, mostly rural, is based in the various ecological zones of the continent, with more than half in the most populous countries – Nigeria, Ethiopia, Egypt, DRC, South Africa, Tanzania, and Kenya. A large proportion of Africa's urban population resides in informal, unplanned settlements on the periphery of towns and cities. Thus rural and urban modes of life in each African country are blended in such ways as to portray the complex mosaic of Africa's multi-cultural heritage. In the Americas, Australia and New Zealand, the remnants of pre-colonial inhabitants are restricted in "reservations" neither viable nor comfortable. It is in these "reservations" that these "indigenous" remnants strive to eke a living and preserve their worldview, overwhelmed by Western "civilization". The resilient struggle of these vanquished peoples is lucidly described by such authors as George Tinker in USA; Leslie

Sponsel in Hawaii; David Suzuki in Canada; Ralph Folds in Australia, Maricel Mena López in Colombia, Carlos Gradin in Brazil, and so on.

John Grim in his book *Deep Ecology* has described the important role of "indigenous" insights in ecological rehabilitation. It is worthwhile, however, to appreciate the marginal space and slot of such Indigenous communities in the dominant, mainstream Euro-American culture. Even in the reservations it is difficult for indigenous communities to live in accordance with their cultural heritage, since these communities are not sovereign powers with authority to determine their own present and future. In Tropical Africa both the rural and the urban areas are populated mostly by Africans, except the expatriates, diplomats, investors, tourists, journalists and researchers who regard their original homes to be elsewhere. The strategy for ecological rehabilitation in Africa differs greatly from that in countries where the carbon footprint is far beyond the global mean of 4 metric tons per capita.

Rural peasants and pastoralists, wherever a majority, are the ones upon whom sustainable ecological rehabilitation will in the long term depend, as emphasized by Prof. Debal K. Singharoy regarding peasant movements in India. In countries where settler colonialism triumphed, the words "native" and "indigenous" have acquired derogatory and condescending connotations, in contrast with "civilized" and "modern". Robert Chambers challenges such denigration in his book *Whose Reality Counts: Putting the First Last* (London: ITDG), 1997).

According to the *UN 2014 Urbanization Prospects Report,* Africa and Asia are the homes of nearly 90% of the world's rural population. Comparatively, the urbanized populations of northern America are (82%); Latin America and the Caribbean (80%); and Europe (73%). Rural populations are much more vulnerable to the impact of extreme weather variations, owing to an underdeveloped infrastructure. One of the impacts of climate change on rural areas is expected to be prolonged droughts and excessive precipitation when the rains finally come.

The Red Cross and other relief agencies are becoming increasingly preoccupied with relief operations for such rural victims of ecological deterioration. For example, the World Meteorological Organization (WMO) warned that the 2015 El Nino rainfall of October–November 2015 would be excessive, and that rural populations should evacuate in advance. In response to this warning, governments of Tropical Africa had to divert their resources to prepare for this ecological eventuality. Such weather variability is expected to be more extreme and more frequent as a result of climate change accelerated by industrial pollution. In 2014 the level of urbanization in eastern Africa was 25%; western Africa 44%; northern Africa 51%; and southern Africa 61%. Some African nations have very low levels of urbanization, such as Burundi 12%; Uganda 16%; Malawi 16%; Chad 22%; and 21% in Swaziland.

The vulnerability of Africa and its inhabitants is summarized in the *Fourth Assessment Report* (AR4) released by the UN Intergovernmental Science Panel on climate change:

> Africa is one of the most vulnerable continents to climate change and climate variability, a situation aggravated by the interaction of 'multiple stresses', occurring at various levels, and low adaptive capacity (high confidence). Africa's major economic sectors are vulnerable to current climate sensitivity, with huge economic impacts, and this vulnerability is exacerbated by existing developmental challenges such as endemic poverty, complex governance and institutional dimensions; limited access to capital, including markets, infrastructure and technology; ecosystem degradation; and complex disasters and conflicts. These in turn have contributed to Africa's weak adaptive capacity, increasing the continent's vulnerability to projected climate change.
>
> *(UN-IPCC 2014, 9.2.2, 9.5, 9.6.1)*

Statistically, the ecological vulnerability of Africa and its inhabitants is quantified in Table 12.1, showing that more than 70% of the continent's area is hyper-arid, arid, semi-arid, or dry sub-arid. Thus Africa's population of more than one billion lives on about 8,895,000 km² (an area about 90% of the USA) out of this continent's total 30,065,000 km². This is an area about 90% of the USA. The available arable land in Africa continues to shrink, and available fresh water supply also continues to decrease per capita owing to reduced precipitation. This ecological deterioration is as a result of global warming, attributed largely to unprecedented industrial pollution – particularly the emission of excessive carbon dioxide. Africa's orbital position makes this continent the worst affected, although Africans are the least responsible. From the perspective of applied ethics, the *Earth Charter* (1992) challenges all people and nations of good will to respond on the basis of the principle of *Common but Differentiated Responsibilities and Respective Capabilities* (CDRC). Africa has the least means to respond to this ecological crisis.

The UN World Conference on the Human Environment, 1972

World Environment Day was proclaimed at Stockholm, Sweden, during the UN World Conference on the Human Environment on 5 June 1972, in recognition of the ecological crisis that had already become evident. Since then, responses have been multi-faceted. Politicians deliberate on it from the perspective of power and influence within and between nations, taking for granted the old adages that "Might is Right" and "Strong never does Wrong". Scientists research this crisis on the basis of another maxim: "Fact no matter What!" The Intergovernmental Science Panel of United Nations Framework Convention on Climate Change (UNFCCC) publishes annual reports containing scientific facts about the state of the global environment, the fifth of which was released on 5 November 2014. Some of the main features of this UNFCCC Fifth Assessment Report (AR5) are the following: A new set of scenarios for analysis across Working Group contributions; dedicated chapters on sea level change, the carbon cycle and climate phenomena, such as monsoons and El Niño; and broader treatment of impacts, adaptation and vulnerability in human systems and the ocean; much greater regional detail on climate change impacts, adaptation and mitigation interactions; inter- and intra-regional impacts; and a multi-sector synthesis, risk management and the framing of a

Table 12.1 Area per aridity zone by sub-region for Africa★

Sub-region	Hyper-arid	%	Arid	%	Semi-arid	%	Dry sub-humid	%	Total (30,065,000 km²)
Northern Africa	4,736	81	640	11	410	7	43	0	5,829
Western Africa	2,363	33	1,465	20	1,278	18	514	7	5,620
Central Africa	0	0	6	0	66	2	144	4	216
Eastern Africa	878	14	1,670	27	1,768	28	767	12	5,083
Southern Africa	96	2	823	13	2,579	42	924	15	4,422
Africa Total	8,072	27	4,604	16	6,100	21	2,392	8	21,170
									70.41%

Note: ★Area numbers are in thousands of km².

Sources: Corbett J. D., R. F. O'Brien, E. I. Muchugu and R. L. Kruska. Data exploration tool: a tool for spatial characterization. CD-ROM and User's Guide, 1996; UNDP/UNSO. Aridity zones and dryland populations: an assessment of population levels in the World's drylands. UNSO/UNDP, New York, 1997.

response (both adaptation and mitigation), including scientific information relevant to Article 2 of the UNFCCC referring to the stabilization of greenhouse gas concentrations in the atmosphere at a level that would prevent dangerous anthropogenic interference with the climate system.

Through the series of reports released by the UNFCCC Intergovernmental Science Panel, on the basis of available data, recommendations are published on what needs to be done towards *precaution, mitigation* and *adaptation* in response to the factors attributed to rapid climate change – especially in the twentieth century. From a business perspective investors respond with yet another maxim: "Business before Sense!" Responses of Transnational Corporations to the series of UNFCC Intergovernmental Science Panel recommendations have tended to be tempered with concern against possible reduction in *profits* in the short term. Thus implementation of some of the recommendations has been delayed by concerns of the business sector regarding profitability.

Since the 1970s academic discourse in Europe and North America has tended to take as normative the maxims of "postmodernism" and "secularism", with religious and moral discourse in retreat. Yet, value considerations are indispensable in politics, business and science. The theme of "Development" is relevant in all disciplines, including religion and ethics. Thus the UN proclamation of 5 June as World Environment Day had religious and ethical implications, irrespective of the moral and religious dispositions that any person may hold or propagate.

In 1972 the book *Limits to Growth* was published cautioning that the rate at which natural resources were being spent was much higher than the rate of population growth and the possible rate of replenishment. The authors suggested that ways and means were urgently needed for managing natural resources more efficiently. Thirty years later, in 2002, the authors wrote a sequel, titled *Limits to Growth: Thirty Years After,* observing that their predictions and cautions were more than accurate, and pleading for more concerted efforts to manage the environment more efficiently and effectively to avoid ecological collapse. Such cautions have been made many times, but politicians are often more concerned about remaining in power than heeding cautions from experts. Likewise, investors are more interested in maximizing their profits, than in heeding cautions about the consequences of their investments. In view of the widening gap between the haves and the have-nots – both nationally and internationally – the most vulnerable will, apparently, have to fend for their own selves without having to wait for patronizing donations from the affluent. Patronage and partnership are not synonyms.

The 1972 World Environment Conference eventually led to the UN Climate negotiations for reduction of greenhouse gases, especially carbon dioxide and methane. The least industrialized nations looked forward to the Kyoto Protocol as a diplomatic mechanism towards just ecological management of global ecology. By 2011, at the 17th Conference of the Parties in Durban, South Africa, it was clear that the industrialized nations were interested neither in taking responsibility for their historical emissions, nor changing course with regard to industrialization. Outcome of the 21st Conference of the Parties (Paris, 2015) re-confirmed this refusal to take immediate remedial action, with implementation postponed beyond the year 2020 and without any guarantee that the Paris commitments will be honored.

The debate continues, but the least industrialized nations, particularly those in Africa, are on their own, bearing the brunt of ecological disaster for which they are least responsible. African proverbial wisdom teaches that when elephants fight, it is the grass that suffers. Africa's rural communities continue to suffer the consequences of environmental degradation caused mostly by others. But in the industrial nations, it is business as usual, to maximize profits and increase the luxuries that consumerism cherishes.

The ecology of Africa from a global perspective

Africa as a continent has the greatest diversity of flora and fauna, owing to its geostationary orbit striding the Equator, the Tropics of Cancer and Capricorn. The orbit of planet Earth in relation to the Sun is such that one half of the continent of Africa is south of the Equator while the other half is north of it. Three oceans and two seas surround Africa: Atlantic Ocean (west); Indian Ocean (east); Southern Ocean (south); Mediterranean Sea (north) and Red Sea (north-east). Africa is the only continent bordering two other continents at one location – the Isthmus of Suez (Europe and Asia). Owing to Africa's topography, the continent has within it all the climate zones of the world, from permanent glaciers on mountaintops to barren deserts; from extreme cold to extreme heat; from high humidity to extreme drought. Sir John Houghton, former Chief Meteorologist of UK and former Chairman of the UN-FCCC Intergovernmental Science Panel on Climate Change, authored the book *Global Warming: The Complete Briefing,* first published in 1997. The fourth edition (2010) contains real-time photos of Africa, taken from space, clearly illustrative of the ecological crisis that Africa and Africans currently face, which future generations will have to inevitably inherit at greater expense than they can afford. The following ecological features characterize Africa's topography.

a) Mountains: Africa has six mountain ranges, four in Eastern Africa (Mount Kilimanjaro; Mount Kenya; Ethiopian Highlands and Ruwenzori Highlands); one in South Africa (Drakensberg); two in western Africa; and one in North Africa. Two of the highest mountains in the world are in East Africa – (Mount Kilimanjaro in Tanzania; Mount Kenya in the Republic of Kenya). Africa also has two large deserts, the Sahara north of the Equator and Kalahari in the southern hemisphere. The largest inland lake in the world is in Africa, namely, Lake Nyanza–Victoria shared by Kenya, Tanzania and Uganda. The Great Rift Valley passes through Africa with the largest inland section passing though Kenya, Tanzania, Uganda, Malawi and Mozambique. These mountain ranges are the "water towers" for this fragile continent. However, the most of permanent glaciers have melted during the twentieth century, greatly reducing the volume of the rivers that originate there and the underground aquifers also. The consequence of this ecological disaster is devastating for the communities dependent for livelihood on these rivers.

b) Rivers: The Nile is longest river in the world flowing for 6,843 kilometres (4,248 miles) northwards for from the East African highlands to drain at its delta in Egypt, with a catchment covering twelve countries: Egypt, Sudan, South Sudan, Ethiopia, Uganda, Kenya, Tanzania, Burundi, Rwanda, the Democratic Republic of Congo (DRC) and Eritrea as an observer. The largest waterfall in the world (Victoria Falls) is along the Zambesi River, on the border between Zambia and Zimbabwe. Victoria Falls is more than a kilometre long; its drop is than four hundred meters. The largest desert lake in the world is in Africa – Lake Turkana in Northern Kenya. The largest salt lake in the world is also in Kenya (Lake Magadi). The longest coral reef in the world is along the shores of Kenya, Tanzania and northern Mozambique. The Congo River catchment has an equatorial rain forest habitat comparable to that of the Amazon in South America, covering eleven countries (DRC, Central African Republic, Angola, Republic of Congo, Zambia, Tanzania, Cameroon, Burundi, Rwanda, Gabon, Malawi). The Congo is the deepest river at 220 meters (720 feet). The second highest waterfall in the world is on Tulega River in South Africa, Kwa Zulu Natal Province (948 meters). The Senegal River is the second longest in West Africa (after the Niger) and the one closest to the western part of the Sahara Desert. Generally, the volume of Africa's rivers has continued to decrease, owing to the melting of glaciers and the decrease in rainfall precipitation – with attendant negative impact on the populations dependent on these rivers and streams. Seasonal flooding from the

Angola highlands has maintained the Okavango Delta in Botswana, but most of its water evaporates across the Kalahari desert.

c) Lakes: Lake Nyanza (Victoria), located in East Africa, is the largest tropical lake in the world by area, shared by Kenya, Tanzania and Uganda. It is drained by the River Nile and supplied by rivers in the three East African countries. Lake Tanganyika, is the second largest in Africa, shared by Tanzania (46%), DRC (40%), Burundi and Zambia. Lake Malawi is the second largest freshwater lake in the world by volume, and the second deepest. The third largest freshwater lake in Africa is Lake Malawi, shared by Malawi, Tanzania and Mozambique. Other freshwater lakes include Nakuru and Naivasha (Kenya); Tana (Ethiopia); Kivu (DRC and Rwanda); Albert, Edward, (Uganda and DRC); Rukwa (Tanzania); and Mweru (Zambia). The freshwater lakes are a source of livelihood for many communities in Tropical Africa that live within the lake basins. With declining volume on these lakes, the quality of life continues to deteriorate especially for the affected rural communities.

d) Forests: The Tropical Belt of Africa is covered by forest in most countries, except Kenya and Somalia on the east coast – plus Namibia and Angola on the west coast – where the lowlands are mostly arid and semi-arid. Most of these forests have been cleared for agriculture (mainly cash crops for export), urban settlement and also for timber. The impact of deforestation has reduced vegetation, resulting in decreasing precipitation, and the consequent reduction in rain runoff. Water on the riparian valleys continues to dry up, with a corresponding drop in the recharge of aquifers. Logging is big business in Africa's equatorial and tropical forests. Both the demand and the supply ends of this business are outside Africa. The ultimate beneficiaries are outside Africa and the victims are the local African communities. Among prominent African leaders who resonated with the challenge of the Earth Charter are the late Professor Thomas R. Odhiambo, founder of the International Centre of Insect Parasitology and Ecology (ICIPE); Nobel Prize Laureate the late Professor Wangari Maathai, founder of the Green Belt Movement in Kenya; Nobel Prize Laureate Archbishop Emeritus Desmond Tutu of Cape Town, South Africa; and the Nobel Prize Laureate Kofi Annan, former UN Secretary General from Ghana. Afforestation is becoming normative for communities in Tropical Africa. However, it is ironical for these communities to re-forest areas that were cleared to plant cash-crops for export including cocoa, coffee, tea, cotton, pyrethrum, sisal, sugar cane (and such others) for industries abroad, at prices growers could not dictate!

Green economy within the African Union's vision 2063

In 2013 the African Union launched the fifty-year Strategic Plan titled *Agenda 2063: The Africa We Want*. This document, approved by Africa's heads of state, contains the aspirations that Africans wish to enjoy and celebrate during the centenary year since formation of the Organization of African Unity in 1963. Several projects approved for implementation through interstate cooperation directly focus on ecological management, including the following:

a) The "Great Green Wall of Africa". A fifteen kilometer block of forest is to be planted, stretching seven thousand kilometers across the edges of the Sahara Desert from the Atlantic Ocean to the Red Sea, through eleven nations- Senegal, Mauritania, Mali, Burkina Faso, Niger, Nigeria, Chad, South Sudan, Ethiopia Eritrea and Djibouti. The African Union Heads of State endorsed this project in 2007. The government of each country covered by the project is responsible for oversight regarding the planting and management of seedlings in its section of this block. The long-term goal is to create a green buffer for stopping the spread of the Sahara desert southwards beyond this Green Belt. Nobel Prize Laureate the late Professor Wangari Maathai from Kenya was very committed to this project Her leadership in mobilizing communities provided impetus for it.

Implementation of the Africa Great Green Wall Project is with local communities who will be the ultimate beneficiaries for the present and future generations in all the eleven nations to be afforested. Apart from their foreign languages of rule – English, French and Arabic, these communities speak dozens of African languages, and their religious affiliations include African Religion, Islam (both Shia and Sunni), Christianity (Protestant, Catholic, Orthodox, Independent, Evangelical, etc.). The common goal of rehabilitating their ecology, to curb the spread of the Sahara Desert southwards unites them in one endeavor.

b) Sustainable food production and consumption: The Food and Agriculture Organization (FAO), in December 2013, issued a Progress Report on this project, observing that a variety of natural resource management programs underway in some of the countries demonstrated the potential of sustainable land management to boost food security, improve community livelihoods and build the resilience of the land and the people to the changing climate. Thus the FAO recognizes and appreciates the necessity of local and contextual approaches to solving country-specific problems, and applying those solutions in specific local contexts. Any approach applying the "one size fits all" template is doomed to failure, from the outset.

c) Sustainable management of Africa's river basins: The African Union through its various organs and agencies is facilitating inter-state collaborations for sustainable management and utilization of the main river basins in the continent- the Congo; the Niger; the Nile; the Okavango; Zambezi. This facilitation is multi-faceted, to ensure optimal intergenerational benefit for both the local communities and the national economies.

d) Hydro-electric power for national and regional consumption: Several projects are planned along the main river valleys that serve as national borders. These projects will provide electricity on the basis of agreed commitments.

e) Photo-voltaic farms for generation of electricity: The African Union is also facilitating regional collaboration in the generation of Carbon-free electricity from photo-voltaic panels, to be fed into national grids for both national and trans-national consumption. Kenya is promoting this technology for both rural and urban supply of electricity

f) Production of electricity from wind turbines: The Tropical position of Africa is favorable for all-year supply of electricity through installation of energy-efficient wind turbines. Kenya, South Africa and Tunisia are leaders in this strategy. The power is fed into the national grid, reducing the cost of transmission and installation in remote locations.

g) Geo-thermal generation of electricity: A few African countries have sites that are suitable for generation of electricity from geo-thermal steam. Kenya is one of the leading nations in the use of this technology.

Innovative community-based initiatives

Rural communities in Tropical Africa are taking responsibility to rehabilitate their local environments through various methods of rainwater harvesting. The following are some examples:

a. The Utooni Development Organization (UDO), which constructed 1,600 sand dams, dug 1,600 kilometres of terraces and planted one million trees over a period of 35 years between 1978 and 2013. UDO won the 2014 UNDP Equator Award, as the most accomplished sustainable community-based organization for environmental resilience and rehabilitation. www.utoonidevelopment.org/

b. The Christian Impact Mission International (CIM) has transformed the lives of individuals and communities in Eastern Kenya through the harvesting of rainwater from roofs and roads. The water is collected in polythene tanks and polythene sheets for use in vegetable and fruit cultivation on the Yatta Plateau. www.ipsnews.net/2012/09/men-and-women-farming-together-can-eradicate-hunger/

c. The Kenya Wildlife Conservancies Association (KWCA) is a landowner-led national membership organization representing community and private conservancies in Kenya. The Association works with conservancy landowners and regional associations to create an enabling environment for conservancies to deliver environmental and livelihood bene-fits. Its mission is to be the forum where landowners have a unified voice, share experiences and actively participate in protecting and benefiting from wildlife. The long-term goal is a future where wildlife and communities benefit from a network of functional conservancies that complement state protected areas. Its strategy is to change how Kenya's wildlife and wild places are managed as well as strengthen people's rights to manage and benefit from nature. http://kwcakenya.com/page/about

d. MKOPA Solar provides photo-voltaic panels to rural households not on the national power grid, enabling families to enjoy affordable lighting, internet mobile phone charging and other electronic necessities at home. The credit facilities are serviced via MPESA, the mobile money exchange invented by Kenyan youth. www.m-kopa.com/

e. The International Centre of Insect Physiology and Ecology (ICIPE), based in Nairobi, Kenya, focuses on sustainable development, to include human health as the basis for devel-opment and the environment as the foundation for sustainability. Working in a holistic and integrated approach through the 4Hs Paradigm – Human, Animal, Plant and Environmental Health – ICIPE aims at improving the overall health of communities in tropical Africa by addressing the interlinked problems of poverty, poor health, low agricultural productivity and degradation of the environment. www.linkedin.com/company/icipe

These innovative initiatives resonate with the works of such scholars as: Calestous Juma – *The New Harvest: Agricultural Innovation in Africa*, (OUP, 2011); Jesse Mugambi – *Fresh Water to Eradicate Poverty* (Norwegian Church Aid, 2006); Robert Chambers, *Whose Reality Counts Putting the First Last* (Intermediate Technology Publications, 1997); Ulrich Duchrow – *Alternatives to Global Capitalism* (Heidelberg: Kairos Europa, 1996); M. Tiffen and M. Mortimore and F. Gichuki – *More People, Less Erosion: Environmental Recovery in Kenya*, Nairobi: ACTS Press and ODI, 1994).

Relevant "empirical science" and "appropriate technology"

At the 1981 UN Conference on New and Renewable Sources of Energy in Nairobi, Kenya, the Governing Council made the following resolution:

> Deeply concerned at the disastrous ecological consequences, which might result in the near future from the excessive use of wood as the sole source of energy for the vast majority of people in a large number of developing countries, ... Urgently appeals to the Preparatory Committee of the United Nations Conference on New and Renewable Sources of Energy to ensure that, in drawing up the provisional agenda for the conference and preparing the draft plan of action to be submitted for adoption, it gives sufficient emphasis and attaches high priority to the uses of new and renewable energy sources which would make it possible to tackle the major problem of fuelwood.
>
> *(United Nations Environment Programme, 1981, Resolution 9/7)*

In subsequent years, efforts were made to popularize "appropriate technology" in the energy sector. However, the adjective "appropriate" was "derogatory" and also "demeaning" from the perspective of the target population – since it was viewed as "technology for the poor and marginalized". Electricity and cooking gas were, and continue to be viewed as, options of the rich who can afford it, while firewood, charcoal and other biomass are relegated to the poor, living mainly in rural areas and informal settlements in the peripheries of towns and cities. Until the notions of "empirical science" and "appropriate technology" are defined as applicable to the whole population, the poor and marginalized will remain excluded from the mainstream of national economies in Africa.

The youth, representing the majority of the population in Africa, can contribute immensely to ecological stewardship, if they are socialized as active participants in the social transformation of their respective societies. Science as "organized knowledge" is not value-free. Designers of research projects determine the variables and constants, core and peripheral factors, and the utility of the results. When such results are popularized outside the cultural context for which they were intended, the consequence is distortion of reality, especially when assessed from the perspective of those to whom the research is exported. Such has been the tragedy of research and technology originating from other cultures and dumped in Africa as "aid", or sold for use in cultural contexts for which it was not originally intended.

There is no culture-free science; no culture-free technology. When tools designed in and for a particular culture are exported to another culture, the latter users have to adapt those tools to their cultural context. Some adaptation is successful, while most of it is not. Technological "globalization" is ideologically misleading, since it remains foreign for the cultures that adapt it and homely for the culture for which it was originally designed. The mobile phone in Kenya is an interesting example. In a country with a per capita GDP of USD1.358 in 2014, more than three-quarters of the population own at least one mobile phone, which is used for both communication and for money exchange in lieu of formal banking. The M-Pesa platform, in which "M" stands for "mobile and *pesa* is Swahili for "money," was originally designed by a young Kenyan, to increase the opportunities for mobile phone money transfers. For many lower-income Kenyans living in rural contexts, where both banking and landline telephone were inadequate, mobile phones were unavailable or dysfunctional for the majority of the population. Because of this initiative in 2015 M-pesa in 2015 transacted 42% of Kenya's GDP! When this insight is applied to ecological stewardship, Africa's younger generation in each nation will have to design and apply scientific, technological and culturally attuned innovation to meet respective needs and contexts. Appropriateness can be accurately determined only contextually – not globally. What then, is the fate of the "globalization" agenda? The trade walls and wars in the WTO and other international forums indicate that "globalization" is "buzz word" heavily loaded with assumptions, but devoid of any culture-specific meaning – except for those who peddle it for their own advantage.

Implications

Demographically, in the current decade Africa is the youngest continent, with a median age of 19.1 in 2012. This demographic profile will shape Africa's future. With regard to the African heritage and ecological rehabilitation the following implications can be discerned:

a.   The history of African peoples (pre-colonial, colonial and post-colonial) to a large extent determines how the citizens of a particular nation respond to the challenges arising from environmental degradation. The one-century struggle against Apartheid in South Africa is instructive.

b.   Land tenure policies greatly influence the response of communities to ecological degradation and rehabilitation. Many of the conflicts in Tropical Africa seem to be political. In the final analysis, however, they have to do with distribution of natural and human resources – Kenya; Sudan; Nigeria; Cote d'Ivoire; Zimbabwe – the list is long!

c.   There is great dissonance between proposed "global" solutions and "local" perceptions, especially when "donor" agencies impose "global" solutions to "local" problems. The "Aid Business" reacts and lurches from disaster to disaster. But the victims always live at the edge of existence. This is one of the ironies of the international "Development" agenda.

d.   The hope for successful ecological stewardship in Africa is in endogenous innovation, provided that it passes the test of scientific rigour and contextual applicability. An example is A Rocha Kenya, rehabilitating the Mangrove forest on the shores of the Indian Ocean, and at other locations with different flora and fauna.

e.   Importation of gadgets designed for other cultures is unavoidable. However, the application of those gadgets outside the culture of origin will depend on the cultural importance and utility of those gadgets for the importing culture. A dramatic example is the mobile phone in Kenya, which today is a money transaction tool for low-income subscribers, in addition to its "global" function as a phone.

f.   These insights are applicable as much to ecological stewardship, as to any other challenge faced by cultures that import gadgets – for whatever purpose.

Conclusion

In conclusion, it is worthwhile to return to the challenge aptly expressed in the *Earth Charter,* emphasizing the inevitable but constructive tension between the global and the local; between Fact and Value; between Past and Future; between Despair and Hope; between Oppression and Liberation; Destruction and Reconstruction; between Self-pity and Self-affirmation. The choice is ours. No one has the right to choose on our behalf:

> The choice is ours: form a global partnership to care for Earth and one another or risk the destruction of ourselves and the diversity of life. Fundamental changes are needed in our values, institutions, and ways of living. We must realize that when basic needs have been met, human development is primarily about being more, not having more.

Kofi Annan's address, as UN Secretary General, to the UN World Summit at Johannesburg in August 2002 is also aptly relevant:

> We should no longer imagine either that one fifth of humanity can indefinitely enjoy prosperity, while much larger numbers live lives of deprivation and squalor, or that patterns of production and consumption which destroy the environment can bring us lasting prosperity. The issue is not environment versus development, or ecology versus economy. It is how to integrate the two.

The ecological rehabilitation of Africa is a challenge for the entire "global community". It ought not to be a burden for Africans to bear, particularly in view of the fact that the natural and labour resources extracted from Africa have benefited nations and peoples outside this continent for centuries. Africans have a great role in this challenging task. But they cannot, and ought not, to bear the burden alone. It is the challenge for the whole of humankind. The three principles of precaution, mitigation and adaptation remain valid, but their application must, as a matter of necessity, be differentiated.

References

Books

Annan, K. "A Chance to Secure Our Future", Address to the UN World Summit on Sustainable Development, Johannesburg, August 2002, in *UN Chronicle,* 1 Sept. 2002.

Birch, C. and Paul, D. *Life and Work: Challenging Economic Man*, UNSW Press, 2003.

Chambers, R. *Whose Reality Counts: Putting the First Last?* London: ITDG, 1997.

Duchrow, U. in his book *Alternatives to Global Capitalism*, Heidelberg: Kairos Europa, 1996.

Hallman, D. G. *Spiritual Values for Earth Community: Updated Edition*, Geneva: Risk Book Series, 2012.

Houghton, J. *Global Warming: The Complete Briefing,* 4th Edition, Cambridge University Press, 2010.

Juma, C. *The New Harvest: Agricultural Innovation in Africa*, OUP, 2011.

Meadows, D., J. Randers and D. Meadows. *Limits to Growth: The 30-Year Update,* Chelsea Green Publishing, 2004.

Mugambi, J. N. K. *God, Humanity and Nature in Relation to Justice and Peace,* Geneva: WCC, 1987.

Mugambi, J. N. K. and G. Kebreab. *Fresh Water to Eradicate Poverty,* Oslo: Norwegian Church Aid, 2006.

Rasmussen, L. *Earth Community Earth Ethics: Ecology & Justice,* Maryknoll, New York: Orbis, 1997.

Tiffen, M., M. Mortimore and F. Gichuki. *More People, Less Erosion: Environmental Recovery in Kenya,* Nairobi: Africa Centre for Technology Studies and London: Overseas Development Institute, 1994. Available at: www.odi.org/sites/odi.org.uk/files/odi-assets/publications-opinion-files/4600.pdf (accessed 17 March 2016).

African Union and UN Documents

African Union, *The Great Green Wall for the Sahara and the Sahel Initiative.* Available at: www.fao.org/docrep/016/ap603e/ap603e.pdf (accessed 17 March 2016).

UN – Intergovernmental Panel on Climate Change (IPCC): *Fourth Assessment Report* (AR4), New York: United Nations, 2007. Available at: www.ipcc.ch/report/ar4/ (accessed 17 March 2016).

UN – Intergovernmental Panel on Climate Change (IPCC): *Fifth Assessment Report* (AR5), New York: United Nations, 2014. Available at: www.ipcc.ch/report/ar5/wg1/ (accessed 17 March 2016).

UN – *Report of the UN Conference on the Human Environment,* Stockholm, 5016 June 1972. Available at: www.un-documents.net/aconf48-14r1.pdf (accessed 17 March 2016).

UN – Habitat, The State of African Cities 2014: Re-imagining Sustainable Urban Transitions. Available at: http://sd.iisd.org/news/un-habitat-calls-for-re-thinking-african-approaches-to-urbanization/ (accessed 17 March 2016).

UN – Department of Economic and Social Affairs, Population Division (2014). World Urbanization Prospects: The 2014 Revision, Highlights (ST/ESA/SER.A/352).

UNEP – AMCEN, *Atlas of Our Changing Environment* (Africa), New York, 2008.

UNEP Governing Council Resolution 9/7 of 28 May 1981.

13

ASIA

An indigenous cosmovisionary turn in the study of religion and ecology

Dan Smyer Yü

Asia is the home of two-thirds of the world's indigenous peoples. The total estimated indigenous populations of Asia are said to be around 260 million (IWGIA), mostly found in Siberia, South Asia, Southeast Asia, and Southwest China. Unlike their counterparts in the premodern era who relatively had more autonomy, the contemporary indigenous peoples of Asia are facing multiple distresses regarding the loss of their ancestral lands, the deprivation of their rights for self-determination, and the external exploitation of their natural resources. Modern nation-states often socially and economically marginalize indigenous populations. For instance, Karen in Burma are rarely given land and citizenry rights (Horstmann 2015, 130). China has fifty-five officially recognized ethnic minority groups. All of them except Hui (Muslims) are indigenous populations; however, the Chinese state does not officially recognize their indigenous status. Like the Chinese state, the Bangladesh government does not recognize its ethnic minorities as indigenous peoples but gives them marginal status as "tribes" and "minor races" (Dhamai and Chakma 2015, 314).

Due to their social marginalization, they have often been portrayed with a set of terms judgmental and alienating, such as "backward," "primitive," and "marginalized" marking them as an Other against the backdrop of modern progress. Scholars commonly recognize that indigenous peoples, including those in Asia, have been subjects of various civilizing projects (Harrell 1995, 3–36) and development aid programs. The modernized world is only beginning to recognize that many indigenous modes of being have preserved living knowledge that might prove to be invaluable in the current era confronted with problems that challenge the sustainability of sentient flourishing, ecological integrity, environmental health, and interspecies ethics.

In such a global context, I write this chapter as "a *radical rethinking* of the myth of progress and humanity's role in the evolutionary process" and "the *creative revisioning* of mutually enhancing human-earth relations" (Tucker and Grim 2001, xvii, xxii) in the growing study of indigenous religion and ecology in Asia. The added emphases are meant to spur a reorientation of localized indigenous ecological knowledge toward a pan-human approach to the current environmental crisis arising from multiple anthropocenic impacts, e.g. biophysical, climatic, atmospheric, and eco-psychological. While I revisit an important body of literature concerning the recent trends in the study of religion and ecology in Asian contexts, I wish to reconceive "the indigenous" or "indigeneity" not as an exclusive concept denoting peoples and practices outside those societies that politically and economically dominate their surrounding regions and nations or the rest of the world. Instead, I intend to re-identify "the indigenous" as an inclusive term with two significances. First, it encompasses peoples who possess place-specific but pan-humanly intelligible knowledge of the earth as a living being in ethical and spiritual terms, and those who no longer live on their native land due to migration but retain their ancestral memories of such animated human-earth relationship. Second, while we continue to advance the science of religious ecology advocated by Grim and Tucker (2014, 35–41), such reconceptualization of the indigenous allows us to exercise the study of traditional ecological knowledge not merely as a modern scientifically objective endeavor, but more critically as an empathetic understanding of the indigenous cosmovisions and ecological knowledge with a humanistic orientation (Berkes 1999; 2004; and Berkes and Davidson-Hunt 2006).

Rethinking of ecological elementals

Scholarly documentations and interpretations of Asian origin myths and indigenous ecological practices are undoubtedly abundant (Eliade 1978; Rappaport 1979; Leeming 2010); however, scholars' in-depth analyses of the inherent connection between religious cosmovision, human ecological behaviors, and environmental sustainability are rather a recent trend. This awareness has steadily arisen along with the publications on world religions and their ecological implications, such as those of Tucker and Williams (1997), Richard Foltz et al. (2003), and Roger Gottlieb (2006). However, publications on indigenous religions and ecology are relatively scarce. Grim's *Indigenous Traditions and Ecology* (2001) and Gottlieb's *This Sacred Earth* (2004) stand out as the leading collections of new perspectives. Both scholars express their profound care for the wellbeing of the earth and moral concern for the ongoing environmental crisis. The geographical coverage of both volumes is worldwide. The difference is that Grim focuses on the lifeways of indigenous peoples while Gottlieb on both world religions and traditional religions with short research essays and mythological narratives. While these two landmark publications have laid a foundation for the study of indigenous religion and ecology, Asian indigenous traditions proportionally remain on the margins of these two volumes. The indigenous traditions that most frequently appeared in these two volumes and elsewhere are those of the Americas and Australia.

In this regard, whether in Asian contexts or elsewhere, the study of indigenous religion and ecology is a matter of interdisciplinary pursuit inclusive of all indigenous traditions. The question then is how and where we reposition Asian traditions with their peers around the world. The reason why I raise this question is because I notice a sustained trend of accumulating Asian cosmovisionary narratives without enough complex identification of their mechanisms for sustainable modes of living relevant for the contemporary societies. For instance, Eliade's *From Primitives to Zen* (1978) and Leeming's *Creation Myths of the World* (2010) are the most impressive volumes collecting origin stories around the world. Mircea Eliade's copious works on origin myths laid a foundation for our systematic understanding of indigenous cosmovisions. The presence of the sacred is the center of Eliade's inquiry into the cosmological meanings of this world in connection with the invisible worlds of gods, spirits, and supernatural beings. His approach is consistently shown as an exegetical bridge between the world of the profane and the world of the sacred. Yet its gravity appears to lean toward a string of synonymous terms, such as "the ultimate reality" and "archaic humanity" (1959a: 3–5). All these conceptual terms point to one direction – the sacred as the primal mover of this world.

It is obvious that we do not lack documented narratives of indigenous cosmovisions. However, in our effort to identify their ecological significances, it should be pointed out that these collections of indigenous origin stories appear to emphasize passages that connect the separate worlds of the sacred and the profane but with little discussion on their ecological implications. Their approach is mostly centered on religion without including its relationship with ecology. This is where I wish to fill in the gaps by identifying the materiality of indigenous cosmovisions in ecological terms.

Gottlieb's volume reflects the growing trend of triangulating religion, human, and nature. He takes a step forward to emphasize the multifunctions of religion as the explanatory system regarding the origin of nature and as the medium connecting human existential values with nature (Gottlieb 2004, 54). What should be applauded is that Gottlieb and his contributors give their interpretive priority to the eco-logic of the sacred as the foundation of environmental sustainability (Sponsel and Natadecha-Sponsel 2004, 142). In the meantime, it should be also pointed out that this religion-based eco-logic seems to appear in the sacred realm only; thus, it does not extend itself to the sustainability of all spheres of natural and built environments. This prompts me to post these questions: Is sustainability limited to sacred sites? Are we scholars reflexively constrained in our culturally trained subjectivity in terms of the dichotomization of the sacred and the profane? Can we go beyond such dichotomy to address the linkages between indigenous cosmovisions and sustainable modes of living?

To address these questions, I propose a deeper understanding of what I call ecological elementals that are the building blocks of the earth and creative life forces that make the life-cycles of all species possible. These ecological elementals have been animated in many living cultures and remembered in the cosmological folktales of indigenous peoples. In my ethnographic work with Tibetans, I recorded a folktale concerning the earth as a supernatural living being whom humankind is dependent upon:

> Sadeg Doche is a god who lives inside the earth. The literal meaning of his name is 'big-bellied master of the earth'. Sadeg Doche does not reside in a particular part of the earth but wanders from one end of it to the other. The earth is his body. It has blood, flesh, and bones just like the components of a human body. No one would fail to protect his body...So when we fell forests on the surface of the earth or dig deep into the mountains we anger him. Every one of us is made out of earth, water, fire, and wind—the four elements. When we anger him, he mobilizes these elements as our enemies. This is seen when a fire swallows up an entire forest or a sudden earthquake turns what is on the earth's surface into a pile of ashes or a vast body of water. It is because Sadeg Doche shakes up the earth when he is angry.
>
> *(Smyer Yü 2014, 483–484)*

This narrative could be read as a fairytale that have little to do with our existential state of being; however, from the eco-religious perspective, they contain many expressions of the ecological elementals, such as water, earth, wind, and fire. These elementals are the basic composites of the earth and the human body. They could be seen as matters of inanimate nature from the perspective of modern physics; however, the indigenous cosmovision tells us otherwise: the earth and the humankind are two sentient species, which are bonded with each other based on mutual respect and affective expressions. It is precisely these expected moral and emotional fabrics that weave together a sustainable human–earth relationship. It is also the conceptual space for me to inform readers of the advantages and limitations of the comparative and anthropological approaches to indigenous cosmovisions and their ecological implications.

Comparative approach

In my case study of the Tibetan mountain community, I began with the comparative approach to focus on the mountains as sacred sites distanced from humans. I equated the verticality of the mountains with the potency of their sacredness as some of my peers. To many of them, "UP is the realm of the gods, from up the first kings often come, UP the ancestors reside; UP is the pure world" (Michaels 2003, 17). Thus, sacred mountains in Tibet are seen as pillar-like high grounds reaching into the heavens where the gods reside. In this sense, the sacred is vertically other-worldly and thus it is implied that anything this worldly is its opposite. This perspective seems to coincide with the Judeo-Christian dichotomized image of the cosmos, that is, the upwardness of God and the downwardness of the profane human realm. The apex of the vertically impressive mountain is therefore regarded as an "opening" (Eliade 1959b, 26) bridging the human realm with the gods but serving as a boundary separating each from the other.

I find that Eliade's idea of "opening" is best understood in horizontal rather than vertical terms in the Tibetan context, meaning that a mountain's perceived degree of sacredness is not necessarily in direct proportion to its height. It could also be understood as a mode of relationship rather than a demarcated physical place that contains the perceived sacredness. It consists of an ethically and spiritually interdependent relationship for all parties involved. Such relationship then signifies the communality of the earth, deities, and humans. It saturates a shared environment and a common inner realm. It is best to understand that the sacred in this regard is horizontally felt among humans.

When I situate Eliade's concept of sacred sites as openings in my case study, I see them as multidirectional connectivities with gods, the earth, humans, and other sentient beings, not just with gods alone. They signify an intersentient communication. The center of this multilayered communication is the local spirits and deities embodied in the surrounding landscape. If the sacred sites of monotheistic religions are the sites of theophany or disclosure of God above, the sacred mountains in Amdo represent what I call eco-theophany, as they and their deities choose to dwell in the local landscape and intimately bond with both humans and non-humans. The eco-theophanic nature of many Tibetan mountain communities lies in the lively traffic of natural, supernatural, and spiritual powers. The materiality of these powers is often manifested as wealth/poverty, health/sickness, fertility/infertility, and other bifurcated pairs contingent upon the presence or absence of the equitable relationship between the gods and humans. In this sense, the eco-theophanic landscape is an "energy zone" (Ivakhiv 2001, 228) of geological, ecological, sentient, and spiritual origins; thus the communality of gods, the earth, and humans is the defining force for a community of interbeing (Grim 2001, xxxvii).

For humans in this case, their mode of existence fits Eliade's idea of the "double." According to Eliade the "double" nature of human dwelling on earth signifies that the terrestrial and the celestial worlds are a linked pair comprising the visible and the invisible or the human and the divine, as he writes, "The creation is duplicated" (Eliade 1959a, 7). Its invisible blueprint is manifest in the physical architecture of the world, which is the ultimate hierophany commemorating the first mythical–physical act of the divinity (Eliade 1959a, 27–28). Through Eliade's conceptual perspective, I see the entwinement of the earth, the gods, and humans in the Tibetan case. The sacredness of the earth is no longer one-sided but reflects a multisided connectivity especially regarding the foundational elements of the earth including the intangible forces of the divinity, the cosmos, and human psyche. Herein, I also wish to point out that Eliade's collections of origin myths (Eliade 1963, 1978) show the animated ecological elementals, e.g. earth, water, and fire. However, in the comparative study of religion, these ecological elementals are mostly seen as the background of divine creative acts. Therefore, their ecological function largely remains unseen to the modern anthropocentric mind.

Anthropological approach

The Tibetan folk story of *Sadeg Doche* cited above was a part of my interview with an elderly Tibetan medicine man and a tantric yogi in the Amdo area, currently Qinghai Province of China. In his narrative, the earth is obviously sentient; however, my initially social scientifically trained subjectivity tended to negate its sentience with these analytical options: (1) the earth is not sentient but is perceived as the body of the deity *Sadeg Doche*; (2) the moral of the story is not so much about the animated nature of the earth but pertains to humans' self-centered, utilitarian needs for natural resources; so the preservation of forests and minerals in the name of *Sadeg Doche* reveals humans' inherent motivation for self-preservation; and (3) the scientifically recognizable part of the story is the four elements, namely earth, water, fire, and wind, and yet this could only qualify as an archaic science compared with modern science. To interject: my social scientific reflexivity revealed the systemic arrogance toward an ancient human eco-religious cosmovision. It negated the spiritual and emotional relations of the earth, humans, and invisible deities.

To justify my disciplinary viewpoint, I thought of adopting Roy Rappaport's paired models known as the operational model and the cognized model. The former is concerned with "the assumptions and methods of the objective sciences, in particular the science of ecology" (Rappaport 1979, 97) – whereas a cognized model signifies "a people's knowledge of their environment and of their beliefs concerning it" and "encodes values" (Rappaport 1979, 97, 101). In retrospect, both models reflect two ways of classifying cultural knowledge, namely the etic (non-native) and the emic (native). Such classification acknowledges a given local knowledge system but, at the same time, separates the native from the social scientist as if the latter belongs to a suprahuman category not susceptible to his or her own culturally trained worldview. The dichotomization of the native and the social scientist reflects the anthropologist's culturally conditioned subjectivity. It separates "scientific knowledge" from "native knowledge." The former is founded upon the latter and yet the latter only appears to be a subject matter to the inquiry of and representation by "the objective sciences;" thus, the native view is considered secondary in the arena of modern scientific discourses of environment, ecology, and sustainability. In this regard, a radical, inclusive social scientific approach to the study of indigenous religion and ecology is in demand.

A scientific cosmovision of indigeneity and traditional ecological knowledge

With this said, I wish to propose a scientific cosmovision of indigenous ecological knowledge, which is intended to expand the interpretive frameworks of different disciplines for more empathetic understanding of indigenous knowledge systems. Herein, the scientific cosmovision mostly refers to a science of religion and ecology that not only studies the micro cases of eco-religious practices of a given society but also bears ethical responsibilities for identifying the mechanisms of sustainable living embodied in them.

Science in this context does not exclusively refer to modern science but is rather a neutral term that includes indigenous knowledge systems. Most of indigenous societies did not have the word "science" but have managed to translate it from Western languages. This does not mean that they do not possess their own versions of scientific practices of natural and human knowledge in oral, written, or landscape marking methods. At the same time, to embrace indigenous knowledge systems with the proposed scientific empathy, it is also essential to recognize the culturally specific, indigenous roots of modern science. For instance, the ancient Greek root of modern science tells us that science was known as *epistēmē* or knowledge. *Epistēmē* does not stand alone but is rooted with *phusis* and *psuché* or nature and soul. From this indigenous European perspective, nature and soul both are primeval creative forces entwined with each other (Plato 2000, 239). In this regard, such a knowledge system is a science of nature and psych. Both mutually embody each other. In this process, the creative force in nature is recognized and revered as the inner principle of the cosmos. This indigenous root of modern science undoubtedly resonates with its peers elsewhere in the world. For instance, *reg-gni* in Tibet and *medleg* in Mongolia both refer to knowledge acquisition systems via living experience and education.

By pluralizing sciences with their indigenous roots and different orientations, I also wish to emphasize the meanings of what I call *indigenous experiences* and *indigenous memories* rather than continue to dwell on redefining who indigenous peoples are in ethnic, racial, and developmental terms. My premise is that everyone on earth is indigenous or a descendent of an indigenous people; indigenous eco-cultural values and practices are never static but change over time; and our ecological and spiritual empathy for environmental health and sustainable modes of being is predicated upon the antecedence that we are an environed species. Thus, an indigenous experience in the ecological sense is understood either as a directly lived experience in an ancestrally defined environment or as an empathetic resonance with that experience but based on one's associative ancestral memories of similar experiences retained in one's oral or written cultural history.

In my ethnographic study of the Tibetan village as forementioned, the meaning of being indigenous has evolved with changing historical and religious circumstances and conditions. As the village is located in the far-eastern part of Amdo, its religious history has evolved with two strands of changes. One is Tibetans' overall conversion process from Bonpo to Buddhism since the seventh century (Rambo 2008, 215). Another is the historical encounter between Tibetans, imperial China, the Mongol empire, and the Islamic world. If I look at the religious history of the region from a static and isolated perspective, the village could not be qualified as an indigenous community; however, the actual practice of Buddhism among villagers rather points to the presence of pre-Buddhist gods and spirits. The Buddhist conversion process is not a clear-cut shift from one religion to another; instead, it is a hybrid process with the combination of both Bon and Buddhist influences (Smyer Yü 2011, 58) or a syncretic meeting point of two religions with their respective indigenous origins.

For instance, mountain gods had long been in existence in pre-Buddhist Tibet. They were regarded as supernatural beings; however, after undergoing Buddhist conversion, they have retained their supernatural power but are now subject to their human counterparts' Buddhist ritual power that is purported to harness their supernatural power for worldly needs, e.g. prevention of illness and insurance of harvests. On the level of local cultural history and ancestral memory, the local mountain deities are all descendants of Ode Gungyal, the original Tibetan mountain deity. Mt Ode Gungyal is located in southern Üzang in the current Tibetan Autonomous Region. The name 'Ode Gungyal' is a pseudonym of Mutri Tsempo, the second king of ancient Tibet, who, according to the historical record, lived from 345 to 272 BCE. From the historical perspective, many mountain deities in Tibet are both gods and humans, with human ancestors immortalized by being incarnated as deities. In this sense, the 'bloodline' of mountain deities is entwined with the ancestral line of humans. Local knowledge is then a set of complex relationships between gods, nature, and human history. Humans entrust their ancestral memories to the gods, and the mountainous landscape is the shared dwelling space of gods and humans. The local environment and human livelihoods are sustained within an equitable but constantly negotiated relationship between these elementals of human dwelling (Smyer Yü 2011, 62–69).

The spiritualized nature of the Tibetan traditional ecological knowledge is not unique but also found in other Asian indigenous communities. In East Malaysia, the Penan landscape embodies not merely ecosystems, natural sustenance, human history, and cultural changes, but also the supernatural presence of gods and spirits in the eco-geological features of their rivers and forests (Brosius 2001, 140–148). In Piers Vitebsky's ethnographic study of the Eveny in Siberia, he finds that the human existence is deeply entwined with both domesticated and wild reindeers. The reindeer often mediates the human–earth relationship via shamanic visions besides the utilitarian provision humans receive from the reindeer. The partnership of humans and the reindeer is lodged so deeply in their existential and spiritual relationships that human consciousness of the earth continues to express itself through the symbol of the reindeer (Vitebsky 2005, 212–230).

The cosmovisions, which are documented in the cases of the Penan landscape, the partnership of the Eveny and the reindeer, and the Tibetan folk story of *Sadeg Doche*, are similarly found in the Vedic tradition in India, revealing the earth's animated mode of being in which "the cosmos is seen as a great being" (Dankelman 1995, 46). This Vedic consciousness continues to find its presence in contemporary rituals, such as fire and agni rituals in India. Likewise, the living nature of the earth is also found in how the Altaians in Siberia regard the Katun River and their forest as living beings. When they cross or drink from the river and when they hunt or pick medicinal herbs from the forest, they ask permission and utter prayers (Klubnikin et al 2000, 2000–2001). These case studies of contemporary indigenous eco-religious practices attest to the vitality of their cosmovisions even under adversarial social conditions.

These indigenous traditional ecological knowledge show both place-based manifestations and universal ethical and spiritual orientations. The earth, gods, humans, and non-humans are in "a negotiated order" (Gray 1999, 73). It displays ecological elementals as both natural creative matters and spiritual forces (Posey 1999, 4). These forces constitute the earth and sustain its continuity and the lives of its myriad residents. Traditional ecological knowledge in this sense is concerned with what Fikret Berkes calls "a knowledge-practice-belief complex" (Berkes 1999, 13) and what Grim characterizes as "the interbeing of cosmology and community" (Grim 2001, xxxvii). The shared point of both phrases is that there exists complex interdependence and mutual embodiment of the sensible order of things found in both the modern scientific perspective and the spiritual and affective order of things from the ancient but continuously present indigenous scientific perspective.

The recognition of how all aspects of life are entangled with one another activates what Grim calls "an environmental imagination" and "deeper moralization" (Grim 2001, lv). The former unites indigenous ecological practices with the global environmental concern, while the latter enacts "a creative behavior that not only responds to the concerns of place, knowledge, and sovereignty of indigenous peoples, but also collaboratively explores visions of flourishing life" (Grim 2001, lv). Thus, for the beginning of a new, holistic science of eco-religious cosmovision, the separation of modern science and ancient indigenous science needs to end, not only in Asia but also elsewhere in the world.

This new scientific cosmovision is already underway thanks to the collective effort of scholars and the environmental concerns of the public. I thus summarize it in eco-spiritual and applied terms: From the perspectives of environmental ethics (Tucker and Grim 2001, 2) and spiritual ecology (Sponsel 2012, 13), this new scientific cosmovision is a science of holistic awareness or inner ecological civilization (Dankelman 1995, 46) committed to relearning and experimenting with new meanings and applications of the eco-spiritual wisdom of both ancient and modern origins in our renewed appreciation of the earth as a living being antecedent to the births of diverse geological formations, ecosystems, and civilizations. From an applied perspective, this new scientific cosmovision pertains to a body of diverse ecological worldviews and their culturally embodied practices and subsequently yielded knowledge with a teleological orientation toward the restoration and the preservation of natural and human heritages as ways of sustaining the health of the earth and its affective bond with humankind and all other species.

References

Berkes F. (1999) *Sacred Ecology: Traditional ecological knowledge and resource management*, Taylor & Francis, London.

Berkes F. (2004) "Rethinking Community Based Conservation", *Conservation Biology* 18:3, 621–630.

Berkes F. and Davidson-Hunt I. (2006) "Biodiversity, traditional management systems, and cultural landscapes: examples from the boreal forest of Canada" *International Social Science Journal* 58:187, 35–47.

Brosius J. P. (2001) "Local knowledges, global claims: On the significance of indigenous ecologies in Sarawak, East Malaysia", in Grim J. ed. *Indigenous Traditions and Ecology: The interbeing of cosmology and community*, Harvard University Press, Cambridge 125–157.

Dankelman I. (1995) "Culture and cosmovision: Roots of farmers' natural resource management", in Oglethorpe J. ed. *Adaptive Management: From theory to practice*, (Sui Technical Series Vol. 3) World Conservation Union, Gland: Switzerland 41–52.

Dhamai B.M. and Chakma P. (2015) "Bangladesh", in Mikkelsen C. ed. *The Indigenous World 2015*, Eks-Skolens Trykkeri, Copenhagen 314–320.

Eliade M. (1959a) *Cosmos and History: The myth of the eternal return*, Harper Torchbooks, New York.

Eliade M. (1959b) *The Sacred and the Profane: The nature of religion*, Harper & Brothers, New York.

Eliade M. (1963) *Myth and Reality*, Harper Torchbooks, New York.

Eliade M. (1978) *From Primitives to Zen: A thematic sourcebook of the history of religions*, HarperCollins, London.

Foltz R. C., Denny F. M. and Baharuddin A. eds. (2003) *Islam and Ecology: A bestowed trust*, Harvard University Press, Cambridge.

Gottlieb R. ed. (2004) *This Sacred Earth: Religion, nature, environment*, Routledge, London.

Gottlieb R. ed. (2006) *The Oxford Handbook of Religion and Ecology*, Oxford University Press, Oxford.

Gray A. (1999) "Indigenous peoples, their environments and territories", in Posey D. A. ed. *Cultural and spiritual values of biodiversity*, United Nations Environment Programme, Nairobi, Kenya 61–138.

Grim J. (2001) "Introduction", in Grim, J. ed. *Indigenous Traditions and Ecology: The interbeing of cosmology and community*, Harvard University Press, Cambridge xxii–lxiv.

Grim J. and Tucker M. E. (2014) *Ecology and Religion*, Island Press, Washington.

Harrell S. (1995) "Civilizing projects and the reaction to them", in Harrell S. ed. *Cultural encounters on China's ethnic frontiers*, University of Washington Press, Seattle 3–36.

Horstmann A. (2015) "Secular and Religious Sanctuaries: Interfaces of Humanitarianism and Self-Government of Karen Refugee-Migrants in Thai-Burmese Border Spaces", in Horstmann A. and Jung J.H. eds. *Building Noah's Ark for Migrants, Refugees, and Religious Communities*, Palgrave Macmillan, New York 129–156.

International Work Group for Indigenous Affairs (IWGIA), Asia (www.iwgia.org/regions/asia). Accessed 1 June 2015.

Ivakhiv A. J. (2001) *Claiming Pilgrims and Politics at Sacred Glastonbury and Sedona Ground*, Indian University Press, Bloomington.

Klubnikin K., Annett C., Cherkasova M., Shishin M., and Fotieva I. (2000) "The sacred and the scientific: Traditional ecological knowledge in Siberian river conservation", *Ecological Applications* 10:5, 1296–1306.

Leeming D. (2010) *Creation Myths of the World: An encyclopedia*, ABC-CLIO Inc, London.

Michaels A. (2003) "The sacredness of (Himalayan) landscapes", in Gutschow N., Michaels A., Ramble C. and Steinkellner E. eds. *Sacred landscape of the Himalaya*, Austrian Academy of Sciences Press, Vienna.

Plato (2000) *Laws* trans. Jowett B. Prometheus Books, Amherst.

Posey D. A. (1999) "Introduction: Culture and nature – The inextricable link", in Posey D. A. ed. *Cultural and Spiritual Values of Biodiversity,* United Nations Environment Programme, Nairobi, Kenya 3–16.

Rambo C. (2008) *The Navel of the Demoness: Tibetan Buddhism and civil religion in Highland Nepal,* Oxford University Press, Oxford.

Rappaport R. A. (1979) *Ecology, Meaning, and Religion,* North Atlantic Books, Berkeley.

Smyer Yü D. (2011) *The Spread of Tibetan Buddhism in China: Charisma, money, enlightenment,* Routledge, London.

Smyer Yü D. (2014) "Sentience of the Earth: Eco-Buddhist mandalizing of dwelling place in Amdo, Tibet", *Journal for the Study of Religion, Nature and Culture* 8:4, 483–501.

Sponsel L. and Natadecha-Sponsel P. (2004) "Illuminating darkness: the monk-cave-bat-ecosystem complex in Thailand", in Gottlieb R. ed. *This Sacred Earth: Religion, nature, environment,* Routledge, London 134–144.

Sponsel L. (2012) *Spiritual Ecology: A quiet revolution,* Praeger, ABC-CLIO, LLO, Santa Barbara.

Tucker M. E. and Ryuken D. W. (1997) *Buddhism and Ecology,* Harvard University Press, Cambridge.

Tucker, M. E. and Grim J. (2001) "Series Forward", in Grim J. ed. *Indigenous Traditions and Ecology: The interbeing of cosmology and community,* Harvard University Press, Cambridge, xv–xxx.

Vitebsky P. (2005) *The Reindeer People: Living with Animals and Spirits in Siberia,* HarperCollins, New York.

14

PACIFIC REGION

In search of harmony: Indigenous traditions of the Pacific and ecology

Mānuka Hēnare

Harmony and economics of traditional religion are key themes of this chapter on Pacific ecology, with trees as the case study and the impact on ecology. On large and small islands, atolls, and other places, coconut trees and large canopy kauri trees provide humans with food, shelter, and identity. These identities connecting trees, cultures, and ecologies have shaped the cultures of the peoples of the Pacific (Tui Atua, 2007). Our primary focus in this chapter is the Māori, my people, of Aotearoa.

Māori are the *tangata whenua*, the people *of* the land, and people who *are* the land, of Aotearoa-New Zealand. Māori have settled and developed the land since ancestors from East Polynesia arrived some 1,000 years ago. Notwithstanding the Aotearoa sojourn, the genetic and genealogical taproots are much older. Before becoming Māori of Aotearoa the earlier Pacific ancestors were described by linguists as Austronesians, meaning, first, a language family, and second, the peoples and cultures of the many islands and languages of Island South East Asia to Madagascar. Austronesian peoples entered the Pacific some 3,500 years ago after a 40,000 year sojourn in South East Asia (Feinberg and Macpherson, 2002: 101, 106). Over their time of constant occupation of Aotearoa, Māori developed a distinctive traditional Polynesian culture and economy, wrapped in its own religion and spirituality. Religious concepts such as *tapu*, *mana*, *mauri*, *hau*, *wairua* are not autochthonous to Aotearoa, but rather to Island SE Asia and atolls and islands of *Te Moana Nui a Kiwa*, the Great Ocean of Kiwa. Today this is known as the Pacific Ocean.

Oral histories inform us that Māori are people of the Pacific Ocean—the Great Ocean of Kiwa. Kiwa is one of the children of Sky Father and Earth Mother and is responsible for the domain of the oceans (Best, 1982). Pacific Island cultures emerged during thousands of years of constant habitation on small islands, atolls, and reefs that are spread over thousands of kilometres of an oceanic world and its multiplicity of ecosystems and species diversity (Rappaport, 1979: 4–5; South Pacific Regional Environment Programme, 1992: 7, 9). In this environment, distinctive human cultures developed and are often referred to as Micronesians, Melanesians, and Polynesians. Māori of Aotearoa are Polynesians.[1]

Archaeologists, geneticists, and oral traditions tell us that the first human inhabitants of Aotearoa were East Polynesians who brought with them distinctive Polynesian cosmology and mythology, including the concepts and beliefs of *tapu*, *mana*, *mauri*, *hau*, and *wairua* (Davidson, 1987: 171, 22). The genetic trail is some 19,000 years duration (O'Connell and Allen, 2004).

Further religious developments took place after settlement in Aotearoa as the founding settlers quickly adapted to the totally new environment, which contrasted with the earlier habitation on smaller islands and atolls of East Polynesia, and Island South East Asia

Cosmovision

Māori cosmology is the particular way that Māori view the world in terms of the ultimate metaphysical principles relative to their material environment. Cosmology is a theory, which in the Pacific is a mix of oral tradition and modern Asian and Western sciences. The creation story of the universe is the Pacific cosmogony

Traditional Māori religion is a belief in spiritual beings with its own order of reference that is meaningful and real to the people (Geertz, 1973; Tylor, 1871). The cosmogony refers to the creation stories and beliefs, and the genealogical links between the seen and unseen worlds (Schrempp, 1992: 4, 55, 114). Māori tradition views the cosmos as being dynamic rather than static and that the world of the living was derived through the genealogical links from the creative powers referred to in the cosmogonic stories (Irwin, 1984). Thus, in describing Māori cosmology, the narrator is at the same time characterising the socio-political processes of the culture.

The cosmic religious worldview of Māori is as old as the culture itself and constitutes a philosophy, which is a love of wisdom and search for knowledge of things and their causes (Williams, 1983: 235). At the heart of this view of the creation process is an understanding that humanity and all things of the natural world are always emerging, always unfolding (Prytz-Johansen, 1954: 40–47). Within this knowledge and enlightenment seeking framework, Anne Salmond explains that Māori and other peoples of the Pacific have "their own ideas on how relations between people and between people, earth and sea must be conducted" (Salmond, 1997: 176).

Worldview and cosmos

Creation accounts are the foundations upon which Māori build and maintain a distinctive cosmological, religious philosophy and metaphysics. The basis of a Māori philosophy of vitalism is the idea that all things in creation, whether material or nonmaterial, contain a life-force that is independent of the thing itself. This life-force emanates from the original source of life itself, namely the Supreme Being, Io-matua-kore, Io who has no parents.

Worldview, ethics, values, morals, and associated cultural practices are integral components of Māori ancestral legacy that preserve both unity and identity with roots in and continuity with the past. They are the signal of where Māori are in the present. According to oral tradition, indigenous worldviews transmit certain crucial features (Marsden and Henare, 1992). First, myth and legends are neither fables nor fireside stories; rather, they are deliberate constructs employed by the ancient seers and sages to encapsulate and condense their views into easily assimilated forms of the world, of ultimate reality, and of the relationships between the Creator, the universe, and humanity. Worldview, then, is at the heart of Māori culture, touching, interacting with, and strongly influencing every aspect of it.

Philosophically, Māori do not see themselves as separate from Te Ao Mārama, the natural world, being direct descendants of Earth Mother and Sky Father (Royal, 2015). Thus, the resources of the earth do not belong to humankind; rather, humans belong to the earth. While humans as well as animals, birds, fish, and trees can harvest the bounty of Mother Earth's resources, they do not own them. Instead, humans have user rights, which Māori have recorded in cosmic and genealogical relations with the natural world (Marsden and Henare, 1992).

Oral language is full of the very best of metaphor and symbolism, which alludes to Māori history and fundamentals of cosmological thought. Māori history starts in the time before creation, and progresses through to the birth of the mythical and original homeland, called Hawaiki—a place distant in time and space, which is the link with the spirit world. In this sense, Hawaiki is both a cosmic place and a physical place on earth.

In Māori mythology the creation of the universe begins with the Supreme Being Io-matua-kore and continues through Ranginui, the Sky Father, and Papatūānuku, the Earth Mother. These two were lovers locked in an age-long embrace, during which they had many children. The offspring lived between them, becoming the spirit beings and caretakers of the sea, winds, forests, wild foods, crops, and humanity; the progenitors and personification of the world and its environment as we know it. After living in continuous darkness and being totally inhibited by their parents, the children separated Rangi and Papa, creating a world of light. In doing this, Whenua Hawaiki, the New World, the original homeland of the Māori, came into being. Thus the primal parents and their children are the source of all life, and everything in the world, including ideas and language, has a life essence that maintains its existence.

Philosophically, the sundering of the parents and the concomitant burst of light into the cosmos is the spark that started life for plants, fish, birds, and people. According to Māori narratives, the cosmos started with a burst of primal energy. Like a wind it swept through the cosmos bringing freedom and renewal. Once established in the new milieu, the power could be called upon by humans and transmitted through ritual pathways into receptacles, such as stones or people (Salmond, 1997).

This account of the origins of the cosmos, the world, nature, and humanity are at the heart of Māori worldview. It provides a sense of unity, identity, and continuity with the past. Everything in nature has its own special *tapu,* which is derived from the primary *tapu* of the particular progenitor. In this way kumara, a sweet potato, participates in the *tapu* of Rongo, persons of Tū, fish and reptiles of Tangaroa, the winds of Tāwhirimatea, birds and forests of Tāne, and fernroot of Haumia (cf. Shirres, 1982: 42). Māori conceive of the universe as a two-world system in which the material proceeds from and interacts with the spiritual. Primacy, however, rests in the spiritual sphere.

Kupe and Kuramārōtini

Let me take you back to around 925 CE, the time when the great Polynesian trader explorer, Kupe, and his wife, Kuramārōtini, voyaged the wild Pacific seas filled with creatures and sea demons to discover Aotearoa and many other islands. Travelling over 8,000 kilometres from the North East Pacific, from the earthly Hawaiki, in their ocean going *waka,* or canoe, *Matawhaorua,* they set their sights on a long, large white cloud. Kuramārōtini, excitedly pointing, named the land beyond—Aotearoa: Land of the Long White Cloud.

Imagine what they sighted upon their arrival: a large, complex ecosystem and environment alien to their homeland of Hawaiki. Spanning 1,600 kilometres from north to south, a land covered in tangled native forest, dense and dark, alive with an abundance of unknown birds, insects, and lizards. Sometimes impenetrable to humans, tall trees towering above shrubs and younger trees, shading ferns and mosses on the peaty ground illuminated faintly by the shards of sunlight through the foliage.

Large trees, later identified and named by other generations of Māori, provide the canopy covering the main broad-leafed species. In the coastal areas they sight for the first time the ancestral native trees, particularly the *pohutukawa* (*Metrosideros exceisa*), all much revered today, and, along the muddy margins of the coast, mānawa or mangroves, which can have a life span of a million years.

The ancestors of Kupe were coastal dwellers and never lived within the forest, but they used its plants for food and medicines, and hunted the birds that lived within the bush. They believed it was the home of *atua*, spiritual beings, a wondrous place that necessitated caution, where an unsuspecting warrior could get lost and be swallowed up, never to be seen again. Conversely, when Anglo-European settlers arrived, they found the bush gloomy, frightening, and irritating, clearing vast areas to create open pastures.

After circumnavigating the islands of Aotearoa, Kupe and family return to Hawaiki, recounting the wonders of this far land. In time, their descendants migrated to the new land to settle according to both patrilineal and matrilineal kinship groups known as *whānau*, *hapū*, and *iwi*. The historical link to the landscape was established and consummated within these early encounters and migrations and became the foundations upon which the kinship system prospered. The land is seen, not as a commodity, but as a part of the individual and tribal self. Names given in the founding times remain important today for local communities who identify themselves with the history of their ancestors.

The importance of the land

Māori see the land as a material manifestation of the spiritual, and a person's identification with their tribal land provides both spiritual and physical nourishment. Indeed, the Māori word *whenua* means both 'land' and 'placenta'. The land is as important as the placenta that nourished them when they were in the womb. It is a force that relates to and interacts with Mother Earth's life essence (J. Hēnare, 1981, 4 July).

The environment provides people with signs of and signals from the spirit world. Omens are seen in natural phenomena, such as lightning, the presence of certain birds, fish, or animals, the condition of forests, or particular plants. Every kinship group identifies particular locations as the dwelling place of guardian spirit beings, called *kaitiaki* and *taniwha*. Kaitiaki are frequently identified with specific animals into which the soul of an ancestor has passed. Each community has its own named kaitiaki. Taniwha are creatures who dwell in pools, rivers, or the ocean, or sometimes they take many forms of inanimate objects such as stones, trees, or logs.

The landscape itself was the setting for the cosmological and mythical dramas of creation. The landscape has its own story of origins. The mountains, rivers, lakes, and rocks were created in specific acts, frequently involving atua and ancestors of a heroic age. However, in terms of the Māori worldview, people and the natural world are in a state of harmony or balanced equilibrium towards each other.

Kauri: a canopy tree of significance

According to Māori philosophy and understanding of the spiritual world, cosmos, and the natural world, all trees come from the spiritual world and over time are gifted by spiritual ancestors to the natural, material worlds and humanity. The Kauri tree is among the most ancient in our part of the Universe. Its antecedents appeared during the Jurassic period, between 190 and 135 million years ago. It is thus a *taonga tuku iho*, a treasured gift of the ancestral spiritual world of the Supreme Being, Io Matua Kore, followed through aeons to Ranginui and Papatūānuku, Father Sky and Mother Earth, and their child Tāne. All Kauri and other trees attest to the wonder of evolution, the ability of life to adapt to unexpected challenges, and to perpetuate itself over vast periods of time. Rooted securely in the earth, Kauri, like other trees, reaches for the heavens.

Kauri, as do other trees, has a *tinana*, rendered as its body; it is imbued at its birth with its own tapu, meaning its potentiality to be a remarkable tree, and it is the tapu of Tāne that gives to the Kauri its primary mana, its authority and status as a *rangatira* of the forest. The Kauri is a child of Tāne, Rangi, and Papa, and ultimately of Io Matua Kore. From its conception it receives its own *mauri*, or life-force, from Io Matua Kore who inspired its name, Kauri. In Māori metaphysics, the name is closely connected with the very being of things of creation. The action of Io Matua Kore by giving the name established its tapu, mana and wairua, and in so doing its identity. It is from this action the Kauri takes its unique *ahua* or form.

It is the mauri that defines the nature of the Kauri, thus determining its specific characteristics. The mauri also binds the tinana to its wairua which is the spiritual kaitiaki or guardian that guides and protects it throughout its long life. Each Kauri has its own *hau*, another life-force of Io Matua Kore which guides its capability to be productive as a Kauri tree. It is this productivity that gives sustenance and life to associated flora and fauna. Furthermore, the production of its unique male and female cones is the resilience expected of such a tree. The death of the Kauri occurs when its mauri separates itself from the tinana, for various reasons, thus releasing the wairua of the tree to return to its spiritual source. In this ethical view of life and of the natural world, all trees such as Kauri come from the spiritual world and are gifted to the cosmos, the natural material world, and humanity.

This combination of unique spiritual, genetic, and biological attributes enables each Kauri to grow according to its nature to be a *rākau rangatira*, a canopy tree of significance. Over its full life of 500 to 2,000 years it becomes a *kāinga*—a habitat and home to a succession of creatures and other plant species. It thus protects and sustains specific birdlife such as Kaka, Kereru, and other birds; its own family of insect life; and associated shrubs such as tree ferns, Nīkau palms, orchids, vines, and flowers. Finally, it creates continuously its own *pā-kāinga*, a habitat around itself nurturing other native and endemic tree species of the ecology, namely, *Rimu, Tōtara, Miro, Pūriri, Taraire,* and *Kohekohe*. It is in its nature to live a very, very long life, and grow to immense proportions, both width and height. According to tradition throughout its life cycles each Kauri is protected by its own wairua, its spiritual guardian from Io Matua Kore, and kaitiaki tangata, the human guardians.

The demise of the Kauri and its forest companions over the last 200 years is a consequence of a variety of attacks on each of its individual attributes and also its spiral or matrix of attributes, whether by nature, human, animal, or disease. We are watching the near destruction of a sacred species. Who hears the pain and agony of a dying Kauri tree?

The loss of the Kauri to *Māoritanga* is akin to the loss of a rangatira or outstanding leader who has died. Historically and scientifically, Kauri trees are significant to the ecology of Aotearoa forest, particularly in the warmer zones. With the potential demise of the Kauri tree there is a likely demise also of other flora and fauna that have symbiotic relationships between them and the tree. It is the loss of the lifeforce, or the vital essence, of the tree that matters.

Religion, philosophy, ethics—a philosophy of vitalism

Within the culture and its language a religion emerged. With its sense of the holy, beliefs, rituals, and ethics, it continues to be significant in the lives of Māori. A cosmic outlook is at the core of the Māori religious and philosophical worldview (M. Hēnare, 1999: 52). Philosophically, Māori and other Pacific peoples are vitalists, acknowledging daily in ritual prayer, speechmaking and song an understanding of *te tangata*, the human person, and creation. For Māori philosophical anthropology, 'life is an absolutely originary phenomenon, irreducible to matter', and is traced to a Supreme Being (Mondin, 1985: 26–27). It follows that mechanistic explanations of life make little sense in Māori thought. The traditional religions of the Pacific are not found in any written text or a set of dogmas, but is found in a close study of *Māoritanga*, meaning the total way of life itself. Both the religion and philosophy have within them a discernible set of ethics and forms of virtuous behaviour, which are embedded in the language and sketched as follows:

- *te ao mārama*, the cosmos, seeking enlightenment;
- *mauri*, life essences in all things material and non-material, vitalism;
- *tapu*, the state of being and potentiality, sacred, holy;
- *mana*, power, spiritual authority and concern for the common good;
- *hau*, the spiritual source of obligatory reciprocity in relationships and economic activity;
- *te ao hurihuri*, change and tradition;
- *wairuatanga*, spirituality, the recognition that all things in creation have a spiritual dimension;
- *whānau*, the extended family as the foundation of society;
- *whanaungatanga*, belonging to and maintaining the kinship system;
- *kotahitanga*, solidarity;
- *manaakitanga*, quality care, kindness, hospitality; and
- *kaitiakitanga*, the guardianship of creation.

All these are recognized as a membrane of divine and human virtues of traditional contemporary life. They are tangible signs of a person's or group's mana, or well-being and integrity. Mana is recognized for its active manifestations and is always closely linked to the powers of the spiritual ancestors. It is a quality that cannot be generated for oneself; rather it is generated by others, and is bestowed upon individuals and groups. To understand mana is to gain a pathway into the Māori, Polynesian, and Austronesian worldviews.

The life force

The belief and respect given to the mauri, the life force, and life itself, is deep in Māori consciousness. Mauri is the life force of all things inanimate and animate, and the basis to the wellbeing of the people and the tribe. The natural world, no less than humans themselves, proceeds from the spiritual. Both the human person and all created things have material and spiritual aspects. The spiritual essence or life force is known as mauri. Trees, forests, and birds, the ocean and its creatures, and gardens and the work of human hands, all have their own mauri, which maintains their existence. It is the mauri of people and things that sorcerers sought to destroy.

Mauri can be established for such things as a new house and it is often in the form of a stone. This mauri embodies the life force, the mana (power) of the house. Its purpose is to assist in the fulfilment of the hopes and aspirations of those belonging to it (Barlow, 1991: 128). The oceans, rivers, and forests also have their mauri and this is the life force of these resources. When the mauri is endangered and stocks have become depleted through carelessness or overuse, it can be restored through rahui, ritual restriction, and karakia. The restored mauri can ensure that depleted food supplies, such as fish, shellfish, or birds, can become abundant again (Barlow, 1991: 83; cf Ranapiri quoted in Best, 1909: 439; Gudgeon, 1905: 128).

Closely associated with mauri is hau, glossed as the very essence of vitality (Barlow, 1991:

83; Best, 1909: 439; Gudgeon, 1905: 128). There would appear to be two connected but important distinctions to be made about hau. The hau of things is different from the hau of persons. When applied to things, hau and mauri often seem to be one and the same. Both can be glossed as life principle (Gathercole, 1978: 338). Thus there was a life principle of a forest or sea. Due to the tapu and mana of these aspects of the environment, maintenance of the hau of a forest, a river, or the sea, is considered an imperative by Māori. This is because of the association with food supplies.

Prior to fowling or fishing or other forest activities, *tohunga* or spiritual leaders would check the appropriate mauri because it caused birds, fish, or trees to be in sufficient supply. This enabled the people to have food supplies or other materials for daily living. *Karakia* would be given by the tohunga and others, and later, after catches or resources were taken, another ritual would take place. This was *whangai hau*, feeding the hau, which entailed offerings to the forest or sea. Often the first catches would be returned to their source or put aside for ritual use. In these ways, both the hau of the forest or sea products, and the mauri, may return to the forest or sea. That is, in this reciprocal exchange, the mauri and the hau return to the original source of the life force (Best, 1909: 439).

The power to act on the world is to some extent at least a spiritual quality, referred to as mana. While human beings have intrinsic psychic powers or *ihi*, which enables them to act decisively in a natural way, mana by contrast is a uniquely spiritual force that comes from the atua. However, mana is itself a manifestation of the power of being, or existing. The recognition of that existence, and of its potentiality for creative power, was expressed in the concept of tapu, namely, a being with the potentiality for power (Shirres, 1982: 46).

Ecological economics

Māoritanga is the way in which Māori ensure their customary knowledge, values, and principles are adhered to in an appropriate modern context. We have responsibility to the environment and community, and must balance the growth of the Māori economy and compliance to Treasury.

A Māori worldview finds kinship with ecological economics (Costanza et al., 2015: 119–20), the proposition being that economies exist in the ecology, and not the other way around. Thus forestry can be considered as part of nature capital. Further research that integrates ecology and economics and humanity's impact on the natural world; historically and in the future, will in this way lead to a greater understanding of both the natural and economic value of the trees and the forest and its specific integral part in the ecological system of Aotearoa. Furthermore, such research will identify its economic values of significance.

In traditional Māori economics, that is, the Economy of Mana, there are two significant understandings of economy and the values that inspire its productivity. First, the economy is embedded in society and the values of that society inform the economy. The Economy of Mana, also referred to as *he whenua rangatira*, is inspired by the worldview of the first Aotearoa-New Zealanders and its four well-beings—spiritual, environmental, kinship, and economic. The second understanding is that the Economy of Mana is embedded in the ecological system that sustains it. These two beliefs integrate methods for understanding and promoting regional resilience and transformations relevant to the survival, sustainability, and productivity of the forests of Aotearoa.

In this framework of the economies of Mana embedded in society and ecology, trees are part of the natural capital of nature itself in the habitats. Forests are structurally complex, with many different tree species per hectare. These ancient economies are significant, first to the ecological economics of each region, and second to the kaitiaki, or guardian function of the kinship family systems of the regions. The demise of ecology of the forest and the economy will lead to the diminution of the Ecological Economy of Tai Tokerau Forestry and to the stock of natural capital of trees. The Kauri trees were a particular interest to earlier English explorers like Captain James Cook in the 1700s and later explorers who saw the Kauri as a tree of resilience and strength. Hundreds and thousands of feet of Kauri timber was cut and sent to be used as masts and spars in the rebuilding of the British Navy fleet, which were to prove a decisive factor in Lord Nelson's defeat of the French in the Battle of Waterloo.

The above understanding of the nature of trees and forests is evidence of the flow of valuable ecosystem goods or services into the future. Furthermore, a stock of trees and associated flora and fauna ensures a flow of future trees as well as the flora and fauna, thus creating a flow of sustainable nature capital. This is natural resilience in transformation. Trees as natural capital also provide services such as recycling wastes, water catchment, erosion control, and a haven for other species, such as birds, insects, and other plants. According to ecological economic thinking, the flow of services from ecosystems requires these systems to function as a whole and thus the structure and diversity of the systems are important components of natural capital. Each tree is thus individually and collectively significant.

The loss to the ecological economy of Aotearoa and the loss to the kaitiaki function of Māori are separately and collectively significant. As stated, all trees have value in themselves and contribute to both the Ecological Economy of forestry and the Economy of Mana, and enhance the total sustainable future of the cosmos, natural world, and humanity. Should the forests suffer through an overreliance on one dominant species and forests of diverse species as we have known over the millennium, the Ecological Economy of Aotearoa forests suffers and its natural capital is greatly diminished. In turn, the ability of the kaitiaki, both spiritual and human, to carry out their function as caretakers are profoundly diminished also. The restoration of Kaitiaki capabilities is an agenda of the current times for the Pacific-Austronesian peoples and economies of mana.[2]

Notes

1   The categories of Polynesian, Melanesian, and Micronesian are not Pacific indigenous peoples' classifications.
2   This chapter has been adapted from earlier work: (M. Hēnare and Kernot, 1996; Henare, 2001). My thanks to Amber Nicholson at the University of Auckland Business School for her contributions in this chapter.

References

Barlow, C. (1991). *Tikanga Whakaaro: key concepts in Māori culture*. Auckland, NZ: Oxford University Press.
Best, E. (1909). Māori Forest Lore: Being Some Account of Native Forest Lore and Woodcraft, as also of many Myths, Rites, Customs, and Superstitions Connected with the Flora and Fauna of the Tuhoe or Ure-wera District. *Transactions of the New Zealand Institute*, 42, 434–481.
Best, E. (1982). *Māori Religion and Mythology: Being an account of the cosmogony, anthropogeny, religious beliefs and rites, magic and folk lore of the Māori folk of New Zealand*. Wellington: Govt. Printer.
Costanza, R., Cumberland, J. H., Daly, H. E., Goodland, R. J. A., Norgaard, R. B., Kubiszewski, I., and Franco, C. (2015). *An Introduction to Ecological Economics* (Second edn). Boca Raton, FL: CRC Press.
Davidson, J. M. (1987). *The Prehistory of New Zealand* (New edn). Auckland, N.Z.: Longman Paul.
Feinberg, R. and Macpherson, C. (2002). The Eastern Pacific. In A. Strathern, P. J. Stewart, L. M. Carucci, L. Poyer, R. Feinberg, and C. Macpherson (eds.), *Oceania: An introduction to the cultures and identities of Pacific Islanders*. Durham, NC: Carolina Academic Press.
Gathercole, P. (1978). Hau, Mauri and Utu: A Re-examination. *Mankind*, 11(3), 334–340.

Geertz, C. (1973). *The Interpretation of Cultures; selected essays.* New York: Basic Books.

Gudgeon, W. E. (1905). Māori Religion. *The Journal of the Polynesian Society,* 14(3(55)), 107–130.

Hēnare, J. (1981, 4 July). *Address to Auckland District Law Society.* Photocopy of typescript.

Hēnare, M. (1999). Sustainable social policy. In J. Boston, P. Dalziel, and S. S. John (eds.), *Redesigning the Welfare State in New Zealand: Problems, policies, prospects* (pp. 39–59). Auckland: Oxford University Press.

Hēnare, M. (2001). Tapu, Mana, Mauri, Hau, Wairua: A Māori Philosophy of Vitalism and Cosmos. In J. A. Grim (ed.), *Indigenous Traditions and Ecology: The Interbeing of Cosmology and Community* (pp. 197–221). Cambridge, Massachusetts: Centre for the Study of World Religions Harvard Divinity School.

Hēnare, M. and Kernot, B. (1996). The spiritual landscape. In J. Veitch (ed.), *Can Humanity Survive?: The world's religions and the environment.* Auckland: Awareness Book Co.

Irwin, J. (1984). *An Introduction to Māori Religion: Its character before European contact and its survival in contemporary Māori and New Zealand culture.* Bedford Park, South Australia: Australian Association for the Study of Religions.

Marsden, M. and Hēnare, T. A. (1992). *Kaitiakitanga: A definitive introduction to the holistic world view of the Māori.* Typescript.

Mondin, B. (1985). *Philosophical Anthropology: Man, an impossible project?* (M. A. Cizdyn, Trans.). Rome: Urbania University Press.

O'Connell, J. F. and Allen, J. (2004). Dating the colonization of Sahul (Pleistocene Australia–New Guinea): a review of recent research. *Journal of Archaeological Science,* 31(6), 835–853. doi:10.1016/j.jas.2003.11.005

Prytz-Johansen, J. (1954). *The Māori and his Religion in its Non-ritualistic Aspects: With a Danish summary.* København: E. Munksgaard.

Rappaport, R. A. (1979). *Ecology, Meaning, and Religion.* Richmond, California: North Atlantic Books.

Royal, T. A. C. (2015). Te Ao Mārama – the natural world. Retrieved from www.TeAra.govt.nz/en/te-ao-marama-the-natural-world (accessed 9 March 2016).

Salmond, A. (1997). *Between Worlds: Early exchanges between Māori and Europeans, 1773–1815.* Auckland, N.Z.: Viking.

Schrempp, G. A. (1992). *Magical Arrows: The Māori, the Greeks, and the folklore of the universe.* Madison, Wisconsin: University of Wisconsin Press.

Shirres, M. P. (1982). Tapu. *The Journal of the Polynesian Society,* 91(1), 29–51.

South Pacific Regional Environment Programme. (1992). *The Pacific Way: Pacific island developing countries report to the United Nations Conference on Environment and Development.* Noumea, New Caledonia: South Pacific Commission.

Tui Atua, T. T. T. E. (2007). In Search of Harmony: Peace in the Samoan Indigenous Religion. In T. T. T. E. Tui Atua, Tamasailau M. Suaali'i–Sauni, Betsan Martin, Mānuka Hēnare, Jenny Plane Te Paa, and T. K. Tamasese (eds.), *Pacific Indigenous Dialogue on Faith, Peace, Reconciliation and Good Governance* (pp. 104–112). Apia, Samoa: Alafua Campus, Continuing and Community Education Programme, University of the South Pacific.

Tylor, E. B. (1871). *Primitive Culture: Researches into the development of mythology, philosophy, religion, art, and custom* (Vol. 1–2). London: J. Murray.

Williams, R. (1983). *Keywords: A vocabulary of culture and society* (rev. and expanded. ed.). London: Fontana Paperbacks.

15

NORTH AMERICA

Native ecologies and cosmovisions renew treaties with the earth and fuel indigenous movements

Melissa K. Nelson

We have treaties with creation. We have treaties with the fish, we have a treaty with
the rice, with that lake ... Creation doesn't give a second chance; we can't renegoti-
ate again. Protect the land, live with the land, not off of it.

> (Michael Dahl, White Earth Chippewa Tribal member at public hearing opposing the
> Enbridge Sandpiper oil pipeline, 2015)

Indigenous ecological consciousness persists throughout Indian Country in a variety of diverse
and beautiful ways. From on top of the world at the Arctic Circle of Nunavut on Turtle Island
(North America) to the tip of South America at Tierra del Fuego, Native Peoples assert their
rights to land, culture, and a healthy future for generations to come and for all of the ecological life that creates sacred homelands. Honoring the Prophecies of the Eagle and the Condor,
the Prophecy of the Seventh Fire, and many other revered teachings, the Native Peoples of the
Americas are in a time of renewal and resurgence regarding their "treaties with creation" –
honoring reciprocal relationships with their inter-species kin, landscapes, and ancestral places
in a time of great challenge.

This chapter focuses on the Indigenous traditions of North America, specifically the cosmovisions and native ecologies of my Anishinaabeg heritage of the Northern Great Lakes regions
that promote ethical relations with the more-than-human world. The Anishinaabeg and other
Native American traditions include sophisticated spiritual values and lifeways that bring
together law, religion, art, science, and governance in a holistic vision and practice of regeneration. In addition to the Anishinaabeg tradition, there are examples from the great
Haudenosaunee (Six Nations Iroquois) peoples of the Northeast. These strong nations and
confederacies have shared their worldviews widely and are today using the Traditional
Ecological Knowledge (TEK) and ethical wisdom of their cosmovisions to address some of the
most-pressing ecological problems of our time: water scarcity and biodiversity loss that
contribute to food insecurity and overall loss of life and well-being. Integral to these revitalization efforts is an understanding that people and place, nature and culture, biodiversity, and
cultural diversity, are inextricably linked and must be addressed together, holistically, not in a
fragmented way. The resurgence of Indigenous cosmovisions and lifeways are helping address
the roots of our ecological crisis and are being included in contemporary social movements
such as the Idle No More, Climate Justice, and Indigenous Rights movements.

Anishinaabeg cosmology and cosmovision

The Anishinaabeg cosmology and cosmovision are based on our sacred stories, *aadizookaan*, original instructions and ancestral practices that help us maintain and renew our relationships with all of creation. Our creation stories have many versions and two main messages that stand out: 1) if people do not cooperate, express gratitude for the gifts of life, and get along sharing together, Creator will alter the world to start over fresh again (i.e., a great flood, fires, earthquakes); and 2) after the great flood, it took inter-species cooperation between a number of animals, most significantly the turtle, muskrat, beaver, and loon, to help Nanaboozhoo, our Trickster culture hero, re-create the world from water and produce Turtle Island (I discuss the Earthdiver concept later). Within this story and other related creation stories one can see the important values of human and inter-species cooperation and the understanding that to survive and thrive living on this Earth, we must all do it together—plant, animal, bird, fish, human, and more.

We also find four major concepts in the Anishinaabeg cosmovision that articulate this holistic vision and practice of Earth guardianship and inter-species harmony: Gitche Manitou (Great Mystery), Mino-bimaadaziwin (the Good life), Trickster consciousness, and the Earthdiver.

Gitche Manitou

In our language, Anishinaabemowin, this sacred concept is very difficult to translate but it is generally translated to mean "great mystery" or "great spirit." This energy is not personified or represented by a thing even though it is an animate noun. It is something alive and an oral form because of its inherent dynamic spiritual energy. As I have said elsewhere, "A manitou is most easily and often translated to mean 'spirit…'(Nelson 2013). Ojibwa elder, Basil Johnston, states, 'Depending on the context, they knew that in addition to spirit, the term also meant property, essence, transcendental, mystical, muse, patron, and divine'" (Johnston 1995).

Accordingly, *Gitche Manitou*, the Great Spirit, the Great Mystery, is ultimately unknowable but is a powerful source of life and creation that is with us and available to us at all times for guidance, power, and peace. But in order to access Gitche Manitou's power we must follow the teachings from creation time that show us how to respect, honor, and celebrate the gifts of life. These values are often communicated in other stories such as the teachings of Little Boy and the Seven Grandfathers. In this story Little Boy goes on a journey to seek out truth and find out what life is all about. Along his journey he encounters different animals (Grandfathers) that teach him different life lessons. From the wolf he learns humility; from the bear, bravery; from raven, honesty; from beaver, wisdom; from turtle, truth; from buffalo, respect; and from eagle, love (National Museum of the American Indian exhibit). All of these values (and animals) must work together to create balance and harmony. Again we see the theme of inter-species cooperation, which from a modern ecological perspective would be noted as part of population biology, food webs, and ecosystem dynamics. The ecological function of these animal species reflects, from an Anishinaabeg worldview, other significant traits and symbolic values for human life, such as the clan system. We humans too are simply another species originating from and imbued with Gitche Manitou. To know our specific purpose and responsibilities and to be able to share and interact with other life forms in a reciprocal way is key to creating a diverse and concordant co-existence.

Bimaadaziwin

This word is made of an adverb *mino* meaning, "good, nice, well" and an animate verb *bimaadazi* meaning "a way of being or life in time and space" and is generally translated to mean "living a good life." This is an Anishinaabeg concept of sustainability and means living in a good and respectful way: "Baamaadziwin motivates people to go beyond being good and just 'to being servants, devoting ourselves to making a difference in all that has occurred and may still be occurring within our respective communities and environment.' This includes restoring the balance of our shared natural environment and of all inhabitants who are dependent upon a robust ecosystem" (Holgren et al. 2015).

In another Anishinaabeg interpretation of this foundational concept, Winona LaDuke interprets mino-bimaadaziwin as "continuous rebirth" (LaDuke, 51). Inherent in this concept of mino-bimaadaziwin is the notion of rebirth and regeneration as an obligation to perpetuate and reproduce life. Consequently, the continuous implementation and revitalization of mino-bimaadaziwin is essential for a healthy, abundant life for humans and all beings with whom we share planet Earth.

Other tribes such as the Seneca Nation of the Haudenosaunee also have similar concepts to mino-bimaadaziwin. David Bray, a Seneca farmer and educator, talks about the importance of *Jöhehgöh*, the life sustainers" as critically important to his nation ("Nourishing the Body" 2014). Like the emphasis on "continuous rebirth," *Jöhehgöh's* emphasis on "life sustainers" shares the importance of the reproductive capacities of humans and nature. One of the most significant and potent expressions of these life sustainers are the Three Sisters of Native American agriculture—corns, beans, and squash. These food plants are so important that they are referred to as sisters and are essential sustainers of life for the people and the nation. Their significance goes back to their creation story when Sky Woman fell through a hole in the sky where a great tree grew and landed on the top of Turtle, who was swimming on a great sea surrounding the Earth. Under Sky Woman's fingernails were the seeds of these Life Sustainers, which had been scraped from the roots of the Sky Tree as she fell. With the help of Earthdivers, such as muskrat, otter, swan, and other creatures, Sky Woman was able to plant these Life Sustainers in fresh dirt and dance corn, beans, and squash into life. This dance is sometimes known as the "Old Time Women's Dance" and a line of it states, "Women and our life sustainers are as sisters" (Fenton, 23). There is also a Three Sisters Wampum Belt that depicts the importance of this symbiotic relationship and its power in the agricultural economy and ceremonial cycles of the Seneca (Kelsey 2014, 71).

Both *mino-bimaadaziwiin* and *Jöhehgöh* are two North American Indigenous spiritual concepts that emphasize continuing, sustaining, and regenerating the life-enhancing principles of the Earth and cosmos in all aspects of individual and collective life—physical, emotional, mental, and spiritual—from food gathering and farming to governance and ceremony. These concepts are not historic, intellectual artifacts but living, dynamic life principles and instructions for contemporary Native communities to manifest and sustain a good life for all of creation, now and for the future.

Trickster consciousness

Another key concept is that of the trickster. This term has many meanings and interpretations depending on the nation, tribe, and language, but it has a particular focus for the Anishinaabeg. This has been theorized most notably by the great Ojibwe scholar-intellectual, Gerald Vizenor. "Trickster consciousness" is a term that he coined and it has become a neologism for Native American and Indigenous Studies today, as well as in Indian Country (Nelson, 2008, 288). In true fashion of the Trickster, this term is not easily defined but it emerges from the stories of *Nanaboozhoo*, the Ojibwe Trickster culture hero. Most simply put, Trickster consciousness "defies singularity, monocultures, and completeness" and embraces paradox and contradiction.

Trickster consciousness can facilitate a shift in paradigms by mediating supposedly contradictory forces (Vizenor 1990). The significance of Trickster consciousness is that it is an integral part of many Native American and Indigenous worldviews and it helps support cognitive pluralism, that is, the ability to embrace multiple ways of knowing and being simultaneously.

When consciousness is fluid, open, and dynamic as opposed to rigid, myopic, or apathetic, humans learn more, adapt, and embrace change. Ojibwe have always relied on dreams, emotion, insight, and imagination to embrace deeper understandings of reality and provide greater vision. These trans-rational abilities characterize Trickster consciousness and are key skills for survival and resilience that Native Peoples have relied on for millennia. This type of cognitive dexterity and emotional intelligence is also well known in the Haudenosaunee Law of Peace. In this system, the Six Nations Confederacy utilizes the complex checks and balances of the clan system. That is, clan mothers and faithkeepers maintain traditional peace-keeping and seasonal rituals for peace and unity amongst communities. The Iroquois commitment to a "clear, unified mind" in the midst of diversity and contradiction stands as an ethical counterbalance and acknowledgement of trickster ambiguity in the world (Mohawk, 2008, 48).

Trickster consciousness is not only a signature trait of the cosmovision of the Anishinaabeg but also of many Indigenous traditions of North America. It has great relevance to our current global climate crisis, where the ability to adapt to sudden, unexpected, and radical environmental and social change will mean survival or not, for the Earth community. Clearly Trickster consciousness is an important Native American cultural value and practice, which may serve as a compelling idea for humanity in these tumultuous and challenging times.

Earthdiver

The Earthdiver is a widespread creation motif in North America and Eurasia. This includes stories of Northeast nations like the Anishinaabeg and the Haudenosaunee but also many others such as California Indian tribes. The Anishinaabeg have an Earthdiver creation story whereby a group of creatures try to figure out how to dive deep into the vast, primal sea to find some earth or mud with which to re-create the world. "In every version of the Earthdiver story, it is never a human being who dives down to bring up the earth. Loon, Muskrat, Pike, Beaver, Duck—any creature with the powerful ability to move between the world of water and the world of air plays that role" (Bruchac 2003, 67). This process of Earthdiving includes discussion, negotiation, and competition as the various animals try to outdo each other holding their breath to reach solid earth under the endless waters. It is finally the modest muskrat who retrieves soil at the bottom of the ocean and place it on turtle's back to create Turtle Island.

In some versions, muskrat sacrifices his life; in others, Creator or Nanaboozhoo take pity on him and blow new life back into him to reward him for his efforts. This story shows the importance of sacrifice to help others and of working together. It is through inter-species cooperation and support from animals that the Anishinaabeg sought stability from out of the trickster ambiguity of life. Gerald Vizenor uses this mythic story as a metaphor for modern mixedblood identity. He believes that many of us are modern earthdivers; we are "mixedbloods, Métis, tribal tricksters and recast cultural heroes, the mournful and whimsical heirs and survivors from that premier union between the daughters of the woodland shamans and white fur traders" (Vizenor, 1981, xvi). Earthdivers navigate trickster consciousness in that they honor *Gitche Manitou* and manifest *mino-bimaadaziwin* in both historic and modern contexts. Yet these ancestral values are often difficult for contemporary Native Peoples to embody due to centuries of colonial disruption.

Disruption of treaties with creation

There was a deliberate disruption of pre-contact and post-contact treaties by colonial forces that instituted systems of cultural genocide on Native Americans. The impacts of colonialism included forcing Native Peoples off of their ancestral lands, which meant they could not practice and implement their ecological and spiritual values and honor their treaties with creation. This profound loss of ancestral lands and disruption of our Earth-honoring practices means we have not been able to uphold the original instructions of Gitche Manitou. We have been cut off from our Life Sustainers by this violent disruption of our treaties with the more-than-human world. Many species have disappeared; many ecosystems are damaged. Sacred waters and medicines have been desecrated, and the overall well-being of life for Native Peoples and the lands has greatly diminished. Guilt and shame about not being able to fulfill our sacred covenants has led to mental ailments that Apache psychologist Eduardo Duran calls "the soul wound" (Duran 2006). Healing this soul wound goes hand in hand with healing the Earth from the ravages of consumer capitalism and what Leanne Simpson calls the mindset of "extractavism" (Klein 2013).

Asserting native values in contemporary movements

Native American leaders in the Idle No More, Climate Justice, and Indigenous Rights movements work toward rebuilding "healthy, just, equitable and sustainable communities" (Idle No More). The United Nations Declaration on the Rights of Indigenous Peoples, article 24 states, "Indigenous peoples have the right to their traditional medicines and to maintain their health practices, including the conservation of their vital medicinal plants, animals and minerals" (United Nations Declaration on the Rights of Indigenous Peoples [UNDRIP]). Article 25 acknowledges the profound spiritual relationship and responsibility Indigenous peoples have to the land: "Indigenous peoples have the right to maintain and strengthen their distinctive spiritual relationship with their traditionally owned or otherwise occupied and used lands, territories, waters and coastal seas and other resources and to uphold their responsibilities to future generations in this regard." Supported by our original instructions, TEK, UNDRIP, and other modern laws and tools, Native Peoples are regaining their rights and responsibilities to protect Mother Earth. Recent actions, victories, and struggles portray this effort and show how modern Native American and First Nations communities continue to assert their cosmovisions and traditional knowledge and practices.

Idle No More

The Idle No More movement started in Canada in December 2012 and has become one of the largest mass, Indigenous grassroots movements in the world. Its vision is clear and simple: "Idle No More calls on all people to join in a peaceful revolution to honour Indigenous sovereignty, and to protect land, and water" (Idle No More). Its goals include resisting neo-colonialism in all its forms and it was spawned in large part to protest efforts by the Canadian government to dismantle First Nations sovereignty.

In an effort to protect Indigenous treaty land and water rights in Canada, First Nations Chief Teresa Spence protested a legislative overhaul, which would have radically changed the protections of forests and waterways in Canada making them extra vulnerable to extraction. Chief Spence went on a six-week hunger strike in a tipi in the nation's capital of Ottawa in the middle of winter in 2013. It was in large part out of this outrage of the complete disregard for First Nations rights and consultation that Chief Spence and four other Native women from Saskatchewan co-founded the "Idle No More" movement.

Cree, Ojibwe, and Métis, all Anishinaabeg language communities, have been leaders of this movement utilizing the strength of Indigenous cosmovisions as "original instructions." Round dance flash mobs, blockades, campaigns, teach-ins, summits, walks, and court cases are all used to stop fracking and oil extraction, protect treaties and water rights, defend native lands, and protect the sanctity of First Nations and Indigenous women. These are the main actions and strategies used by the Idle No More movement to assert sovereignty. Song, dance, and prayer are integral components of all of these actions with great reverence for community Elders and Knowledge Holders as well as youth leaders. Most importantly, the drum continues to be used to awaken the people and create unity (Vennum 1982). Anishinaabeg writer and activist Leanne Simpson shares:

> People within the Idle No More movement who are talking about Indigenous nationhood are talking about a massive transformation, a massive decolonization. A resurgence of Indigenous political thought that is very, very much land-based and very, very much tied to that intimate and close relationship to the land, which to me means a revitalization of sustainable local Indigenous economies that benefit local people.
>
> *(Klein 2013)*

Climate Justice

Connected to the Idle No More and larger Indigenous rights movements is the climate justice movement, a social and environmental justice effort to prevent damage and prepare for impacts caused by climate change, especially to poor, Indigenous, and disadvantaged communities, and make the primary polluters and causes of climate change, such as petroleum industries, responsible for their actions. The climate justice movement draws on law and policy reform, and nonviolent protest and marches. These tactics stem from a deep commitment to peace and justice rooted in Indigenous values. For example, according to the Indigenous Environmental Network's (IEN) "Four Principles of Climate Justice," the fourth principle is "Living in a Good Way with Mother Earth." It states that,

> Climate justice calls upon governments, corporations and the peoples of the world to restore, revaluate and strengthen the knowledge, wisdom and ancestral practices of Indigenous Peoples, affirmed in our experiences and the proposal for 'Living in a Good Way,' recognizing Mother Earth as a living being with which we have an indivisible, interdependent, complementary and spiritual relationship.
>
> *(IEN)*

Again, we see this contemporary movement being infused with the traditional teachings of mino-bimaadaziwin, "living the good life," and other native teachings of reciprocity, and care for Mother Earth.

Climate Justice is on most everybody's mind these days as the Earth continues to experience unprecedented changes in air, land, and ocean temperatures, super storms, migrating species, insect infestations, melting ice, eroding communities and flooded islands, and ongoing shifts in the atmospheric weather patterns that creates droughts and flooding, among other unprecedented changes, in different places.

In many ways humanity is facing these unprecedented climate changes for the first time. In other ways, Indigenous peoples have had to deal with catastrophic change over centuries of colonization and millennia of Earth changes. As Potawatomie botanist, Robin Kimmerer has shared, "I remember an elder once saying that we have protected our traditional knowledge against so many assaults and that one day the whole world would need it. In this time of accelerating climate change and the Age of the Sixth Extinction, we know that traditional teachings of care for the land and water, of respect for the living earth are more critical than ever" (VanPool 2015). Kimmerer and others have also stated that when Native Americans were forcibly relocated from cool maple forests to southern, humid woodlands; or from coastal areas to inland dry plains; or from great basin deserts to temperate rainforests they experienced "climate change" in a very real and significant way. They had to use the strength of their traditional teachings of resilience to adapt quickly to a radically different environment in order to survive. So Native Peoples who have been relocated to different landscapes and ecosystems certainly have strong experiences of climate change that can be helpful adaptation strategies for all people today.

Indigenous Rights Movement

The global turn to Indigenous rights is a massive, multifaceted movement fueled by the 2007 passing of UNDRIP, an extraordinary human rights declaration that hundreds of Indigenous leaders worked on for over thirty years. It is both a flagship and Rosetta stone for Indigenous peoples worldwide and gives ethical, political, and legal power to Native communities across the globe.

In terms of protection of land, territories, and resources a study from my own Turtle Mountain Chippewa Reservation in North Dakota brings this discussion into personal territory. This community, in an effort to protect their lands and waters, preemptively banned fracking on all Reservation lands. In the Western side of North Dakota, where the abundant and lucrative Bakken shale formation lies, the oil industry has boomed and busted with oil extraction. The main way this industry extracts oil is through the destructive process of fracking. In this hidden, subterranean process a mixture of water, sand, and chemicals are pumped deep down into the shale deposits in the earth below at high pressure, "fracturing" the rock and releasing the oil to flow for extraction. A group called "No Fracking Way Turtle Mountain Tribe" has led the effort here. The main concern has been the protection of water from chemical contamination. In their Tribal Resolution signed by the Chairman and Secretary on November 2011, one of the lines reads, "The Turtle Mountain Band of Chippewa is responsible for protecting Mother Earth from any pollutants that may cause harm to its citizens, land, water, and air" (Resolution 2011). In an effort to protect Mother Earth and uphold our commitments to Gitche Manitou and mino-bimaadaziwin, the Turtle Mountain Chippewa Nation became the first tribe in the US to "prohibit in perpetuity" any fracking on or near Reservation lands.

Conclusion

Clearly, Native American cosmovisions and spiritual concepts and values are infusing modern Indigenous ecological movements for flourishing life. As Anishinaabeg, our guiding principles of respect and honor for *Gitche Manitou* are alive and well as we work to manifest the instructions of the Seven Grandfathers: humility, respect, bravery, honesty, wisdom, truth, and love.

Additionally, we strive for mino-bimaadaziwin, or living the good life, for ecological sustainability and cultural regeneration and rebirth. We utilize our understanding of being Earthdivers from our creation stories, and how we must continue to dive deep, metaphorically, for solid earth in the midst of chaos and imbalance on the surface of things. Additionally, we understand Trickster consciousness as a dynamic and fluid way to embrace complexity, paradox, and change. We use these gifts of trickster to help ourselves and others adapt and find resilient tools as we confront the climate crisis, economic disparities, food and water shortages, and ongoing assaults on human and Indigenous rights. So also the Six Nations of the Haudenosaunee have their Life Sustainers, Wampum Belts, and understanding of reciprocity and the Great Law of Peace. These gifts fuel their contemporary work to protect their ancestral lands, waters, and cultural rights. Something similar can be said of the 560 Native American nations in the United States today.

Interestingly, both the Anishinaabeg and the Haudenosaunee are "border" nations, meaning their traditional territories were politically bisected by the US/Canadian nation state boundary. Being in the heart of Turtle Island, these nations often work on both sides of that political borderline. As we learned from the overview of the Idle No More movement, First Nations in Canada are actively resisting the neocolonialism of extractivism and the violation of their rights and are pursuing justice based on their traditional values and in solidarity with other Indigenous peoples. According to the government of Canada, there are 617 First Nation communities in Canada (First Nations People of Canada).

In part due to the success of the Idle No More, Indigenous Rights, and Climate Justice movements in Canada, the US, and globally, and the egregious action of former Prime Minister Harper toward First Nations, Canada voted out Prime Minister Stephen Harper in the October 2015 election and voted in Justin Trudeau, the second youngest prime minister to be elected in Canada. Trudeau, in his first month of office, vowed to change and improve the Canadian governments relationship with the Indigenous peoples of Canada. In his letter to the new minister of Indigenous and Northern Affairs, Carolyn Bennett, Prime Minister Trudeau had this to say:

> As Minister of Indigenous and Northern Affairs, your overarching goal will be to renew the relationship between Canada and Indigenous Peoples. This renewal must be a nation-to-nation relationship, based on recognition, rights, respect, co-operation, and partnership. I expect you to re-engage in a renewed nation-to-nation process with Indigenous Peoples to make real progress on the issues most important to First Nations, the Métis Nation, and Inuit communities – issues like housing, employment, health and mental health care, community safety and policing, child welfare, and education…
>
> To support the work of reconciliation, and continue the necessary process of truth telling and healing, work with provinces and territories, and with First Nations, the Métis Nation, and Inuit, to implement recommendations of the Truth and Reconciliation Commission, starting with the implementation of the United Nations Declaration on the Rights of Indigenous Peoples.
>
> *(Minister Indigenous and Northern Affairs Mandate Letter 2015)*

Having survived what Eduardo Duran calls "the soul wound" and what Ojibwe scholar Lawrence Gross calls "post-apocalyptic stress disorder," Native Americans in North America continue to heal themselves (Gross 2003). They also work hard to heal the land from damage and protect it from extractive industries. This rise of the Indigenous mind is being voiced by native communities and being heard by national leaders. This tenacity, strength, and power within the Indigenous movement is fueled by cosmovisions that emulate and make present the procreative power of stars, winds, volcanoes, and seeds.

As the great Mohawk midwife Katsi Cook has stated, "women are the first environment." The Shoshone spiritual teacher Corbin Harney stated that, "we *are* the environment" (Cook 2003; Gonzales and Nelson 2001). Living with the truth of these profound understanding and the wisdom of our original instructions, it is clear that Native American communities are mending and renewing our treaties with creation and will be here for the next seven generations and beyond.

References

Blaeser, Kimberly. 2012. *Gerald Vizenor: Writing in the Oral Tradition*. Norman: University of Oklahoma Press.

Bruchac Joseph. 2003. *Our Stories Remember: American Indian History, Culture and Values through Storytelling*. Golden: Fulcrum Publishing.

Cook, Katsi. 2003. Cook: Women are the First Environment. Indian Country Today Media Network.com: http://indiancountrytodaymedianetwork.com/2003/12/23/cook-women-are-first-environment-89746; December 23, 2003.

Dahl, Michael quoted in Winona LaDuke's "The Thunderbirds versus the Black Snake." *Earth Island Journal*, special issue on "Indigenous Resistance and Restoration, guest editor Melissa K. Nelson, Volume 30, No. 3 (Autumn 2015).

Fenton, William Nelson. 1998. *The Great Law and the Longhouse: A Political History of the Iroquois Confederacy*. Norman, University of Oklahoma Press.

First Nations People of Canada, www.aadnc-aandc.gc.ca/eng/1303134042666/1303134337338

Gonzales, Tirso and Melissa K. Nelson. 2001. "Contemporary Native American Responses to Environmental Threats in Indian Country. In *Indigenous Traditions and Ecology – The Interbeing of Cosmology and Community*. Edited by John Grim, Center for the Study of World Religions, Harvard University Press.

Gross, Lawrence. 2003. Cultural Sovereignty and Native American Hermeneutics in the Interpretation of the Sacred Stories of the Anishinaabe. *Wicaso Sa Review* 18, no. 2, 127–34.

Holgren, Marty, Stephanie Ogren, and Kyle Whyte. (2015). "Renewing Relatives: One Tribe's efforts to bring back an ancient fish." *Earth Island Journal*, special issue on "Indigenous Resistance and Restoration, guest editor Melissa K. Nelson, Volume 30, No. 3 (Autumn).

Idle No More Manifesto www.idlenomore.ca/manifesto

Indigenous Environmental Network http://perspectives.apps01.yorku.ca/2010/12/15/four-principles-for-climate-justice/

Johnston, Basil. 1995. *The Manitous: The Spiritual World of the Ojibway*. New York, HarperCollins.

Kelsey, Penelope Myrtle. 2014. *Reading the Wampum: Essays on Hodinohso:ni' Visual Code and Epistemological Recovery*. Syracuse: Syracuse University Press.

Klein, Naomi. 2013. "Dancing the World into Being: A Conversation with Idle No More's Leanne Simpson. Yes! Magazine! March 5. www.yesmagazine.org/peace-justice/dancing-the-world-into-being-a-conversation-with-idle-no-more-leanne-simpson

LaDuke, Winona. 1992. *Minobimaatisiiwin*: The Good Life. *Cultural Survival Quarterly* 16, No 4, Winter, 69–71.

Minister Indigenous and Northern Affairs Mandate Letter, http://pm.gc.ca/eng/minister-Indigenous-and-northern-affairs-mandate-letter

Mohawk, John, 2008. "Clear Thinking: A Positive Solitary View of Nature." In *Original Instructions: Indigenous Teachings for a Sustainable Future*. Edited by Melissa K. Nelson. Rochester, VT, Bear & Company.

National Museum of the American Indian (NMAI) exhibit, www.nmai.si.edu/explore/exhibitions/item/?id=530

Nelson, Melissa K. 2008. "Mending the Split-Head Society with Trickster Consciousness." In *Original Instructions: Indigenous Teachings for a Sustainable Future*. Edited by Melissa K. Nelson. Rochester: VT: Bear and Company.

Nelson, Melissa K. 2013. The Hydromythology of the Anishinaabeg: Will Mishipizhu Survive Climate Change or is he Creating It? In *Centering Anishinaabeg Studies: Understanding the World through Stories.* Edited by Jill Doerfler, and Niigaanwewidam James Sinclair and Heidi Kiiwetinepinesiik Stark. Michigan State University Press.

"Nourishing the Body, Honoring the Land": https://vimeo.com/100924901

Resolution TMBC627-11-11, Nov. 29, 2011, Turtle Mountain Band of Chippewa Indians.

UNDRIP/United Nations Declaration on the Rights of Indigenous Peoples www.un.org/esa/socdev/unpfii/documents/DRIPS_en.pdf

UNESCO. 2015. "Resilience in a Time of Uncertainty: Indigenous Peoples and Climate Change" (http://Indigenous2015.org/).

VanPool, John. 2015. Q&A with Robin Wall Kimmerer, November 3. Citizen Potawatomie Nation. www.potawatomi.org/news/top-stories/1854-q-a-with-robin-wall-kimmerer-ph-d

Vennum, Thomas, Jr. 1982. *The Ojibwa Dance Drum: Its History and Construction.* Smithsonian Folklore Studies No. 2, Washington DC, Smithsonian Institution Press.

Vizenor, Gerald. 1990. Trickster Discourse *American Indian Quarterly* 14, no. 3 Summer, 277–87.

Vizenor, Gerald. 1981. *Earthdivers: Tribal Narratives on Mixed Descent.* Minneapolis, University of Minnesota Press.

16

ARCTIC

Ontology on the ice: Inuit traditions, ecology, and the problem of categories

Frederic Laugrand

Many Indigenous groups populate the Arctic regions: Yup'it, Inupiat and Dene in Alaska, Siberian groups in Russia, Athabaskan and Inuit in Canada, Kalaallit in Greenland, Sami in Northern Scandinavia, to name a few. In this paper based on long-term research with my colleague Jarich Oosten, I will focus on the Inuit groups from Northern Canada where a wide variety of subgroups can also be distinguished. Nevertheless, I'll make a few references to other Arctic Indigenous groups, especially Yup'it and Cree, when connections can be made, or in order to discuss some theoretical points.

For thousands of years Inuit from the Arctic regions have survived by subsistence hunting. They developed sophisticated techniques that allowed them to hunt for land and sea game, and an extensive knowledge of the land and its animals. Inuit society was completely oriented to hunting, and the beliefs and practices of the past all focused on sustaining a flow of life that allowed animals to be captured and used so Inuit could survive. Once Euro-Americans appeared in the Arctic in the sixteenth and seventeenth centuries, and later the Hudson's Bay Company, this lifestyle transformed considerably. In the twentieth century, most Canadian Inuit also converted to Christianity, leaving behind many rules and ritual injunctions connected to the hunting practises, and shamanism went underground (see Laugrand and Oosten 2010).

In the second half of the twentieth century the Canadian administration began to take control in the north. The move of the Inuit to permanent settlements terminated their nomadic existence. The administration often considered the hunting mode of existence an anachronism and wished to integrate Inuit into a wage economy. Inuit were aware of this negative attitude towards their mode of existence. The relocations in the 1950s and the killing of the dogs in the 1970s were viewed as attempts to destroy Inuit people and their culture (see Laugrand and Oosten 2014).

In the last decades, Inuit were also faced with the developments of external bans on whaling and the fur boycott. Although Western companies had made fortunes with whaling and the fur trade for centuries, these activities were declared immoral and prohibited by the time Inuit were beginning to take control of their own destiny. Even today it is hard to explain to Inuit why furs should be boycotted when the trade in leather from the animals slaughtered in the Western bio-industry is perfectly legitimate (see Wenzel 1991).

Today, whereas climate change has become a political concern (Stuckenberger 2007; Fienup-Riordan 2010), new animals such as the polar bear and, in certain regions, the caribou, are now subjects of a very limited hunt. In this context, the management of wildlife remains one of the central political matters in Nunavut. Inuit still reject or distrust the approaches and methods of biologists and bureaucrats (Nadasdy 2003). In the US as well as in Canada, the administration has tried to protect wildlife by implementing quotas, which caused great resentment among indigenous peoples. To some extent, Inuit never understood or believed in Western ecology. Inuit conceive the management of wildlife as the foundation of the relationship between people and land. They do not possess the land but say that they are themselves possessed by the land. They hunt animals and are upset by any unnecessary killing. A few years ago, Philipp Qipanniq (quoted in Laugrand and Oosten 2014), an elder from Igloolik, made a statement that illustrates well the problem all over the Inuit lands, "There should be no restriction on what a hunter can hunt. (...) From time immemorial Inuit never wasted any of the game animals that they caught. They never killed game animals and just left them." Inuit are critical of sport fishing. They are more open to polar bear sport hunting, since this activity not only provides a lot of money for families, but also facilitates sharing, which maintains social relationships. (Wenzel 2008, Wenzel and Dowsley 2008).

Barnabas Peryouar from Qamanittuaq explained that Inuit still do not understand biologists as they never grew up with caribou all their lives (quoted in Mannik 1998: 177–8).

According to Peryouar, caribou are not on the verge of disappearance but fleeing from human activities, and especially mining activities: "Mining, the sound of airplanes, and the foul smell of the mining stuff has made it so the caribou can't roam around the tree line any more" (quoted in Mannik 1998: 176).

This distrust with respect to biologists is less marked in Alaska but in Canada, it also applies towards ecologists as this is illustrated by the following event that took place in Mittimatalik. During the early fall of 2008 about 629 narwhals were trapped by the winter freeze-up of the sea ice and doomed to starve. With the authorization of the Department of Fisheries and Ocean, the Nunavut Wildlife Management Board and local elders, the Mittimatalik Hunters and Trappers Organization spent two weeks killing the narwhals, pulling them out of the water to harvest the meat. This massive catch gave the hunters an occasion to teach the younger generations how to kill and harvest whales and it soon became the largest bounty in decades. An Inuk from Mittimatalik stated on a blog: "We thank the Creator for giving us the animals as we thank Him every time we harvest something. (...) We hope you learn that Inuit respect the animals, and nothing will go to waste. We are so thankful to our Creator for giving us all the meat we can use (...)!!!"(see Laugrand and Oosten 2014)

But this successful catch resulted in a bitter debate between Inuit leaders and an environmentalist group, the Sea Shepherd Conservation Society. The society's leader, Paul Watson, called the killings a war crime. Watson described the killings as a "bloody massacre". Such a conflict illustrates well the different views Inuit and Western peoples have on ecology and animals. It is clear that from the Inuit perspective, animals can be respected and protected in many ways but that hunting is in itself a way of respecting animals. Through hunting, human society can take shape, and the well-being of animals as well as human beings depends on it. This awareness that animals depend on human beings and human beings depend on animals is a key point in Inuit cosmology. Being an Inuk is not just an identity constructed by people as in Western naturalism; it is also provided by the animals themselves. Similarly, knowledge of animals is by no means based on theoretical perspectives and management technologies, it is based on personal experiences, observations and interaction with animals.

In this paper, I will first recall some key elements of Inuit hunting traditions that also appear in other Arctic societies, and secondly discuss some theoretical problems with respect to Inuit animism, a category that is often used to qualify their religious and social systems.

1. Inuit hunting traditions in the Canadian Arctic: a socio-cosmological order

a. Sila and nuna

Inuit and Yup'it emphasize a close connection between the weather (sila), the land (nuna), the sea, animals, and human behavior (see Laugrand and Oosten 2010; Fienup-Riordan 2010; Krupnik et al. 2010). If human beings misbehave, the weather, the land, or the animals will retaliate. The relation between people and the world is embedded in a moral order according to which all actions are reciprocated by these cosmological agencies that missionaries associated with deities known as Sedna (the sea woman) or Sila (the sky), or Taqqiq (the moon). But Inuit did not have such a concept of deity.

The earth holds the deceased as it is covered with graves. It is also populated by many non-human beings, such as dwarves (inugarulligait), mountain spirits (ijirait), etc. Many places thus require respect. Like the land, sila (the sky) cannot be owned by people but only shared. Sila embraces a wide range of related categories, such as the sky and the weather, but also reason and the world and its order. Sila inua was considered to be a powerful person in certain areas of the Arctic also known as Naarjuk, the giant baby. Sila's power is destructive and a force to be feared and bad stormy weather is associated with it.

Today, many of these beliefs still hold, but God has also been incorporated as a new entity able to discipline people for their misbehaviors. Yet, the notion of "nature" and that of "traditional knowledge" are still very inadequate notions for Inuit and mostly relevant in modern societies (Descola 2005). When Inuit decided to implement the notion of Inuit qaujima-jatuqangit (IQ) to replace and broaden the concept of Inuit traditional knowledge, they raised the fact that this Inuktitut term is much better as it encompasses more fully the notion of Inuit knowledge, including social and cultural values, practices, techniques, emotions, beliefs, language, and world view. By defining IQ as "a knowledge from the past that is still useful", Inuit indicated clearly that the dynamic aspects of IQ are essential. Thus IQ is not static, not something abstract and separated from the context in which it is produced, but always related to the present.

b. Making a good hunter

In the past, knowledge of the seasons and the weather, and knowledge of the animals, was by no means sufficient to make one a good hunter. Young men were made good hunters by elders who could pronounce powerful words or use *arnguat*, amulets. Amulets might be worn by the mother on behalf of her son or by the young hunter himself. In certain places, Inuit would become good hunters by playing games such as pretending to be an animal. Someone would act like a polar bear or a walrus, and others would try to harpoon him. Children would first receive toys and later real hunting tools, such as bows and arrows, which would enable them to shoot their first ptarmigans and other small prey. After that they would be taught to go after caribou and seals and their first catches would be celebrated.

In many accounts of first catches, the young hunter does not actually kill the prey, but only contributes to the hunt. The first catch should be distributed and consumed completely and immediately. The young hunter should also abstain from it (Imaruittuq, quoted in Oosten, Laugrand and Rasing 1999: 40).

A first catch ritual often had marked sexual connotations. The midwife (*sanaji* or *arnaliaq*) who had delivered the boy, was considered the maker of the boy, and she/he would receive a part of the first catch. Uqsuralik, an elder from South Baffin related, "*Pijaqsaijuq* is the word for giving boys the skills to be a good hunter. …We would have the boy undo the tie of the pants so the boy would become a really good hunter. … That's how we made boys into very skilful hunters" (quoted in Briggs 2000: 37).

Once the first catch had been made, the boy had opened a connection with the game. Today, first catches are still often celebrated and their patterns differ considerably among families (see Pernet 2013). These type of practises can also be found among the Cree.

The technique of hunting can also be modelled on the act of copulation. An elder from Naujaat, Tungilik, was advised to breathe like he were copulating when he was about to make a kill at a breathing hole of a seal. Thus, there is a suggestion that hunters relate to prey as men do to women, symbolically connecting the acts of killing game and having sexual intercourse with women. Interestingly, Robert Brightman (1993: 127) observes that among the Rock Cree, "sex is a metaphor for hunting, and women are metaphors for animals" (see also Henriksen for the case of the Innu 2009: 41).

c. Treating the animals with respect

Treating the animal with respect is a central topic among most hunting societies of the Arctic. Among the Inuit, a hunter can only catch game if the game itself consents to the kill. But in contrast to an Ingold statement (Ingold 2000: 13–14), this cannot be reduced as a discourse or an allegory as many rules were supposed to be followed. Inuit assumed that even the best hunters will be unsuccessful if animals decide to avoid them. But if human beings showed respect to an animal, behaved themselves in accordance with the rules of the ancestors, maintained the important separations between land game and sea game, between everything connected with hunting and all things concerned with human birth, menstruation, and death, the game might allow itself to be killed. A hunter should also show an attitude of humility, imploring the game to come to him, and to show itself to him. Finally, the hunter should show great joy at the catch and respect the body of the animal once it was killed. Not killing the animal would imply slighting it. If the game did not feel welcome, it would not show itself again.

In the past, respect for animals was organized by many rules. Some basic principles applied to all animals but also varied for each animal. This variation is obviously even wider if we consider other Arctic societies. Today, Inuit hunters still emphasize that any animal, and even mosquitoes, should not be abused, ridiculed, or killed without need. Unnecessary suffering should also be avoided. In addition, hunters should never brag about their abilities nor make fun of animals as those could easily avenge themselves and many stories illustrate this.

Offerings played an important part in hunting. They could be made to the deceased, to the owners of the game or to the animals themselves. Thus, the owners of the caribou demanded offerings of soles and lines before the beginning of the caribou season (Peck, quoted in Laugrand et al. 2006: 388–89), and in the *kiversautit*, a small offering was made to the sea woman before the beginning of the seal season. These small offerings were required for successful hunting. Offerings were often made to the deceased people so that they would give game. Kappianaq (quoted in Laugrand and Oosten 2014) from Iglulik recalled how he was encouraged to offer:

> The act of *tunillainiq* [offering] was encouraged, especially to the grave sites [of those] who had passed away at the time when they were craving for food that was not easily gotten. (…) If I caught a species that he had craved for, I would go and offer a small piece of the meat to the gravesite, or just throw it in the direction of the land where this person was buried.

Today, many of these practises continue but they are difficult to observe, as this was also noticed by Adrian Tanner (1979) in his famous book on Mistassini Cree hunters. Many Inuit hunters also do not offer meat anymore, but they pray to God before going out hunting (see also the case of the Innu who celebrate Saint-Ann feast before going hunting in the summer). In certain communities, elders think that animals are no longer respected sufficiently. They complain about the caribou carcasses left on the land and the bones found around the houses in their communities, claiming more space and opportunities to educate the younger generation. They also blame the mining companies that are too noisy and scare the animals and feel very uncomfortable with the idea of "managing animals", feeling often entirely lost with the new rules and quotas.

d. Sharing and distributing the meat

Sharing, however, was and still is one of the fundamental values of Inuit and many other Arctic indigenous societies. It has been extensively studied not only by Saladin d'Anglure (2000) in Canada, but also by Nutall (2000) in Greenland. Among Inuit, sharing remains a marked feature of all first catch celebrations. Sharing involved hunting partners, relatives, namesakes, and sometimes even the whole community, as in the case of whale hunts. Patterns of sharing varied in each local area, making difficult any general conclusion. Once an animal is killed, its remains have to be used and shared. The game is incorporated into society, respected, honoured, and used. Thus, the obligation to give maintains the flow of game that nourishes the community and enables human beings and animals to reproduce.

The hunter gained prestige by success in hunting, his wife by distributing the meat generously. Emile Imaruittuq (quoted in Oosten, Laugrand and Rasing 1999: 84) also observed, "People who are stingy about food, are probably always hungry. If you are a sharing person, you know you are always going to get more meat. The meat is going to be replenished immediately" (quoted in Oosten, Laugrand and Rasing 1999: 137). As a consequence, there was little need to take action against stingy people. The game would retaliate in its own way. Stingy people would be poor hunters.

Today, the resuming of whale hunts since 1996 greatly illustrates this importance of hunting and sharing for Inuit. Whale are praised for their maktak, their skin, and huge portions are sent by plane to neighbouring communities. Elders pointed out that without whales nutrients, they feel that part of their bodies is missing (see Laugrand and Oosten 2014, chapter 10).

2. Ontology on the ice: the problem of categories

a. The predicament of the hunter

Tivi Etok, from Kangiqsualujjuaq, Nunavik, related the story of Alluriliik that expresses the predicament of the Inuit hunter very well. He describes the case of a hunter who once refused to hunt a walrus with tiny tusks, finding it too small to be killed. The walrus who continued calling the hunter, shouting at him "Harpoon me, as I would like a drink of water", finally got mad and, rebuffed, started shouting to his big herd: 'He does not want us!' And, according to the story the whole herd finally fled on hearing such offence. But the situation soon got worse as the caribou also heard the walrus' words, and fled the area. And, soon, all the animals abandoned the area for many years. (Weetaluktuk and Bryant 2008: 187–88)

Killing game is thus not only a necessity to survive, but it is also an obligation placed upon the hunter by the animals themselves. Hunting is not a matter of choice, but a moral obligation one cannot escape. Only by hunting can Inuit as well as animals prosper. The hardships and sufferings entailed in hunting and the risks involved in killing the animals must all be accepted. Human society itself is made possible by consuming and sharing the game animals.

The modern worldview teaches us that human beings resemble animals. The shamanic perspective informs us that animals resemble human beings, they just have a different body (Viveiros de Castro 1998). Like human beings, they have a *tarniq,* a miniature image or shade, and sometimes they can appear as human beings, because they have an *inua,* its human person or owner. Yet, they all resemble human beings in different ways. In Inuit culture we find many traditions about beings that might look like human beings but were not real humans, *inuunngittut,* such as dwarves, giants and *ijirait* (mountain spirits). These entities are better called non-social beings as they are watching humans and can be quite violent against them. One might have sexual relationships with them, but usually such relations were not supposed to last. Thus, Inuit did not have a universal notion of humanity, but a differentiated one that allowed for other beings that might look like Inuit, but were not really Inuit. Stories warned people to be cautious in becoming too close to these non-Inuit beings.

According to Descola, the common ground of all the entities populating the world is not the humans as a particular species, but humanity as a condition (Descola 2005: 30). As we have seen, Inuit are hunters, with all the moral implications this condition implies. They never feel uncomfortable with killing and eating animals when they need them. They look for game and start their exchanges as soon as the animal has been killed. Their morality is complex and not without contradictions. To some extent it could be summarized by the injunction, "Kill only what you need and share it with others." The game that offers itself, however, should not be refused, and that may sometimes result in killing more animals than are actually needed. But once the game is killed, the spoils should be shared. The animals are not afraid of being killed, but of being treated in a disrespectful way.

Inuit are aware that the social nature of human beings cannot be taken for granted. Only by observing the rules and rituals of their ancestors can people preserve their human and social nature. A person is not a human being by nature, but by being a moral and social person in terms of the ideas and values of his society. The human moral condition is defined in relation to the parties that sanction the existence of human beings, notably the animals, non-human beings such as the as sea woman and the ancestors. The ancestors are primarily the deceased namesakes. In the past, Rasmussen (1929: 58–59) reported, that everyone on receiving a name would receive with it the strength and skill of the deceased namesake. But whereas the deceased namesakes will support the hunter who derives his skills and abilities from them, the punishments for transgressions usually come from the animals and their owners, not the deceased. In particular, the animals will no longer offer themselves to be killed.

It is quite clear that Inuit attribute awareness, intelligence, language, and feelings to animals (Descola 2005). Inuit assume animals have their own communities where they live in human appearance and follow their own rules. Rasmussen (1929: 269) presents a story called "The Owls That Talked and Lived Like Human Beings". Thus, animals are social in their own terms, but their sociality is irrelevant in human society. Here they do not acquire a social status, and they remain anonymous beings to be killed by the hunter. Hunters should kill animals, and animals prosper when they are killed by hunters. But there is no such thing as a social contract between social parties. The hunters may bring small offerings to the deceased or to the owners of the game or even to the animals themselves, but success in hunting is never ensured.

b. Animals as non-human persons?

Already in the early 1990s, Ann Fienup-Riordan qualified animals as non-human persons, implying that even though animals are not humans, they are persons. She emphasizes that animals can reveal themselves in human form once they have been skinned (see Fienup-Riordan 1990: 169–70), but this visual aspect does not imply that they are human beings.

Fienup-Riordan (1994: 159, 188) argues that humans and animals are conceived as incomplete without the other, and she emphasizes reciprocity. According to Fienup-Riordan (1994: 48–49),

> The differentiation of persons into humans and nonhumans was for Eskimo peoples at the foundation of social life … The essential relationship in the Arctic is between humans (male or female) and animals … The ritual process creates the passages between worlds as cultural rules set the boundaries between them. Food sharing and gift giving constitute the core of Yup'ik social life. During the annual cycle of ceremonies, dead humans and animals were gradually drawn into living society, feasted and hosted, and finally sent away again.

Animals can be killed and eaten, because they are not human and do not have a social identity. They remain anonymous. In the myth of Arnaqtaaqtuq, human beings can be reborn as animals, but if this happens, they do not acquire a social identity. It can only be recovered when they are born again as human beings and receive a name. The *atiq,* name, connects one to a deceased namesake and firmly positions one in a social network. However, animals will return time and again to the same hunter who killed them and treated them with respect. They are not connected to their ancestors, but to the hunter who kills them.

In Inuit societies, animals have consciousness and awareness, but they are different from human beings in that they lack names that connect them to their ancestors. It is this relationship that provides human beings with a social identity and the lack of this relationship makes it impossible for animals to be part of human society (see Laugrand and Oosten 2014). Animals are anonymous beings who can only participate in human society by allowing themselves to be killed. Through death they become part of society, as their meat is shared and eaten and their skins are transformed into the furs that enable human beings to survive. The killing of game makes the existence of society possible, as the animal becomes an object of social exchange through sharing.

c. Keeping the correct distance and the problem of the gift

The non-social nature of animals requires that human beings keep their distance from animals and not become too close to them. In the Iglulik we find an extensive discourse on bestiality as a relation that has to be avoided. Thus, human beings should not become too intimate with animals. The dog is the animal that comes closest to human beings, as a hunting companion that only survives in human society. But Inuit always maintain their distance to dogs and traditionally never allowed them in their living space. Sometimes other animals, such as bears, could be adopted and kept in the igloo for some time. Inuit were aware that such a relation could not last, and when the animal grew too big, it was set free again. Similarly, many Indigenous people in the Arctic considered the bear as a companion for hunting (see also Feit 2000), and Inuit often indicated that they learned their hunting techniques from polar bears.

If the correct distance between hunters and animals is maintained, animals will offer themselves to be killed and human beings will be able to survive by killing their prey. It is tempting to interpret this in terms of a gift, in accordance with the magnificent scheme developed by Mauss (1989) in his study of the gift. The notion of the gift is certainly important in Inuit hunting. Simionie Akpalialuk from Pangnirtung stated, "A whale gives of itself: it's an animal that you feel fully about, it's something that gives itself up to you and that's an important thing in our beliefs, that's still very strong" (quoted in Freeman et al. 1998: 42). But in this case the gift of the animal only implied that it made itself available. The hunter still had to capture it and take its life.

The notion of the gift does not exclude an awareness of the violence that is done to an animal in killing it. The killing and preparation of the meat as well as the preparation and sewing of skins were complex operations fraught with danger. The rules of respect were intended to safeguard these processes. Mauss was fascinated by the notion of the gift and the idea of a counter-gift, postulating that each gift required a counter-gift. This may apply to relations within society where human beings are by definition social beings, but it is rather doubtful whether it can apply to beings who lack a social nature. Animals and their owners have to be induced to make the game available by observing ritual injunctions, making small offerings or tempting and seducing their prey. If all this fails, shamans have to force these beings to supply human beings with the food they need and coerce them into continuing to allow animals to be killed.

Regardless of the small gifts that may be given to the owners of the game at the beginning of the hunting season, an asymmetrical relationship continues to exist between the owners of the game (the sea woman, the *tuurngait,* or God) and human beings. Human beings cannot reciprocate with a gift of equal value to the prey. They remain indebted to these transcendental agents. By giving animals, these owners not only enable human beings to survive but also enable them to become social beings by sharing the catch. The owners of the animals and the animals themselves observe whether the prey is handled correctly by the hunter and his wife, whether it is shared in the camp and how it is celebrated in the feasts. The sharing of meat and goods results in more game. Inuit give to each other and their generosity maintains the flow of life. The small gifts to the owners of the game and the spirits of the dead open the path for the game. But in those domains people cannot gain status as a giver of meat or goods. The abundance of game depends on the generosity of the owners. As long as the Inuit respect the game and observe the rules the flow of game continues. But that offer should not be ignored or slighted, as it is not a simple gift. Once the hunting starts, the animal will try to escape the hunter, and the hunter will need all his skills and resourcefulness to make the kill. And he may also fail. There is no such thing as a free gift.

Conclusion

Inuit and Yup'it are good examples of Arctic animism. In most Arctic societies, hunting remains a mode of life and humans are always maintaining relationships with animals, deceased and non-human beings. Shamanism is also often a common institution, the shaman being the master of transformation, but also the one entitled to solve problems in the domain of health, hunting or social relationships. Tanner and Brightman for the Cree; Fienup-Riordan for the Yup'it all provide excellent case studies describing and analysing human–animal relationships, showing the contrasting principles of reciprocity and domination. Ingold and Descola rightly point out the need to go beyond the nature versus culture dichotomy whereas lately, Viveiros de Castro (1998) introduced a very complex debate with the notions of multinaturalism and that of perspectivism, suggesting that the body is the primary locus of perception and that any point of view is located in it.

At the same time, Arctic traditions are all very different from each other, with important variations in languages, categories, local histories, mythologies, etc. These differences should be better acknowledged if we wish to understand indigenous perspectives. Among Inuit, traditions also differ at a regional, local, and family level. But all over the Arctic, some general features can be identified. One of them is certainly the fact that animals are not seen as a "finite resource". If human beings behave properly, respect them, and share their meat, if they are good predators, animals are rather seen as "an infinitely renewable" resource to use Fienup-Riordan's expression (Fienup-Riordan 1990: 167). They thus always remain partners and prey. As for change in ecology, as the Yup'it put it, "The world is following its people" (Fienup-Riordan 2010).

References

Briggs, Jean L. (ed.), 2000. *Childrearing Practices*. Interviewing Inuit Elders 3. Iqaluit: Nunavut Arctic College.

Brightman, Robert. 1993. *Grateful Prey: Rock Cree Human-Animal Relationships*. Berkeley: University of California Press.

Descola, Philippe. 2005. *Par-delà Nature Culture*. Paris: Gallimard.

Feit, Harvey A. 2000. "Les animaux comme partenaires de chasse: Réciprocité chez les Cris de la baie James". *Terrain* 34: 123–42.

Fienup-Riordan, Ann. 2010. "Yup'ik perspectives on climate change: "The world is following its people". *Études Inuit Studies*, 34, 1: 55–70.

Fienup-Riordan, Ann. 1994. *Boundaries and Passages: Rule and Ritual in Yup'ik Eskimo Oral Tradition*. Norman: University of Oklahoma Press.

Fienup-Riordan, Ann. 1990. *Eskimo Essays. Yup'ik lives and how we see them*. New Brunswick: Rutgers University Press.

Freeman, Milton R., Lyudmila Bogoslovskaya, Richard A. Caulfield, Ingmar Egede, Igor I. Krupnik and Marc G. Stevenson. 1998. *Inuit, Whaling and Sustainability*. Walnut Creek, CA: AltaMira Press.

Henriksen, George. 2009. *I Dreamed the Animals: Kaniuekutat: The Life of an Innu Hunter*. New York: Berghahn Books.

Ingold, Tim. 2000. *The Perception of the Environment. Essays in livelihood, dwelling and skill*. London and New York: Routledge.

Krupnik, I., Aporta, C., Gearheard, S., Laidler, G.J., Kielsen Holm, L. (eds.) 2010. *Siku, Knowing Our Ice*. Springer.

Laugrand, Frédéric and Jarich Oosten. 2010. *Inuit shamanism and Christianity*. Transitions and Transformations. Montreal: MQUP.

Laugrand, Frédéric and Jarich Oosten. 2014. *Hunters, Predators and Prey. Inuit Perceptions of Animals*. Oxford, New York: Berghahn Books.

Mannik, Hattie. 1998. *Inuit Nunamiut: Inland Inuit*. Altona, MB: Friesen Corporation.

Nadasdy, Paul, 2003. *Hunters and Bureaucrats. Power, Knowledge and Aboriginal State Relations in the Southwest Yukon*. Vancouver: University of British Columbia Press.

Nutall, Mark. 2000. Becoming a hunter in Greenland. *Études Inuit Studies*, 24–2: 33–45.

Oosten, Jarich, Frédéric Laugrand and Willem Rasing. (eds.) 1999. *Perspectives on Traditional Law*, Interviewing Inuit Elders 2. Iqaluit: Nunavut Arctic College.

Pernet, Fabien. 2013. "Inuguiniq. La construction de la personne dans les rites de passage des Nunavimmiut", PhD dissertation. Quebec City: Université Laval.

Rasmussen, Knud. 1929. *Intellectual Culture of the Iglulik Eskimos,* vol. 7 (pt. 1) of *Report of the Fifth Thule Expedition 1921–24.* Copenhagen: Gyldendalske Boghandel.

Saladin d'Anglure, Bernard. 2000. Pijariuniq: Performances et rituels de la première fois. *Études Inuit Studies* 24, 2: 89–113.

Stuckenberger, Nicole. (ed.) 2007. *Thin Ice. Inuit Traditions within a Changing Environment.* Hanover: Hood Museum of Art.

Tanner, Adrian. 1979. *Bringing Home Animals: Religious Ideology and Mode of Production of the Mistassini Cree Hunters.* London: C. Hurst.

Viveiros de Castro, Eduardo. 1998. "Cosmological Deixis and Amerindian Perspectivism". *Journal of the Royal Anthropological Institute* 4: 469–488.

Weetaluktuk, Jobie, and Robyn Bryant. 2008. *Le monde de Tivi Etok: La vie et l'art d'un aîné inuit,* Québec: Multimondes, Institut Culturel Avataq.

Wenzel, George. 1991. *Animal Rights, Human Rights: Ecology, Economy and Ideology in the Canadian Arctic.* Toronto: University of Toronto Press.

Wenzel, George. 2008. *Sometimes Hunting Can Seem Like Business: Polar Bear Sport Hunting in Nunavut.* Edmonton: Canadian Circumpolar Institute Press.

Wenzel, George, and Martha Dowsley. 2008. "'The Time of the Most Polar Bears': A Co-management Conflict in Nunavut", *Arctic* 61(2): 177–89.

17

LATIN AMERICA

Indigenous cosmovision

Miguel Astor-Aguilera

This chapter surveys native ontologies in Spanish-speaking Latin North to South America with a special Mesoamerican focus. This chapter's author is an anthropologist, specializing in Mesoamerican archaeology and ethnography, descended from Mexican Tarascan-Purepecha lineage and married to a Maya woman. Presented herein are three key ontological foci: one, indigenous Latin American cosmovisions anchor on worldviews intertwined with cosmology; two, indigenous Latin American cosmovisions, pre-Columbian to the present, conceptualize persons, whether animal (human and non-human), plant, and inanimate objects in potential kin-like relationship to one another; three, local ecologies are interwoven within indigenous cosmovisions where each "interbeing person," animate and inanimate, organic and inorganic, participates in seen and unseen manner within relational communities composed of myriad entities.

Spanish colonized Latin America

Indigenous cosmovisions in Latin America exhibit a relational world where causation is expressed through powerful breath-like essences comprising heat, air, and moisture. Indigenous Latin American cosmovisions share core commonalities; however, differences arise due to the size of North and South America and the diversity of its cultures and languages. Latin North America (Mexico and Central America) covers approximately 2,473,343 square kilometers. Latin South America (Argentina, Bolivia, Chile, Columbia, Ecuador, Paraguay, Peru, Uruguay, and Venezuela) covers 8,824,675 square kilometers. Indigenous Latin North America has approximately 24.3 million people and indigenous Latin South America 26.9 million people belonging to hundreds of distinct populations. Hundreds of aboriginal American languages have become extinct since European contact, due to colonial genocide and cultural practice extermination; yet, close to a thousand indigenous languages are still spoken (C.I.A. 2015).

Indigenous Latin American cosmovisions typically have no formal churches; however, an estimated 28 million of the varied Latin American indigenous, especially those with large traditional-conservative populations, have adapted ancient cosmovision templates within Christian traditions. Countries with the largest indigenous Latin North American populations are Mexico (15.7 million) and Guatemala (5.8 million) and in indigenous South America are Peru (13.8 million) and Bolivia (6 million). In Mexico, the largest indigenous population are the

Maya (2.5 million) and in Peru the Quechua (3.25 million) with Mayan and Quechuan language-family speakers respectively spread over various countries (C.I.A. 2015).

Northeastern Asian to indigenous American cosmographies

Pre-Columbian Americans left imagery, sculpture, burials, monumental buildings, and codices that academics continue to assess. Known as *Indios*, "Indians," contemporary indigenous populations of Latin North to South America, descend from the first inhabitants of the "New World." Also called Amerindians, indigenous Americans are the First American Nations whose ancestors are gauged as first arriving in the Americas approximately 33,000 years ago. Genetics and physiological similarities indicate links between First Nation peoples and Eastern Asians crossing a shallow channel, the ninety-kilometer wide Bering Strait (Beringia), linking Alaska and northeastern Siberia, Asia.

A human fossil at Lapa Vermelha (11,000 BC), Brazil, for example, resembles Ainu or South Pacific Islander populations similar to skeletal material from Kennewick (9,000 BC), Washington, United States (Moore 2014, 80). Other Paleo-American (early inhabitant) skeletons also indicate physical similarity with 7,000 BC Ainu peoples and northeastern Asians. Beringia has been exploited since 4,000 BC by *Inuit-Eskimo* fish and sea-mammal hunters living and trading on both sides of the Bering Strait and within its islands. Historical Alaskan Inuit, in addition, are known to have traded deep within the North American eastern interior and far southward toward British Columbia. Indigenous trading involves cosmovision-informed conversation and migrations between northeastern Siberia and Alaska, as recently as 2,000 BC, indicate why core similarities are known to exist between Siberian, Ainu, and indigenous American cosmovisions.

Indigenous Latin American ecological perception

Aboriginal American cosmovisions, pre-Columbian to post-European, vary dependent on local ecologies. Maize, beans, and squash, due to interlinked agriculture at time of Iberian contact, dominated much indigenous American cosmovision (Kehoe 2006, 14). What is eaten, and its procurement, affects economy, politics, and ontology (Astor-Aguilera 2015). Indigenous Latin Americans do not differentiate between economics, politics, and religion and have no discrete native categories/vocabulary to separately describe them (Astor-Aguilera 2010). The environment determines plant and animal lifecycles that in turn influence human perception of food and ecology (Astor-Aguilera 2015).

Earth's blood

Some indigenous American sites, since the Early Archaic (8,000 BC), contain red ocher (Moore 2014, 122) and/or vermilion cinnabar covered human skeletons. Iron-rich ocher and cinnabar have a flesh-and-blood hue indexing inanimate matter as sentient. Ocher is found in "veins" within mounds, hills, and mountains perceived as kin-like persons. Aboriginal Americans, since 8,000 BC, built artificial mounds, hills, and mountains that they often colored from yellow-to-red-to-beige. Pre-Columbian cosmovisions were not codified, either temporally or spatially, so there is no certainty as to what pre-European contact peoples actually thought in terms of religion; however, mound-associated material record conventions indicate ecological regeneration patterns.

The Sun and cosmovision

Wild foods and animals reappear predictably throughout the seasons and, as used by humans, become inextricably linked to indigenous cosmovisions. Until 7,000 BC, aboriginal American hunter-gatherers moved from one wild food to another, shifting from camp-to-camp, according to seasonal shifts around large homelands. Annual treks following animals and plants were seemingly linked to the sun's apparent movement in the sky from an earthbound human's perspective that predictably varies throughout the year in relation to Earth's eastern and western horizons. Tracking seasonal shifts apparently spurred cosmological explanations since circumambulations throughout the homeland seemed as if living within a blue-green bowl-like world (García-Zambrano 2001, 352; Kehoe 2006, 127, 132; Montejo 2001, 182). Within this bowl-like ecological world exist plants and animals, including humans, relying on each other to flourish. Within Latin American cosmovisions there is a power-within-the-world that all things, organic and inorganic, partake of which potentially leads to sentience and animacy (Montejo 2001, 187). Objects in this interlinked universe are potentially powerful as they share vital ancestral essences spread throughout the world.

Ancestors and ecological power

Mummification developed around Peru/Chile 7,000 years ago. Chinchorro "Red Mummies" (2,200–1,700 BC) were red ocher painted, with open eye slots and open mouth indicating sight and speech, exhibiting a non-bilateral cosmovision of life and death. Red-painted Chinchorro skeletons index life—an idea transferred to signify the same for red-painted objects and buildings—both practices of which are common in indigenous Latin America. At Paloma, Peru (5,000–3,000 BC), for example, one mummy was buried with a red pigment filled mussel shell and, at Santa Ana–La Florida, Ecuador (2,600 BC); a coil-shaped lithic wall spiraled inward creating a uterus-like hearth containing human bone and greenstone beads. Next to the hearth were two skeletons accompanied by greenstone artifacts, ceramics, and a strombus conch shell. The ceramics depicted snakes and two scenes of a human emerging from a spondylus scallop shell (Moore 2014, 226).

Shells for indigenous Americans index fertility per female genitalia. Snakes, being phallic shaped, are also a common fertility motif found throughout the aboriginal Americas. A stairway, with sexual connotations, leading into a chamber within a sunken plaza at Kotosh, Peru (1,200 BC), for example, is painted red with a snake entering. Central Andean sunken rooms are similar to early Chaco Canyon, New Mexico (500 AD), kivas representing masculine elements entering the uterine Earth. Andean sunken room floors were often layered with yellow-to-red clays and the female indexing shells. Aboriginal American women were highly esteemed and at Real Alto, Ecuador (2,600 BC), a large mound supported a structure housing a woman's skeleton within a lithic-lined pit. Another large mound at Real Alto had feasting evidence alongside the dead, a common practice throughout indigenous Latin America.

Feasting, fertility, and sustainability

Mesoamerican maize-corn was traded down to South America and seemingly exchanged for cacao pods and orchid vanilla beans. Cacao, per its oval pod and brownish-red seed, and vanilla, per its female genitalia alluding flower, represent female fertility. Tender corn also has fertility connotations in that its kernels, with milky-white fluid, are linked to semen (Astor-Aguilera 2010, 39). Food connects to fertility and aboriginal American cultigens like pumpkin, pecans, gourds, sunflowers, squash, millet, amaranth, beans, tomatoes, cotton, agave, chili peppers, rubber sap plants and trees, potato, yams, peanuts, manioc, avocado, pineapple, and cacao were all linked to indigenous cosmovisions.

Tobacco is also significant within indigenous Latin American cosmovisions. Native tobacco, being quite potent, is used in prognostication rituals and for healing diplomatic discussions. Mind-altering botanicals used in indigenous Latin American rituals are mushrooms, peyote, and ayahuasca. Maize significance to indigenous Americans, however, has few competitors and is observed from southern Canada to central Argentina. Maize anchored death-to-life regeneration concepts are conspicuous in indigenous Latin America and are currently associated in hybrid form with *gracia*, grace, as taught by Roman Catholics. Indigenous grace, however, differs from Christian belief as unmerited favor given by God as a divine blessing. Grace in Maya communities is based on ecological respect anchored by reciprocity and diplomatic favor. While Western ideology seeks to dominate nature (Sale 1990; Thomas 1983), indigenous American cosmovisions center on ecological actions drawing out political grace-and-favor where the environment responds in kind to human behavior (Montejo 2001, 187).

The colonial Maya adopted Catholic grace when speaking of the green maize stalk's ecological potency (see Haly 1992, 288 note 62; Roys 1967 [1933], 107). Maize is important in indigenous Latin American cosmovisions because it feeds the masses (see Florescano 2002, 145); however, maize needs humans to survive since it cannot regenerate by itself, encased in its dense husk, and therefore needs human assistance to flourish. To the Maya, maize represents male–female fertility, indexing humanity (Montejo 2001, 187), and is therefore often depicted as a feminized human male. As a child, this chapter's author was given a "baby corn" with flowing silk hair to play with as if a human doll by his Tarascan grandmother, this while new green corn grew in the maize field in similar manner to the Guatemalan Ch'orti' per Stuart (2011, 226).

Maize to the Maya is *not* a deity as often assumed. The "first fruiting corn," *Hun Yée Nal* in Mayan, is represented as a person with a body and face; however, per McAnany (2008, 224), "the maize deity is no more than the omnipresent fruiting corn stalk." The so-called Maize God is simply the green stalk adorned with corn and long flowing leaves dancing in the breeze (see Florescano 2002, 140–142, 191). A healthy green and fruitful corn stalk swaying in the breeze is described person-like by the Maya as *le nalo tan u okot,'* "the maize stalk is dancing." For indigenous Latin American cosmovisions, their behavioral environment requires personal interaction *not* requiring *nor* having gods but person-like beings representing ecological powers (see Astor-Aguilera 2010; Fischer 2001, 194). For the Maya, dealing with the environment is similar to engaging ones' kin and neighbors (Meskell and Joyce 2003, 105). Ecological power for the Maya, like other indigenous Americans (Morrison 1990, 418–419), is central to their non-polar binary and non-deistic cosmovision. What matters to the Maya is ecologically related power *not* belief and faith per theology. Indigenous Latin American ecological notions are based on an "interbeing" variably termed "Amerindian Perspectivism" (Viveiros de Castro 1998, 471). For the Maya, environment relations focus on the pragmatic consideration that humans need ecological balance to survive. Environmental power here, however, is *not* ethically neutral and responds in personal manner to human actions (Astor-Aguilera 2010).

Every "thing" is potential kin

Sedentariness leads to villages established in different ecological locales varying season-to-season. Unlike industrial city life, village cooperation fosters day-to-day relationships responding to common needs through social diplomacy with immediate to extended kin. Villagers narrate how people deal with their flora and fauna and abiotic environment. Ancient influences, especially from the "Olmec" peoples of Mexico (1200 BC to 200 AD), have had tremendous effect on Mesoamerican cosmovisions. The Olmec are known for imagery linked to powerful animals like jaguars, eagles, and serpents, as well as the regenerating aspects of plants (see Joralemon 1971). Similar to Olmec imagery, extensive northwestern South American (1767–200 BC) motifs represent caimans, raptor birds, serpents, large felines, and human-like beings associated with these predatory creatures (Quilter 2005).

Pre-Columbian cosmovisions were linked to rituals that often included human and non-human animal bone in context with red-to-yellow colored sands and/or clays (Kehoe 2006, 27, 32–33). An Olmec tomb at La Venta (900 BC), for example, had two children's red painted skeletons with a male or female figurine each along with jadeite green stone pendants and beads. The male had a serpentine colored stone 'celt' and the female had a jadeite stingray tail, jadeite frog, and jadeite human hands. Red ocher indexes blood while green stones signify ecological fertility. There are variable practices within indigenous Latin America cosmovisions; however, crucial aspects remain similar. Powerful images, for example, manifest elite kinship with the sun and other environmental phenomena. Indigenous Latin Americans understand celestial bodies, due to their sky-movements, as sentient beings having volition capable of affecting Earth. Being related, what affects the sky affects Earth and vice-versa.

European religious conquest

Europeans and their descendants, from the colonial period onward, have slowly eradicated much indigenous Latin American knowledge. Although there are no known pre-Columbian prophets or ancient inscriptions elaborating religious philosophy, indigenous elders have in recent years been uniting in pan-native movements to strategize against their dominant nation-states. Indigenous Latin American environmental concerns are politically based on ecological diplomacy. Indigenous Latin American rituals focus on regeneration within an experiential perspective where the whole universe is understood within reciprocal relational relationships (Montejo 2001, 192).

Indigenous Latin American cosmovisions differ from the Western due to their holistic cosmological nature interlocking socially with everything that exists. Indigenous Latin Americans seek knowledge through prognosticative healer specialists whose *knowing* acquires "medicine" to heal a particular person and/or situation. A pragmatic caretaking of Earth ensures its survival; however, this ideal is difficult to achieve in practice since *all* humans use Earth and its resources. In present-day Maya villages, for example, there is continual debate, with no clear consensus, as to how much game should be taken from the forest as animal populations continue to decline due to modern technology, such as rifles, facilitating hunting.

The natural and supernatural

A primary characteristic of indigenous Latin American cosmovision is the intersubjective communication with non-human animals, plants, water, inorganic matter, such as rocks and crystals, celestial bodies, and even meteorological phenomena such as clouds and lightning. Pre-Columbian cosmovisions are difficult to unravel within Western ontological frameworks since a particular tree, rock, stream bend, or cave locale can be a site for communication and negotiation. Westerners, however, do not tend to accept non-animal volition per pre-Columbian Americans where many of their descendants' veneration of non-human beings is reciprocal and relational rather than deistic worship.

Ecology, serpents, and "gods"

Assumptions that indigenous Latin American peoples worshipped natural forces as deities and/or gods are notoriously ill determined (Betanzos 1987 [1551], 31; Coe 1973, 10; 1973 [1950], 76–81; Florescano 2002, 191; Graham 2011:301; MacCormack 1991, 107, 109 note 72; Urton 2008; Van Stone 2010:3, 39, 123); indeed, there are no native concepts/words to discuss theological ideas describing transcendent supernatural beings (Arnold 1999:xv; Astor-Aguilera 2009; Holbraad 2009, 433; Nicholson 1973, 89). Snake images, for example, do not evidence serpent worship but only that snakes are respected. Within indigenous America, the phallic-shaped serpent is perceived as regenerative per its growing back its tail (actually a legless lizard), peeling back its prepuce-like old skin, coiling next to watering holes, and penetrating uterine-like Earth holes. Indigenous Latin Americans also respect large cats, predatory birds, deer, wolves, coyotes, foxes, badgers, and sea creatures if near the ocean. Respected in like manner to powerful animals are the sun, fire, moon, light, shadow, wind, storms, lightning, stars, planets, and earthquakes.

Proper names of animals and plants

Westerners see themselves not as animals but as superior to them as well as dominant over plants and Earth. In indigenous cosmovisions, however, humans are not different from animals or even, depending on context, from plants or their environment. The Maya, for example, live in a reciprocal worldview in which favors and return expectations mark their relations with the environment. One manner of potentially communicating intersubjectively with non-human animals and inanimate objects is through indigenous American hunting rituals (Morrison 2000, 27). Non-human persons are named and addressed as if human-persons constituting an "inter-being of cosmology and community" (see Grim 2001, xxxvii).

Non-human beings are here seen to be as conscious as humans (Ingold 2011, 175) and, for this reason, indigenous Latin Americans tell stories about human-like talking animals and plants. Also addressed respectfully here is inanimate matter and, therefore, when addressing rain and lightning one speaks of and to rain and lightning as a kin-like person and same with a rock, tree, etc. Person-like beings in indigenous Latin American environments are addressed as kin because *everything* is potentially extended family. Caretaking Earth requires respectful engagement since consequences to the contrary are fraught with social problems just as within human diplomatic relations.

Cosmovision continuity

Many indigenous Latin Americans, though transformed through adaptation, have retained their ancestor's cosmovision. What we know about aboriginal life at contact mostly stems from European chronicles. We know more about the Mexica ("Aztec"), Maya, and Inca than other pre-contact peoples because the Spanish conquistadors extensively chronicled their conquest. Since conquistador documentation was *not an ethnographic endeavor*, the colonial Spanish assist but also heavily distort our comprehension of indigenous thought. Filtered through Catholic lenses, Spanish clergy's goal was *not to comprehend and arrive at natives' cultural meaning* but to understand them enough in order to subjugate them and supplant native cosmovisions with Christianity. Catholic friars noted that indigenous Americans made offerings to significant ecological features (García-Zambrano 2001, 352); however, they missed seeing that the natives' ontologies did not fit their Judeo-Christian template of worshipping supernatural divinity (see Montejo 2001, 175–176, 178–179).

Throughout the pre-Columbian Americas, indigenous peoples viewed the world as imbued with potent forces with which humans could interact. Exhibiting the pre-Columbian anthropocene, very heavy stones, some weighing tons, were moved many kilometers from distant mountains to building sites. The stones were chosen for their quality, color, and/or cleaving qualities indicating potential power. In South America, natives made offerings to mountains whose snow melted into springs and rivers in reciprocity for good weather and fertile grounds. Colonial friars zealously placed their churches and icons where indigenous Americans made offerings in trying to superimpose their Catholic hierophanies; however, natives created hybrid poly-ontological cosmovisions that remained engaged with their sentient environment.

The search for universal religious meanings within cultures foreign to Judeo-Christian-Muslim theology can obscure the relational process of indigenous cosmovisions. Pre-European contact indigenous populations, and many of their descendants, *do not* conceptualize the supernatural or categorize in other polar binary opposites (see Geertz 1966, 4; Houston et al. 2006, 98, 179; Klass 1995, 25–33; Knab 2004, 109; Montejo 2001, 176; Renfrew 1994, 48; Schaefer and Furst 1996, 24; Sharer 1994, 513–14). Indigenous worlds are inclusive rather than exclusive (Furst 1995, 173–84; Klass 1995, 26, 30). Indigenous cosmovisions formed by interlocking complementary dualities are holistic and emphasize mutuality rather than opposites such as supernatural/natural, sacred/profane, and divine/secular (Astor-Aguilera 2009, 2010). Indigenous Latin cosmovisions emerged from speech relating to a non-theology-based living ecology. Per Schaefer and Furst (1996, 12), indigenous cosmologies "are often called 'animistic,' if not 'nature worship,' but it might be more accurate to call them 'ecological'."

Aboriginal ontologies undergird indigenous Latin-American cosmovisions. The word *camay*, in Quechua, for example, connotes a personal causality expressing itself. Similarly, in Mayan, the phrase *k'aax na'ach'*, is used by Maya ritualists when tethering a non-visible being, or when it tethers itself willingly, to a body-object so it can better express itself. Tethering here applies to humans, animals, plants, ceramics, rocks, water sources, and even meteorological phenomena (Astor-Aguilera 2010, 161–162). The Mayan word *k'ul* references tethered potency such as the fireball in the sky we call the sun as well as people, objects, structures, and land features that have intense character (Astor-Aguilera 2010, 22). The Mayan *k'ul* is similar to the Quechua notion of *huaca*, namely, potent places that one should address with kin-like respect.

Objects as kin

In indigenous American cosmovisions, the sun's heat connects powers circulating in the world held temporarily in various strengths by individual human or non-human entities. This is similar to the *Orenda* concept of the Iroquois, *Wakan* of the Sioux, *Sila* of the eastern Inuit, and *Teotl* of the Mexica (Furst 1995, 177–178). The Sun's heat resonating on Earth is associated with cosmic breath entering objects in differing quantities lending them powerful essences. Environmental potency can be ritually reciprocated if sentient and manifesting volition within its energy-like being. All things have the potential for agency and communication within indigenous Latin-American cosmovisions (Astor-Aguilera 2010).

Being "religious" in aboriginal America

Latin indigenous rituals set in ecological contexts is practical and focused on political action within local environments. What we regard as religious is as much secular within aboriginal America. Being religious, throughout Judeo-Christian-Muslim traditions, relates to piety, worship, and fear/love of the divine as supernaturally derived. Progressively from the colonial period, indigenous Latin Americans have used Western religious terms, in speaking European languages, to defend their aboriginal practices and sovereignty. The use of non-indigenous words, often employed in ideological rhetoric, evokes political reactions that cannot be achieved through the use of non-Western terminology. The circular construction of academic knowledge in which assertion stands as proof affects how indigenous elders speak about their cosmovisions within European language-based Western educational systems and Christian churches. Conservative indigenous Latin Americans, however, base their actions on integrated sociality maintained through interactions with sentient personal beings. Within indigenous cosmovisions the past is in the present and this characteristic is subtle to conspicuous depending on degree of Christian acculturation.

Indigenous Latin American cosmovisions stress a social relational debt-and-merit based reciprocity. What we assume to be indigenous "sacrifice" and/or "penance" is exchange based social respect and diplomatic acknowledgement of variable knowledge, power, and skills with other sentient beings (Bird-David 1999, 73; Klass 1995, 28–29; Morrison 1992, 207; 2002, 38, 42–43). How much social power is attributed to other beings, visible or invisible, is dependent and exhibited within relational debt-and-merit-based behavior. Offerings here are not benevolent actions, that is, gifting is not done out of altruistic behavior but due to diplomatic acts seeking balanced relationships.

Debt-and-merit-based actions entail a constant giving and receiving due to reciprocal relationships presupposing conflict since Earth can be a wonderful and abundant place to live but it is also harsh and deadly. Reciprocation acknowledges that one needs to maintain a diplomatic caretaking role of Earth so it in turn remains "emotionally" balanced (Read 1998). To say, within indigenous cosmovisions, that the world is *alive* is more than just our metaphorical acknowledgement that Earth is a "living planet."

Indigenous Latin Americans potentially communicate with a wide assortment of trees and plants with some having medicinal properties and sentient healing roles. Some pre-Columbian to colonial depictions of trees and plants have open mouths exhibiting communication. Indigenous Latin Americans can potentially communicate with ancestors through animals and plants since they are not in a separate otherworldly realm/dimension but are within and on Earth and sky. In indigenous cosmovisions the ancestors are always present and quotidian life makes sense day-in and day-out in maintaining a dynamically balanced Earth.

End discussion

Indigenous languages use idioms and metaphors deployed in daily practice that are difficult to translate into European speech focused on theology and/or science. When speaking metaphorically in English of mountains having body parts as in the *foot, shoulder,* and/or *head of the mountain,* Westerners are not literally stating that mountains have feet, shoulders, and/or heads; however, for indigenous Latin Americans who relate to their environment interactively these *are* literally embodied terms. Aboriginal Latin-American cosmovisions evolved in very close ecological social relationships focused on invisible person-like beings surrounding their communities. Indigenous American cosmovisions constitute a behavioral ecology that is interrelational and intersubjective and some indigenous peoples living a conservative agrarian lifestyle have retained relational forms of communication with Earth.

Indigenous cosmovisions are not homogenous and especially at present since First Nation peoples are continually adapting to Western technology and Christian theology; however, ontologies remain, temporally and spatially, that are core in similarity. Indigenous Latin

Americans communicate and negotiate with organic and inorganic beings in their world as human-like persons with which they enter into caretaker responsibilities. Indigenous people may be drawn to ecologically sensitive nation-state political parties, such as the *Partido Verde*, Green Party, of Mexico but their primary motivation is not the same as liberal environmental movements within industrial capitalism. Indigenous cosmovisions and resulting ecological practices are pragmatic and not separate from economic or other personal concerns anchored in daily life.

Ritual reciprocity with local environments *does not* separate humanity and nature as suggested in Western environmental slogans per "*Save the Earth*." Saving Earth to indigenous Latin Americans is practical since, not living in urban cities detached from the "outdoors," balancing the environment means maintaining their way of life. Volition and causality, always personal, is generated in relation to non-human kin-like persons that live in the day-to-day behavioral environment of indigenous Latin Americans. The practical quotidian focus of indigenous Latin American cosmovisions is to live within nature, be it visible or invisible, since they are one and the same and nothing exists outside of it.

References

Arnold, P. 1999. *Eating Landscape: Aztec and European occupation of Tlalocan*, University of Colorado Press, Niwot.

Astor-Aguilera, M. 2009. Mesoamerican communicating objects: Maya worldviews before, during, and after Spanish contact, in Pugh, T. and Cecil, L. ed. *Maya Worldviews at Conquest* 159–82, University Press of Colorado, Boulder.

Astor-Aguilera, M. 2010. *The Maya World of Communicating Objects*, University of New Mexico Press, Albuquerque.

Astor-Aguilera, M. 2015. Native religions of the Americas, in Deming, W. ed. *Understanding the religions of the world*, Wiley Blackwell, West Sussex.

Betanzos, J. 1987 [1551]. *Suma y narración de los Incas* Atlas, Madrid.

Bird-David, N. 1999. 'Animism' revisited: personhood, environment, and relational epistemology [comments], *Current Anthropology* 40 67–91.

C.I.A. 2015. *World factbook* Central Intelligence Agency, Washington.

Coe, M. 1973. The iconology of Olmec art, in Easby, D. ed. *The Iconography of Middle American Sculpture*, Metropolitan Museum of Art, New York.

Fischer E 2001 *Cultural Logics and Global Economics: Maya identity in thought and practice*, University of Texas Press, Austin.

Florescano, E. 2002. *The Myth of Quetzalcoatl* trans. Hochroth, L., Johns Hopkins University Press, Baltimore.

Furst, J. 1995. *The Natural History of the Soul in Ancient Mexico*, Yale University Press, New Haven.

García-Zambrano, A. 2001. Calabash trees and cacti in the indigenous ritual selection of environments for settlement in colonial Mesoamerica, in Grim, J. ed. *Indigenous Traditions and Ecology: The interbeing of cosmology and community* 351–375, Harvard University Press, Cambridge.

Geertz, C. 1966. Religion as a cultural system, in Banton, M. ed. *Anthropological Approaches to the Study of Religion*, 1–46, Tavistock, London.

Graham, E. 2011. *Maya Christians and their Churches in Sixteenth-Century Belize*, University Press of Florida, Gainesville.

Grim, J. 2001. Introduction, in Grim, J. ed. *Indigenous Traditions and Ecology: The interbeing of cosmology and community*, xxii–lxiv, Harvard University Press, Cambridge.

Haly, R. 1992. Bare bones: rethinking Mesoamerican divinity, *History of Religions* 31(3) 269–304.

Holbraad, M. 2009. Ontology, ethnography, archaeology: an afterword on the ontography of things, *Cambridge Archaeological Journal* 19(3) 431–441.

Houston, S., Stuart, D. and Taube, K. 2006. *The Memory of Bones: Body, being, and experience among the Classic Maya*, University of Texas Press, Austin.

Ingold, T. 2011. *Being Alive: Essays on movement, knowledge and description*, Routledge, London.

Joralemon, D. 1971. *A Study of Olmec Iconography*, Dumbarton Oaks, Washington.

Klass, M. 1995. *Ordered Universes: Approaches to the anthropology of religion*, Westview Press, Boulder.

Kehoe, A. 2006. *North American Indians: A comprehensive account*, Pearson, Upper Saddle River.

MacCormack, S. 1991. *Religion in the Andes: Vision and imagination in early colonial Peru*, Princeton University Press, Princeton.

McAnany, P. 2008. Shaping social difference: Political and ritual economy of Classic Maya royal courts, in McAnany, P. and Wells, E. eds. *Dimensions of Ritual Economy*, 219–248, Emerald, Bingley.

Meskell, L. and Joyce, R. 2003. *Embodied Lives: Figuring ancient Maya and Egyptian experience*, Routledge, New York.

Montejo, V. 2001. The road to heaven: Jakaltek Maya beliefs, religion, and the ecology, in Grim, J. ed *Indigenous Traditions and Ecology: The interbeing of cosmology and community*, 175–195, Harvard University Press, Cambridge.

Moore, J. 2014. *A Prehistory of South America: Ancient cultural diversity on the least known continent*, University Press of Colorado, Boulder.

Morrison, K. 1990. Baptism and alliance: the symbolic mediations of religious syncretism *Ethnohistory* 37(4) 416–437.

Morrison, K. 1992. Sharing the flower: a non-supernaturalistic theory of Grace *Religion* 22 207–19.

Morrison, K. 2000. The cosmos as intersubjective: Native American other-than-human persons in Harvey, G. ed. *Indigenous Religions: A companion*, 23–36, Cassel, London.

Morrison, K. 2002. *The Solidarity of Kin: Ethnohistory, religious studies, and the Algonkian-French religious encounter*, State University of New York Press, Albany.

Nicholson, H. 1973. The late pre-Hispanic Central Mexican iconographic system, in Easby, D. ed. *The Iconography of Middle American Sculpture*, Metropolitan Museum of Art, New York.

Quilter, J. 2005. *Treasures of the Andes: The glories of Inca and pre-Columbian South America*, Duncan Baird, London.

Read, K. 1998. *Time and sacrifice in the Aztec cosmos*, Indiana University Press, Bloomington.

Renfrew, C. 1994. The archaeology of religion in Renfrew, C. and Zubrow, E. ed., *The ancient mind: elements of cognitive archaeology*, 47–54, Cambridge University Press, Cambridge.

Roys, R. 1967 [1933]. *The Chilam Balam of Chumayel*, University of Oklahoma, Norman.

Schaefer, S. and Furst, P. 1996. Introduction, in Schaefer, S. and Furst, P. ed., *People of the Peyote: Huichol Indian history, religion, and survival*, 1–25, University of New Mexico Press, Albuquerque.

Sharer, R. 1994. *The Ancient Maya*, Stanford University Press, Stanford.

Stuart, D. 2011. *The Order of Days: The Maya world and the truth about 2012*, Harmony, New York.

Urton, G. 2008. The body of meaning in Chavín art, in Quilter, J. and Conklin, W. ed. *Chavín: Art, architecture and culture*, 215–216, Cotsen Institute, Los Angeles.

Van Stone, M. 2010. *2012: Science & Prophecy of the Ancient Maya*, Tlacaél, San Diego.

Viveiros de Castro, E. 1998. Cosmological deixis and Amerindian perspectivism, *Journal of the Royal Anthropological Institute* 4(3) 469–488.

PART IV

Regional landscapes

Introduction: *Willis Jenkins*

Religions have geographies, and those geographies are often landscapes of plurality, hybridity, and interaction. This section focuses on four regions whose cultural ecologies are shaped by multiple religious inheritances. These four chapters are of course merely exemplary of an approach rather than globally representative; an entire volume could be dedicated to geographical treatments of religion and ecology. Yet these four regions are uniquely significant, each bringing into view distinctive elements of possible landscape interactions.

Two of these regions are famously and irreducibly plural. An important feature of what makes India's landscape sacred is the stunning variety of connections between nature and religion. Christopher Chapple's essay identifies resources from that variety for confronting ecological pollution, including: Yoga practices, Jain spirituality, Sikh ideas, and the legacy of Gandhi. China is the land of the Three Teachings (Daoism, Confucianism, Buddhism), to which have been added two more officially recognized religions: Protestantism and Catholicism. Avoiding an "administrative" approach to the official diversity, James Miller begins from geographic spaces in order to illuminate distinctive sensibilities of place. His chapter organizes around mountains, rivers, and coasts, showing how traditional interpretations of each comes into tension with modern dynamics.

Two other regions are shaped by traumatic conflicts with colonial religion that gave rise to new religious formations, new patterns of inhabiting a landscape, and to transformations of European Christianity. Anna Peterson's chapter on Latin America shows how the encounter of indigenous and Christian religiosities continues to shape the landscape and inform liberationist political ecologies. Melanie Harris's chapter on religion and ecology in the African diaspora works with an environmental imagination developed from the experience of environmental loss and trauma. She shows how work on racial justice and reparations does not distract from ecology but in fact interprets its contemporary meaning, and how womanist thought can cultivate a pluralist cultural epistemology needed for interconnectedness.

18

INDIA

Christopher Key Chapple

India, a vast region of more than one billion people extends one thousand miles from the Himalayas in the north to Kanyakumari at the southern tip of the peninsula and another thousand miles, from the Pakistani border to the hills of Assam. Home to one of the world's oldest civilizations, environmental problems have reached immense proportions in India. The air and the rivers are profoundly polluted. By some estimates, the Delhi region has the worst air quality in the world. Long stretches of the Yamuna and Ganges Rivers are no more than industrial effluent. Despite the work of many action plans, urgent triage must be enacted throughout the Indian subcontinent for the sake of human, animal, and plant well-being. This chapter will explore messages from Hinduism and Jainism, Gandhi's legacy, and Sikhism that can help provide alternative inspirations for redressing the problems of ecological degradation.

Hinduism and Jainism offer a personal ethic for dealing with the conflicts that arise in regard to resource management. Gandhi interpreted the *Bhagavad Gita* as calling for an inner change, a mastery over one's lower urges in service of the welfare of all. In the Sikh tradition, the *Guru Granth Sahib* calls for a recognition of God in all things, resulting in a deep appreciation for creation. These traditions of India recognize the sacred within the world and suggest that each individual must strive for inner balance in order to ensure enduring calm and economic sanity.

Hindu spirituality and sustainability

The foundations of the Hindu faith can be traced to the *Rig Veda*. Composed more than 3,500 years ago, this collection of more than one thousand chants gives praise to the powers of the universe, from the heat of fire to the glorious power of human speech, from the magnificence of the sunrise to the flowing of India's great rivers. In addition to including primal, archetypal myths such as the battle between the god Indra and the drought-bringing dragon Vrtra, the Vedas also praise the pinnacle of human accomplishment, referred to as *rita*, the attainment of artistry and ritual and fulfillment within life itself.

One of the best-known hymns, the *Purusha Sukta* (X:190), contains a message that holds a model for thinking about the human–cosmos relationship. The hymn correlates the human body with the far-flung regions of the universe:

The moon was born of his mind; of his eyes, the sun was born;
From his mouth, Indra and fire; from his breath, wind was born;
From his navel there was the atmosphere;
From his head, heaven was rolled together;
From his feet, the Earth; from his ears the directions.

By identifying body parts with heavenly bodies, the heavens and the Earth, and the elements of fire and wind, sanctity is given to both self and cosmos. Like the moon, our mind reflects and changes. Without the light of the sun, we cannot see. Our mouth proclaims our intentions and desires, and like the God of war, Indra, allows us to stake our claim in the world. Each breath we take generates and relies on the circulation of air. Our head pulls upward; our belly gathers us toward the center and allows us to expand; our feet anchor us to the earth. Our ears stabilize us within the space of the four directions. Through this vision, each human being finds a place of importance within the cosmos, signaling a continuity between the human person and his or her place in society and within nature.

A later Vedic passage, the *Prithivi Sukta*, from the *Atharva Veda*, gives praise to human reliance upon Mother Earth for sustenance:

O Mother Earth!
You are the world for us and we are your children.
Empower us to speak in one accord,
steer us to live in peace and harmony,
and guide us in our behavior
so that we have cordial and gracious relationships
with all other people.

(16)

The earth provides the context for human flourishing. The Vedic peoples recognized her importance and called for acknowledgement of her beneficence:

We venerate Mother Earth, the sustainer and preserver
of forests, vegetation, and all things that are held together firmly.
She is the source of a stable environment.

(27)

The text also calls upon humans to protect and preserve nature's resources:

O Mother Earth!
May our bodies enjoy only water that is clean.
May you keep away from us all that is polluted.
May we always do only good deeds.

(30)

The composers of the text see a reciprocal relationship between human kind and the Earth; if humans care for the Earth, the text hints, a covenantal agreement might be formed wherein Mother Earth will extend her protection:

The Earth is adorned with many hills, plains, and slopes.
She bears plants with medicinal properties.
May no person oppress her and may she spread prosperity for us all around.

(2)

Whereas the Hymn of Purusha in the *Rig Veda* emphasizes the interconnectivity between humans and the five great elements, the posture taken in the *Prthivi Sukta* evokes feelings of humility and awe at the power of Mother Earth. By recognizing and enhancing qualities of "truth, strength, artistry, ferocity, dedication, fervor, effulgence, and sacrifice (1)" that sustain the Earth, humans can hope to become worthy of her blessings.

Another genre of Hindu spiritual literature, the epic, spins tales of war and conquest. However, the teaching of non-violence (*ahimsa*) has been declared in the great Indian epic, the *Mahabharata*, to be the highest religious teaching (*dharma*): *Ahimsa Paramo Dharmah*. Several chapters of the eighteenth book are dedicated to non-violence, suggesting that the highest human achievement lies in being able to experience others as oneself and oneself as others. Such a person is said to surpass even the accomplishment of the gods and goddesses in heaven, who still live in a dualistic state:

> Even the gods are bewildered at the path
> of the one who seeks the abode of no abode,
> who sees all beings as oneself,
> as not different from oneself.

(Mahabharata XIII:14.8)

By moving one's identity to a place of virtual homelessness, and by not attaching primary importance to the ego, the practitioner of non-violence attains great peace, leading to the highest dharma:

> When one strides among others seeing them as oneself,
> others also fall in step and see the self as well.
> This is to be followed indeed in the world of living beings;
> by this all dharma is taught.

(Mahabharata XIII:14.10)

Similarly, *Yoga Sutra* II:35 states that "When in the presence of one established in non-violence, there is the abandonment of hostility." By adopting an attitude of similitude with others, a sense of calm and peace can prevail. The Hindu concept of sacrifice (*yajna*) suggests the cultivation of a vision of self that sees action as moving toward a goal beyond selfishness. The self extends to include the other, facilitated through rituals that develop a sense of interconnectivity. Each of the elements listed above: the five great elements of earth, water, fire, air, and space, become part of a recognition of the essentials required for human flourishing. The icons and exemplars as set forth in the array of Hinduism's gods and goddesses provide encouragement and direction for developing and maintaining stability and happiness.

With the advent of global consumer economies, and the co-opting of human happiness by market forces and the thrills and skills of marketing, Hindu spirituality, which emphasizes contentment with the bare essentials, is in danger of being eclipsed in its homeland. Yoga, an aspect of Hindu spirituality originally codified by the scholar Patanjali around 300 CE, has itself entered the global marketplace of ideas and spiritualities. If approached prudently, however, Yoga might help worldwide remembrance that happiness cannot be found through the accumulation of things, but through appreciation and cultivation of simplicity. Beyond the ritual traditions of India, Yoga offers a pan-Asian modality for reflecting on the human condition that offers various entry points into sustainable values. Yoga begins with a series of ethical values (*yama*) adhered to by Buddhists and Jainas as well as Hindus: non-violence, truthfulness, not stealing, sexual restraint, and non-possession. As an ecological ethic, non-violence might be employed to help slow the pace of habitat destruction and the extermination of species around the planet. Truthfulness might encourage the better dissemination of knowledge about the harmful effects of resource exploitation, such as the destruction of forests, monoculture agriculture, and global warming. Deforestation, pollution, and water wastage might be interpreted as theft of the earth's bounty, encouraging people to refrain from such rampant and rapacious activities. Controlling sexual impulses might help slow the rate of population growth. By minimizing their possessions, people might grow accustomed to living on the bare necessities of life, a requisite step toward sustainability.

The other stages of Yoga might also be helpful in developing a spirituality of sustainability. The cultivation of positive observances (*niyama*) can help reground the self within modes of self-reliance and serve as an antidote to the all-pervasive consumer message that true happiness lies in shopping. These observances include searching for and developing purity, contentment, austerity, self-reflection, and an attitude of grateful devotion. By engaging the body in deliberate movement (*asana*) and purposeful breathing exercises (*pranayama*), a centering and inwardness (*pratyahara*) can emerge. As described in Patanjali's *Yoga Sutra* one can then develop great powers of concentration, meditation, and a state of unitive awareness known as *samadhi*, allowing a restructuring of one's being away from familial and cultural constructions toward a point of total freedom (*kaivalyam*) and an automatic positioning of one's actions in accord with the needs of the universe (*dharma megha samadhi*). Newly interpreted in light of environmental distress, Yoga practice can help inspire a life of cheerful sustainability characterized by healthy habits and meager consumption of resources.

Jainism: a living universe

Jainism, which originated in India more than 2,500 years ago, holds that the world has always been present and will exist forevermore. The Jaina approach to the world is different from the views held by her sister faiths of Hinduism and Buddhism. In contrast to Hinduism, Jainism does not advocate sacrificial activity such as the killing of animals to propitiate wrathful deities. Jainism does not posit an underlying unifying state of consciousness, such as Brahman, but insists upon the individual integrity of each and every soul, from beginningless time into an infinite future. Unlike Buddhism, Jainism posits the reality of a Self or Soul that holds the potential to attain its unique state of freedom. Buddhism teaches emptiness of self and other; Jainism teaches about a universe filled with innumerable living souls.

The Jaina approach to the natural world is simultaneously respectful and cautious, and above all driven by moral concerns. First, according to Jaina ontology, the world is suffused with life forces (*jiva*) that merit protection. Hence, the lives contained in particles of earth, drops of water, rays of light, gusts of wind, as well as micro-organisms, plants, and animals must be acknowledged and, to the greatest extent possible, not harmed. Each has entered its particular form due to the materiality of karma. One must exert caution because each harmful action causes the karma surrounding one's own soul to thicken and darken, obscuring the radiant consciousness of the soul and blocking its ascent to freedom. The following passage from the *Acaranga Sutra*, the oldest extant Jaina text (ca. 350 BCE), gives a sense of how the Jainas regard life to pervade all aspects of the natural world:

As the nature of (human beings) is to be born and grow old, so the nature of plants is to be born and grow old... As (humans) fall sick when cut, also that (tree) falls sick when cut; as (the human) needs food, so that (plant) needs food; as the (human) will decay, so that (plant) will decay; as the human is not eternal, so that (plant) is not eternal... As this is changing, so that is changing. One who injures plants does not comprehend and renounce sinful acts. The one who does not injure plants comprehends and renounces sinful acts. Knowing (those plants), a wise person would not act sinfully towards plants, nor cause others to act so, nor allow others to do so. The one who knows the causes of sin relating to plants is called a reward-knowing sage.

(Acaranga Sutra I:1.5.6–7, Jacobi, 101)

David Haberman has written about the affection for trees felt in Varanasi, where he notes that the sentiment can be extended to "tree worship worldwide: trees have not only been commonly thought of as animate beings but also as powerful divine beings who when approached in a respectful manner offer in return life-enhancing benefits to human beings" (Haberman 2013, 57). In the Jaina context, the divinity of the tree would envision the future possibility that the tree, which is divine like all other souls, might take human birth and enter the path toward freedom from all karmic constraints.

The Jaina universe thus conceived becomes a moral universe. In order to advance toward freedom, one must develop impeccable adherence to a moral code in order to purge all impedimentary karmas. The practices of Jaina morality liberate the soul while simultaneously engendering peace in the world. Jainism suggests that surveying the inner landscape as inseparable from the outer landscape can heighten an individual's sense of connectivity and responsibility. Feeling intimacy with life in its many forms, from microbial to heavenly, can prove instructive and corrective. According to contemporary Jaina leaders, most notably the late Acarya Tulsi and Indian Supreme Court Judge L.M. Singhvi, the Jaina account of the universe contains incentives that can rekindle the resolve needed to make substantive lifestyle changes.

Singhvi's 1990 Jain Declaration on Nature frames the core principles of Jainism as dialogue partners in the emerging discourse on religion and ecology (see Chapple 2002, 217) The Declaration states that Jainism presents an ecological philosophy and summarizes various aspects of the faith in light of its particular attention to nature. The first part discusses Jaina teachings on non-violence, interdependence, recognition of multiple perspectives, emphasis on equanimity, and commitment to compassion, empathy, and charity. The second section provides a synopsis of Jaina biological categories, in which all life forms stand within a hierarchy of ascent from elemental beings, microbes, plants, and insects, to thinking creatures like reptiles and mammals. In this pan-ethical universe, life constantly moves from one form to the next, and each category of being rises or falls according to its karmic interactions with others. The third and final part highlights the Jain Code of Conduct as exemplary for bringing about environmental justice. Key aspects include the restatement of the five Jaina vows of non-violence, truthfulness, not stealing, sexual restraint, and non-possession, the history of Jaina kindness to animals, the Jaina advocacy of vegetarianism, the teachings on restraint and avoidance of waste, and, finally, the value of charity in the tradition.

The Gandhian legacy: Narayan Desai, Vandana Shiva, and M.C. Mehta

Mahatma Gandhi (1869–1948) took the yogic and Jaina values of non-violence (*ahimsa*) and holding to truth (*satyagraha*) onto the world stage. He forever changed the course of world history by throwing off the shackles of European colonialism non-violently, a process that spread infectiously throughout the rest of Asia and Africa. His methods have been used with success in the United States, the Philippines, the countries of the former Soviet Union, and, most recently, in the Middle East and Africa and hold great lessons to be learned as the world confronts the ravages of climate changes and species decimation. The austerity of Gandhi's methods, combined with technological ingenuity, may pave a path from uncertainty to sustainability. For example, many leaders of the Occupy movement of the 2010s invoke Gandhian methods, emerging Gandhian modes of spirituality, and can help provide meaning and hope, giving purpose to the quest for a sustainable planet. Drawing from an underlying ethic of non-violence, this spirituality will offer new definitions of human well-being, body-based forms of spiritual practice, and the transformation of food production and patterns of food consumption toward a more sustainable model.

Mahatma Gandhi suggested that by returning to the bare necessities of life, India could disentangle itself from the ravages of the British colonial system. He spun his own thread and wove his own clothing. He led salt marches that helped spare people from taxation and reliance on imported goods. India's response to the difficulties of an enmeshed global economy was well-served by the Gandhian creed and adage: "There is enough in this world for every person's need, but not enough for every person's greed."

Narayan Desai, now in his 90s, serves as Vice Chancellor of Gujarat Vidyapeeth, the university that Mahatma Gandhi established in Ahmedabad, India. Desai, following the footsteps of his father, served as Gandhi's personal secretary during the last month before Gandhi's assassination. Desai continues to lecture and write about Gandhi's core message: holding to truth in a spirit of non-violence. He states that we need to overcome our all-too-human greed and fear with love and cooperation. The plague of greed and fear can be found throughout the world today. Debts, individual and governmental, have been poorly managed with dire consequences. Signs of greed can be found throughout the world in obesity and other food-related crises, and in lumbering, energy guzzling cars, trucks, and overbuilt houses. Likewise, fear can be seen in a deepening split between political parties, between rich and poor, and between religions who ironically all preach peace and love. All these tensions can lead to mistrust and the potential for violence. No political figure today will dare suggest that citizens should alter their lifestyle and reduce their consumption willingly for the sake of the common good. Desai offers a radical personal solution to the world's ills: live within your means, occupy your hands with creative work, and be open to the ideas and views of others. Even when engaged in world travel, Desai, following the model of Gandhi, spins tuffs of cotton into thread, adding to his spool each day. After several weeks or months, he weaves his thread into homespun cloth and wears only self-crafted clothes. All the students and faculty at his university do the same, engaging head, heart, and hands. This simple act helped India cast off its colonial oppressors. Gandhians today continue to espouse the ideals and realities of self-control, moving toward self-sufficiency and self-respect.

Gandhi-inspired activist Vandana Shiva has become one of the world's most outspoken critics of globalization and has provided a trenchant critique of the "patenting" of traditional ways of knowledge for economic gain by corporations. Her activism in the realm of seeds serves as a paradigm for exposing the excesses of the human attempt to manipulate nature. She takes a broad historical view in developing her analysis of "enclosure" or the marketing of what once was held in common, to be commoditized and controlled by industrial and commercial forces.

Shiva's advocacy of sustenance economy would require a shift from a corporate model to one in balance in nature, valuing relationships over and above commodities. She has taken up food production as her primary arena for activism. Shiva (2005) writes:

I launched a national program to save seed diversity in farmers' fields. We call it Navdanya, which literally means nine seeds... The farmers [had] become mere consumers of corporate seed. This excludes the farmer from the critical role of conserver of genetic diversity and development of seed. It robs farmers of their rights to their biological and intellectual heritage... Navdanya wanted to build a program in which farmers and scientists relate horizontally rather than vertically, in which conservation of biodiversity and production of food go hand in hand, and which farmers' knowledge is strengthened, not robbed.

(92)

The corporatization of agriculture, including the patenting of seeds and the mass marketing of herbicides, pesticides, and petroleum-based fertilizers, has created a horrific situation, turning farmers away from organic practices and into debtors. Shiva, citing government statistics, notes that "According to India's National Crime Bureau, 16,000 farmers in India committed suicide during 2004. During one six-month span in 2004, there were 1,860 suicides by farmers in the State of Andhra Pradesh alone" (120). During the first decade of the twenty-first century, 250,000 farmers committed suicide in India (Shiva 2011). This great tragedy has received scant attention from the international press. Shiva advocates for individual farmers and their families, and has helped urge legislation to help India's rural poor.

In 1996, Supreme Court lawyer M.C. Mehta won the Goldman prize for his advocacy on behalf of protecting the Taj Mahal. For years, the industrial grime generated by the city of Agra had degraded the once-gleaming marble of the famous building. Through his efforts, several industries relocated and a stringent program for reducing air pollution was enacted, including a ban on non-electric vehicles within several hundred yards of the Mughal masterpiece.

On April 1, 2001, Mehta achieved another landmark victory. On that day, all public transport vehicles in the Delhi capital region converted from gasoline, diesel, and kerosene fuels to compressed natural gas. The air quality improved right away. The new buses, taxis, and autorickshaws are brightly painted in yellow and green, celebrating the light of the Sun and the bounty of the Earth.

M.C. Mehta has developed a training camp to equip lawyers from throughout South Asia to face the challenges presented by environmental conditions. At his training facility, Eco Ashram near Rishikesh, a dozen young brilliant lawyers from Nepal, India, Bangladesh, Bhutan, Sri Lanka, and Pakistan were taught by Mehta as part of a six-week residential course in 2006 supported by the Ford Foundation. Everyone lived in straw huts, shared simple food, and attempted to keep warm, even when pelted by Himalayan hail. The lectures on industrial pollution included heartbreaking tales of miscarriages, birth defects, and early death.

M.C. Mehta has acknowledged that India's traditional environmental wisdom can provide a valuable resource for moving the public will to support improving air quality, dealing with water pollution, and taking the necessary steps to slow global warming. He commented that "the elements of nature such as air and water are being overtaken by greed. Greed has overtaken us, leaving us under the cloak of greed. For people wearing the cloak of greed, it is very hard to come out and see real life. With correct understanding they will be in a position to respect law, and enforce it. Then we have sustainable elements" (Mehta: 2006). Like other Gandhian-inspired activists, Mehta sees personal change as key to changing society.

Sikhism and love of nature

The Sikh faith, one of the world's newest religions, originated in the Punjab area of northern

India in the fifteenth century. Its founder, Guru Nanak (1469–1539), developed a system of devotion that seeks to honor the infinite force beyond time (*akal*) that empowers all reality. The tenth and final guru, Gobind Singh (1866–1708), established the *Guru Granth Sahib* as the revered sacred text to guide Sikhs in their spiritual life. He also urged Sikhs to identify themselves as an easily recognizable community through the eschewal of cutting their god-given hair, and, for the men, the wearing of a turban.

Several contemporary Sikhs have expressed concern for the environment and have offered ecological interpretations of their religious texts. Surjeet Kaur Chahal argues that "the Sikh Gurus have emphasized a sacramental understanding of the natural world. God is immanent in nature—nature is sacred. Man is enjoined to live in harmony with nature" (Chahal 2015, 88). In the same vein, Surjit Jolly has written that "God did not confer absolute power on man to control and dominate nature. The human race is not an alien species foisted upon this planet to dominate and exploit it, but rather is an integral part of nature itself linked to the rest of creation by indissoluble bonds" (Jolly 1989, 282).

The *Guru Granth Sahib* praises the marvels of the physical world as sufficient proof for the existence of God:

> Marvelous, the multiplicity of creation. Wonderful, their distinctions.
> Marvelous, creation's form. Wonderful, its variety.
> Marvelous, the motion of air. Wonderful, the waters.
> Marvelous, the [power of] fire.
> Air is the vital force. Water, the progenitor.
> The vast earth, the mother of all.

> *(GGS 463)*

Many passages in the *Guru Granth Sahib* affirm that wisdom can be found in the ways of nature. Nature provides instruction and example for living a life mindful of God.

Another Sikh scholar, Amarjit Singh, states that "The Gurus were…sensitive to…socio-natural environments." He correlates the flaring forth or big bang that started the universe with the *Guru Granth Sahib* account of the origin of the universe through God's will. Reminiscent of Pierre Teilhard de Chardin and Thomas Berry, Amarjit Singh celebrates the evolution of life as an unfolding expression of this creation moment. Amarjit Singh sees a continuity of God and soul in all creation, alluding to the presence of God in each form of life (Amarjit Singh 2002, 200).

The Sikh teaching on reincarnation, according to Amarjit Singh, resembles the recycling of life forms and evolution of the biosphere as articulated by Vernadsky, the famed geochemist who described the earth as a self-generating, self-regulating system. For Amarjit Singh, the complexity and magnificence of the process *is* God. Homeostasis, the emergence of perfectly niched, self supporting, and yet interdependent ecosystems, as well as biodiversity, constitute one way of communicating the profound mystery of life.

In the Sikh tradition, God pervades all things; all things are considered to be sacred. Life, grounded in the elements, takes many forms, from insects to fish and reptiles, birds, and mammals. Guru Arjun Dev (1536–1606) praises the many forms of life as follows:

> For several births you became a worm and a moth.
> In several births you became an elephant, a fish, and a deer.
> For several births you became a bird and a snake.
> In several births you were yoked as a horse and an ox.

Meet the Lord of the universe. This is the time to meet.
After a long time this human body is gloriously fashioned.

(quoted in Amarjit Singh 2002, 207)

This long process of birth and rebirth culminates in the human, the state of existence in which the universe becomes aware of itself. The great responsibility of being human requires acknowledgment of the magnificence of all forms of life and, in the case of Sikhism, the creative, nameless, all-pervading reality. In summary, "The Sikh gurus have emphasized a sacramental understanding of the natural world. God is immanent in nature—nature is sacred" (Chalal 2015, 88).

Conclusion

The Hindu, Jaina, and Sikh traditions of India emphasize effect, the cultivation of feeling-tone as integral to religious experience. Each tradition describes and in its own way values and praises the material world. For Hindus, a sense of deep connection arises, when one sees the connections between the human body and the realms of the cosmos. For Jains, once one recognizes the pervasiveness of life, deep feelings of care emerge, leading to carefulness in all one's activities. For Sikhs, observation of the natural world and reflection on one's own deep memory of past lives convinces one of the sense of kinship we hold with all living beings. These respective theologies support the pathways chosen by contemporary activists in India. Like Gandhi, who promoted non-violence and truthfulness as the keys to societal change through personal transformation, India's activists rely upon inner strength through which they inspire others to take up the good work of lifestyle change and legislative advocacy.

One can only hope that the optimism of some theologians will prevail. Professor Sumathy (2014) suggests that:

> In Vaishnavite ecotheology there is reverence, awe, and gratitude for life with its characteristic unity, balance, difference and connectedness—life that centers on divine grace which is compassionate, all inclusive, empowering and showers the gift of unconditional love. This theology enjoins one to behave responsibly toward fellow human and non-human beings and to live a simple life in harmony with the Earth.
>
> *(162)*

David Haberman has written that "Trees are... regularly thought to be embodied forms of divinities who have a strong physical and relevant presence in the world. The most common response to an awareness of the divinity in or as a tree is worship" (Haberman 2013, 184). One can only hope that the worship of the tree can lead to concerted, sustained protection not only of the tree, but the entire system that supports the well-being of trees: the soil, the air, the water, and more.

With irony and sadness, our own awareness of self and nature has brought a tremendous destructive capacity to a terrifying crescendo. Blessedly, the human species has created an awareness of this tremendous destructive power and is beginning to develop an ecological conscience.

Each of us needs to ask the question: How much is enough? How much food? How much entertainment? How large a car? How large a house? From a Gandhian perspective, these questions must be posed repeatedly, relentlessly. The dharma traditions of India suggest that we can and must learn to live within limits, that we must seek inner peace, not outward acquisition.

By appreciating the glories of nature as celebrated in the *Prithivi Sukta* of the Hindus, the *Acaranga Sutra* of the Jainas, and the *Guru Granth Sahib* of the Sikhs, common sense may prevail, leading to a lifestyle that not only sustains but also fosters human flourishing.

References

Chahal S. K. (2015) "Sikhism and environmental ethics" in Siddhartha ed. *Religion, Culture, and the Ecological Crisis,* Meeting Rivers, Bangalore 87–91.

Chapple C. K. (1993) *Nonviolence to Animals, Earth, and Self in Asian Traditions,* State University of New York Press, Albany.

Chapple C. K. and M. E. Tucker. (2000) *Hinduism and Ecology the Intersection of Earth, Sky, and Water,* Harvard University Press, Cambridge.

Chapple C. K. (2002) *Jainism and Ecology: Nonviolence in the web of life,* Harvard University Press, Cambridge.

Chapple C. K. (2008) *Yoga and the Luminous: Patanjali's spiritual path to freedom,* State University of New York Press, Albany.

DeNicolás A. T. (1976) *Meditations through the Rig Veda: Four Dimensional Man,* Nicolas-Hays, York Beach.

Desai N. 2008 Lecture at Loyola Marymount University, 9 October 2008.

Dwivedi O. P. and Chapple C. K. (2011) *In Praise of Mother Earth: The Prthivi Sukta of the Atharva Veda,* Marymount Institute Press, Los Angeles.

Haberman D. (2006) *River of Love in an Age of Pollution: The Yamuna River of Northern India,* University of California Press, Berkeley.

Haberman D. (2013) *People Trees: Worship of trees in Northern India,* Oxford University Press, New York.

Jacobi, Hermann (1884) *Jaina Sutras,* translated from the *Prakrit. Part I: Acaranga Sutra, Kalpa Sutra.* Sacred Books of the East, 22. Oxford University Press. Reprinted: Delhi: Motilal Banarsidass, 1980.

Jolly S. (1989) "Sikhism and the environment", in Dwivedi O. P. ed. *World religions and the environment,* Gitanjali, New Delhi 282–295.

Mehta M. C. (2006) Interview with the author, Dehra Dun, India, 12 December 2006.

Samathy U. (2014) *Vaishnavite ecotheology,* Authorspress, New Delhi 2014.

Sarandha (2011) *Yamuna: Reflections on a river lost,* Vitasta Publishing, New Delhi.

Shiva V. (2005) *Earth Democracy: Justice, sustainability, and peace,* South End Press, Cambridge.

Shiva V. (2011) Lecture at Loyola Marymount University, 1 November 2011.

Singh A. (2002) "The glimpses of environmental and evolutionary biology in Sri Guru Granth Sahib" in Ram Bishnoi K. and Ram Bishnoi N. *Religion and Environment,* vol. II, Guru Jambheshwar University Publication, Hisar, India 194–212.

19

CHINA

Landscapes, cultures, ecologies, religions

James Miller

As the third largest country in the world, China has a vast geographic diversity: arid deserts and snow-capped mountains of Xinjiang in the far West; the unique landscape of the Qinghai–Tibetan plateau, source of the Mekong, Yangzi and Yellow rivers; the rich alluvial plains of Sichuan that provide much of China's food; the northern grasslands of Inner Mongolia; the stunning, golden hues of the loess plateau, source of much of China's coal and minerals; the central province of Henan, which harbors a vast treasure trove of China's ancient civilization, and is now home to over 96 million people; the densely populated coastal regions of Zhejiang, Fujian and Guangdong, now the base of much of China's manufacturing wealth; and the south-western province of Yunnan, bordering Laos, Vietnam, Thailand and Myanmar, home to much of China's biodiversity and as well as 26 of China's 55 recognized ethnic minorities.

In addition to its geographic size, China is also the world's largest country by population, currently standing at 1.37 billion, approximately 19% of the world's total. Of these, approximately 91% belong to the dominant Han ethnic group, with the remaining 9% divided among China's official minority nationalities. China's religious diversity matches its physical and ethnic diversity: China currently recognizes five official religions: Daoism, Buddhism, Islam, Catholicism and Protestantism. This administrative classification depends chiefly on an understanding of religion imported from the West via Japan in the late-nineteenth century. In this view, religions are distinguished in terms of people's affiliation to institutional organizations, a model of religion derived from the European experience of ecclesial belonging. While it is common for Christians to distinguish themselves in terms of the church they "belong to," this model of religious adherence is far from the norm in China's religious history. Religious life in China has often centered on local temples owned collectively by village communities, or on Buddhist or Daoist pilgrimage sites operated by monastic lineages. Attempts to organize these activities into formal religions have usually been sponsored by the state in an attempt to organize, classify and monitor religious activities. After the Communist revolution in 1949, for instance, the multiplicity of Daoist lineages, city temples and mountain retreats were brought under the administrative oversight of a single, overarching body, the Chinese Daoist Association. Similarly, all Protestant denominations were forcibly "ecumenized" into a single "patriotic" association. These social-organizational dynamics were not motivated by internal religious demands to unify, but were the result of political demands of the modern nation state.

Such demands did not originate with the Communist party, but had begun earlier in the

Republican period (1912–1949), during which leaders sought to unify China as a single, modern, nation state in part by replacing the diverse, diffuse and local local networks of social power with national, patriotic associations (Duara 1991). In this way, local religions that had formed around ancestral veneration, local gods and goddesses were deprecated as "cultural relics," or reclassified and absorbed into the formal, state organizations of Buddhism and Daoism. At the same time practices associated with China's Confucian heritage were not included as part of this classification scheme. Consequently "Confucianism" and the widespread practices of ancestral veneration, including annual tomb-sweeping are not commonly understood as belonging to a "religion." Indeed Confucianism is not officially part of China's "religious" landscape despite the fact that, from an anthropological point of view, many aspects of Confucianism can be understood as religious.

Rather than understanding religion administratively, this chapter presents an overview of China's religious scene from the perspective of its diversity of the geographic spaces in which it takes place, with an understanding that the result is meant to be illustrative, rather than comprehensive. Five key spaces function as this chapter's organizing themes: the Middle Kingdom; rivers; coasts; mountains; and margins.

The Middle Kingdom

The Chinese word for China, Zhongguo 中國, is commonly translated as the "middle kingdom," though in all likelihood this term was originally understood in the plural, referring to the kingdoms that occupied central China in the period of disunity known as the Warring States (475–221 BCE). The term has come to represent a key feature of imperial China's social imagination of itself as occupying the central space within a cosmic frame, bounded on each side by "barbarian" nations, a square earth sitting under a rotating circular canopy of stars, spread out like an umbrella held up by a central *axis mundi*. The imaginative scheme of centre versus periphery functions as a key organizing principle not only of early Chinese geography, but also in terms of religion and culture. From within this scheme, the world's peoples are divided into two basic categories: "Zhongguoren 中國人," or people from the central kingdom(s); and "waiguoren 外國人," people from the outer kingdoms. The earth is imaged as a three by three square, with China occupying the central location. This imagery is repeated throughout Chinese art, architecture and city planning, with the three by three or nine by nine squares symbolizing the full extent of the cosmos and China's central place within it. The capital cities of Beijing, Xi'an and Nanjing, for instance, were laid out as square, walled cities. In Beijing, the imperial palace complex sits at the centre, and at the centre of this lies the "purple forbidden city" (*zijincheng* 紫禁城), and at the centre of this the Hall of Supreme Harmony with the imperial throne.

In this scheme, the king or emperor occupied the key location at the apex of the society of people from the middle kingdom, and therefore possessed the sacred duty of uniting heaven, earth and and all humanity. The Chinese character for king 王 is three horizontal lines one above the other, bisected vertically by a single line. According to traditional interpretation, the three horizontal lines represent the earth at the bottom, the heavens at the top and humanity in the middle. These three realms are united in the person of the king, the single vertical line that touches all three. In this way the king, or emperor, functioned so as to produce the "unity of heaven and earth" (*tianren heyi* 天人合一), the state of optimal flourishing among the three realms of the cosmos, thus bringing about wealth and prosperity for all.

This geographic schema was thus also the foundation for the state religion, which refers to the official religious life of the emperor conducted on behalf of the people. Tourists today can see the vestiges of this in Beijing's most recognizable landmark, the Temple of Heaven, where the emperor traditionally conducted animal sacrifices on behalf of the state. The architecture of the site symbolizes the traditional conception of the universe, with the circular temple, signifying heaven, located on a square platform, signifying earth. Only the emperor was able to undertake such sacrifices on behalf of the people, and this ritual performance was the chief way in which the sacred geography of heaven, earth and humanity underwrote the divine authority of the emperor himself.

The cosmic pattern of heaven, earth and emperor concretely symbolized in the imperial architecture of the capital city was also reproduced in the social imagination of the nation's geography itself. Five sacred mountains, also known as marchmounts, were designated as marking the boundary points of China's north, south, east, west and center. These mountains were the location of further imperially sponsored rites to promote the harmony of heaven, earth and humanity. In this way the state religion functioned as a kind of religious ecological mechanism, focussed on the body of the emperor himself. This system was fully set out in a grand, unified cosmology synthesized by Dong Zhongshu (179–104 BCE), which became the basis of state Confucianism.

In this view, the emperor functioned as the sacred linchpin of the social, agricultural and cosmic order, ensuring that all three realms work together. Such a system also imposed an obligation upon the emperor's person to constantly bring the three realms into harmony. Since the natural world was in constant transformation according to the seasons, the emperor also made corresponding changes to his life, wearing certain colors, and undertaking certain seasonal activities (see Miller 2012).

In this way the sacred geography of central capital and compass-point marchmounts was also paralleled by a sacred seasonality, east corresponding to spring, west corresponding to fall, and so on. Space and time were co-ordinated together in a single overarching cosmology focused on the body of the emperor himself. The geography of the traditional Chinese empire cannot therefore be fully understood without reference to the inner landscape of the body in which the solid *yang* structures of bones, sinews, flesh and organs correspond to mountains; and the fluid *yin* dynamics of *qi* (subtle breath or vital force) corresponded correspond to the flow of water through China's landscape.

Mountains

China's five sacred mountains are usually known in English as the five marchmounts, or mountains in the "marches" or border regions of China. They occupy key positions in the sacred cosmography that underpinned the imperial order. Corresponding to the four directions, plus the centre, the marchmounts symbolized and marked out the cosmic limits of the empire. They also functioned as tutelary deities who protected the Middle Kingdom from external threats (Verellen 1995). They originated in an earlier set of four marchmounts associated in the *Zuozhuan* with "barbarian" tribes, and more specifically their leaders who acted as a buffer between the Chinese ruling house and foreign powers (Kleeman 1994: 227). By the Han dynasty, this scheme of four mountains was absorbed into the cosmological system of five phases, colours, direction, etc., with the addition of a fifth, central mountain, Mt Song in Henan. The four mountains that previously marked the borders of the empire by now lay well within Han territory and, during the reign of Emperor Wu of the Han, came under the direct patronage of the Emperor (Kleeman 1994). In so doing the mountains were fully integrated into the Han cosmology with the body of the emperor as the supreme node joining heaven, earth and human beings in a single, coherent, system. The Han empire, constructed on the subjugation and pacification of marginal peoples thus fully incorporated these border lands into Middle Kingdom by incorporating the sacred marchmounts into the sacred space governed by the emperor himself.

The emperor's duty was to offer blood sacrifices at the mountains in order to propitiate the tutelary deities, ensuring success and prosperity for the dynasty. The trouble and expense of such rites meant that the most elaborate and sumptuous, the Feng and Shan sacrifices, were performed only five times in the Han Dynasty (Bokenkamp 1998: 384). These rituals were performed at the foot of Mt Tai, the sacred mountain of the East, in present-day Shandong province. Due to this imperial patronage, the Eastern marchmount achieved pre-eminence among the five marchmounts, and to this day temples in its honour exist across China.

The religious traditions of Daoism and Buddhism also vied with state Confucianism for a claim over these spaces in an attempt to strengthen their relative position within the empire. Sometimes these traditions occupied the same space at the same time. At other times, as the fortunes of religions rose and fell, the mountain spaces inscribed by one religious tradition would be overlaid by a new one in a complex rewriting of ritual space. For example, in his study of the Southern marchmount (Nanyue 南越), known as Mt Heng 恆山 in Hunan, James Robson (1995: 230) writes that "the sacredness of Nanyue was continually produced and reproduced throughout history by different religious traditions whose discourses and attempts to define the sacredness of the mountain were at times in direct competition." The Daoist intellectual Sima Chengzhen, for instance, persuaded the Tang emperor Xuanzong to recognize the five marchmounts as the "terrestrial abodes of Daoist 'perfected ones' (*zhenren* 真人)." In so doing the emperor recognized the Daoist claim to imperial sites, thereby strengthening its position vis-à-vis Buddhism. This imperial recognition led to the gradual imprint of Daoist religious activity on the five marchmounts, and vied with a competing Buddhist layer that had been established on Nanyue since the mid-sixth century CE.

Mountains were not only significant in China's religious heritage as contested sites of Imperial, Buddhist or Daoist patronage. They were also the residences of gods, recluses and unusual fauna, and the source of rare flora sought by alchemists in their quest for transcendence or "immortality." Campany (2001: 127) notes in his study of the alchemist Ge Hong (283–343) that such seekers valued natural materials that were "hard to obtain, and located in barely accessible places;" and secondly, materials that had an unusual appearances, being "visually and morphologically anomalous, straddling taxonomic boundaries" (128). The combination of difficulty of access and strange appearance went hand in hand with their numinous qualities. In this way it can be said that the natural world is not in some way "flat" or "democratic" (see Miller 2008: 32), but rather possesses a hierarchy of power, accessibility and strangeness. Alchemists such as Ge Hong emphasized the value of these rare and powerful materials, believing they contained the power of transcendence when ingested. Just as the mountain has a roughly pyramid shape, the natural world itself can be understood by analogy as a pyramid in which the rare and valuable substances are the hardest to reach and fewest in number.

The Daoist fascinations with mountains as repositories of rare and precious substances also gave rise to the association between mountains, recluses and the revelation of religious texts. Daoists travelled to sacred mountains in search of techniques of meditation, teachings from Daoist masters and the transmission of Daoist texts revealing secret traditions of meditation and self-transformation. The Chinese term for mountain cave or grotto (*dong* 洞) also became the word we roughly translate as "canon" meaning a selection of religious texts. Mountain grottoes can thus be understood as locations for the revelation of sacred mysteries, whether through the intense meditation of the recluse, the transmission of oral teachings from a Daoist transcendent, or the initiation into an esoteric text. As Verellen (1995: 271) notes, the grotto can be understood by means of a close homophone *tong* 通, meaning to penetrate or connect, and Daoist cosmography came to envision an interconnected network of "grotto heavens and blessed places" (*dongtian fudi* 洞天福地) that were deemed particularly auspicious sites for engaging in Daoist cultivation.

In addition, the altar space erected by the Daoist priest for the performance of rituals is also envisaged as a cosmic space bounded by the five marchmounts, with the priest at the middle. The image of the mountain is replicated over and again: the cosmic space of Daoist ritual is bounded by mountains; the body of the priest is imaged as a mountain; inside his body in the grotto-chambers of his organs dwell the spirits of the cosmos (Schipper 1993: 91–93). The network of mountains within mountains and grottoes connecting to grottoes functions as the basis for an economy of cosmic power in which the vital breath or *qi* flourishes and circulates, pervading the myriad dimensions of the cosmos, interpenetrating organs, caves, bodies and mountains in the ceaseless exchange of energy: life begetting life, inner begetting outer, physiology and geography interwoven in a dazzling, mysterious and endless overflowing of vitality.

Rivers

China's physical landscape is defined by its hydrological cycle in which waters emerge from the Qinghai–Tibet plateau in the West, flow East through the three great rivers, Yangzi, Mekong and Yellow, and pour into the sea. This West to East flow can be understood through the Chinese term "Dao" meaning Way or Path, but also denoting a fluid vector by which the processes of life are never static but always in motion. These processes of life, or "ten thousand things" (*wanwu* 萬物) include all things, human, animal, vegetable and mineral. All are composed of or shaped by the flow of water, the streaming Dao that is the basic vector of the Chinese landscape (Miller 2006). As the *Daodejing* notes (ch. 4):

> The Dao is empty [empties], yet using it it does not need to be refilled.
> A deep spring (*yuan*)—it seems like the ancestor of the myriad living things.
>
> *(Quoted in Allan 1997: 76)*

Here Dao is envisaged as the flood of liquid vitality from which all phenomena emerge. As the first chapter of the *Daodejing* mentions, this Dao is not a thing with a name or form, but acts generatively to give birth to all the phenomena of the natural environment. In Wittfogel's (1957) thesis of the "hydraulic state," taming this flood of life was tantamount to gaining political power. Indeed, there is no doubt that one of China's major early technological achievements was the construction of the Dujiangyan irrigation system (267–256 BCE) in present-day Sichuan province. A weir across the Min river regulates the flow during the spring floods, directing the flow into a network of irrigation channels that to this day provides water for 50 cities and irrigates 672,000 hectares of farmland (Miller 2013). Today this dam is regarded as a feat of "Daoist" engineering in which the flow of water is not blocked completely but productively distributed to promote the fertility of the landscape. As Miller (2013) notes the weir constitutes a concrete expression of the Daoist concept of *wuwei* 無為 variously translated as "non-aggressive" or "effortless" action because "rather than damming the river completely the site employs a weir and irrigation system to channel and regulate water's natural power." A Daoist temple on the site to this day memorializes the architect Li Bin.

The second sense in which water functions as a category of religio-cultural life in China is in the concept of *fengshui*, literally "wind and water," the cultural practice by which houses, tombs and other human structures are located to take advantage of the nature's fluid powers, channelling good fortune, health and happiness to the earth's human inhabitants, both living and deceased. The natural ecology of plants and trees is here directly connected to the religious ecology of spirits, ancestors and descendants. When the land flourishes, the ancestral spirits will be at peace and this economy of cosmic power will contribute to the flourishing of the family lineage and the future prosperity of the clan.

According to Coggins (2014), although the dominant method of Han Chinese expansion was the deforestation and reconfiguration of the landscape to support agriculture, fengshui forests and temple forests emerged as protected wilderness spaces valued by monastic leaders and local village communities for non-economic reasons. He writes (2014: 15): "Corporate groups in lineage villages had additional reasons to preserve certain forests and groves, reasons that transcended immediate economic concerns and reflected a profound regard for their own long-term viability." This concern he traces to a seminal fengshui text, the *Book of Burial* by Guo Pu (276–324). The text notes:

> The *Classic* says, *qi* rides the wind and scatters, but is retained when encountering water. The ancients collected it to prevent its dissipation, and guided it to assure its retention. Thus it was called feng shui (wind/water). According to the laws of feng shui, the site that attracts water is optimal, followed by the site that catches wind … Terrain resembling a palatial mansion with luxuriant vegetation and towering trees will engender the founder of a state or prefecture.
>
> *(Trans. Field 2001: 190)*

"Attracting water" and "catching wind" may be understood as taking advantage of the natural fluid dynamics of physical and subtle energies, or *qi,* which animate the landscape and the body. The construction of water channels and preservation of "fengshui forests" may thus be understood as part of the Chinese attempt to take advantage of, without using up, the physical and subtle *qi* of the natural landscape. This would produce not only natural, biological fertility, but also socio-cultural fertility in the continuity of lineage from deceased ancestors to the as-yet unborn descendants. This "continuity of *qi*" functioned not only ecologically to bring the human world into dynamic correlation with the natural ecosystems and hydrological cycles, but also historically in the production of genealogical narratives by which Chinese communities are traditionally organized.

Coasts

Traditional scholarship on Chinese religions divides gods into local and national categories. Local gods have their specific tutelary domains and are worshipped only by people living in those particular geographic areas. National gods, such as Guan Di, the Jade Emperor, or the God of Wealth, can be found throughout the country. Local gods, conversely, are worshipped only in specific regions. Prominent among these regional deities is Mazu 媽祖 (Matsu) a goddess associated with the South China sea whose temples are found throughout the coastal provinces of Fujian and Guangdong, and also Hong Kong, Macau and Taiwan. According to tradition, Mazu was a girl who lived in the late-tenth century who was renowned for her assistance to seafarers. She was posthumously deified and attracted a wide cult throughout the southern China coastal area in the Ming dynasty. Over the past few centuries she has become one of the most popular local deities in China.

Devotion to Mazu is widespread throughout South East China's coastal areas because of her association with seafarers and fishermen. She can be thought of in bioregional terms, corresponding to the Southern China Marine Ecoregion as identified by the World Wildlife Fund (WWF), that is, the sea area between Taiwan, mainland China, Hong Kong and Macau. Her worship emerges from the engagement of peoples with fish, coastlines, tides and the sea. Out of this complex of social, economic and ecological interaction developed a powerful bioregional religious tradition. Typically, Mazu temples are located in strategic coastal sites, and her statues watch over the marine activities of local seafarers. Indeed, residents of Macau attributed the fact that they escaped the 2003 Severe Acute Respiratory Syndrome (SARS) crisis that gripped Hong Kong to the prophylactic powers of the enormous Mazu statue that they had recently erected. Now Mazu is beginning to take on new political responsibilities as a symbol of harmonious relations between Taiwan and the mainland. A huge emerald statue of Mazu, valued at US$28.25 million, arrived in Taiwan from the mainland in December 2011. Both religious and political dignitaries attended the reception ceremony for the Mazu statue (Taipei Times 2011). Mazu's bioregionalism thus opens her up to the possibility of being exploited for political ambitions, as a symbol of the unity of people on both sides of the Taiwan straits. Mazu's significance thus demands analysis from a complex of religious, ecological and political perspectives.

Margins

The Chinese religious imaginations of nature in mountains, rivers and coasts may alternatively be understood in terms of center and periphery, or valleys and hills (see Weller 2014). This figure of center and edge is replicated throughout the multiple imaginations of nature in Chinese tradition, from food practices, garden design, to urban planning and even China's Great Wall. Of particular interest in the contemporary period is the multiple imaginations of nature that take place at the borders of China's land mass, home to extraordinary cultural and ecological diversity.

In the southern province of Yunnan, for instance, the concept of "holy hills" among the Dai ethnic minority has preserved fragments of old-growth rainforests from massive deforestation and replacement with rubber plantations (see Zeng 2012). At the same time, the traditional complex of religion and ecology among the Blang people is being rewritten as the people transform their indigenous agricultural practices through the development of a cash economy based on growing highly lucrative Pu-er tea (see Miller and An 2013).

More significant from the point of view of securing China's continued access to water is the fate of China's nomadic Tibetan people in the area of the Qinghai–Tibetan plateau, the source for China's three major rivers systems. Since 2005 these herders have been subject to forcible resettlement, known euphemistically as "ecological migration" (Qi 2014). The reason for this resettlement was to create a national nature preserve in this area so as to help preserve water supply downstream. Alarm bells rang in spring 1996 when for the first time in thirty years the water supply in the upper reaches of the Yellow River was cut off. In 1997 the interruption lasted 226 days and no water flowed along 706 km of the river (Qi 2014, 182). As a result of the drastic measures undertaken by the central government, the culture and religion of these nomadic peoples has been changed forever, attracting widespread criticism in the Western media (Jacobs 2015). Ecologists are uncertain as to whether the resettlement will have any positive effect upon preserving the water supply. Nonetheless it is clear that at the margins of China's fragile environment the stakes are enormous. Without Tibet's pristine waters, the lives of millions of ordinary Chinese people may be at severe risk. As a result of China's massive modernization and development, the traditional complex of ecology, culture and religion that has persisted for centuries at China's margins will likely soon disappear.

Conclusion

China's central government recently announced plans to create a new megaregion "Jing-Ji-Jin" comprising the previous cities of Beijing and Tianjin and the less-developed regions of Hebei province in between. The new region will be approximately the size of Kansas and will have a population of over 100 million people (Johnson 2015). Similar long-term plans are under way for the Pearl River Delta region, combining the cities of Hong Kong, Macau, Shenzhen, Guangzhou and Dongguan. As China builds hyper-dense megacities in order to house, feed and provide employment for its massive urbanizing population, it is clear that the traditional networks of religion, landscape, ecology and environment will undergo unprecedented transformation. As President Xi Jinping builds China's connections westwards in a new Silk Road stretching towards central and southern Asia, this transformation will encompass China's traditionally poorer, more marginal and ethnically diverse regions and religions just as much as its rich eastern coastal provinces.

References

Allan, S. 1997. *The Way of Water and the Sprouts of Virtue,* State University of New York Press, Albany.

Bokenkamp, S. R. 1998. A Medieval Feminist Critique of the Chinese World Order: The Case of Wu Zhao (r. 690–705), *Religion* 28, 383–392.

Campany, R. F. 2001. Ingesting the Marvelous: The Practitioner's Relationship to Nature, According to Ge Hong in Girardot, N. J., Liu, X. and Miller, J. eds *Daoism and Ecology: Ways with a Cosmic Landscape,* Harvard, Cambridge, 125–147.

Coggins, C. 2014. When the Land is Excellent: Village Feng Shui Forests and the Nature of Lineage, Polity and Vitality in Southern China in Miller, J., Smyer Yü, D. and van der Veer, P. eds *Religious Diversity and Ecological Sustainability in China,* Routledge, Abingdon, 97–126.

Duara, P. 1991. Knowledge and Power in the Discourse of Modernity: the Campaigns Against Popular Religion in Early Twentieth-Century China. *Journal of Asian Studies* 50.1, 67–83.

Field, S. 2001. In Search of Dragons: The Folk Ecology of Fengshui in Girardot, N. J., Liu, X. and Miller, J. eds *Daoism and Ecology: Ways with a Cosmic,* Landscape Harvard, Cambridge.

Jacobs, A. 2015. China Fences In Its Nomads, and an Ancient Life Withers, *The New York Times,* July 11, 2015 www.nytimes.com/2015/07/12/world/asia/china-fences-in-its-nomads-and-an-ancient-life-withers.html (accessed 15 July 2015).

Johnson, I. 2015. Chinese Officials to Restructure Beijing to Ease Strains on City Center, *The New York Times,* www.nytimes.com/2015/07/12/world/asia/china-beijing-city-planning-population.html (accessed 15 July 2015).

Kleeman, T. F. 1994. Mountain Deities in China: the Domestication of the Mountain God and the Subjugation of the Margins, *Journal of the American Oriental Society,* 114.2.

Miller, J. 2006. Daoism and Nature in Gottlieb, R. ed., *The Oxford Handbook of Religions and Ecology,* Oxford University Press, New York 220–235.

Miller, J. 2008. *The Way of Highest Clarity: Nature, Vision and Revelation in Medieval China,* Three Pines Press, Magdalena.

Miller, J. 2012. Nature in Nadeau, R. L. ed., *The Wiley Blackwell Companion to Chinese Religions,* 349–368.

Miller, J. 2013. Monitory Democracy and Ecological Civilization in the People's Republic of China, in Trägårdh, L., Witoszek, N. and Taylor, B. eds, *Civil Society in the Age of Monitory Democracy,* Berghahn Books, Oxford 137–148.

Miller, J. and An, J. 2013. 现代化程中布朗族的宗教与生态 Religion and Ecology in the Modernization of the Blang People, in Su F ed. 文化多元与生态文明 *Cultural Diversity and Ecological Civilization,* Minzu University Press, Beijing 353–364.

Qi, J. 2014. 'Ecological migration' and cultural adaptation: A case study of the Sanjiangyuan Nature Reserve, Qinghai Province in Miller, J., Smyer Yü, D. and van der Veer, P. eds, *Religious Diversity and Ecological Sustainability in China,* Routledge, Abingdon, 181–193.

Robson, J. 1995. The Polymorphous Space of the Southern Marchmount [Nanyue], *Cahiers d'Extrême-Asie* 8.1, 221–64.

Schipper, K. M. 1993. *The Taoist Body* University of California Press, Berkeley.

Taipei Times 2011. Emerald Matsu statue lands at Taichung Harbor Dec 15, 2011 http://taipeitimes.com/News/taiwan/archives/2011/12/15/2003520791 (accessed 15 July 2015).

Verellen, F. 1995. The Beyond Within: Grotto-Heavens (Dongtian) in Taoist Ritual and Cosmology *Cahiers d'Extrême-Asie* 8.1, 265–90.

Weller, R. P. 2014. Globalizations and Diversities of Nature in China, in Miller, J., Smyer Yü, D. and van der Veer, P. eds *Religious Diversity and Ecological Sustainability in China,* Routledge, Abingdon, 147–163.

Wittfogel, K. A. 1957. *Oriental Despotism,* Yale University Press, New Haven.

Zeng, L. 2012. Cultural Transformation and Ecological Sustainability among the Dai people, in Xishuangbanna *Sustainable China,* www.sustainablechina.info/2012/03/26/coping-with-change-rapid-transitions-faced-by-the-dai-in-xishuangbanna/ (accessed 15 July 2015).

20

LATIN AMERICA

Anna Peterson

Like other "regional landscapes" chapters, this one presents the religious landscape of a particular part of the world, in this case Central and South America and the Spanish-speaking Caribbean. I do not attempt here to provide a comprehensive survey of either the history or contemporary situation of Latin American religion, tasks which are the subject of several other scholarly books (see Lynch 2012; Gonzalez and Gonzalez 2007; Peterson and Vasquez 2008). Instead, I concentrate on the religious ideas, histories, movements, practices, and institutions that are most important for the study of religion and ecology. I do not assume any familiarity with Latin American religion, so this chapter will be of primary interest to scholars and students who wish to understand the basic contours of religion in the region. However, I hope it will also prove helpful for Latin Americanists who have not paid special attention to the intersection of religion and environmentalism. By highlighting the distinctive features of this interaction in Latin America, the chapter contributes to the larger comparative context developed by this book and its aim of providing a global perspective on both the study of religion and ecology and the role of religion in environmental movements.

In order to make the chapter as helpful as possible for those studying religion and ecology, I focus on four central themes. I begin with the study of Latin American religion, explaining how its institutional location, theories, methods, and major concerns differ from the study of religion in other regions. Next, I turn to the historical context of Latin America's religious geography, including a short overview of pre-Columbian religion, a discussion of the European conquest and the colonial period, and important subsequent developments. The third theme is contemporary religious pluralism, including the diversification of the Roman Catholic church, the growth of Protestantism, and the shifts in indigenous and African-based traditions. Last, I examine the role of religious ideas and groups in contemporary environmental movements and politics, with an emphasis on the ways both environmental movements and religious participation differ from religion and ecology interactions in the US and elsewhere.

Studying religion in Latin America

Identifying who studies Latin American religion and how can help us understand the themes that are important for religion and ecology, the methods and sources that are used, and the theoretical perspectives that dominate. Religious studies is not the dominant discipline in the study of Latin American religion, and North American scholars are joined by numerous Latin Americans and Europeans in defining the field. As a result, the themes, methods, and theories that dominate the study of world religions in the US, particularly the drive to identify and compare distinctive features of religion, are not central to scholarship on religion in Latin America. On the other hand, Latin Americanists pay attention to some issues that receive little attention in the field of religion and ecology in the US, especially political and economic factors.

These differences emerge in part from disciplinary background. Most people studying Latin American religion, including scholars from Latin America, the US, and Europe, are historians, anthropologists, sociologists, or political scientists. Social scientists have concentrated on topics such as church–state relations, the role of religion in social movements, and ethnographies of particular communities (especially indigenous indigenous and African-based). Recent work has also paid attention to religious pluralism, the rise of evangelical Protestantism, and the interactions between religion, gender, and race. Religion and ecology, as a distinct subfield, has not received much systematic attention, although there have been excellent studies of the role of religious ideas and communities in social movements, including those focused on environmental protection and sustainability.

Historical context

The religious history of the Latin America began long before Europeans arrived. Unfortunately, the conquest led to the destruction not only of many native communities but also of many important religious texts and structures. As a result, there are many gaps in scholarly knowledge of pre-conquest religions. Still, the existing resources provide a fascinating picture of many aspects of religious life and practice, especially for the great civilizations of these regions – the Incas in the Andes and the Mayans and Aztecs (or Mexica) in Mesoamerica. In these cultures, religion did not constitute a separate realm, distinct from other aspects of personal and collective lives. Religious practices and ideas permeated power relations among people; daily activities such as farming, hunting, and childcare; attitudes toward physical and mental health; and interactions with the natural world. While generalizations about "Native American" cultures or religions miss many important differences, it is possible to identify some common themes for pre-colonial indigenous traditions in the Americas. One of the most important, and one that is especially relevant to scholars of religion and ecology, is the mutual dependence between human and divine beings. This reciprocity was expressed in sacrifice, rituals related to planting, harvesting, fishing and hunting, human rites of passage, changes of season, lunar and solar cycles, and other natural patterns that structured people's daily personal and collective lives.

While all people in all places depend on nature, indigenous peoples in the Americas often made their relationships with the non-human world particularly explicit in their rituals and narratives. Mutuality characterized human relationships both with the sacred and with non-human animals, plants, and landscapes. This is evident in the Quiche Mayan story of human creation from corn, told in the sacred text *Popol Vuh*, as well as in the intricate balance between creation and destruction negotiated in Mesoamerican rituals of sacrifice (see Read 1998). A similar mutuality characterized Andean understandings of *huacas*, the mountain gods that demanded human offerings of food. After the conquest, native Andeans, the gods became hungry and restless, because the people no longer "fed" them (Silverblatt 1987, 185). While the conquest and forced conversion to Catholicism altered the relationship between Andean people and the *huacas*, it did not destroy it entirely. Many contemporary Andean communities, especially in the highlands, continue to make offerings and engage in other practices that "feed" the *huacas*, fulfilling the human side of this reciprocal relationship (see Nash 1993; Taussig 1983). In many native myths, not only landscape features but also non-human animals play important roles in religious belief and practice, as deities and as fellow residents of sacred and mundane space.

Beliefs and practices regarding the natural world mark one of the sharpest distinctions between Amerindian cosmologies and those of European colonizers, who had a more instrumental view of nature, often based on culturally particular ideas of human superiority and dominion (see Sale 1990; Thomas 1983). In part because of these ideas, the *conquistadores* wrought unprecedented destruction of human lives and cultures and of nature. Some of this destruction was unintended, the result of native vulnerability to European diseases, such as smallpox. These diseases killed indigenous people in huge numbers, up to 90 percent of the population in many areas. Further deaths resulted from forced labor in mines, plantations, and other European enterprises and military conflicts. In addition to the human costs, the conquest brought massive destruction of Latin American ecosystems. Colonizers came from a Europe that, as Kirkpatrick Sale argues, "was estranged from its natural environment and had for several thousand years been engaged in depleting and destroying the lands and water it depended on, and justifying that with one or another creed or conviction" (Sale 1990, 74). The dominant European approach to nature, as a source of wealth to be exploited, contrasted sharply with the emphasis on reciprocity that characterized many indigenous worldviews. In the new world, Europeans encountered vast resources in mines, forests, and waterways, which they exploited with the help of forced indigenous labor and, later, African slaves. In addition to these destructive attitudes, settlers brought with them animals and plants, both domestic and wild, that wreaked much the same havoc on New World ecosystems as European diseases did on Native American people.

While Europeans were on the whole extremely destructive, some explorers appreciated the beauty of the startlingly new landscape they encountered in the Americas. European evaluations and descriptions of the animals, trees, flowers, waterways, forests, and mountains of this new world were infused with religious assumptions, fears, and hopes. Some found the New World to be a paradise, as evidenced in the lush descriptions of early explorers from the Amazon to Florida. Other Europeans believed it a hell and considered its human and nonhuman inhabitants to be demons and monsters (de Mello e Souza 2004, 10–13, 39–43).

Inevitably, both European and indigenous religiosity changed in the distinctive cultural, political, and ecological landscape of the new world. Interactions among indigenous, Catholic, and, in many places, African traditions generated a distinctively Latin American form of Catholic Christianity, often described as syncretic or hybrid. While there were many regional variations, some important common features of this popular religion included a practical emphasis on rituals, festivals, and personal devotions, especially to Mary and other saints. The central role of saints was influenced by indigenous beliefs, especially the emphasis on reciprocity and sacrifice. David Carrasco writes that pre-Columbian understanding that "life is a sacred gift from the deities for which humans are responsible to nature" shaped Mesoamerican attitudes toward the saints (Carrasco 1990, 152). The saints mediated believers' relationships not only with God but also with landscapes, animals, and other natural features. Many natural features, especially mountains or rivers, were incorporated into myths and rituals. Almost all villages have patron saints and annual festivals that emphasize the link between religiosity and local identity. Further, apparitions of the Virgin Mary or Jesus often occur in significant natural places, such as the mountain Tepeyac associated with the Virgin of Guadalupe. This place-based religiosity continued an important feature of most indigenous traditions in the Americas, and continues to play a role in many environmental social movements today. For example, for many indigenous people, the preservation of traditional culture is inextricable from the protection of natural places that serve as both sacred sites and resources for continuing long-standing practices such as hunting, fishing, and subsistence farming.

The central role of religion in contemporary indigenous movements is just one manifestation of the pervasive importance of religion in Latin American political life. In the colonial period, church and state were intertwined, providing each other with mutual justification as well as resources. At the end of the colonial period, this alliance shattered, as the church sided with conservatives loyal to the crown, against the mostly republican and secular independence movements that swept throughout the region in the early 1800s. The Catholic Church remained both dominant and conservative in the post-independence era, until a host of political, religious, and social changes in the mid- to late-twentieth century dramatically changed both the religious landscape and religion's role in the larger Latin American landscape. These include growing diversity within and among religions, collaboration and competition among religious groups, and increasing global connections and mobility. All these changes have important implications for the relations between religious groups and environmental movements and for religiously influenced attitudes and values regarding nature.

In addition to indigenous traditions, Latin American popular Catholicism was also influenced heavily by African religions, especially in the regions that had large enslaved populations, primarily on the Atlantic Coast, from Brazil north to the Caribbean. Both enslaved and free Latin Americans of African descent helped create new religions, incorporating some Catholic and even indigenous elements into ritual structures taken from West African, especially Yoruba, traditions. Contemporary religions such as Candomblé and Umbanda in Brazil and Santería and Vodoun in the Caribbean result from centuries of syncretism and creativity. These traditions are often pragmatic in character, oriented toward addressing health, relationship, or financial difficulties. They share with many indigenous traditions an emphasis on the need for reciprocity in interactions between humans and divinities (Murphy 1995). This mutuality is embodied in rituals, many of which incorporate non-human animals. In Candomblé, for example, exchange and consumption are not only guiding metaphors but practical realities in religious life (Johnson 2002). This reflects a very different understanding of the human relationship to nature than the dominant European Catholic vision of a hierarchical cosmos, in which only the institutional church mediates people's relations to God. African and indigenous emphases on reciprocity and exchange entered into popular Catholicism to create a distinctive understanding of the ways people should and do interact with the non-human world.

Contemporary religious pluralism

One of the most important features of Latin American religious life since the mid-1900s is increased diversity, including both diversification within Catholicism and the emergence of alternative religious traditions. Most Latin Americans remain Catholic, at least nominally, and Catholicism today provides a variety of options in terms of pastoral and liturgical style as well as moral and theological priorities. The most important streams within the church are progressive Catholicism, best known for its ideological expression in liberation theology, and the Catholic Charismatic Renewal, which emphasizes the Holy Spirit, healing, and an active, participatory worship style.

The Catholic movement with most relevance to environmental concerns is progressive Catholicism, which emerged in the wake of the second Vatican Council (1962–65) to challenge the conservative, pro-establishment position of most church institutions, officials, and ideologies. Progressive Catholicism includes pastoral, institutional, and liturgical elements, as well as liberation theology, which became well-known in Europe and North America as well as throughout Latin America. In its early stages in the 1970s and 1980s, theologians focused on defining the liberationist perspective in regards to the church, Jesus, and political ethics, especially in the context of dictatorship that many nations faced. In more recent years, other issues have become important, including race and gender and also the environment. As I will discuss below, Catholic attention to environmental problems has focused on social justice and human needs, although ecocentric attitudes also enter into many statements and movements, especially as Latin American environmentalism – like Latin American religion – becomes more diverse and more global.

The Catholic church remains a very big tent, and many people who adhere to indigenous and African traditions also identify themselves as Catholics. In Brazil, for example, people who frequent Umbanda or Candomblé centers often baptize their children and marry in the church. The same is true for many Santería practitioners in Cuba, Puerto Rico, and other parts of the Spanish-speaking Caribbean. By any measure, however, African-based religions are vibrant and influential aspects of the religious and cultural arenas throughout Latin America, particularly Brazil, Colombia, Venezuela, and the Caribbean. While these traditions are very diverse, they share some common features, several of which are especially relevant for scholars of religion and ecology. In contrast to Western traditions, most African-based religions assert that everything in the universe, from natural phenomena to humans and spirits, is animated by a vital force, known as *Axé* in Portuguese and *Aché* in Spanish. In this non-dualistic worldview, diverse entities exchange energies in order to maintain balance and harmony. Like many indigenous cultures, African-based religions understand the natural and spiritual worlds as characterized by balance and mutual interaction, creating an interrelated cosmos that requires humility, respect, and reciprocity (Murphy 1995; Seligman 1995). Humans should engage in religious practices such as prayers, sacrifices, and special celebrations, to the ancestors and other spirits. This worldview suggests a modest and respectful attitude toward nature, although there have not been many official statements to this effect. This is at least in part because African-based religions lack the central authority and textual emphasis of Roman Catholicism and mainline Protestant churches.

The same decentralized and practical (non-textual) approach characterizes indigenous religions in Latin America. With the greater cultural and political openness following the end of most dictatorships by the 1990s, indigenous groups have joined other social sectors in building both a stronger public identity and political organizations. Some of the most active native groups are Mayans in Guatemala and southern Mexico and Amazonian groups in Brazil, Colombia, and Peru, although there are native political movements in many other areas, including some nations, like Chile and El Salvador, in which native populations have been relatively small and quiet.

The largest Christian minority in Latin America is Pentecostalism, a stream within Protestantism that focuses on baptism in the Holy Spirit and a direct personal relationship with God. Pentecostal and other evangelical Protestant churches also emphasize conversion, and often are seen as politically withdrawn or conservative. For these reasons, the growth of evangelical and especially Pentecostal Protestantism in Latin America was fraught with political conflict. As progressive Catholicism became tied to progressive social movements, Pentecostals sometimes appeared as allies of repressive regimes and their US benefactors. Today there is much greater political diversity among Latin American Protestants, but many secular and Catholic progressives continue to view Protestants as individualistic, spiritualized, and apolitical. This has contributed to the relatively low participation of Protestants as both leaders and lay activists in environmental movements compared with Catholic and indigenous communities.

Religion and the environment

Environmental concern and activism have grown rapidly in Latin America in the past few decades, especially in the wake of the 1992 Rio conference on sustainable development. However, environmental advocacy in the region differs in important ways from environmental movements in North America and the US. While Latin American environmentalists are diverse, in general they pay more attention to political and economic matters, especially poverty and inequity, than do mainstream environmentalists in the north. Many focus on urban factors, such as air and water quality, and even more traditional "ecological" concerns, such as protecting forests and preserving natural areas, are placed in the context of social justice, economic subsistence, and human rights to basic resources. Ricardo Navarro, a prominent environmental activist in El Salvador, explains that "to speak of ecological problems is to speak about social problems, it is to speak about poverty, about relations of power, at a national and international level, and it is thus 'to get mixed up in politics' [*meterse en política*], as some would say" (Navarro 1990). Until poor people are able to provide for themselves and their families, as Navarro points out, they will continue unsustainable practices, cutting down trees for firewood and planting seeds on exhausted soils (see Peterson 2005, 65). Many Latin American environmental advocates have also worked in leftist movements for human rights, social justice, and democratization, experiences that bring distinctive political aims and activist styles.

Many Latin American environmental campaigns appear more "anthropocentric" than many European and North American movements. Practical goals focus less on species diversity and wilderness protection, and there is little discussion of philosophical issues that are important in environmental ethics and eco-theology in the North, such as the intrinsic value of nature. While Latin American activism does not coincide exactly with any particular stream in US environmentalism, it has more in common with social ecology than deep ecology or land ethics. This human-centered approach coincides with that of many religious groups, which often emphasize values such as stewardship rather than ecocentrism. Environmental concerns have been incorporated into a broader Catholic humanist position, which contends that God created all of nature in order to serve human dignity and the common good, not individual profit. This assertion of the social purpose of created goods suggests a strong connection between environmental and social problems. This model is exemplified in Roman Catholic statements on the environment, such as John Paul II's message for the 1990 World Day of Peace, titled "Peace with God the Creator, Peace with All of Creation." Describing a connection between social and environmental problems that is common in Latin America, John Paul (1990) asserts that "the proper ecological balance will not be found without directly addressing the structural forms of poverty that exist throughout the world. Rural poverty and unjust land distribution in many countries, for example, have led to subsistence farming and to the exhaustion of the soil. Once their land yields no more, many farmers move on to clear new land, thus accelerating uncontrolled deforestation, or they settle in urban centres that lack the infrastructure to receive them." Interest in the environment continues to be strong in the Catholic church, and Francis I – the first Latin American pope – published his encyclical on the environment, *Laudato Si*, in 2015. *Laudato Si* builds on and fits within the long tradition of Catholic social thought, calling on Catholics to care for nature both because it is part of God's good creation and because the common good for humans requires environmental health and sustainability.

Within Latin America, Catholic statements on the environment echo the themes seen in official documents from Rome. Some liberation theologians, including the Brazilian Leonardo Boff, have incorporated ecological concerns into their work. In *Cry of the Earth, Cry of the Poor*, Boff

(1997) makes that connection explicit, arguing that ecological and social justice issues are linked both theoretically and practically, and must be addressed in an integral way. These concerns are echoed in *Laudato Si*, which reflects not so much the possibility that Pope Francis was influenced by liberationist thought (though he probably was) but more importantly the fact that this integration characterizes Catholic social thought more broadly. On an institutional level, a number of local and regional church groups have also become involved in environmental issues. For example, a pastoral letter issued in December 2000 by the Apostolic Vicariate of Petén, Guatemala titled *El grito de la selva en el a o jubilar: Entre la agonía y esperanza* [*The cry of the forest in the jubilee year: Between agony and hope*] asserts that "It is not possible to speak of ecology without taking justice into account. It is not possible to defend the conservation of the forest apart from the advancement and life of the poor." Excessive consumption by some harms both poor people and the natural world, according to this statement. The solution will entail both more restraint in human use of natural resources and a better distribution of the goods they make possible.

In addition to Catholics, indigenous groups have been the most active participants in connecting religion and ecology in Latin America. Native activists emphasize their traditional knowledge of and connection to non-human nature and also their practical ties to the land. Some native movements have highlighted environmental matters, especially the protection of native lands and waters. These issues have been especially important in national politics in Brazil, where Amazonian indigenous communities have cited traditional religious values in their resistance to logging, agriculture, and other uses of their traditional lands by private companies and the government. Some of these movements have received support from local or national Catholic organizations, particularly in Brazil.

Other Latin American religious groups have played much less prominent roles in environmental activism. Among Protestants, the historic churches linked to the World Council of Churches, such as Lutherans, Methodists, Presbyterians, and American Baptists, have been the most active, at least in affirming statements of environmental concern and urging stewardship. However, mainline Protestants are a small minority in Latin America. Most Latin American Protestants are Pentecostals, and many belong to "storefront" churches, small congregations led by an individual pastor, often not affiliated with any larger denomination. Some Latin American churches belong to the Assemblies of God or other Pentecostal denominations (mostly based in the US), some of which have taken an interest in environmental topics. A number of American evangelical groups have called attention to global warming, in particular. With some exceptions, most Latin American Pentecostals remain uninvolved in social and political issues, and even when they do step into the public square, it is usually around issues such as gang violence, rather than ecological problems.

While much of the information about religious involvement in environmental issues is anecdotal, there is some evidence that Catholics and Pentecostal Protestants hold substantially different attitudes about ecological issues (Cooper 2015: 36; Otsuki 2013: 418). Practitioners and researchers involved in sustainable development projects in Latin America, for example, have found that Catholics have far more positive opinions of such efforts. This issue demands additional research, to identify the extent of and reasons for religiously based differences and also to learn how to build wider support for environmental programs among different religious communities. If, as scholars of religion and ecology believe, religion is an important factor in attitudes toward environmental issues, then we need much more information about the causes and implications of these attitudes. We need to ask, for example, if statements from the Vatican primarily influence Catholics, or the stance of local church officials, or deep-rooted cultural attitudes and practices. We also need to learn whether Protestants are less likely to support environmental initiatives because they are preoccupied with spiritual matters, because their pastors discourage activism, or because they hold anthropocentric theological beliefs that find nature to be of little value. In addition, socio-economic differences and other factors that distinguish Protestants and Catholics may be related to environmental attitudes and practices. These and many other questions remain to be investigated, making the intersection of religion and the environment a rich field for further research.

Conclusions

Religion and ecology, as a scholarly field, is extremely diverse, incorporating resources, theories, and questions from both religious studies and environmental studies. Focusing on Latin America complicates the situation even further by adding yet another interdisciplinary field to the mix. However, Latin Americanists have much to add to the study of religion and ecology, and students of religion and ecology in the US and Europe can learn both from experiences in Latin America and from the work of Latin Americanist scholars.

Diverse Latin American religious communities approach environmental matters with a shared emphasis on human needs and, in the case of activists, an insistence that environmental protection must be linked to social justice and other large political issues. This is not to say that nature is not valued in and of itself; there is often a great deal of attachment and appreciation, especially among rural and indigenous people. However, the realities of poverty, economic inequity, and lack of access to basic resources frame both practical and moral approaches to the natural world. People care about the theological and moral foundations of attitudes to nature, but there is little interest – popular or academic – in comparing the "green" elements in various world religions, an issue that preoccupies many US scholars. This does not mean that the latter should abandon their work on these issues, but it does point to the possibility of addressing explicitly political and practical themes.

The conversation, of course, should not be one sided. Latin Americanists will also benefit from engagement with North American and European scholarship on religion and ecology. For example, the extensive work on eco-theology and environmental ethics can help provide theoretical grounding for environmental projects. Latin American theologians, including Boff, have paid some attention to these issues, but the conversation lags far behind the discourse in the US and Europe. As noted above, it is probably not accidental that Catholics tend to support environmental initiatives more than Protestants. Research investigating the historical, ideological, and institutional reasons for these differences – including deeply rooted theological and moral claims – can make an important contribution to the success of future initiatives. In addition, many social scientists do not attend seriously to the role of religion in social movements, including environmental ones; they will benefit from engagement with the growing literature on religion and ecology.

As the study of religion and ecology expands to new parts of the world, we cannot simply apply accustomed methods and theories from North American and European scholarship – we cannot just apply "religion and ecology" and stir. To do so would miss much of what is important and distinctive in religious understandings of nature and religious environmentalism in other parts of the world. Instead, novel settings demand new (to us) ways of understanding, which we can then turn back on our familiar landscapes to see them differently as well.

References

Apostolic Vicariate of Petén. 2000. *El Grito de la Selva en el año Jubilar: Entre la Agonía y Esperanza*. Petén, Guatemala: Vicariato Apostólico de Petén.

Boff, Leonardo. 1997. *Cry of the Earth, Cry of the Poor.* Maryknoll, NY: Orbis Books.

Carrasco, David. 1990. *Religions of Mesoamerica: Cosmovision and Ceremonial Center.* San Francisco: Harper & Row.

Cooper, Natalie. 2015. "Uncovering Resident Perceptions of a New Forest Livelihood: Timber Extraction within the Chico Mendes Extractive Reserve, Acre, Brazil." M.S. thesis. University of Florida, Gainesville.

Francis I. *Laudato Sí: On Care for our Common Home.* https://w2.vatican.va/content/francesco/en/encyclicals/documents/papa-francesco_20150524_enciclica-laudato-si.html

Gonzalez, Ondina E. and Justo Gonzalez, *Christianity in Latin America: A History.* 2007. Cambridge: Cambridge University Press.

John Paul II. 1990. "Peace with God the Creator, Peace with All of Creation." Message for the World Day of Peace. https://w2.vatican.va/content/john-paul-ii/en/messages/peace/documents/hf_jp-ii_mes_19891208_xxiii-world-day-for-peace.html

Johnson, Paul Christopher. 2002. *Secrets, Gossip, and Gods: The Transformation of Brazilian Candomblé.* New York: Oxford University Press.

Lynch, John. 2012. *New Worlds: A Religious History of Latin America.* New Haven, CT: Yale University Press.

de Mello e Souza, Laura. 2004. *The Devil and the Land of the Holy Cross.* Austin: University of Texas Press.

Murphy, Joseph. 1995. *Working the Spirit: Ceremonies of the African Diaspora.* Boston: Beacon Press.

Nash, June. 1993. *We Eat the Mines and the Mines Eat Us: Dependency and Exploitation in Bolivian Tin Mines,* rev. ed. New York: Columbia University Press.

Navarro, Ricardo. n.d. [1990?]. "Presentación del Libro" [Preface]. In Ricardo Navarro, Gabriel Pons, and German Amaya, *El pensamiento ecologista.* San Salvador: Centro Salvadoreño de Tecnología Apropiada [CESTA].

Otsuki, Kei. 2013. "Ecological rationality and environmental governance on the agrarian frontier: The role of religion in the Brazilian Amazon." *Journal of Rural Studies* 32: 411–419.

Peterson, Anna L. 2005. *Seeds of the Kingdom: Utopian Communities in the Americas.* New York: Oxford University Press.

Peterson, Anna and Manuel Vasquez. 2008. *Latin American Religion: Histories and Documents in Context.* New York: New York University Press.

Read, Kay Almere. 1998. *Time and Sacrifice in the Aztec Cosmos.* Bloomington: Indiana University Press.

Sale, Kirkpatrick. 1990. *The Conquest of Paradise: Christopher Columbus and the Columbian Legacy.* New York: Alfred A. Knopf.

Seligman, Rebecca. 2014. *Possessing Spirits and Healing Selves: Embodiment and Transformation in an Afro-Brazilian Religion.* New York: Palgrave-MacMillan.

Silverblatt, Irene. 1987. *Moon, Sun, Witches: Gender Ideologies and Class in Inka and Colonial Peru.* Princeton, N.J.: Princeton University Press.

Taussig, Michael. 1983. *The Devil and Commodity Fetishism in South America.* Durham, NC: University of North Carolina Press.

Thomas, Keith. 1983. *Man and the Natural World: Changing Attitudes in England 1500–1800.* London: Penguin.

21

AFRICAN DIASPORA

African American environmental religious ethics and ecowomanism

Melanie L. Harris

In this chapter the work and ideas of African, African American, and Black theologians, ethicists, environmental geographers, sociologists, literary artists, activists, and scholars are referenced as leaders shaping the discourse. That is to say, environmental justice discourse is diverse, dialogical, and interdisciplinary. Likewise, African American religious environmental religious ethics, and ecowomanism are also diverse, dialogical, interdisciplinary, and interreligious. This important theoretical framing builds on three important hallmarks of third wave womanism that I have articulated in my previous work, *Gifts of Virtue, Alice Walker and Womanist Ethics* (Harris, 2010). These hallmarks include: 1) the global links across the African diaspora and its peoples that recognize connections between worldviews, ethics, and cultures, as well as the parallels and common struggles experienced by African peoples as a result of globalization and the impact of climate change; 2) the necessity for interdisciplinary approaches to help womanist religious thought engage a variety of discourses; and 3) the importance of interreligious dialogue in womanist and ecowomanist discourse, thus breaking new ground and providing new entries into interfaith and interreligious dialogue about earth justice.

In sum, this chapter is written from a third-wave womanist theoretical frame that features contemporary womanist discourse, (discourse that centers the theological, religious, and ethical reflection of women of African descent and their interdisciplinary perspectives on religion) and shares dialogue across the African diaspora making "global links." In addition the chapter engages interreligious dialogue and interdisciplinary methods that help to acknowledge, embrace, and examine, the rich complexity of womanist religious thought.

There is an implicit assumption and myth in the traditional environmental movement that African Americans and other communities of color have little interest in earth justice and that their numbers have been sorely lacking in the movement. The rationale often given to support such a claim is that African Americans and other communities of color are too often engaged in battles of racial justice in an age of white supremacy. Facing the impact of unjust police brutality evoking fear into the rhythm of one's everyday existence, the constant pressure of having to prove the humanity and innate dignity of African peoples, and the reality of living in a culture wherein black women's bodies continue to be subjected to rape, shame, and even death at the hands of those who exercise social dominance is difficult "death dealing" and life transformative work (Townes 1998). In the midst of the urgency of climate change however, many in activist communities have to fight for social justice in new and unexpected ways.

Considering the significance of the unjust treatment and exploitation of African Americans and the history of race and white supremacy in America, a scholar concerned about responsible cultural production of knowledge must ask the question: Why is African American environmental history missing from the discourse of religion? This question might prompt additional questions such as: Has it been left out of the discourse intentionally? If so, why? Is the absence of African American environmental history an intentional move to reinforce particular racially biased beliefs pertaining to myths about the lack of African American intelligence, ability, humanity, or ethical sensibility? Is the absence designed to devalue the religious and social protest thought that emerges from African American life, thus problematizing categories in Western discourse used to ignore these realities? Or rather is the absence of African American environmental history simply a problem of neglect? An unconscious "innocent" move of unknowing, perhaps based on an implicit bias that seems to repeatedly, systematically, and normatively ignore the history of race (and racism) in America and its connections to the environment?

However ill founded or naïve the reasons for the absence of attention to African American environmental history (as well as the history of other peoples of color in the environmental movement) in the move towards recovering this history—and correcting the meta-narrative that does not always acknowledge the significance and presence of this field—one must also be careful not to further silo African American environmental history as it is being developed. This move could result in the unhelpful effect of marginalizing all history about African Americans and the environment into one "subfield" casting it into a lesser, or seemingly less valuable discourse. It is important to acknowledge the important critical and creative thought necessary to develop theory and praxis in these fields. It is also important to engage the politics of the cultural production of knowledge in a way that insists on equity and historical accuracy and serves to subvert the dominance of a false meta-narrative that lacks racial and social consciousness but claims to be operating with good intentions.

In other words there is an assumption within the discourse of environmental studies, and more specifically within the subfield of environmental religious ethics, that the history of African Americans, the Civil Rights Movement, and such contemporary social movements as *#BlackLivesMatter* have little in common with concerns for earth justice. This assumption must be debunked if environmental ethics is to function in a true mode of justice. Social justice is connected to reflections on environmental degradation and should be considered vital for mainstream talk in religion and ecology.

This move not only widens the scope of environmental ethics but also invites the inclusion of theoretical frames that are non-Western and influenced by African, Asian, or Indigenous cosmologies. The inclusion of such cosmologies and religious worldviews sheds new light on the perils of colonialism and its connection to environmental degradation. It helps us to recognize that African American environmental history is different and offers unique contributions to the wider discourse because it demands that the remnants of what African theologian Edward P. Antonio (2004) calls "colonial ecology," be acknowledged. Too often traditional discourse in religion and ecology leaves out the important conversations that blend ethics, social justice, religion, racial consciousness and ecology together. The development of African American environmental history, religious history, and ecowomanism suggests that these conversations must take place.

Having explained why and how African American environmental history, religious history, and ecowomanism engage religion and ecology, I now turn to a specific question. In light of the significant contributions and templates for social justice organizing that have emerged from the Black Lives Matter movement, are there specific anti-racist reparations paradigms that can be translated for environmental justice paradigms and ecological reparations work? In this first section I introduce African American Environmental Religious Ethics and describe ecowomanism. The existence of these two fields fills a void in the traditional environmental movement and argues that particularly when it comes to conversations about religion and ecology, African Americans and many communities of color have long been shepherding and guiding an important discourse that links social justice to earth justice by using an environmental justice paradigm.

Secondly, building upon the third wave womanist hallmark that recognizes global links and connections about religion, ecology, and gender across the African diaspora, this essay explores how a frame for ecological reparations can be shaped from an ecowomanist African diasporic perspective. That is, I discuss parallels and intersections between racial and ecological reparations using an African disaporic lens. I conclude the chapter with remarks about the importance of uncovering African American religious and indigenous eco-wisdom and environmental history for the development of ecological reparations in the face of climate change.

African American Environmental History

To date, there is limited focused research on the question of why African American Environmental Ethics has been left out of the discourse of religion and ecology (Jenkins 2013, 206–211). In this section I will focus on the history of the African American Environmental Justice movement in connection with the Black Church and the social protest tradition. This information provides a counter-narrative to the meta-story of the environmental movement that too often leaves out the histories and contributions of women and communities of color. This exploration begins by pointing to an important origin point in the work of Dr. Martin Luther King Jr. that makes the link between social and earth justice impossible to ignore.

As Dianne D. Glave (2005) highlights, the writings, preaching, and oratorical genius of Martin Luther King Jr. offered much more than a theoretical and theological foundation for Black Liberation Theologies and contemporary perspectives on religion and the African diaspora. King's work of social justice, and in particular his radical stance on love, the beloved community, and racial and economic justice opened the doors for scholars, artists, everyday folk, and activists to get involved and raise consciousness about the gap between the haves and the have-nots. Interwoven within King's Christian understanding of redemptive suffering, divine justice, and unconditional love, was a deep commitment to non-violence and a belief in the innate dignity of all. King believed that if white Americans and people around the planet could overcome the myth of white racial superiority—if they could see and accept African Americans as fully human, as equally deserving of rights, humane treatment, and opportunity—their own morals would convince them to change public policy and laws that degraded persons of African descent. But as he records in many writings, speeches, and letters including the *Letter From A Birmingham Jail*, King was often disappointed with white moderates whose political stance stopped short of taking a stance for justice and ending racial segregation. Too comfortable with the social power that white privilege presented to them, King became untrusting of white politicians, religious, and community leaders who offered words of faith and reminders of Jesus' virtue of patience, but little action towards breaking the code of institutional racism. Dispirited but still brave, King didn't stop. Moving beyond disappointment—and death threats—King expanded his scope of the civil rights movement to include not only "radical love" and racial justice, but also economic and environmental justice. King's passion for justice erupted past his theology as a Christian, to include the religious orientation and economic status (or lack thereof) of others. That is, King was concerned, deeply concerned about the poor and the oppressed.

This is precisely why King accepted a request to speak and organize on behalf of sanitation workers in Memphis, Tennessee on April 3, 1968 (Harris 2016, 5–14). The Memphis movement exposed the deplorable conditions and environmental health hazards that workers had to face daily, all the while combating racism on the job. King grasped the connections between poverty, individual and institutional racism, and environmental health hazards, and he interpreted these links as threats to justice. These connections are not lost on many African American environmentalists. In fact, it is precisely this awareness of the links between race, class, and environmental health that guides much of the research of scholars in environmental justice (e.g. Bullard 2000; Taylor 2009; Margai 2010). Coming from a variety of disciplines including, sociology, geography, and religion, these scholars also signal an important theoretical frame and approach to investigating environmental injustice, climate violence, and environmental racism. The work of historians and literary artists such as Kimberly K. Smith (2007), Kimberly K. Ruffin (2010), and Camille T. Dungy (2009) also point to the significant contributions to the field of environmental studies coming from people of African descent who have been forced to confront systems of economic, gender and racial oppression.

The environmental justice paradigm links environmental concern with a social justice agenda attuned to the connections between environmental health hazards and race, class, and gender disparities (Taylor 2009). Illustrated in acts of justice modeled by Dr. King, Black church women protesting toxic waste dumping in North Carolina, New Orleans, and New York City, National Association for the Advancement of Colored People (NAACP) leaders, such as Dorothy Height, and contemporary leaders, such as Majora Carter, the environmental justice paradigm provides a foundation for both theoretical and praxis oriented approaches that help expand traditional environmental paradigms of "conservation" or "preservation." This expansion helps the discourse recognize the impact of historical and contemporary realities (such as a rise in racial violence) as well as the cumulative impact that structural forms of environmental injustice can have on *earthling* communities and communities of faith.

African and African American Environmental Religious Ethics (AAERE)

Far from absent, African Americans have been on the journey for justice, and earth justice since—some would argue before—being forcibly brought to North American shores on the slave ships from Africa. While arguments from political theorists and historians might place the origins of African American environmentalism at the start of black agrarianism, during the height of the transatlantic slave trade, or in the midst of Jim and Jane Crow, a disciplinary lens from religion suggests that the origins of African American environmentalism lie in a history much deeper than that: in the heart of African cosmology.

African cosmology connects the realms of spirit, nature, and humanity into one flowing web of life. That is, instead of a hierarchal or dualistic structure, African cosmology functions in a circular manner emphasizing interconnectedness and, in the words of Thich Nhat Hahn (1988), "interbeing." It is important to note that I am intentionally making connections between Asian and African cosmologies and argue that the worldviews and perspectives on earth ethics that emerge from these cosmologies shed new light on environmental justice. These perspectives from communities of color are necessary in the movement as we face down climate change.

As has been argued by religious scholars including John Mbiti (2015), Peter Paris (1994), Barbara Holmes (2002), and others, an African cosmological perspective drastically changes one's perspective of earth care or environmental justice because in fact a being, or human being, is understood to be a vital part, but not the center of the universe. That is, these cosmologies lend themselves towards seeing the earth as fully and intimately connected. "Everything is everything" is one hip-hop phrase whose signified meaning connotes this kind of interconnectedness (Hill 1998). As such, African cosmology also promotes a kind of innate ethical message to care for the planet. In addition to the ancestral spiritual connections that must be honored according to many African religious traditions, an ethical mandate to care for the earth is often communicated from African cosmologies because of the interconnectedness. To care for the earth is to care for the self and vice versa.

This kind of cosmology serves as a base from which African American Environmental Ethics and Ecowomanism grow. Both these fields honor and operate in ways that evoke the same kind of interconnectedness in African cosmology and understand this to be a central value (such as the value of community) in womanist thought, and in African and Black life. The community, and herein, the earth community is important to keep in balance. Honoring this balance is also reflected in relationships between human and non-human beings. That is, African cosmology deeply shapes the ethical mores of how one ought to care for the earth, and sheds light on how individual humans ought to act towards the earth and each other in communion. This kind of "earth ethic" that emerges from African cosmology has deeply influenced forms of black liberation theology and womanist religious thought. This can be seen most clearly in the work of eco-theologians, Theo Walker (2004), Karen Baker-Fletcher (1998), and Emilie M. Townes (2006). Of particular note, Delores S. Williams' essay, "Sin, Nature and Black Women's Bodies" (1993) (a foundational resource for ecowomanism) argues that there is a connection between the logic of domination present in white supremacy, colonial Christianity, and sexist ideology that sanctioned the objectification of Black enslaved women's bodies during slavery (and since). Drawing upon the work of several pioneering womanist thinkers, and theologians, James H. Cone (2001) draws direct links between black liberation theology and ecological justice:

> The logic that led to slavery and segregation in the Americas, colonialization and apartheid in Africa, and the rule of white supremacy throughout the world is the same one that leads to the exploitation of animals and the ravaging of nature…People who fight against white racism but fail to connect it to the degradation of the earth are anti-ecological—whether they know it or not. People who struggle against environmental degradation but do not incorporate in it a disciplined and sustained fight against white supremacy are racists—whether they know it or not. The fight for justice cannot be segregated but must be integrated with the fight for life in all its forms.
>
> *(2001, 23)*

Recognizing the links within the logic of oppression that have negatively impacted peoples of color, and peoples of African descent throughout the diaspora, Cone's argument points us back to the important source that African cosmology is for developing the discourse of black liberation theology, religion, and ecology, and especially ecowomanism.

Ecowomanism

Ecowomanism is critical reflection and contemplation on environmental justice from womanist perspective and more specifically from the perspectives of African and Indigenous women. It links a social justice agenda to environmental justice and recognizes the similar logic of domination at work in parallel oppressions suffered by women and the earth. That is, just as women of color have historically suffered multiple forms of oppression including racism, classism, sexism, and heterosexism, so has the earth suffered due to debasing and devaluing attitudes about the earth that suggest that non-human beings (earth) have less status and worth than human beings. Regrettably, this kind of anthropocentric attitude has resulted in many human cultures adopting hierarchal values and dualisms that have resulted in the Anthropocene: a new geological epoch and many would argue a new climate age resulting from the impact human life and lifestyle has had on the planet.

As a discourse that blends methods from environmental studies and womanist religious thought and ethics, ecowomanism offers unique gifts to the wider field in that it highlights the contributions that women of African descent and Indigenous women have made to the environmental justice movement. Ecowomanism also examines how their theological voices and ethical perspectives contribute new strategies of earth justice. Noting the paradoxical historical relationship that African American women have had with the earth, (i.e. being named as property the same as the earth, and devalued in similar ways as the earth) in the midst of surviving a white supremacist society, ecowomanism pays attention to the complex subjectivity of African American women. The approach uses multilayered analysis to investigate earth justice in a way that honors the bio-diversity of the earth, while also honoring varied particular experiences of African American women as human beings living in the earth community. Using intersectional analysis when approaching earth justice, ecowomanism is a lens that contributes solutions to the ecological crisis that we are living in today, by connecting the agenda of social justice with earth justice; or in other words by developing a theological liberationist social justice agenda—moving towards race, class, gender, and sexual justice, while also lifting up earth justice.

As noted in the work of celebrated cultural geologist Florence Margai (2010), sociologist Dorceta E. Taylor (2009), and others, the connection between race and class and environmental health disparities can no longer be ignored. As a contribution to this discussion, contributing a religious perspective, ecowomanism joins the conversation inviting reflection on the inner lives of women environmental activists, the spiritual practices of those who practice environmental justice, as well as the communities of faith and encouragement that instill and reinforce their values for earth justice with hope.

Making connections: African religion, ecology and the diaspora

Before moving forward to discuss more about the contributions of ecowomanism, it is important to nuance the understanding of African cosmology and an ecowomanist adoption of it. Most thinkers would consider it a risk to ground a concept and method such as ecowomanism into a cultural frame, rather than a historical frame. One disadvantage for example of claiming that womanist race–class–gender analysis is central to ecowomanist method and approaches is that it places ecowomanism in a center of a debate about identity politics at best and exhibits the impact of colonial ecology at worst. That is, without the presence of racial hierarchies, meta-narratives that leave out communities of color, and other structural oppressions, some argue there would be no need for a womanist or ecowomanist approach. Black and womanist theologies and ethics have been interpreted as reactionary responses to traditional (read: white) theologies and therefore suspect in that their theoretical base may in fact be based on the same theory used by the oppressors. Using the "master's tools to dismantle the master's house," to paraphrase Audre Lorde (1984), is a difficult quandary and debate that many black liberation and womanist theologians, ethicists and theorists have engaged.

Another contentious element in ecowomanist theory involves its adoption of an African cosmological frame. Antonio (2004), Amenga-Etego (2011) and other African Traditional Religion (ATR) scholars argue that Western religious, anthropological, and even cultural lenses have misframed the relations between ATRs and ecology in problematic ways. This is important because womanist ethics has often been labeled a Western enterprise and ecowomanism could be (mis)understood the same way. Although I do not agree with this characterization and maintain that womanist religious thought and ecowomanism both attempt to be shaped by non-dualistic and non-hierarchal thinking, it is worth noting this very important debate.

Many scholars of African traditional and Indigenous religion argue that ATRs are too often misunderstood and misinterpreted due to a colonial impetus within most Western lenses. For example, Antonio (2004) warns of the Western colonial influence on ecological discourse by problematizing the assumptions that many religious thinkers in the West make about African people and worldviews. In Antonio's view, the claim that Africans are closer to nature and therefore their worldviews have within them practices that are more earth-honoring than European practices and theologies of nature is wrong and constructed on a false premise. This Western assumption is at best culturally unaware and at worst racist.

Antonio critiques malformed assumptions often made by white and European environmentalists and exposes their deep investment in a logic that attempts to control Africans and African worldviews. Noting how the logic of domination has woven through the history of colonization and crippled the earth, Antonio points to the eerie similarity between colonial moves to control the bodies of Africans and the body of the earth. He asks whether the white and European assumption that Africans are closer to nature and their move to point their scholarly arrows back towards African religious roots for answers about climate change is authentic, or an odd act of repentance for the construction of colonial ecology. His essay prompts ecowomanist scholars to ask the question forthrightly: Is the move back to African religion, the adoption of African cosmology, and the over-romanticizing of the very nature, culture, and values of Africanness (i.e. blackness) actually an odd act of repentance—a conceptual move towards some form (however misshapen) of ecological reparations?

In the frantic urgency of climate change, environmentalists of every discourse are often left scrambling, ready to pounce and exploit the very blackness, and African cosmological connection embedded in African indigenous religions for the sake of saving their own (colonial) ecological home (or their "ecological souls"). As the history of colonialism proves, there is a connection between exploiting African peoples and their lands, and this logic of domination has had a major impact on our current climate crisis. Colonization is one of the reasons for climate injustice; therefore, social justice is directly related to climate or earth justice. Social justice is earth justice. Earth justice is social justice.

Dialogue with theologians and African religious scholars, such as Antonio, turn the ecowomanist's gaze into herself for self-reflection and critique. Similar to the methodological move of self-reflection gleaned from the work of womanist ethicist Stacey Floyd-Thomas (2006), Antonio's point invites ecowomanism to more closely analyze its structures and moral foundations. By starting with African cosmology as a base from which to honor African, African American, and Indigenous voices of women and their contributions to environmental justice is ecowomanism casting itself into the long shadow of colonization? Is it practicing colonial ecology?

I say no. Ecowomanism honors its African cosmological roots as a valid and authentic epistemology of how to be an ethical earthling, living on the planet today. As many African peoples' values are shaped by their religion and religious practice, so too are ecowomanism's values shaped by the identity as descendants of Africa, carrying embodied and actual indigenous roots. To honor African cosmologies—and the theories and practices of environmental justice that are informed by them—is an act and method of resistance in that it joins the postcolonial move to expose the remnants of colonial ecology and dismantle this by infusing the field with true ecowomanist epistemology: womanist earth stories, African American agriculture knowledge, and African ethical world views that model interconnectedness and help us honor the earth and live faithfully on the planet with others.

Conclusion: "I can't breathe"

Following the death of Eric Garner—an African American man who was killed on July 17, 2014 in New York City at the hands of a police officer who used an illegal chokehold to pin him down to the ground, ignoring his pleas for life, saying " I can't breathe, I can't breathe"— I, like many environmental activists, took a step back. #*Blacklivesmatter* had grown into a movement and many scholars, activists, teachers, and religious leaders were taking to the streets to protest the rise in racial violence. Clearly black women and men were being targeted. Calling upon the wisdom of Ida B. Wells, whose *A Red Record* exposed the horrors of white supremacist lynching of black women and men across the United States for three gruesome years, and examining the roots of white supremacy and its connections to forms of Christianity that normatively ignores social justice as a base from which to do theology, I began to take a long view of the work of justice being done in the green movement. The autopsy performed by the coroner in New York revealed that Eric Garner not only suffered from the impact of the illegal chokehold forced upon him by the officer. Eric Garner also had asthma. A consequence of the environmental health hazard of air pollution in Staten Island, Eric Garner, like thousands of children, and adults living in and breathing in non-clean air, suffered not only because of racially motivated violence, but also because the air in his community robbed him of a normal quality of life: the right to breathe clean. The crescendoing call that many of us are hearing now is begging us to use an environmental justice paradigm, to see the connections between social justice and environmental justice, to hear these important earth stories. As paradoxical as they may be, African American environmental history and herstory has never been so important. Its riches, contributions, and stories and solutions have never before been so valuable for the work of justice.

References

Amenga-Etego, R. M. (2011) *Mending the Broken Pieces: Indigenous religion and sustainable rural development in Northern Ghana* Africa World Press, Trenton.

Antonio, E. P. (2004) "Ecology as experience in African indigenous religions", in Thomas, L. E. ed., *Living Stones in the Household of God*, Fortress Press, Minneapolis, 146–157.

Baker-Fletcher, K. (1998) *Sisters of Dust, Sisters of Spirit: Womanist wordings on God and creation*, Fortress Press, Minneapolis.

Bullard, R. D. (2000) *Dumping in Dixie: Race, class and environmental quality*, Westview Press, Boulder.

Cone, J. (2001) "Whose earth is it anyway?", in Hessel, D. and Rasmussen, L. eds., *Earth Habitat: Eco-justice and the church's response*, Augsburg Fortress, Minneapolis, 23–32.

Dungy, C. T. (2009) *Black Nature: Four centuries of African American nature poetry*, University of Georgia Press, Athens, GA.

Floyd-Thomas, S. M. (2006) *Deeper Shades of Purple: Womanism in religion and society*, New York University Press, New York.

Glave, D. D. and Stoll, M. (2005) *African Americans and environmental History*, University of Pittsburgh Press, Pittsburgh.

Hanh, T. N. (1988) *The Heart of Understanding: Commentaries on the Prajnaparamita Heart Sutra*, Parallax Press, Berkeley.

Harris, M. L. (2010) *Gifts of Virtue, Alice Walker and Womanist Ethics*, Palgrave Macmillan, New York.

Harris, M. L. (2016) "Ecowomanism: An introduction", *Worldviews: Global Religions, Culture, and Ecology* 20:1, 5–14.

Hill, L. (1998) "Everything is everything," song from *The Miseducation of Lauryn Hill*, Ruffhouse Records.

Holmes, B. (2002) *Race and the Cosmos: An invitation to view the world differently*, Bloomsbury T&T Clark, New York.

Jenkins, W. (2013) *The Future of Ethics: Sustainability, social justice, and religious creativity*, Georgetown University Press, Washington, DC.

Lorde, A. (1984) "The masters' tools will never dismantle the master's house", *Sister Outsider*, Crossing Press, Bel Air, California.

Margai, F. (2010) *Environmental Health Hazards and Social Justice: Geographical perspectives on race and class disparities*, Earthscan, Washington, DC.

Mbiti, J. (2015) *Introduction to African Religion*, 2nd edn, Waveland Press, Long Grove, IL.

Paris, P. J. (1994) *The Spirituality of African Peoples: The search for a common moral discourse*, Fortress Press, Minneapolis.

Ruffin, K. K. (2010) *Black on Earth: African American ecoliterary traditions*, University of Georgia Press, Athens, GA.

Smith, K. (2007) *African American Environmental Thought: Foundations*, University of Kansas Press, Lawrence.

Taylor, D. E. (2009) *The Environment and the People in American Cities, 1600s–1900s: Disorder, inequality and social change*, Duke University Press, Durham.

Townes, E. M. (1998) *Breaking the Fine Rain of Death: African American health issues and a womanist ethic of care*, Wipf and Stock, Eugene.

Townes, E.M. (2006) *Womanist Ethics and the Cultural Production of Evil*, Palgrave Macmillan, New York.

Walker, T. Jr. (2004) *Mothership connections: A black Atlantic synthesis of neoclassical metaphysics and black theology*, State University of New York Press, New York.

Williams, D. S. (1993) "Sin, nature and black women's bodies", in Adams, C. J. ed. *Ecofeminism and the sacred*, Continuum, New York, 24–29.

PART V

Nature spiritualities

Introduction: *John Grim*

The contributions in this section ask fundamental questions: what does it mean to describe an experience of the natural world as spiritual? If spirituality characterizes a meaningful relationship with the natural world in what ways does it relate to environmental and social responsibility? These and other questions have become increasingly widespread in discussions of nature religion, paganism, and nature spirituality.

There has been a resurgence in recent decades in contemporary forms of ancient European paganism. That recovery can be credited in part to widespread honoring of experiences in nature as sacred, to a lack of attention to nature in traditional Western religions, as well as admiration of the rituals and religious traditions of indigenous peoples. Underlying many of these discussions is the term "animism," which carried connotations of primitive belief in spirits frozen in a timeless traditionalism. Often used to separate traditional tribal peoples from the monotheistic religions, the category of animism was also used to underscore theological arguments against syncretistic influences on the Abrahamic traditions. In these arguments, any borrowing from pagan "animists" was seen to subvert scriptural revelations from a divine transcendent source.

With the rejection of those facile prejudicial readings and a growing enthusiasm for nature experiences, many practitioners and scholars have productively revisited animism. Most importantly, there is a recognition that the early introduction of the term was used broadly to define religion, not simply the spiritual practices of traditional peoples. This reevaluation has served to establish animism as beginning a theoretical language for describing nature spiritualities, as well as a correction to mis-readings of indigenous religions as indiscriminately projecting spirituality onto the natural world.

This reexamination of animism neither begins with doctrinal beliefs nor lingers on any particular cultural representations of spirits. Rather, it has opened inquiries about pagan and other community relations with a world that is alive with persons-in-the-world having will, intention, and power. Paralleling these pagan nature spiritualities as relational, there are forms of nature mysticism in the institutional religions that acknowledged a divine immanence in the natural world. These ranged, for example, from the simple elegance of Julian of Norwich in the fourteenth-century English Catholic church who saw the universe in an acorn in her hand, to the poetic brilliance of Kabir in fifteenth-century northern India who drew on Islam and Vaishnavite Hinduism to express an experience of the natural world as manifesting the divine body. Thus, some nature spirituality cannot be wholly separated from the institutional religions. However, these nature mysticisms were often considered anomalous or marginal in those religions.

Efforts to reach beyond the human cultural sphere into the community of life on Earth largely characterize nature spiritualities. So human experiences of nature may resist or even subvert anthropocentric perception; mystical perspectives may be used to affirm ontological experiences of a unitive, or a difference-in-identity, or a devotional experience in which a divine reality infuses all dimensions of the nature spirituality, especially the "self."

Similarly, a philosophical or scientific perspective may be evoked to explain nature religion using collective evolution as a frame or ecological interdependence as an interpretive whole. In these hermeneutical contexts the human cannot be easily separated out as an independent causal agent. Rather, the human may be seen as one biotic entity among a collective evolutionary scene having other animal, plant, and mineral participants. Finally, esoteric thought provides another perspective in which elemental constituents in pagan and nature spirituality, such as food, soil, water, air, or fire are recognized as causal agents that work on the human interior shaping one's "soul." Broadly speaking, then, the chapters in this section bring us into consideration of nature spirituality as an intellectual endeavor, a form of environmental activism, and a personal path of heartfelt commitment to meaning in the world.

22

PAGANISM AND ANIMISM

Graham Harvey

It used to be straightforward to say "Paganism is a nature religion". One conference about "Contemporary Paganism" (at the University of Newcastle upon Tyne in 1994) was followed by another on "Nature Religions Today" (at the University of Lancaster in 1996) debating similar topics and with a significant overlap in the list of speakers and participants. Since then, the referents and resonances of the three words—Paganism, nature and religion—have become more difficult to be clear about, especially when collocated. The reasons for this are closely entangled with changing and contested uses of the term animism in both popular and scholarly contexts. What was once a synonym of religion, emphasizing "belief in spirits" (Tylor 1871) and thus prioritizing metaphysics, has become more closely allied to nature than to supernature. It speaks of interactions with the world rather than of its transcendence, and of ontology rather than epistemology. Precisely these dynamic and ongoing contests over the uses of "Paganism" and "animism" encourage and reward further critical reflection on what ecology means in relation to religion and/or spirituality, and vice versa.

Nature religion as one label for Paganism

Catherine Albanese (1990, 2002) gave the phrase "nature religion" the task of revealing that American history is replete with cultural and religious movements that have nature at their heart. "Nature religion," in this Albanesean frame, is not a single religion in which nature is sacred. Rather, it is a "useful analytical abstraction" (Beyer 1998: 11), directing attention towards an unsteady amalgam of ideas and practices that can be recognized by those looking for commonalities between people who emphasize orientations towards the natural world. It is, therefore, a useful tool for initiating and propelling debate.

Pagans and scholars of Paganism made "nature religion" their own term. It has linked Paganism to broader currents in contemporary society (e.g. indigenous traditions and Emersonian Transcendentalism) whilst simultaneously distinguishing it from religions that may be said to emphasize other tropes than nature (e.g. salvation or enlightenment). It is hard to conceive of an introductory publication or talk about Pagans or Paganism that does not make nature central. This rhetorical move matches the performative centrality of nature among Pagans. For example, they celebrate festivals that honor relationships between the earth, sun and moon. They value narratives, rituals and experiences patterned by solstices, equinoxes, new and full moons and the turning of seasons. Embodiment, physicality and materiality have been cherished among Pagans for longer than they have been critical terms for interdisciplinary scholarly debate. Nonetheless, this is far from a monochrome picture. Shifting definitions of both "Pagan" and "nature" resonate together to illuminate significant engagements between the phenomena they are intended to identify.

In the nineteenth century, poets and novelists began to reclaim the adjective "pagan" from those who used it pejoratively. Under the influence of Romanticism they resisted the notion that bodily desires and needs were antithetical to proper religion or respectable morality. They imagined rural spaces to be better places than the fast growing cities in which to achieve self-knowledge, world-appreciation and other new or recovered virtues. The notion that "pagan" was derived from Latin references to countryside and its "down to earth" inhabitants further valorized these notions. (It is more likely that in ancient Rome *pagus* meant something like "local administrative district" or "parish" than that it referred to rural locations or peasant communities.) Under similar inspiration, wilderness was created by imagining less urban spaces to be pure and pristine—energetically ignoring or romanticizing millennia of dwelling and shaping by indigenous and non-metropolitan communities.

In the early- to mid-twentieth century, those who explicitly disseminated the self-designation "Pagan" blended these Romantic currents with esotericism (which found nature to symbolize transformative truths or human "interior" processes) and varied forms of naturism (e.g. the celebration of natural places and of nudity). Under the influence of Margaret Murray's faulty histories of "witchcraft" (1921, 1929, 1933; Hutton 1999), these Pagans claimed to be making public a previously secretive fertility religion that had survived centuries of Christian witch hunts. Seasonal festivals that were supposed to have supported the fecundity of the world were remade as means of energizing the pursuit of what later became known as "personal growth." Thus a putative fertility religion fed the popularization of esotericist traditions (Hanegraaff 1996) and made their ideas and, more especially, their ritual practices available to a wider public inspired by Romantic views of "wild" places. The emergence of this new Paganism not only coincided with but was made possible by shifting ideas about and responses to "nature." The continuing evolution of Paganism was and remains braided with changing understandings of human relations with the larger than human world and cosmos.

Nature different animisms

Alongside changing associations of Paganism and nature, in the same period (from the mid-nineteenth century to now) there has been a diversification of animism—both as a term and as worldviews or lifeways. Edward Tylor (1871) adopted "animism" as a catchword for his claim that the essential, definitive characteristic of religion was "a belief in souls and spirits." He theorized that religion began as a misinterpretation of experiences in which people postulated the existence of non-empirical entities in order to explain dreams of the deceased or feverish visions of distant places. He expected better science to replace such theories with more rational knowledge of the workings of the world and the nature of humanity. His understanding of animism has had considerable (if uneven) impact. However, it is not always noticed that he defined religion as "belief in spirits." Despite the title of his book, he did not intend to suggest that only the allegedly "primitive" religions are animistic. Religion, rather than religions or sub-classes of religion, is animism for Tylor. Nonetheless, in a quasi-Tylorian twist, animism has most often been imputed to those who are not monotheists. It is, for example, sometimes used to contrast West Africans who identify with what are otherwise called "traditional African religions" rather than with Islam or Christianity. (The evident weakness of the notion that people

must belong solely within one clearly bounded religion is a legacy of official theologies that polemicize against "syncretism" as if sharing and permeability were not pervasive cultural habits.)

Alongside these Tylorian or quasi-Tylorian animisms, another animism emerged in multi- and inter-disciplinary debates towards the end of the twentieth century. It has proved far more helpful in understanding the lived realities of various religious and cultural phenomena than older approaches to animism. It does not always contest, let alone reject, "belief in spirits," but it rarely finds either believing or spirits to be generative of debate or understanding. Many of those engaged by the "new" animism[1] have been significantly influenced by Irving Hallowell's writing about Anishinaabe ontology, behavior and worldview (especially see Hallowell 1960). This animism entails the understanding that the world is a community of persons, only some of whom are human, but all of whom deserve respect (Harvey 2005; 2013a). Locally appropriate ways of demonstrating respect towards other species vary considerably but typically match the forms of etiquette deemed appropriate towards human elders and other respected persons. Appropriate gifts for elders (e.g. tobacco among indigenous North Americans; beer among West Africans) may also be correct for other-than-human persons. Being silent so as to pay attention to elders may be matched by practicing silence in the woods that are home to many of our other-than-human neighbors (Gross 2014: 55–79). The ubiquity of the word "respect" in indigenous discourses illustrate how pervasive and widespread these modes of acting well are among those who understand the world to be a larger-than-human community. In these and other ways, this animism points to non-anthropocentric understandings of nature and of religion.

The most significant contrast between the scholarly approaches of Tylor and of "new" animism researchers was already signaled by Hallowell. His discussion of an elder's reply to his question, "are *all* the stones we see around us here alive?" (Hallowell 1960: 24, emphasis original) is frequently cited. This question is not totally dissimilar to Tylor's effort to understand how an indigenous person can imagine or experience putatively inanimate objects as alive. However, the elder's reply pointed Hallowell and his successors in a new direction: "No! But *some* are." As the elder and Hallowell unpack this seemingly enigmatic answer it becomes apparent that the proper indigenous question is not "is this alive?" but "how should we show our respect?" or even "what gifts are exchanged between persons?" Animism is not an attempt to understand vitality but a life-long effort to improve relationships among persons (across species boundaries and responsive to differences of species, age and other characteristics).

Implicit in these summary ideas about this "new" animism is the understanding that terms like "culture" and "nature" become problematic. Albanese recognized this even as she discussed "nature religion" among Algonquian peoples, including the Anishinaabe, among whom

> Western cosmological notions, particularly the categories of "nature" and the "supernatural," do not fit, represent, or explain the Ojibwa cosmos.
>
> *(Morrison 2002, 57)*[2]

To understand why "nature" fits uneasily within Pagan categories for speaking of the world, it will be helpful to consider Pagan esotericism and animism more closely. This will provide a foundation for arguing that "nature" is a problem for "nature religions."

Animisms within Paganisms

All kinds of animism have nourished the creation and development of Paganism. Stronger or weaker tensions between them are responsible for at least some of the diversification of

Paganism. Put another way, different worldviews and/or "theories of nature" that might be labeled animism have shaped particular aspects of Paganism.

According to Antoine Faivre, "living nature" is one of four "fundamental characteristics" the simultaneous presence of which are required for something to be "included in the field of esotericism" (Faivre and Needleman 1993: xv). "Living nature" can refer to notions of a cosmos replete with living beings (some of whom might be called "spirits") or to a planet imbued with consciousness and/or agency. The inheritance of this esoteric and Tylorian form of animism is evident in some of the earliest and most long-lasting trends in Paganism. For instance, Pagan rituals often invoke elementals, the four winds, guardians of the cardinal directions, tree or river spirits, and other such beings. For some, the earth is a single if complex living being, Gaia or Mother Earth, with intentions, desires and needs that shape what happens to all who live on or in her. If "living nature" is evident in such practices and understandings, what needs clarification is the implications of the edgy and fluid interplay between esotericism and "new" animism. Two vignettes of Pagan events related to food might be enlightening.

Lughnasad vignette

In a British pasture field somewhere near ancient sacred sites, away from non-participating others, a Druid group has gathered to celebrate Lughnasad, a grain harvest festival. They have camped here for up to a week already and most will stay for a few more days. Episodes of conviviality and learning have structured the camp's days and nights so far. But this day has involved a quite serious mood, almost entirely anticipating the camp's main ritual. Subgroups have prepared the physical location, practiced appropriate songs or ritual acts. One small group has gone to a nearby organic barley field, offered greetings and gratitude and then harvested grain. They have baked bread in a replica Iron Age style oven. Finally, everyone has donned robes and regalia, floral wreathes or other signs of festivity. Led by senior Druids and musicians, a procession slowly forms a circle. The four cardinal directions and associated elements, guardians or powers are greeted by pre-selected officers or volunteers. A ritual drama of the harvest is presented—mixing entertainment into the serious business of focusing attention. A speech is made about Lughnasad and its place in the "wheel of the year" marked by eight seasonal festivals. This states, as if it were not contentious, that the purpose of Lughnasad celebrations is to reflect on who we have become in the months since Spring when we set out our intentions for the year. The bread and drink symbolize the joy of fulfilled hopes or satisfied desires, and the harvest of improved lives. The speech introduces the high point of the celebration: the sharing of loaves of bread and cups or horns of barley wine or mead around the circle. Some participants make offerings to the earth of portions of bread or libations from the cups or horns. Finally, ritualized farewells to the quarters lead to the end of the ritual and departure from the circle.

Midwinter vignette

At the edge of a village in the southwest of England, an educational center with re-constructed ancient style buildings is the venue for an explicitly "animistic gathering." Its express purpose is to develop rituals in which gratitude can be expressed towards those other-than-human persons who are treated as our food. Permission is also sought from them to treat them in this way. Approximately sixty people gather for a long weekend at the winter solstice. They sleep in a Viking-style long hall or an Iron Age-style roundhouse. They share the preparation and consumption of meals. Conviviality is enhanced by music making, story-telling and entertainments led by group members. A sense of unity is already established by clarity about the purpose of the event and by the increasing percentage of people who participant annually. The main day of this "Bear Feast" is devoted to the eponymous bear. Elements of an emerging liturgy of songs and chants drawing on arctic and sub-arctic traditions of varied antiquity (e.g. the Finnish *Kalevala*) are introduced so that everyone is prepared to participate fully. Taboos on speaking of the bear are encouraged and alternative names (e.g. "Honey Paw" or "our honored guest") are suggested. Before dawn the bear is ritually hunted. At this point, it takes the double form of a Siberian black bear skin gifted to the feast organizers and a fancy-dress style costume worn by one of the group. The ebb and flow between game, role play, spectacle and serious ritual are hard to convey. Once the bear gives up life to the hunters and gratitude is expressed, much of the day is spent preparing for the evening feast. This takes place in a large earth lodge with stepped seating that begins at ground level and descends to a central fire. A hole in the turfed roof not only allows smoke to escape but will play an important role later at night. Locally sourced venison and vegetables are shared by a community open to varied dietary choices from omnivorous to vegan. After the meal, songs and speeches continue the theme of expressing gratitude and seeking permission for the taking and consuming of other-than-human lives. The bear (in its Siberian skin form) presides, surrounded by gifts. It then ascends through the smoke-hole "as if by magic" and accompanied by a chant about the "golden thread" by which it is drawn back into the otherworld in which all life originates. The group then step outside to see the bear in the form of the circumpolar great bear constellation. Usually the sky is clear at this point of the ritual even if the rest of the night has been rainy or snowy. The participants remark on this and give thanks that this extra vision is granted.

Esotericism and animism mingling and purifying

In these two events, as in many other Pagan celebrations, there is no pure or bounded form of either esotericism or animism. However, abstractions or distillations from each can cast light on the ways in which they shape experiences and groups. They also convey something of the ways in which "nature" and human movement through and within the world may be perceived.

In the Lughnasad ceremony, esotericism and animism both mingle and confront each other. The ritualist privileged to speak about the festival concisely summed up an esotericist understanding: grain, bread and drinks symbolize, represent or otherwise stand for "inner" realities. They are sacramental, outwardly expressing inner and transcendent truths. In this case, the intentions and ambitions of the participants are evoked as a kind of harvest and the ceremony invites people to recognize, acknowledge and celebrate themselves. Although it was not made explicit, the "self" in question was largely an individualized and privatized one of intentionality, agency and increased "self understanding". However, at least some participants evinced surprise that nothing was said about the actual grain, actual bread and actual drinks. Neither these nor any of a range of possible deities or spirits responsible for good harvests were thanked. Tylorian and quasi-Tylorian animists might have expected the invocations of the elements, directional guardians and/or the "Goddess and God" or "Great Mystery" to feed from the opening establishment of ritual space into this high point of the ritual. "New" animists might have expected grain and consumable "persons" to be addressed personally. In fact, some participants knelt (without prompting) to offer portions of bread or libations to the earth, not only honoring divine or spiritual beings but (also or instead) honoring the actual grains who had given up their lives. Importantly, however, even the most animistic participants are unlikely to have entirely rejected the esotericist trend. They too may have gladly considered the ways in which their ambitions for the year had been met. Their "self," however, was a relational one, an in-between, "I-and-Thou" pluralist personhood defined by relationships. It was a selfhood engaged with "nature"—or whatever the larger-than-human world should be called—in a ceremony that (regardless or alongside of the esotericism) served to celebrate relationality.

In turn, although the selfhood of the "new" animistic Bear Feast was explicitly relational, some participants wanted to "look within". For example, in conversations about global climate change some said that "change must begin within". They rejected eco-activism and confrontational politics on the grounds that expressing negativity perpetuates problems. Their preferred path to positive change is individualistic ("change begins within me"). These are, most likely, minority views among Bear Feast animists but they resonate with a more esotericist notion of the ways in which putative interiority ("soul" or "psyche") ripples outwards into activity if not activism. A more evident tension and/or mingling at the Bear Feast is that between the "old" (Tylorian) and "new" animisms. Is the bear a "spirit," a member of a metaphysical class of beings or is it a citizen (albeit an unusual one) of the more everyday world in which different species encounter one another as potential foods or potential consumers? The language and performance of participants suggests that an answer to such questions is still under consideration and that diverse views will continue.

In short, both vignettes involve Pagans who engage with both "real" and "metaphysical" entities. Bread and wheat, and their "spirits" (for want of a better word) are addressed and honored. Other-than-human persons who provide food and symbols for humans are involved. However, in the Bear Feast the focus is on relations with other-than-humans while the Lughnasad vignette is marked as more esotericist by the ritualist's emphasis on the "inner harvest" idea. Esoteric and various animistic understandings are a matter of emphasis not absolute contrasts in these interactive events.

Pagan ecologies

Pagans have not made theology a central pursuit or a cause for division among themselves. Their words about deities are spoken in ceremonies rather than in creedal statements. Much the same could be said of ancient and contemporary polytheistic or animistic traditions. Theology—deity talk—is an aspect of the ways in which people speak about and to the wider world. Ecology or "talking about home, our community" takes priority. Deities, hedgehogs, humans and others are members of this ecology that fuses with cosmology. This is one sense in which Paganism is a "nature religion," a religion in which forms of ecology take precedence. Paganism provides inspiration and encouragement to try different ways of engaging with "nature." Those who adopt the habit of celebrating sunrises, full moons, ancient woodlands, wild hills and bracing waves are likely to embrace attitudes that contest anthropocentrism. Multiple shades of "green" are evident among them, from the "light green" of those who donate to environmental causes to the "dark green" of radical protest (Taylor 2010) and the earthy green of those turning to organic gardening and composting as everyday Pagan practices. None of this means that Pagans spend all their time discussing ecology or environmentalism. Despite expectations, Pagans are no more likely to be eco-activists than are members of other religious or social groups. However, the frequency with which Paganism is encountered among eco-activists suggests at least an elective affinity for some. Similarly, Pagan ecology may offer an eco-therapy in which people re-connect themselves with the wider world in outdoor activities. To judge by internet search returns, eco-therapy is a growing movement that includes those who wish to promote mental health, mindfulness, physical exercise and holistic well-being (e.g. Harris 2015). Again, this is not to suggest that Pagans are particularly concerned to promote their religion as a form of therapy or even that they are all aware of the claims made for the beneficial outcome of spending time outdoors. Rather, it is to indicate an elective affinity with a contemporary cultural ambience in which human well-being is braided back into the wider-than-human world.

It is possible to hear the resonances of everything from Romanticism and Transcendentalism to new animism and anarchism in these varied forms of Pagan ecology. In them, we might discover new or deeper implications of the notion that Paganism is a "nature religion." Nonetheless, nature, religion and spirituality remain difficult terms. Precisely their collocation in provocative or pregnant phrases like "nature religion" indicates that all the phenomena they are utilized to address have changed.

Religion, spirituality and the difference that "nature" makes

David Shorter's (2015) discussion of spirituality in relation to Native American cultures and peoples is helpful in pursuing an understanding of nature, religion and spirituality. He rejects the term "spirituality" because it privileges beliefs about non-empirical or "supernatural" realities and beings (souls or spirits). It has a history that defines it in opposition to materiality (e.g. embodiment, animality, artifacts and location). It is most unhelpful in relation to indigenous religions and cultures when it reinforces the marginalization of religion from politics and everyday life. This trajectory began within the early modern European "Wars of Religion" which re-organized societies as "Nation States" (Cavanaugh 1995). For this to work, religion could not be allowed to continue to shape transnational identities larger than the citizenship required of modern people. It could only be about private beliefs, almost a hobby. To identify people as "spiritual" became a way of indicating their acquiescence (deliberate or otherwise) to this marginalization of religion. However, against all expectations, as the modernization project expanded from northern Europe to become the globally dominant culture, religion survived. Alongside its acquiescent "spiritual" or "liberal modern" form, religion continued both in a "fundamentalist modern" form (in which people refuse to separate religion from politics or private from public) and in an "everyday" form in which reality has never been disenchanted or humanity divorced from the larger-than-human community (Orsi 2012; Harvey 2013b).

Paganism illustrates these forms of religion. The Pagan adaptation of esotericism could have remained firmly within the realm of privatized and individualized believing. "Belief in spirits," especially those found in "natural" places, could have continued as an eccentricity among otherwise productive and consumerist citizens. Meanwhile, the more confrontational Pagan eco-warriors and wilderness protectors speak and act politically from deeply held worldviews that seem to demand a contest with the destructive ideology of the anthropocene. Finally, an everyday religion that utilizes ritual and story-telling as a means of interacting with an enchanted, animate and relational world inspires increasing numbers of Pagans to try new ways of expressing gratitude to consumable beings. Again, these are only rarely entirely bounded and discrete practices or traditions. Often there is an ebb and flow through and among the varieties, especially as Pagans engage with others in events or networks that emphasize particular modes of being religious.

Alongside the changing forms of Paganism are changing notions of "nature." Indeed, it becomes difficult to know which changed first: the uses of the term "religion" or the uses of "nature." In Tylor's theory, "nature" implicitly refers to an inert world that is wrongly understood by pre-scientific, religious humans to act intentionally and communicatively. Quasi-Tylorian animists (data about whom fed Tylor's more grandiose argument about all religion) found "spirits" within and among trees, rocks and rivers. They are the archetypal believers in a supernatural realm beyond the natural one. Whether we need to believe in such believers (to extrapolate from Bruno Latour's 2010 challenge) is doubtful. Once we pay attention to religion as interactions with the world, the ("new") animists become more interesting. What was once named "nature" in the Euro-originated world becomes visible as a thoroughly cultural world of multi-species interrelationship, a realm in which every event is at least potentially a personal action (Pflug 1992).

It is possible to think that the world is "all nature" (with "human nature" as a subset) or that it is "all culture" (as intentionality, humor and artifice are discovered among more species). If we play the more radical move of rejecting both sides of this dualistic construction, we will need to find alternative ways to speak of the world and ourselves. Religion has played a significant part in both culture (epitomizing either what is distinctive about humans or about some humans over against others) and, more recently, in nature (as cognitivists have naturalized religion as an evolutionary mechanism or byproduct). However, the animism of Anishinaabe and Pagan interactions with the world makes it hard to say "nature religion."

Conclusion: "religion as ecology" not "nature religion"

The changing meaning of "animism" within Paganism clarifies wider shifts in contemporary understandings of human relations with the world. While the term "nature" remains problematic, its collocation with "spirituality" and "religion" provide rich resources for recognizing and debating contemporary cultures. Catherine Albanese's term "nature religion" strategically placed an odd assortment of phenomena alongside one another. It encouraged contrasts with such categories as "world religions," "new religions," and "indigenous religions," and aided both comparison and contrast with "revealed or scriptural religions." However, the term remains problematic both because it tends to suggest an individualization of religion (aka "spirituality") and because "nature" is inadequately discussed.

Collocating "nature" with "religion," however, at least points towards something more radical. It could encourage careful attention to indigenous knowledges about the inter-relating and interacting of diverse communities of persons. However, since "nature" and "culture" keep speaking of separate realms of human and non-human action, they are unhelpful. In relation to religions, "nature" also imports a contrast with a "supernatural" or transcendent realm of deities and spirits where it misdirects attention from animist and other religious understandings of reality. Nonetheless, "nature religion" has provided one opening to conversations about ontologies and epistemologies, which are alternative to modernity's dualisms and separations. The "new" animism pushes this further by encouraging the re-theorizing of religion as etiquettes and rites of inter-species relationships, mediated by gift exchange and respectful behaviors. Such studies of indigenous and Pagan animism should further the re-examination of *all* religion as everyday acts in an evolving larger-than-human world. If "nature religion" can get us to rethink both "religion" and "nature" then it will have served a valuable purpose in placing religion back into ecology.

Notes

1 Which is only "new" as an academic term, and in comparison to Tylor's coinage.
2 Anishinaabe and Ojibwa are among the self-designations of the same indigenous nation.

References

Albanese, C. L. (1990) *Nature Religion in America: From the Algonkian Indians to the New Age,* University of Chicago Press, Chicago.

Albanese, C. L. (2002) *Reconsidering Nature Religion,* Trinity Press, Harrisburg.

Beyer, P. (1998) "Globalisation and the Religion of Nature," in Pearson, C., Roberts, R. H. and Samuel, G. eds, *Nature Religion Today: Paganism in the Modern World,* Edinburgh University Press, Edinburgh, 11–21.

Cavanaugh, W. T. (1995) "A Fire Strong Enough to Consume the House: 'The Wars of Religion' and the Rise of the State," *Modern Theology* 11.4: 397–420.

Faivre, A. and Needleman, J. eds (1993) *Modern Esoteric Spirituality,* SCM Press, London.

Gross, L. W. (2014) *Anishinaabe Ways of Knowing and Being,* Ashgate, Burlington.

Hallowell, A. I. (1960) "Ojibwa Ontology, Behavior, and World View," in Diamond, C. ed. *Culture in History: Essays in Honor of Paul Radin,* Columbia University Press, New York, 19–52.

Hanegraaff, W. (1996) *New Age Religion and Western Culture: Esotericism in the Mirror of Secular Thought,* E. J. Brill, Leiden.

Harris, A. (2015) "Ecotherapy" www.adrianharris.org/ecopsychology/ (last accessed 2 September 2015).

Harvey, G. (2005) *Animism: Respecting the Living World,* Hurst, London; Columbia University Press, New York.

Harvey, G. (2013a) *The Handbook of Contemporary Animism,* Routledge, New York.

Harvey, G. (2013b) *Food, Sex and Strangers: Understanding Religion as Everyday Life,* Routledge, New York.

Hutton, R. (1999) *Triumph of the Moon: a History of Modern Pagan Witchcraft,* Oxford University Press, Oxford.

Latour, B. (2010) *On the Modern Cult of the Factish Gods,* Duke University Press, Durham.

Morrison, K. M. (2002) *The Solidarity of Kin: Ethnohistory, Religious Studies, and the Algonkian-French Religious Encounter,* State University of New York Press, Albany.

Murray, M. (1921) *The Witch Cult in Western Europe,* Clarendon Press, Oxford.

Murray, M. (1929) "Witchcraft," *Encyclopedia Britannica* 23: 686–8.

Murray, M. (1933) *The God of the Witches,* Oxford University Press, Oxford.

Orsi, R. A. (2012) "Afterword: Everyday Religion and the Contemporary World," in Schielke, S., and Debevec, L. eds, *Ordinary lives and grand schemes: an anthropology of everyday religion,* Berghahn Books, Oxford 146–61.

Pflug, M. A. (1992) "Breaking Bread: Metaphor and Ritual in Odawa Religious Practice," *Religion* 22: 247–58.

Shorter, D. D. (2015) "Spirituality," in Hoxie, F. ed., *The Oxford Handbook of American Indian History,* Oxford University Press, Oxford

Taylor, B. (2010) *Dark Green Religion: Nature Spirituality and the Planetary Future,* University of California Press, Berkeley.

Tylor, E. B. (1871) *Primitive Culture: Researches into the Development of Mythology, Philosophy, Religion, Art and Custom,* John Murray, London.

23

SPIRITUAL ECOLOGY AND RADICAL ENVIRONMENTALISM

Leslie E. Sponsel

Spiritual ecology

Spirituality is undeniably an elemental and often pivotal manifestation of human nature; even atheists, agnostics, and secular humanists may be spiritual (Sponsel 2012, 149–154). A survey of 14,527 new students in 136 colleges and universities in the US during 2003 to 2010, revealed that the majority were spiritual, but not necessarily religious (UCLA Higher Education Research Institute 2010). A survey by the Pew Research Center on "America's Changing Religious Landscape" concluded that since the last project of its kind in 2007, the number of religiously unaffiliated adults has increased by about 19 million. By now there are around 56 million religiously unaffiliated adults in the US, a group more numerous than either Catholics or mainline Protestants (Smith 2015). These two surveys highlight one reason why it is important to consider not only religion, but also spirituality (cf. Gottlieb 2012). (For an especially useful discussion of the distinction between religion and spirituality see Taylor 1991a, 175–178.)

Spiritual ecology is carefully chosen as an umbrella term to designate the vast, complex, diverse, and dynamic arena at the interfaces of religions and spiritualities with environments, ecologies, and environmentalisms. Other authors use the term spiritual ecology as well (e.g., Laszlo and Combs 2011; Merchant 2005, 117–138; Taylor 2007). Elsewhere other labels are used instead. However, they reflect a variant of spiritual ecology and usually a narrower pursuit, such as dark green religion, deep ecology, earth spirituality, earth mysticism, ecomysticism, ecopsychology, ecospirituality, ecotheology, green religion, green spirituality, nature mysticism, nature religion, nature spirituality, religion and ecology, religion and nature, religious ecology, religious environmentalism, and religious naturalism.

Basically, individuals and organizations may pursue spiritual ecology in one or more of three ways: as an intellectual endeavor, a form of environmental activism, and/or a personal path of spirituality (Sponsel 2011, 2012, 2014). In academia, the most significant advances have been made by pioneering leaders like Mary Evelyn Tucker and John Grim in developing the series of conferences in 1996 to 1998 and subsequent edited books on different world religions and ecology at Harvard University and also the Forum on Religion and Ecology (FORE) with its email newsletter (Tucker 2006). In addition, they contributed to the establishment in 1997 of the periodical *Worldviews: Environment, Culture, Religion*, subsequently called *Worldviews: Global*

Religions, Culture and Ecology. Another extraordinary pioneering leader, Bron Taylor, developed the International Society for the Study of Religion, Nature and Culture in 2006, and its official publication a year later called the *Journal for the Study of Religion, Nature and Culture.* Taylor (2005a) is also Editor-in-Chief of the benchmark two-volume *Encyclopedia of Religion and Nature* encompassing 1,000 entries by over 500 authors. A third major contributor in the development of spiritual ecology is Roger S. Gottlieb (2006a) who edited *The Oxford Handbook of Religion and Ecology,* and documented religious environmentalism in his own book, *A Greener Faith: Religious Environmentalism and Our Planet's Future* (Gottlieb 2006b). Gottlieb (2010) also edited the four-volume collection of mostly reprinted articles in *Religion and the Environment.* Of course, many others have made important contributions as well to the development of these two components of spiritual ecology and allied arenas (Sponsel 2011, 2012, 2014). (The second component, a form of environmental activism, will be discussed later).

The third component, spiritual ecology as a personal path of spirituality, often motivates the first and second ones. Furthermore, spiritual experience in nature as a motivating factor is a phenomenon common to spiritual ecology and radical environmentalism. For example, Taylor (1995a, 16) mentions that: "What animates most Earth Firsters! are their own spiritual experiences in nature which convince them of the interrelatedness and sacrality of all life." Likewise, Scarce (2006, 31) observes that: "Radical environmentalism emerges out of an ecological consciousness that comes from the heart—not the head—that has experienced the natural world." (For other cases, see Schauffler 2003.) There are numerous guides to spirituality in nature (e.g., Bekoff 2014; Chase 2011; Hecking 2011). Also, nature can be a catalyst for many positive neurophysiological reactions in the human body, such as lowering stress and blood pressure (Selhub and Logan 2012). Andrew Newberg (2010) is a leader in exploring the new frontier of neurotheology, the neurobiological correlates of religious and spiritual activities. In ecopsychology, reconnecting with nature is a primary prescription for emotional as well as physical health and healing (e.g., Chalquist 2007).

Kinsley identified ten basic principles of spiritual ecology (1995, 227–232). These principles represent points of convergence between spiritual ecology and radical environmentalism as well. Many would resonate also with numerous moderate environmentalists. However, they are best formulated as hypotheses for empirical examination and testing in the context of particular cases.

Secular approaches to environmental problems and issues include environmental sciences, environmental studies, environmental laws and regulations, and the like. Certainly they are absolutely necessary and have achieved significant progress. However, so far they have proved insufficient in resolving many environmental concerns, let alone the environmental crisis as a whole. A multitude of diverse environmental problems, issues, and crises continue from the local to the global levels: new ones are being discovered; many are becoming worse (Schwagerl 2014, Millennium Ecosystem Assessment 2005). Most secular initiatives merely treat the superficial symptoms of specific problems, instead of addressing the underlying root causes as a whole, a primary concern of spiritual ecology. Accordingly, Rabbi Michael Lerner (2000, 138) remarks that: "The upsurge of Spirit is the only plausible way to stop the ecological destruction of our planet. Even people who have no interest in a communal solution to the distortions in our lives will have to face up [to] this ecological reality. Unless we transform our relationship with nature, we will destroy the preconditions for human life on this planet."

The fundamental changes required to turn the environmental crisis around for the better, if not completely resolve it, must involve far greener environmental worldviews, values, attitudes, behaviors, and institutions (Sponsel 2012). No specific religion is considered to be the cause or the solution for the ongoing environmental crisis (cf. White 1967). Instead, among numerous other initiatives, the adherents to any particular religious or spiritual practice are encouraged to experience nature and to pursue introspection in order to develop far greener beliefs and practices (Dobson 1991; Hecking 2011).

The transformation generating greener lifestyles and societies worldwide is sometimes referred to as "The Great Turning" (Hawken 2007; Korten 2006; Macy and Johnstone 2012). Metzner (1999, 171–182) developed the most detailed characterization of this transformation. Many attributes of what he calls the Ecological Age, such as the intrinsic value of nature, are already endorsed and increasingly pursued in various ways and degrees by proponents of spiritual ecology, moderate environmentalism, and radical environmentalism (e.g., Dobson 1991; Pepper 1996). Some individuals consider the impact of spiritual ecology to be potentially revolutionary, albeit in a nonviolent and constructive manner (Korten 2006; Macy and Johnstone 2012; Sponsel 2012).

Spiritual ecology can provide critiques and alternatives regarding the exclusively anthropocentric, dualistic, materialistic, reductionist, mechanistic, utilitarian, consumerist, and economistic worldview that is considered by many to be the major cause of environmental problems and crises from the local to the global levels (Dobson 1991; Hawken 2007; Macy and Johnstone 2012). Among other pivotal points, spiritual ecology challenges the fundamental fallacy infecting industrial capitalism: that unlimited growth is possible on a limited base. That base includes land, water, and natural resources as well as the resiliency capacity of environmental systems to process and absorb waste and pollution (Meadows et al., 2004; Rees et al., 1998). For those pursuing spiritual ecology, nature is a grand cathedral of communal beings, rather than an unlimited warehouse of mere objects to endlessly exploit for profit and greed (cf. Berry 2006, 17).

Spiritual ecology is contributing to the vital necessity of re-thinking, re-feeling, and re-visioning the place of humans in nature (cf. White 1967). Despite many complications, difficulties, obstacles, and uncertainties, spiritual ecology might finally prove to be a turning point in alleviating many environmental concerns. Only future decades will reveal the full extent of its success or failure, although there have already been many important achievements. Spiritual ecology is growing exponentially (cf. Gladwell 2002). A search in Google.com on November 28, 2004, revealed 420,000 results for spiritual ecology. On May 22, 2015, a similar search revealed 3,730,000 results, an increase of almost nine times within a decade. Furthermore, spiritual ecology complements secular approaches to environmental concerns.[1]

There is considerable overlap between spiritual ecology and radical environmentalism, although the two arenas are not isomorphic. Some points of overlap have already been noted, such as spiritual experiences in nature as a major motivating factor, critique of industrial capitalism, and the ten basic principles. Other similarities and differences will become apparent in the next section. Boundaries between these arenas or domains are not always discrete, but it is necessary to distinguish them for heuristic and analytical purposes.

Radical environmentalism

Radicalness may be in the eye of the beholder. Even some establishment personages have been identified as radicals because of their environmentalism, such as Prince Charles (Lorimer 2003, 14) and Pope Francis (Burton 2014). In such cases, radical refers to its original meaning of root, as in the ultimate causes and solutions of the environmental concerns. In this sense spiritual ecology is radical too.

There are two basic types of radical environmentalism. Type 1 fits the common stereotype of individuals and groups illegally disrupting economic activities in response to some environmental concern. Type 2 is far broader and more diverse, encompassing bioregionalism, deep ecology, ecofeminism, primitivism, and social ecology (List 1993). To these, Taylor (1991a,b, 2010) adds ecopsychologists, feminist spirituality, Pagans, Wiccans, green anarchists, anti-globalization protesters, and animal liberation activists, although the last three are related to Type 1 in their tactics. (Also, see the articles on such topics in Taylor 2005a.)

Moderate environmentalism emerged in the US in the late-nineteenth century with organizations like the Sierra Club founded in 1892 (Turner 1993). Radical environmentalism developed much more recently, such as Earth First! around 1980. In many instances this radicalization was a reaction to increasing concern over the gravity and urgency of environmental matters and growing commitment to deep ecology and related perspectives; frustration with the compromising and ineffectiveness of moderate environmentalism; and the anti-environmentalism policies of the administration of President Ronald Reagan in 1981 to 1989. Many radical environmentalists became convinced that the government was owned by corporations, ineffective in dealing with environmental concerns, regressive in its environmental attitudes, and illegitimate in disregarding the interests and rights of nonhuman beings and their habitats (e.g., Scarce 2006, 20–21, 23). Many believe that mainstream environmental organizations have sold out to corporations (see MacDonald 2008). The first type of radical environmentalists, like those of Earth First!, believe that they have no choice but to engage in forceful direct actions, even if illegal, despite risks to their own safety, life, welfare, and income, including imprisonment (Scarce 2006, 14; Taylor 1995a, 17). They assert that the true extremists, radicals, and ecoterrorists are those who commit wanton environmental destruction, termed ecocide (Scarce 2006, 20).

Initiatives by either of these two types of radical environmentalism may involve only dramatic protests or demonstrations on the threatened site or in strategic locations designed to draw media and public attention to an environmental concern. Letter writing, petitions, lobbying government agents, and lawsuits are other tactics of radical as well as moderate environmentalism (Hawken 2007). Here space limitations do not allow more discussion of the second type of radical environmentalism because of its great breadth and diversity (see Merchant 2005; Taylor 2005b). Consequently, most of the remainder of this essay focuses on the first type, and on Earth First! in particular.

In Type 1, subversive direct action concentrates on a local or regional environmental problem or issue. It may deploy ecotage or monkeywrenching; that is, illegally sabotaging the machinery or damaging the property of an organization that is allegedly threatening, degrading, or destroying a particular environment. This uncompromising and militant ecodefense tactic stems from intense moral outrage at what is viewed as a continuous war against nature by rampant and rapacious industrial capitalism. It also stems from deep disillusionment, frustration, and desperation with the limited effectiveness and compromising of moderate environmentalism, and similar limitations of government. For instance, as described below in some cases ecotage by Earth First! has stopped logging in an old-growth forest, whereas letter-writing campaigns and other initiatives of moderate environmentalists over many years failed (Roselle and Mahan 2009, 172–175, 209).

Edward Abbey initially described monkeywrenching and many of the confrontational tactics involved in Type 1 in *The Monkeywrench Gang* (1975). Dave Foreman, a co-founder of Earth First!, characterizes monkeywrenching as nonviolent, not organized, individual, targeted, timely, dispersed, diverse, fun, not revolutionary, simple, deliberate, and ethical. He insists that monkeywrenching is not a general policy of Earth First!, but a matter of personal choice in particular situations (Foreman and Haywood 1993, 9–11). Among the specific tactics of monkeywrenching discussed in the book *Ecodefense: A Field Guide to Monkeywrenching* (Foreman and

Haywood 1993) are removing survey stakes, tree spiking, road spiking, disabling equipment, burning machinery, billboard trashing, spray paint slogans, computer sabotage, avoiding arrest, camouflage, and disposing of evidence. Such actions in defense of nature are based on moral, ecological, and political grounds. The laws of nature transcend those of any government from the perspective of many radical environmentalists (cf. Scarce 2006, 200).

Monkeywrenching can be effective through elevating business costs to the extent that profits are reduced, thereby discouraging further operation. Foreman claims that Earth First! contributed to the empowerment, restructuring, and realigning of the environmental movement (Foreman and Haywood 1993, 14, 16). (Also, see Scarce 2006, 280–281.) Furthermore, he claims that Earth First! significantly influenced ideas and practices in conservation biology (Foreman 2004, 158–161). However, he does not explain such claims.[2]

Regarding ethics, Foreman asserts that monkeywrenchers "are engaged in the most moral of all actions: protecting life, defending Earth" (Foreman and Haywood 1993, 11). While monkeywrenching may be considered by Earth First!ers to be nonviolent, critics view it as violent, although it is pursued against machines and other equipment, rather than against humans or other sentient beings. The overwhelming majority of radical environmentalists are not really terrorists because they are firmly committed to nonviolence in their actions, being extremely careful to avoid causing harm to other humans (Potter 2011, 47; Rosebraugh 2004, 236; Scarce 2006, 13).

Direct action by Earth First! and other Type 1 radical environmentalists like the Earth Liberation Front (ELF) and the Animal Liberation Front (ALF) has involved thousands of incidents of property damage or destruction, even some arsons, yet so far these have not injured or killed a single human being (Potter 2011, 48, 122, 124–126, 184–185; Rosebraugh 2004, 236; Roselle and Mahan 2009, 204–206). By contrast, Foreman was hit by a logging truck; a bomb was planted in the car of two members of the ELF, Judi Bari and Darryl Cherney, seriously injuring them; and a member of Earth First!, David "Gypsy" Chain, was killed by a giant tree felled by a logger (Potter 2011, 84; Rosebraugh 2004, 237–239). In addition, police used violence against radical environmentalists, not only tear gas and pepper spray on peaceful demonstrators, but even rubbing pepper spray with cotton swabs along their lower eyelids (Potter 2011, 56). A former spokesperson for the ELF, Craig Rosebraugh (2004, 236–241) argues that such incidents are terrorism, whereas it is ridiculous propaganda to consider the actions of radical environmentalists to be terrorism (cf. Likar 2011). On the other hand, a Congressman from the state of Washington, George Nethercutt, even called for the death penalty for ecoterrorists (Rosebraugh 2004, 241).

Scarce and others assert that since the mid-1990s, the stigmatization of domestic terrorist, ecoterrorist, and associated rhetoric applied to radical environmentalists has become a tactic to distract attention from the real environmental matters at stake and the real terrorists – the corporations and other collaborators who should be held responsible for degrading and destroying nature (Scarce 2006, 77–78, 181; cf. Likar 2011; Taylor 1998, 2004). Potter (2011, 241–245) argues that such paranoid reactions to radical environmentalism, and sometimes even to environmentalism in general, stem from the belief of opponents that these are a threat to the American way of life, a culture war and a war of values, and a direct challenge to capitalism, modernity, and even civilization itself. Here Potter is referring to both of the two types of radical environmentalism. Potter (2011, 245) asserts: "Their confluence is the redefinition of what it means to be a human being…. At their core, they challenge fundamental beliefs … that human beings are the center of the universe and our interests are intrinsically superior to those of other species and the natural world." In the opinion of some, this is also a threat to deeply held Judeo-Christian beliefs and values in human exceptionalism; namely, that humans are superior to nature and other creatures (Potter 2011, 246). Potter's characterizations apply to spiritual ecology as well.

The variety within Type 1 can be illustrated by a few cases involving old growth forests in the western US, where redwood trees can grow up to 30 feet wide and 35 stories tall plus live for 2,000 years, yet 95% of such forests have been destroyed. In the opinion of many environmentalists, redwoods are ancient noble beings forming a natural cathedral and merit reverence and protection. They view clearcutting such forests as an environmental disaster, crimes against nature, and even sacrilege (cf. Stone 2010).[3]

One of the early cases is John Muir (1838–1914), in many ways a radical environmentalist in his time, who nevertheless founded the mainstream environmental organization called the Sierra Club to preserve western wilderness. Eventually he contributed more than any other individual to the development of the national park system in the US. (Sponsel 2012, 57–63; Turner 1993; Worster 2008). A contemporary case, the Redwood Rabbis, pursued civil disobedience and other tactics like planting tree seedlings to campaign against deforestation and in favor of the protection of the old-growth forests in California. They managed to convince Maxxam and its subsidiary, Pacific Lumber, to establish a forest reserve of 7,470 acres and to apply restrictions on logging in the Headwaters Forest of Northern California, the largest stand of unprotected old-growth redwoods anywhere (Steinberg 2005). A more radical case was the longest tree sit by anyone for 738 days of Julia Butterfly Hill (2000). She succeeded in widely publicizing the issue of the preservation of redwood forests (Scarce 2006, 263). Eventually Hill and her support team from Earth First! obtained from Pacific Lumber/Maxxam Corporation in perpetuity a three-acre buffer zone around the redwood she had bonded with and named Luna. (Also, see Wolens 2000.) Still more radical was the campaign in 1983 of monkeywrenching by Earth First! that prevented the construction of the Bald Mountain Road in the northern boundary of the Kalmiopsis Wilderness Area in the Siskiyou National Forest of southwestern Oregon. Kalmiopsis is one of the oldest continuously forested areas in the west. The road would have opened access to 160,000 acres for clearcutting. Their direct action made the destruction of old growth forest by the Nation Forest Service a national issue and drew attention to Earth First! as a formidable force in environmentalism.[4]

Radical environmentalism may be viewed as a religion in the sense that adherents find ultimate meaning and transformative power in nature (Taylor 1991a, 175). Most radical environmentalists have experienced something mystical or some kind of an epiphany involving a sense of identity, unity, or wholeness with nature (Taylor 1991a, 179–181; 1991b, 231–235). Many of them also embrace elements of religions and/or spiritualities that they believe pursue values and actions that are environmentally friendly, such as Buddhism, Daoism, Indigenous spiritualities, Paganism, Wicca, and Gaian mysticism. They share the conviction that part of the solution to the environmental crisis is in reconnecting with nature and "resacralizing" it, to "listen to the land" through its sacred voices (Taylor 2005b, 1326). Foreman views monkeywrenching as a form of earth worship engaged in by religious warriors to defend the earth (Taylor 1991a, 175). Most Earth First!ers believe that there must be a spiritual awakening as well as better environmental education and profound political change to resolve the environmental crisis (Taylor 1995a, 18). Earth First! literature has engaged deep ecology and spiritual ecology (see Davis 1991). (Also, see Scarce 2006, 31, 35–36, 39, 91, 97, 225–226.) However, Taylor (2005b, 1327) cautions that: "radical environmentalism is plural and contested, both politically and religiously; it is characterized by ongoing controversies over strategies and tactics, as well as over who owns, interprets, and performs the myths and rites." Again, spiritual ecology and both types of radical environmentalism overlap, although they are not isomorphic.

Conclusions

While those engaged in spiritual ecology and radical environmentalism largely agree on many ends, they disagree on means, as do both types of radical environmentalists. Although no quantitative data from surveys have been found, apparently most engaged in spiritual ecology and Type 2 radical environmentalism reject violence of any kind. While adherents of spiritual ecology and Type 2 may pursue civil disobedience for protests, otherwise most reject illegal activities.

Manifestations of spiritual ecology and the two types of radical environmentalism are quite diverse, as are the ways in which they overlap and sometimes even reinforce one another. They have a substantial history and literature, and they continue to flourish and expand. Surely their impact on environmentalism, nature, and society is significant, but only coming decades will reveal just how significant. Moreover, it seems probable that both types of radical environmentalism as well as spiritual ecology will only increase in the future as local and global environmental crises worsen, especially with global climate change.

Notes

1 Surveys of spiritual ecology and related domains include Bauman et al., 2011, Gottlieb 2006a,b, Grim and Tucker 2014, Jenkins and Chapple 2011, Kinsley 1995, Sponsel 2012, 2014, Taylor 2005a, 2010, and Tucker 2006. For anthropological contributions to spiritual ecology see Sponsel 2011.
2 For first-hand accounts by radical environmentalists see Curry 2011, Pickering 2007, Potter 2011, Rosebraugh 2004, and Roselle and Mahan 2009.
3 Chase 1995 and Durbin 1996 provide overviews on environmentalism in western old-growth forests.
4 For ecodefense by Earth First! and similar radical environmentalists see Foreman 1991, Foreman and Haywood 1993, Love and Obst 1972, Rosebraugh 2004, Roselle and Mahan 2009, and Taylor 1995b.

References

Abbey E. (1975) *The Monkeywrench Gang*, Avon, New York.
Bauman W. A., Bohannon R.R. II and O'Brien K. J. eds. (2011) *Grounding Religion: A field guide to the study of religion and ecology*, Routledge, New York.
Bekoff M. (2014) *Rewilding Our Hearts: Building pathways of compassion and coexistence*, New World Library, Novato.
Berry T. (2006) *Evening Thoughts: Reflecting on Earth as sacred community*, Mary Evelyn Tucker ed., Sierra Club Books, San Francisco.
Burton T. I. (2014) "Pope Francis's radical environmentalism" *The Atlantic*, 11 July 2014 (www.theatlantic.com/international/archive/2014/07/pope-franciss-radical-rethinking-of-environmentalism/374300/) Accessed May 21, 2015.
Chalquist C. (2007) *Terrapsychology: Reengaging the soul of place*, Spring Journal, Inc., New Orleans.
Chase A. (1995) *In a Dark Wood: The fight over forests & the myths of nature*, Houghton Mifflin, New York.
Chase, Steven (2011) *Nature as Spiritual Practice* William B. Eerdmans Publishing Company, Grand Rapids.
Curry M. (2011) *If a Tree Falls: The story of the Earth Liberation Front*, Oscilloscope Laboratories, New York (DVD, 85 minutes).
Davis J. ed. (1991) *The Earth First! reader: Ten years of radical environmentalism*, Gibbs Smith, Salt Lake City.
Dobson A. ed. (1991) *The Green Reader: Essays toward a sustainable society*, Andre Deutsch, New York.
Durbin K. (1996) *Tree Huggers: Victory, defeat, and renewal in the northwest ancient forest campaign*, Mountaineers, Seattle.
Foreman D. (1991) *Confessions of an Eco-Warrior*, Harmony Books, New York.
Foreman D. (2004) *Rewilding North America: A vision for conservation in the 21st century*, Island Press, Washington, DC.
Foreman D. and Haywood B. eds. (1993) *Ecodefense: A field guide to monkeywrenching*, Third edition, Abbzug Press, Chico.
Gladwell M. (2002) *The tipping point: How little things can make a big difference*, Back Bay Books, New York.

Gottlieb R. S. ed. (2006a) *The Oxford Handbook of Religion and Ecology*, Oxford University Press, New York.

Gottlieb R. S. (2006b) *A Greener Faith: Religious environmentalism and our planet's future*, Oxford University Press, New York.

Gottlieb R. S. ed. (2010) *Religion and the Environment*, Volumes 1–4, Routledge, New York.

Gottlieb R. S. (2012) *Spirituality: What is it and why does it matter*, Oxford University Press, New York.

Grim J. and Tucker M.E. (2014) *Ecology and Religion*, Island Press, Washington, DC.

Hawken P. (2007) *Blessed unrest: How the largest movement in the world came into being and why no one saw it coming*, Viking Penguin, New York.

Hecking R. J. (2011) *The Sustainable Soul: Eco-spiritual reflections and practices*, Skinner House Books, Boston.

Hill J. B. (2000) *The Legacy of Luna: The story of a tree, a aoman, and the struggle to save the redwoods*, HarperSanFrancisco, San Francisco.

Jenkins W. and Chapple C.K. (2011) "Religion and environment", *Review of Environment and Resources* 36, 441–463.

Kinsley D. (1995) *Ecology and Religion: Ecological spirituality in cross-cultural perspective*, Prentice-Hall, Inc., Englewood Cliffs.

Korten D. C. (2006) *The Great Turning: From empire to community*, Berrett-Koehler, San Francisco.

Laszlo E. and Combs A. eds. (2011) *Thomas Berry, Dreamer of the Earth: The spiritual ecology of the father of environmentalism*, Inner Traditions, Rochester.

Lerner M. (2000) *Spirit Matters*, Hampton Road Publishing Company Inc., Charlottesville.

Likar, Lawrence E. (2011) *Eco-Warriors, Nihilistic Terrorists, and the Environment*, Praeger, Santa Barbara.

List P. C. (1993) *Radical Environmentalism: Philosophy and Tactics*, Wadsworth Publishing Company, Belmont.

Lorimer D. (2003) *Radical Prince: The practical vision of the Prince of Wales*, Floris Books, Edinburgh.

Love S. and Obst D. eds. (1972) *Ecotage!* Pocket Books, New York.

MacDonald C. C. (2008) *Green, Inc.: An environmental insider reveals how a good cause has gone bad*, Lyones Press, Guliford.

Macy J. and Johnstone C. (2012) *Active Hope: How to face the mess we're in without going crazy*, New World Library, Novato.

Meadows D. H., Randers J. and Meadows D. (2004) *Limits to Growth: The 30-year update*, Chelsea Green Publishing, White River Junction.

Merchant C. (2005) *Radical Ecology: The search for a sustainable world*, Second edition, Routledge, New York.

Metzner R. (1999) *Green Psychology: Transforming our relationship to the earth*, Park Street Press, Rochester.

Millennium Ecosystem Assessment (2005) *Ecosystems and Human Well-Being: Synthesis*, Island Press, Washington, DC. (www.millenniumassessment.org/documents/document.356.aspx.-pdf) Accessed May 21, 2015.

Newberg A. (2010) *Principles of Neurotheology*, Ashgate Publishing, Burlington.

Pepper D. (1996) *Modern Environmentalism: An introduction*, Routledge, New York.

Pickering L. J. (2007) *The Earth Liberation Front 1997–2002*, Arissa Media Group, Portland.

Potter W. (2011) *Green is the New Red: An insider's account of a social movement under siege*, City Life Publishers, San Francisco.

Rees W. E., Wackernagel M. and Testemale P. (1998) *Our Ecological Footprint: Reducing human impact on the earth*, New Society Publishers, Gabriola Island.

Rosebraugh C. (2004) *Burning Rage of a Dying Planet: Speaking for the Earth Liberation Front*, Lantern Books, New York.

Roselle M. with Mahan J. (2009) *Tree Spiker: From Earth First! to lowbagging: My struggles in radical environmental action*, St. Martin's Press, New York.

Scarce R. (2006) *Eco-Warriors: Understanding the radical environmental movement*, Left Coast Press, Walnut Creek.

Schauffler F. M. (2003) *Turning to Earth: Stories of ecological conversion*, University of Virginia Press, Charlottesville.

Schwagerl C. (2014) *The Anthropocene: The human era and how it shapes our planet*, Synergetic Press, Santa Fe.

Selhub E. M. and Logan A.C. (2012) *Your Brain on Nature: The science of nature's influence on your health, happiness, and vitality*, Harper Collins Publishers Ltd., Toronto.

Smith G. (2015) "America's changing religious landscape", *Pew Research Center*, 12 May 2015, Washington, D.C. (www.pewforum.org/2015/05/12/americas-changing-religious-landscape/) Accessed May 21, 2015.

Sponsel L. E. (2011) "The Religion and Environment Interface: Spiritual Ecology in Ecological Anthropology", in Helen Kopnina and Elleanore Shoreman, eds. *Environmental Anthropology Today*, Routledge, New York, 37–55.

Sponsel L.E. (2012) *Spiritual Ecology: A quiet revolution*, Praeger, Santa Barbara.

Sponsel L. E. (2014) "Bibliographic essay – Spiritual ecology: Is it the ultimate solution for the environmental crisis?", *Choice* 51:8, 1339–1348.

Steinberg N. (2005) "Redwood rabbis", in B. Taylor ed. *Encyclopedia of Religion and Nature*, Vol. 2, Thoemmes Continuum, New York, 1352–1354.

Stone C. D. (2010) *Should Trees Have Standing? Law, morality and the environment*, Third edition, Oxford University Press, New York.

Taylor B. (1991a) "Earth and nature-based spirituality (part I): From deep ecology to radical environmentalism", *Religion* 31, 175–193.

Taylor B. (1991b) Earth and nature-based spirituality (part II): From Earth First! and bioregionalism to scientific paganism and the new age", *Religion* 31, 225–245.

Taylor B. (1995a) "Earth First! and global narratives of popular ecological resistance", in Taylor B. ed. *Ecological Resistance Movements: The global emergence of radical and popular environmentalism*, State University of New York Press, Albany 11–34.

Taylor B. ed. (1995b) *Ecological Resistance Movements: The global emergence of radical and popular environmentalism*, State University of New York Press, Albany.

Taylor B. (1998) "Religion, violence and radical environmentalism: From Earth First! to the Unabomber to the Earth Liberation Front", *Terrorism and Political Violence* 10:4, 1–42.

Taylor B. (2004) "Threat assessments and radical environmentalism", *Terrorism and Political Violence* 15:4, 173–182.

Taylor B. ed. (2005a) *The Encyclopedia of Religion and Nature*, Vols. 1–2, Thoemmes Continuum, New York.

Taylor B. (2005b) "Radical environmentalism", in Taylor B. ed. *Encyclopedia of Religion and Nature*, Vol. 2, Thoemmes Continuum, New York 1326–1335.

Taylor B. (2010) *Dark Green Religion: Nature spirituality and the planetary future*, University of California Press, Berkeley.

Taylor S. M. (2007) *Green Sisters: A spiritual ecology*, Harvard University Press, Cambridge.

Tucker M. E. (2006) "Religion and ecology: Survey of the field", in Gottlieb R. S. ed. *The Oxford Handbook of Religion and Ecology*, Oxford University Press, New York 398–418.

Turner T. (1993) *Sierra Club: 100 Years of protecting nature*, Abradale/Abrams, New York.

UCLA Higher Education Research Institute 2010 Spirituality in higher education (http://spirituality.ucla.edu/) Accessed May 21, 2015.

White L, Jr (1967) "The historical roots of our ecologic crisis", *Science* 155:3767, 1203–1207.

Wolens D. (2000) *Butterfly*, Doug Wolens, Berkeley (VHS, 80 minutes).

Worster D. (2008) *A Passion for Nature: The life of John Muir*, Oxford University Press, New York.

24

NATURE WRITING AND NATURE MYSTICISM

Douglas E. Christie

"Thinking like a mountain." It has been more than half a century since Aldo Leopold first artic-ulated this idea in his now classic work *A Sand County Almanac* (Leopold 1949, 129–133). Few ideas have resonated as deeply within or had as lasting an impact upon ecological thought as this one. Certainly part of its enduring appeal has to do with its radical, if also playful, upend-ing of our usual sense of subjectivity: who is the thinker? Who is being thought? Leopold's phrase calls to mind Paul Cézanne's famous observation to Joachim Gasquet that "The land-scape thinks itself in me and I am its consciousness." Or, as Thomas Berry has observed: "We are the universe conscious of itself."

Such expressions, playful and imaginative but also serious in their own way, point to the challenge we sometimes face when trying to describe how we encounter, live within, relate to, construct, and respond to the natural world. Especially the profound sense of intimacy and reci-procity, even shared life that often characterizes these encounters. To imagine oneself as capable of "thinking like a mountain," or the landscape as thinking "itself in me" is to entertain the possibility of inhabiting a richly indeterminate and fluid world where subject and object, self and other, inner and outer landscape, ebb and flow together without clear boundaries. In ecological terms, such a space is known as an ecotone, a rich place of exchange between one ecosystem and another, characterized by biological diversity, abundance, and opportunity. But what about our own capacity for such fluidity and reciprocity, especially in the realm of ecolog-ical thought and practice? How does such deep identification and exchange with the natural world happen? Why, in this moment of deepening environmental degradation, does such a sensibility appear so increasingly rare (and, paradoxically, sought after)? And what might it mean—personally, ecologically, ethically, and politically—to retrieve the kind of intimate iden-tification with the natural world that enables us, together with Aldo Leopold, to think like a mountain?

Such questions form an increasingly important part of the contemporary discourse concerning our relationship with and responsibility for the natural world. Often these ques-tions are framed primarily in ethical terms: what kind of responsibility do we have for other living beings? How far does that responsibility go? How should we enact and embody that responsibility, personally and collectively? These are important and necessary questions. Still, as we struggle to answer them, we sometimes find ourselves confronting another dimension of our felt relationship with the living world, something akin to what is sometimes named as spiritual or even mystical: an awareness of the self as capable of becoming immersed, even lost within an immensity far beyond one's capacity to articulate or understand. A sense of the numinous, the wholly Other, the ineffable. One can think of many examples of this kind of experience within those literary traditions devoted to articulating and interpreting the human encounter with the natural world.

In what sense do questions of religion and spirituality come into play in helping us think about our relationship with the natural world? Sometimes, as is the case with formal declarations on the environment on the part of particular religious communities, or in self-consciously theological or mystical reflections on the relationship between human beings and the natural world, the potential role of religious traditions to shape and guide our response to the environmental crisis is clear (Pope Francis 2015; Grim and Tucker 2014). Still, there has also been a longstanding ambivalence in environmental and ecocritical literature toward religion and spirituality, the sense that they have little to contribute to discussions about our relationship to and responsibility for the natural world (Hiltner 2015; Westling 2014). In spite of this, one can sense a shift taking place in relation to the distinctive role that spirituality can play in helping us reimagine our relationship with the living world. Increasingly, spirituality refers not necessarily to spiritual ideas and practices arising out of particular religious traditions (although these also have their place), but the often-implicit and inchoate sense of what William James once describes simply in terms of "the more" (James 1976, 35). Or what poet Czeslaw Milosz calls "the real" (Milosz 1983, 25).

Such language captures something important about the way many contemporary persons approach the question of spirituality. In particular, it reflects a growing feeling that the language of spirituality needs to be translated and reinterpreted continuously if we are to find a meaningful way to express our own relationship with the sacred. This includes a seeking out new ways of expressing our sense of the sacred in this particular historical moment. In relation to the natural world, the efforts to name this new and still emerging sensibility are striking in their range and diversity. Allan Hodder, for example, calls attention to an attitude he describes as a "mindful naturalism," a sense of oneself as so deeply immersed within the rhythms of the natural world that any notion of human identity separate from that larger reality becomes impossible to conceive (Hodder 2001, 66). Fiona Ellis speaks of an "expansive naturalism," a sense of nature as capable of revealing and making present to us a great immensity (Ellis 2014). Timothy Morton notes the importance of an ecological-spiritual sensibility that will enable us to discover "the liminal space between things" (Morton 2014). And Cathy Rigby describes this moment as one in which we are learning to respond to the challenge of "rematerializating religion and spirituality" (Rigby 2014).

These efforts to describe the depth dimension of our relationship with the natural world— in terms that draw upon the language of classical spiritual thought and practice while also revising it in important ways–comprise a critical part of our common environmental work in the present moment. Increasingly, we are recognizing that the way we speak of our relationship with the natural world impacts how we live within it and respond to it. Hence the heightened sense of value we attribute to writing that integrates of natural history observation, personal narrative, philosophical reflection, and social-economic-political analysis, while also making room for a consideration of the deep and abiding hunger we feel for a more intimate relationship with the natural world. A relationship that still, in spite of all the destruction we have wrought upon the natural world, exists and matters. The effort to articulate the character of this relationship and its potential for helping us heal what has been broken has led to a renewed attention to the place of spirituality in our ecological thought and practice. Something very much like this sensibility seems to haunt Leopold's conviction regarding the importance of learning to "think like a mountain."

Rethinking spirituality and nature

What does it mean to describe our feeling for the natural world as having a spiritual character? And why, in a cultural moment when religion has become so problematic and divisive (not least in relation to the question of what it means to cultivate ecological responsibility), do we encounter such strong attention to spirituality and spiritual practice? Is the language of spirituality really helpful here? Or does it distract from the urgent necessity to develop an ethic of care toward the natural world? In other words, do the fundamental questions of who we are in relation to the natural world and why and how we have grown so alienated from it have a contribution to make to the larger project of ethical and political renewal that is so urgently needed in the present moment?

To address these questions, it is important to acknowledge how ambiguously and variously spirituality and spiritual practice have come stand in relation to explicitly religious symbols and traditions. Sociologists and historians of religion have drawn our attention to the extent to which spirituality and spiritual practice are increasingly understood as distinct from and sometimes even antithetical to religious identity and practice (Wuthnow 1998; Roof 1998). The phrase "spiritual but not religious," heard with ever greater frequency during the last twenty to thirty years, is emblematic of this growing divide and hints at why spiritual practice remains so vital even amidst growing alienation on the part of many from religion and religious traditions (Fuller 2001; Schneiders 2003; Frankenburg 2004; Schmidt 2005). However, there is also evidence for the increased presence and influence of spirituality and spiritual practice *within* religious traditions, something that can sometimes contribute significantly to the renewal of those traditions, including the responsibility communities of faith have for the natural world (Jenkins 2008; Pope Francis 2015). Still, the disengagement of spirituality from its traditional relationship with religious communities and religious symbols and the increasing prominence of what Meredith McGuire and others describe simply as "lived religion"—a wildly diverse, deeply personal, often-eclectic and improvised orientation to spiritual practice—have become characteristic features of spiritual thought and practice in this particular historical moment, certainly in many parts of North America and Europe (McGuire 2008). And although not all these emerging forms of contemporary spirituality are ecocentric in character, we are witnessing a renewal and reimaging of spiritual life and practice that offers something potentially important for helping us address some of our most intractable environmental problems.

One of the striking things about this broad cultural shift is the complex way in which religious language and symbols continue to exert their power. In the literary tradition of nature writing, this sometimes means that "trace elements" of religious language or religious symbols appear as part of a narrative account without explicit reference to the particular tradition from which they originated. In other cases, it means that more encompassing spiritual language, such as mystery, awe, the sacred, or the numinous appears as a referent for something of ultimate value within the natural world or crucial to our relationship to the natural world. This is not an entirely new phenomenon. In the tradition of American nature writing, for example, one can already find elements of this eclectic spiritual hybridity in the work of Emerson, Thoreau, and others (Gatta, 2004; Robinson 2004; Hodder, 2001). Still, it has become a striking part of the contemporary discourse about the natural world. And in an historical moment marked by increasingly acute environmental destruction, it has taken on a new poignancy. There is a growing sense that we need a language strong enough to help us articulate both the depth of the loss we are experiencing and the grounds for a renewal of hope. This may sometimes mean drawing explicitly on the symbols that are central to the great religious traditions of the world.

But not always. And it is here that the dense, metaphorically rich, and allusive language of spirituality has an important role to play.

Returning to Leopold's narrative in *A Sand County Almanac*, it becomes clear that we are not simply being invited to wonder about the life of the mountain and what it might mean to understand its complex and delicate ecosystem more clearly and deeply. This would already be something valuable and worthwhile, and a clear advance on our usual habits of thinking of mountains (as well as rivers, forests and other natural habitats) simply as resources, to be used and disposed of for our own purposes. But Leopold seems to suggest that this is not enough, not nearly enough. We must go deeper; we must consider what it might mean to *think like* a mountain: to allow our own subjective awareness to be affected by the life of actual places, to allow ourselves to become implicated in that life, and become responsible for it. It is important to acknowledge that the dawning of this new awareness and the articulation of what Leopold called the "land ethic" was born from a jarring and painful experience of loss in which he himself was complicit: what he came to see as a callous, unnecessary, and ecologically destructive killing of a wolf. But it was the sight of a "fierce green fire" in the dying wolf's eyes that ushered him across the threshold toward a new sense of vulnerability toward and intimacy with the land; it was this that had the strongest and most enduring impact on his understanding of what "thinking like a mountain" might mean. Leopold found himself confronted with the question of whether he was willing to relinquish a narrow and controlling understanding of himself and the world for the sake of something larger, more complex, and more capacious. Whether he was willing to live with an orientation rooted in humility and openness to mystery that could enable him to *identify* with and *participate* in the living world.

Here we see an example of how the narrative accounts that lay at the heart of so much contemporary nature writing pose and demand responses questions of great moral and spiritual significance. For instance: what does it mean to open ourselves in this way to the living world? What value does the effort to do so have within the larger work of cultivating ecological awareness and environmental and social responsibility? These two questions run like a steady pulse through much contemporary discourse about the natural world. And their roots run deeper still. But it is not easy in this post (or trans, or meta) modern moment to interpret or respond to them. Part of the challenge has to do with trying to understand the character and aim of the questions themselves. Should Leopold's words be understood primarily as an invitation to become more deeply *informed* about the ecosystems we inhabit? Certainly this is part of it. But there is also an implied *ethical* obligation: we are being invited to reorient ourselves, even rethink our very subjectivity in relation to the natural world. If mountains (and other sentient beings) do in fact possess subjectivity (however we understand this), then how are we to live in relation to these other subjects? These ethical questions in turn open out onto others that might well be considered spiritual or even mystical in character.

There is a suggestion in Leopold's provocative challenge to "think like a mountain" of something long familiar to readers of mystical texts: an identification with the object of one's attention so deep and encompassing that the boundaries between subject and object begin to blur; a willingness to relinquish your ego identity so completely that you virtually disappear into the life of the Other, even as the life of the Other comes to suffuse your own life. "Undifferentiated unity" is the term often employed among mystical writers to describe such radical identification with the Divine Other. There is nothing explicitly religious or mystical in Leopold's text. Still, much of the power of his brief account lay in his expression of dawning respect for the utter unknowability of the mountain and its wild inhabitants (especially the wolves)—indeed the entire pattern of relationships between and among these different beings that together comprise the life of the place: an ecological awareness rooted in humility and respect. Also in his palpable hunger to identify more deeply with the place and his sense of what such identification might mean for the way we live in relation to such places. So: not a religious account perhaps. But one that invites a consideration of what it might mean to allow ourselves truly feel the immensity and power of the world in which we live and move, and to risk being changed by it. An account that resonates with undeniable spiritual power and meaning—a sense of spiritual transformation that carries within it the potential for a more-enduring ethical commitment.

Nor is Leopold's account unusual. It calls to mind other writers in whose work such questions occupy a prominent place. Edward Abbey, famously skeptical of traditional religion and anything smacking of metaphysical thought, nevertheless dreams of "a hard and brutal mysticism in which the naked self merges with a nonhuman world and yet somehow survives intact, individual, separate" (Abbey 1968, 14, 21). Kiowa writer N. Scott Momaday describes a young woman observing the corn dance at Cochiti, and her dawning awareness that the dancers "were held upon some vision out of range, something away in the distance, some reality that she did not know, or even suspect. What was it they saw? Probably they saw nothing after all, nothing at all. But that was the trick, wasn't it? To see nothing at all, nothing in the absolute" (Momaday 1968 36). And poet Tracy K. Smith recounts her childhood sense of awe at glimpsing the first pictures from the Hubble telescope: "We saw to the edge of all there is—/So brutal and alive it seemed to comprehend us back" (Smith 2011 11).

It is not easy to locate such accounts on any recognizable map of religious thought or practice; or to situate them within a framework of conventional ecological thought. Still, they are important, for they touch into and reflect a fundamental dimension of experience that challenges and sometimes transcends ordinary language and imagery and thought: the ineffable as mystical writers sometimes describe it. These accounts serve as a kind of provocation, drawing us into wondering about that moment when "the naked self merges with a nonhuman world." Or considering a way of seeing that is a kind of 'non-seeing'. Or grappling with what it is to be "comprehended by" the immensity of the universe. Even so, such moments of heightened awareness only tell part of the story. There is also the harsh and ongoing reality of loss and brokenness, the "slow violence" that environmental destruction visits upon the bodies of the poor and marginalized (Nixon 2011). The social context of racism and other forms of structural injustice increasingly informs our perception of what it means for human beings to engage the natural world (Ruffin 2010; Finney 2014). If we wish to understand what it might mean to "think like a mountain" in the current moment, we will need to attend carefully to the entire complex of social, economic, and political realities that inform our experience of nature.

An enduring mystery

What does such immersion look like and feel like, in practice? What does it mean for those engaged in such immersive practice? Part of this work involves learning to look at the living world more closely, learning to notice and describe the myriad, physical details of the world with precision and care. It means coming to *know* the world. But there is also a clear acknowledgment of an irreducible dimension of beings and relationships within the natural world, and of our experience and understanding of them, which resists complete understanding or explanation. It is for this reason, Robert Finch says, that nature writers often try to suggest "a relationship with the natural environment that is more than strictly intellectual, biological, cultural, or even ethical … they sense that nature is, at its very heart, *an enduring mystery*." It is because of this, Finch suggests, that "we must look to the sounds and images of our unedited natural experience for the true source of our emotions, our impulses, our longings—even for the very language of imagination itself" (Finch 1991, 101).

These observations are emblematic of a sensibility one encounters often in the tradition of nature writing, a deep respect for the complexity and intricacy of the living world, but also a certain wariness of the impulse to reduce or explain a reality that ultimately resists explanation. Respect for mystery does not mean mystification however. Nor does it necessarily imply a commitment to a theistic or religious explanation for the origins or purpose of the natural world. But it does suggest a willingness to inhabit a posture of openness, curiosity, and humility in relation to the living world. To pay careful attention to what Finch describes as our "unedited natural experience" and to learn to stand within it with respect and awe.

What do we know? That question, arising at the end of Gary Snyder's "Pine Tree Tops," echoes for a long time afterwards in the silence of the night. It is an important, troubling question. What *do* we know—about the lives of wild animals, about rivers, trees, the wind? How much has our *assumption* of knowledge obscured our vision, prevented us from seeing, feeling, loving what is continually unfolding before us? How much arrogance is wrapped up in that assumption of knowledge?

To struggle with these questions is, I think, to find ourselves confronted with a profound moral challenge. It is the challenge of discovering whether we have the capacity to adopt a posture of genuine humility before the ever-elusive, ever-mysterious natural world. In this, one can understand why Stephanie Mills argues that the question of our relationship with the wild world ultimately "shakes out as a religious question." Or, one might say, "a spiritual question." And the question is this: "Is nothing sacred? Are there no natural phenomena—cells, organisms, ecosystems—before which we might stand in humble awe?" (Burks 1994, 53).

The sense of awe. Here we are close to the very root of what has always characterized intense spiritual experience. Its recurrence in the present moment in the context of our experience of and response to the natural world seems telling and significant. Freeman House, whose book *Totem Salmon* chronicles the painstaking efforts among communities in Northern California to restore the Pacific Salmon run in their rivers and streams, offers an important example of this in his account of what it feels like to draw close to such beautiful, mysterious, and increasingly rare beings.

> King salmon and I are together in the water. The basic bone-felt nature of this encounter never changes, even though I have spent parts of a lifetime seeking the meeting and puzzling over its meaning, trying to find for myself the right place in it. It is a *large* experience, and it has never failed to contain these elements, at once separate and combined: empty-minded awe; an uneasiness about my own active role both as a person and as a creature of my species; and a looming existential dread that sometimes attains the physicality of a lump in the throat, a knot in the abdomen, a constriction around the temples.
>
> *(House 2000, 13)*

It is a *large* experience, he says, for which one must struggle (and in all likelihood fail) to find adequate language. But the language House does find is revealing. It is analogous to the language often used to describe intense religious experience. He calls it an "encounter," a "meeting," a "large experience," whose immensity evokes in him a deep sense of humility as he struggles to find his "right place in it." It pervades his entire being. Later, at some remove from the immediacy of this experience, he reflects on its meaning: "Each fish brought up from the deep carries with it implications of the Other, the great life of the sea that lies permanently beyond anyone's feeble strivings to control or understand it. ... True immersion in a system larger than oneself carries with it exposure to a vast complexity wherein joy and terror are complementary parts" (House 2000, 70).

Who is this Other whom we meet in such moments? Is it the "world" of these luminous beings ("a system larger than oneself ... a vast complexity"—a world that will forever elude our understanding and because of this remains fundamentally mysterious and alluring)? The beings themselves? God? An encounter with salmon, or any living species, House's narrative suggests, is an invitation to consider all of these possibilities. It is an invitation to open ourselves and respond to the mysterious Other with honesty and imagination and, perhaps, faith.

To open ourselves in this way to the mysterious Other, without prejudice or constraint, may well be one of the keys to rediscovering a sense of intimacy with other living beings that was once a common and accepted part of human experience. In such moments, Paul Shepard argues, we find ourselves participating in a kind of "archaic spirituality," rooted in the rhythms of the natural world. It is, he suggests, "The way of life to which our ontogeny was fitted by natural selection, fostering a calendar of mental growth, cooperation, leadership, and the study of a mysterious and beautiful world where the clues to the meaning of life were embodied in natural things, where everyday life was inextricable from spiritual significance and encounter, and where the members of the group celebrated individual stages and passages as ritual participation in the first creation" (Shepard 1982, 6).

It is striking to note how deeply communal this vision of the world is. Many of us, at least in North America and Europe, still suffer from a highly individualized understanding of spirituality. Gradually this is changing. Pope Francis's vision of spirituality in *Laudato Si*, for example, is inclusive of the entire web of life that connects us to every living being—non-human beings as well as our fellow human beings, especially the poor and marginalized who suffer most from our careless attitudes toward the environment. Paul Shepard's vision of spirituality, rooted in ancient human practices, is another: here we encounter an understanding of spiritual practice as deeply embedded within the life of the community, involving rituals of initiation, a daily, intimate immersion with the rhythms of the natural world and clear sense of responsibility for others. The question of whether it is still possible for us to participate in such an archaic spirituality, to absorb it into our consciousness and our lived reality, and to act in the world in response to it is difficult to answer. But one thing seems certain: not even to attempt to do so will almost surely contribute to an even deeper sense of alienation from the natural world than we are currently experiencing. And to live with such alienation and estrangement from the natural world can only hinder our efforts at restoration and renewal. The challenge is to learn to open ourselves to the inescapably biological–carnal dimension of our spirituality, and to respond to the entire non-human natural world as part of a single field in which we ourselves participate.

Conclusion

"Thinking like a mountain." The challenge of taking seriously the life of the world, on its own terms, remains. An even deeper challenge: allowing ourselves to become so identified with the life of the world that we can no longer stand aloof from it, or behave as if what befalls it does not concern us. Becoming (again) part of the world. Recognizing the sacredness of all living beings, the entire living world. Recognizing that "There are no unsacred places/there are only sacred places/and desecrated places" (Berry 2001, 270).

In this moment of acute environmental destruction, this work of simple recognition takes on a new and urgent meaning. As does the work of imagining and describing what happens when we recognize ourselves as participants in and responsible to the natural world, rather than disinterested and disengaged spectators. The upending of narrow models of subjectivity is no small part of this work. But so too is the work of creating a new vision of community within which every subject has a voice. We need more than ever before spiritual practices that will help us to hear the voices of those long neglected, deeply threatened but still precious beings in the natural world, as well as those human voices that have been too often relegated to the margins. The literature of nature is making significant contributions to this project of spiritual and ecological renewal. It serves as a reminder that "thinking like a mountain" can and must become part of an everyday spiritual awareness, reminding us of who we are in the world and how we are to act within it.

References

Abbey, E. (1968) *Desert Solitaire,* Tucson: Arizona: University of Arizona Press.

Berry, W. (2001) "How to be a poet (to remind myself)", *Poetry.*

Burks, D. C. ed. (1994) *Place of the Wild,* Washington, DC: Island Press.

Ellis, F. (2014) *God, Value and Nature,* Oxford: Oxford University Press.

Frankenburg, R. (2004) *Living Spirit, Living Practice: Poetics, Politics, Epistemology,* Durham, NC: Duke University Press.

Finch, R. (1991) Being at Two with Nature, *The Georgia Review,* 45.

Finney, C. (2014) *Black Faces, White Spaces: Reimagining the Relationship of African Americans to the Great Outdoors,* Chapel Hill, NC: University of North Carolina Press.

Francis (Pope) (2015) *Laudato Si: On care for our common home,* Washington, DC: USCC.

Fuller, R. (2001) *Spiritual but not Religious: Understanding Unchurched America,* New York: Oxford.

Gatta, J. (2004) *Making Nature Sacred: Literature, religion, and environment in America from the puritans to the present,* New York: Oxford.

Grim, J. and Tucker, M.E. (2014) *Ecology and Religion,* Washington, DC: Island Press.

Hiltner, K. ed. (2015) *Ecocriticism: The essential reader,* London: Routledge.

Hodder, A. D. (2001) *Thoreau's Ecstatic Witness,* New Haven, CT: Yale.

James, W. (1976) *Essays on Radical Empiricism,* Cambridge, MA: Harvard University Press.

Jenkins, W. (2008) *Ecologies of Grace: Environmental ethics and Christian theology,* New York: Oxford University Press.

House, F. (2000) *Totem Salmon: Life lessons from another species,* Boston, MA: Beacon.

Leopold, A. (1949) *A Sand County Almanac and Sketches Here and There,* New York: Oxford.

McGuire, M. B. (2008) *Lived Religion: Faith and Practice in Everyday Life,* New York: Oxford.

Milosz, C. (1983) *The Witness of Poetry,* Cambridge, MA: Harvard University Press.

Momaday, N. K. (1968) *House Made of Dawn,* New York: Harper and Row.

Morton, T. (2014) "The Liminal Space between Things: Epiphany and the Physical," in Iovino Serenella and Serpil Oppermann, eds. *Material Ecocriticism,* Bloomington, IN: Indiana University Press.

Rigby, K. (2014) "Spirits that Matter: Pathways toward a Rematerialization of Religion and Spirituality," in Iovino Serenella and Serpil Oppermann, eds. *Material Ecocriticsm,* Bloomington: Indiana University Press.

Roof, W. C. (1999) *Spiritual Marketplace: Baby Boomers and the Remaking of American Religion,* Princeton, NJ: Princeton University Press.

Ruffin, K. N. (2010) *Black on Earth: African American Literary Traditions,* Athens, GA.

Schmidt, L. E. (2005) *Restless Souls: The Making of American Spirituality,* San Francisco, CA: HarperSanFrancisco.

Schneiders, S. (2003) "Religion vs. Spirituality." *Spiritus* 3, 163–185.

Shepard, P. (1982) *Nature and Madness,* Athens, GA: University of Georgia.

Smith, T. K. (2011) *Life on Mars,* Minneapolis, MN: Graywolf.

Snyder, G. (1974) *Turtle Island,* New York: New Directions.

Westling, L. (2014) *The Cambridge Companion to Literature and the Environment,* Cambridge: Cambridge University Press.

Wuthnow, R. (1998) *After Heaven: Spirituality in America Since the 1950s,* Berkeley, CA.

PART VI

Planetary challenges

Introduction: *Willis Jenkins*

The problems that organize this section represent the basic difficulty for global ethics: how can humans living in many different cultural and religious worlds develop common responsibility for one shared planet? We inhabit one planet but live in many worlds. This volume dwells on the religious entanglements of environmental problems and ecological relations. Most of the contributors to this book think that, in some way, understanding religious dimensions of ecological relations will help humans more adequately confront shared challenges. Yet religion seems as likely to impede as improve cooperation on planet-wide challenges that face humanity as a species; so why focus on it here?

The contributors to this section would answer that question in different ways, in part because they take quite different approaches to practical problems. Some of the chapters are written by religious studies scholars (water, conservation, food, consumption, and gender injustice), some by social scientists (population, environmental justice), some by natural scientists (climate change, biodiversity), and some by interdisciplinary writers (oceans, animals). Yet, while none of these chapters suppose that religion holds the solution to planetary challenges, they all share a sense that religion (in some form) is more or less unavoidable in understanding and beginning to become responsible for these challenges.

Here are three basic reasons to focus on religion in the midst of planetary challenges. Not all the contributors would hold to all three reasons, or perhaps any one of them in this articulation, but a reader can find these themes emerging from the section as a whole. First, if religious difference is an impediment to addressing planetary problems, that is good reason to critically engage it (see Wyman on population). Second, as the global ecological influence of humanity creates challenges that flow across borders, it is important to understand how those challenges are interpreted by multiple cultural inheritances, and where may lie resources for recognizing shared values or generating new ones (see Peppard on water). Third, some of these problems so pervasively call into question industrial patterns of inhabitation that they throw modern cultures back to fundamental questions – questions that could be called religious in scale (see Lovejoy on biodiversity).

These chapters do not offer comprehensive treatments of the challenge they address; they rather offer one interpretive transect through it. The cultural and religious entanglements identified will make up a significant part of how each challenge is confronted (or not). It is possible that these religious dimensions may be transformed as multiple worlds invent ways to take shared responsibility for planetary systems. In any case, each planetary challenge here poses an important aspect of the human future, religious and ecological.

25

CLIMATE CHANGE

Varieties of religious engagement

Mike Hulme

On the 27 April 2015, several weeks before the Vatican issued Pope Francis' encyclical *On Care For Our Common Home* (Pope Francis 2015), the Cornwall Alliance, an American Christian evangelical coalition, issued an open letter on climate change addressed to the Pope. Whilst commending him for his care for the Earth and for God's children, the letter raised concerns about the quality of some aspects of climate science and about the worldviews underpinning some climate policy advocacy. Interpreting the Bible as mandating a preference for the poor, the authors of the letter concluded that "it is both unwise and unjust to adopt policies requiring reduced use of fossil fuels for energy" (Cornwall Alliance 2015).

Three years earlier on 22 February 2012, Operation Noah, another Christian evangelical coalition, but one based in the UK, had also issued a public statement on climate change, the so-called Ash Wednesday Declaration (Operation Noah 2012). It challenged the church that care for God's creation – and therefore concern about climate change – was foundational to the Christian gospel. Consciously echoing the 1934 Barmen Declaration, which gave coherence and visibility to the emergent Confessing Church during the Nazi regime, Operation Noah claimed climate change to be just such another "confessional issue." Taking inspiration from the same Scriptures as the Cornwall Alliance, they declared, "For our generation, reducing our dependence on fossil fuels has become essential to Christian discipleship."

These two examples spotlight the complex relationship between religion – in this case the Protestant Christian faith – and climate change. On the one hand they clearly show that the questions raised by the idea of human-caused climate change have increasingly come to occupy Christian institutions, theologians and faith-holders (and indeed, as we shall see, those of other faith traditions). But these vignettes also capture something of the diversity of religious engagements with the subject. Though appealing to the same revealed divine authority in the Bible, the Cornwall Alliance and Operation Noah reach radically different conclusions about what constitutes an appropriate response to climate change. The arguments, controversies and calls to diverse actions (and inactions) that have characterized the public (and mostly secular) discourse surrounding climate change are also to be found powerfully at work within religious communities.

In exploring the relationship between religion and climate change, this essay argues three things. First, it makes the case that religious thought and practice is important for understanding how the idea of climate change is given meaning in the contemporary world. There are many ways in which "climate change comes to matter" (Callison 2014), and to do full justice to understanding these processes one needs to study religions. Second, it emphasizes the empirical observation that the meanings attached to climate change, both between and within different religious traditions, will be diverse and at times contradictory (Veldman et al. 2013). The idea that religion could somehow act as a unifying platform to offer a "planetary opportunity and driving force to stay within planetary boundaries" – as a recent conference has suggested (EFSRE, 2015) – misreads both religion and ecology (see Hulme 2009, 142–177). The essay illustrates these two initial arguments by identifying a number of the more salient contact points between religious thought and practice and climate change: cosmologies, beliefs and perceptions, ethics and practices.

Third, the essay concludes by suggesting a number of areas where these tensions in the relationship between religion and climate change seem most acute. And yet I also suggest that deeper, broader and more informed engagement with the world's religions – on the part of scholars, advocates and politicians – is essential to shape the unfolding story of climate change and humanity; to understand how people in all their diversity and fractiousness will come to navigate the physical and cultural force field of climate change.

When did religions discover climate change?

The rhythms of the sky have long been triggers and companions of human thought and ritual. From the frigid north to the torrid tropics they have induced wonder and fear, whilst also offering comfort and assurance. Alongside the experience of intense yet predictable diurnal and seasonal weather cycles, sits the unreliable performance of climate from year-to-year and from generation-to-generation. No two years are the same; the climate of old age seems unlike the climate of youth. While a drought is to be feared; a mere dry season is not. A winter is not an ice age; neither are all storms hurricanes. It is little wonder that human anxieties, hopes and the search for explanation – and hence many of our spiritual longings and theologies – have been bound up with the skies. Religions have found many ways to make sense of these cruel fates, acknowledging our dependence on powers beyond our control and giving thanks for mercies and blessings received. So even as social, cultural and climatic influences combine in complex ways to allow for the emergence of different religious beliefs and practices, religion has "played a crucial role in fostering the ideas that humanity and Nature could reciprocally transform each other" (Barnett 2015, 232). Climate and religion have a long history of interdependence.

And so it is somewhat surprising that there has been such a notable exclusion of religions, and the religious, in the forging of late-modernist accounts of climate change and its multiple causes. For example, the United Nations expert body on climate change – the Intergovernmental Panel on Climate Change (IPCC) formed in 1988 – has managed successfully to prise the idea of climate and its capricious behavior away from its deep historical and cultural anchors. In the IPCC, and pre-dominantly in most public discourse, climate is framed as a physical phenomenon, to be studied using the theories of physics and the tools of numerical simulation models.

In recent years this "purification" (cf. Latour 1993) of climate has begun to change. As the scientized account of climate change and human agency has spread around the world it has run up against deeper narratives and stronger resistance than many scientists – and many politicians and campaigners – might have thought. Climate change turns out to be not just one thing, a thing defined and simulated inside Earth System models. As a hybrid physical–cultural phenomenon, climate change needs to be studied not just by meteorologists, ecologists and economists – the dominant disciplines assessed by the IPCC (Bjurström and Polk 2011) – but by sociologists, anthropologists, philosophers and, importantly, by theologians and religious scholars. The meanings of climate change are multiple and the ways in which people express, represent, engage and resist climate change are too numerous to be controlled by a single dominating perspective, namely science. And many of these diverse reactions have their roots in ancient religious cosmologies, doctrines, traditions and practices.

Whereas formal academic scholarship on climate change and religion is a phenomenon largely of the last decade, the engagement of religious faiths with climate change has a much longer history. For example, as far back as 1988, before the IPCC was constituted, the World Council of Churches launched its Climate Change Program, which was aimed at promoting transformation of social structures and lifestyle choices. The Church of Scotland issued its first assessment of climate change in 1989 – *With Scorching Heat and Drought* (Pullinger 1989) – the Dalai Lama made his first speech on climate change in 1990, during the Kalachakra Initiation at Sarnath, India and in 2000 the Coalition on the Environment and Jewish Life (COEJL) issued a report on *Global Warming: A Jewish Response*. Among the growing number of fora where climate change was debated by religious leaders and scientists, the Oxford Forum on Global Climate Change in July 2002 was particularly significant (Wilkinson 2012). The resulting Oxford Declaration signed by more than 70 leading climate scientists, policy-makers and American Christian leaders recognized that "the Christian community has a special obligation to provide moral leadership and an example of caring service to people and to all God's creation" (Climate Forum 2002).

Why religions matter for climate change

Researchers, policy-makers and leading scientists have recently recognized the importance of religion for understanding how people make sense of climate change and also for identifying meaningful responses to the challenges that are raised. Conversely, climate change also matters for religions, as has been succinctly argued by Anglican Bishop David Atkinson: "the questions posed by climate change reach to the heart of faith: our relationship to God's earth and to each other; the place of technology; questions about sin and selfishness, altruism and neighbour love; what to do with our fears and vulnerabilities; how to work for justice especially for the most disadvantaged parts of the world and for future generations" (Atkinson 2012).

Religious traditions influence the cosmologies of believers and, less directly, unbelievers, which in turn give shape to how people make sense of unsettling changes in their local climatic environments. Major religious faiths also possess substantial institutional and economic resources, as well as possessing significant political power (Grim and Tucker 2014). Arresting climate change is not just beyond the capacities of science; it is also beyond the capacities of the state. As with other non-state actors, such as businesses, cities and NGOs, religious movements and institutions have the mobilizing power to enlist and de-list multitudes of citizens in influential causes. Religious actors are key contributors to political discourses at local, national and international levels and prominent climate activists regularly cite the importance of religious participation in international climate negotiations. Influential climate scientists have publicly called for enhanced collaboration among religious institutions, policy-makers and the scientific community (e.g. Dasgupta and Ramanathan 2014).

Finally, religions give substance and power to social and ethical norms, enhance social capital and valorize certain lifestyles. Many commentators have remarked that climate policies need to tap into intrinsic, deeply held values and motives if cultural innovation and change are to be lasting and effective. As the Alliance of Religions and Conservation observed in 2007: "The emphasis on consumption, economics and policy usually fails to engage people at any deep level because it does not address the narrative, the mythological, the metaphorical or the existence of memories of past disasters and the way out. The faiths are the holders of these areas and without them, policies will have very few real roots." Religious practices can not only ameliorate hardships affecting communal life, but also animate calls for alternative value systems and lifestyles.

Cosmologies

The historian Lynn White Jr., in his influential 1967 essay "The historical roots of our ecological crisis," observed that "What people do about their ecology depends on what they think about themselves in relation to things around them. (It) is deeply conditioned by our beliefs about nature and destiny ... that is by religion" (White 1967, 1205). How people understand the ordering of the natural world, the animate and inanimate agencies at work and the appropriate duties and responsibilities of humans, i.e., their cosmologies, significantly affects how people interpret climate change and make it meaningful. For most people in the world, scientific claims that through their aggregate acts of material transformation and consumption humans are largely responsible for changes in the world's climates are at best partial. Such claims do not engage with or make sense of the spiritual and moral lifeworlds of many people. For others, such scientific claims may conflict more fundamentally with traditional beliefs about agency and causation in the sky (Donner 2007); weather phenomena are believed to be explicitly controlled by the gods.

In his study of climate change beliefs amongst Marshall Islanders in the Western Pacific, the anthropologist Peter Rudiak-Gould shows how the blending of local and Christian cosmologies offer different accounts of blame and agency than would easily be understood in many Western settings (Rudiak-Gould 2012). For others, knowledge about the weather and its effects on local ecology and physical landforms (e.g. glaciers, Allison 2015) are infused with spiritual worldviews. Although not easily fitting the category of 'religious beliefs' (Leduc 2010), such worldviews nevertheless challenge modernist and materialist accounts of agency. For some Hindus, belief in the epoch *Kaliyuga* allows the divinity of the sacred Ganges River to offer reassurance in times of climatic disturbance. Reasonings of blame for climatic misdemeanours and deviations come in many different guises (e.g. Rudiak-Gould 2015), which can blur the lines of rationality assumed to distinguish between early-modern and late-modern cultures. Barnett (2015) shows how early modern scholars deemed the Biblical Flood 'anthropogenic' in the sense that it was divine punishment induced by human sin. This is not so far from the belief expressed by the eco-theologian Michael Northcott when he proclaims "Global warming is the Earth's judgement on the global market Empire and the heedless consumption it fosters" (Northcott 2007, 7). The moral failures of humanity are in each case the cause of global climatic change, whether mediated by God or by "the Earth."

Without understanding the religious and spiritual dimensions of peoples' lifeworlds, climate change communication, advocacy campaigns and policy development and implementation will be deficient. Unravelling and giving salience and credibility to these different accounts of climatic agency and blame is a task of religious studies scholars.

Beliefs and perceptions

The study of climate change is informed by numerous public opinion surveys that seek to capture the extent of popular belief in anthropogenic climate change and about what should be done to tackle it and by whom (Capstick et al. 2015). Popular discourse about religion, publics and politics often reduces the political agency of religious actors to fixed theological positions. Such a view of politics and religion has bifurcated public opinion about the role of religion in addressing climate change. Especially in the United States and in the UK, religious actors are frequently criticized as climate obstructionists.

However, if American religious conservatives tend to be climate "skeptics," for example, recent scholarship suggests that theological beliefs are not the primary motives for such a position (Jones et al. 2014). Hispanic Catholics, Black Protestants and non-Christian/Jewish religious Americans all tend to be more concerned about climate change than the average American; it is only white Protestant and Catholic Americans who are less likely to be concerned. And with respect to attributing the severity of recent climatic disasters to different causes, it is only white evangelical Protestants who are substantially more likely to attribute them to a Biblical "end times" belief than to (human-caused) climate change. The theologies of "white" Catholics and white mainline Protestants do not lend themselves to such interpretative beliefs (Jones et al. 2014).

Many commentators concerned about climate change praise the work of religious advocacy networks, citing the historical role of religion in activating social change with respect to racial justice, poverty alleviation and human rights. Nevertheless, religious attitudes about the scientific consensus on climate change and about what constitutes an appropriate response are decidedly mixed (Taylor 2015). These divergent claims about religion and climate change indicate the need for public discourse that is better informed by the diverse ways religious actors are engaging climate politics (Hulme 2009). Experts in this field are currently charting the myriad responses of religious communities to environmental crises and describing how various religious systems of thought confront the challenge of climate change on different terms (Gerten and Bergmann 2012; Veldman et al. 2013).

Ethics

Beyond cosmology and doctrine, any attempt to understand how religions engage with the idea of climate change must appreciate how different religious faiths reason ethically. In relation to climate change there are three places to start such an inquiry, whether religious or not: (1) What is our responsibility to the non-human world? (2) What is our responsibility to the human other? (3) How should we care for the future? From a Christian theological perspective these inquiries might be framed as questions about "creation care," "neighbor care" and eschatology (Wilkinson, 2010). Different religious faiths might frame these questions differently and within any single faith tradition there will be different interpretative positions. Nevertheless, religious thought can contribute to global cultural dialogue about such generic questions without requiring adherence or identification with any particular religion. I don't have to be a Buddhist to be interested in how Buddhists use their tradition to answer these questions.

We have already seen how within evangelical Christianity the Cornwall Alliance and Operation Noah construct ethical responsibility for the poor in radically different ways. Religious responses to the fossil fuel divestment campaign orchestrated by the social movement "350.org" are another site where differences in ethical reasoning can be studied. For example, the Church of England has about $16 billion of invested capital and in 2014 the Ethical Investment Advisory Group (EIAG) of the Church was charged by the governing Synod to revise its ethical investment policy. This was in light of the divestment movement's challenge to prominent public bodies to withdraw all investments in oil, coal and gas companies. Their response, approved by Synod in July 2015, drew upon a variety of theological traditions and Biblical hermeneutics and recommended selected rather than full-scale divestment. It could be seen as offering a middle way between the more extreme positions of the Cornwall Alliance and Operation Noah. Nevertheless, the paper concluded, "climate change is an urgent ethical issue and ... calls for an urgent response from all parts of society, including investors" (Church of England 2015).

Another example where religious ethics have an important role to play is with regard to the question of climate engineering. The questions raised by these putative technologies of deliberate climate modification relate to the drawing of boundaries between human and non-human entities (for a general treatment of this question see Albertson and King 2010). If climate engineering is thought of as a form of climatic enhancement, then there is a parallel to be drawn with how religious ethics engage the idea of human enhancement (Hulme 2015). Ethical questions about "manipulating nature" and "playing God" come to the fore – questions which different religions traditions approach differently. Clingerman (2015) suggests that one role for theologians in this debate is their ability to offer hermeneutical tools to engage different narratives of climate engineering in constructive ways. Religious thought and belief, according to Clingerman, are to be offered as a resource "to help us to understand the machinations of actual domination of the atmosphere" (Clingerman 2015, 16). This is not dissimilar to the call to think carefully about what metaphors of agency are adopted for environmental action (Jenkins 2005). Religions can again be seen as a cultural resource, helpfully widening the ways in which we think about responses to climate change.

Practices

Religion is relevant for understanding responses to climate change not simply in terms of cosmologies, the shaping of abstract beliefs through formal doctrines or through ethical principles. To think thus would be to succumb to a discredited Enlightenment prejudice about the pre-eminence of abstract reason over embodied action. Perhaps more important for understanding climate change in relation to religion is the way in which religious institutions, communities and practices shape cultural imaginaries and individual behaviors.

There is a considerable literature from sociology and social psychology about how beliefs, values and attitudes work to shape pro- or anti-environmental behaviors (e.g. Gifford et al. 2011). There is also growing interest in how institutions, communities and individuals engage in, or are prevented from engaging in, adaptation actions to reduce vulnerabilities to extreme weather (Adger et al. 2009). Yet not much of this literature, and even then only recently, has examined the specific role of religious networks, practices and rituals in this context. For example, Kuruppu (2009) drew attention to the importance of religion for adaptation of community water resources in Kiribati, particularly in contexts where religion is central to peoples' lives. From a different part of the world, Hesed and Paolisso (2015) showed how amongst African American communities living along the Chesapeake Bay, faith-based knowledge and religiously shaped social networks work to mediate the adaptive capacity of these coastal communities in the face of storm-risk. Other social scientific investigations into how religious practices – from within Buddhism, several variants of Christianity, Hinduism, Islam and traditional indigenous beliefs – are shaping responses to climate change are collected in Veldman et al. (2013).

A focus on religious practices and climate change can also lead to reflections on the idea of virtue (Hulme 2014). Religious responses to climate change involve the community of believers, and since one of the abiding goals of most religious communities is the pursuit of holiness, one can ask how virtue is to be acquired and exercised. Or as Protestant theologian Tom Wright puts it, "How can we acquire that complex 'second nature' which will enable us to grow up as genuine human beings?" (Wright 2010, 220). Wright offers a virtuous circle in which five elements work together to cultivate character: community, stories, scripture, practices and examples. One can recognize in this schema elements that are not unique to Christian tradition. They can be developed in many different religious and secular settings by paying attention to the wisdom of the past, human life stories, rituals that reinforce connectivity and community cohesion that reaffirms individuals' self-worth. For example, the appeal of such a virtuous circle can be found in the writings of the philosopher Alain de Botton and his "Manifesto for Atheists: ten virtues for the modern age" (de Botton 2013).

Conclusion

Different regions and diverse groups of stakeholders understand the threat of climate change according to particular and often distinct religious frames of reference. These religious narratives and rituals shape the nature and credibility of different knowledge claims about climate – what is happening to it and why – as well as shaping individual and communal ethical and social behaviors. Religious faith communities therefore offer "thick" accounts of moral reasoning for acting in the world, in response to climate change as much as in response to other social and ecological challenges. Such an approach sits in contrast to secular calls for mitigation and adaptation that rely upon "thin" global values: widely shared, but culturally non-specific, moral criteria (Wolf and Moser 2011).

But while I would argue that religious engagement with climate change is both necessary and inevitable, it is hardly a panacea for resolving the many deep divisions and dilemmas in our world which climate change reveals. Whilst the Interfaith Statement on Climate Change (Interfaith Summit 2014) or Pope Francis' Encyclical *On Care For Our Common Home,* hold out hopeful visions for common action on climate change, there remain many sources of tension within and between different religious traditions that complicate these hopes. Far from inspiring a replicable or universal response, the world's religions are engaging the idea of climate change for diverse reasons and in divergent ways (Veldman et al. 2013). Where local groups are affected by climate change, communities necessarily respond in vernacular terms consistent with their own religious and cultural self-understandings. "Thin" global values do not fully capture the full range of concerns and commitments expressed by affected communities, for example claims about sacred landscapes, divine causality, ethical responsibility or social solidarity.

Unanswered questions

There is much yet to discover about how religious beliefs, institutions and practices around the world interact with the idea of climate change, and with what effect. A crucial first step towards forging more culturally grounded policy responses is improving public understanding of the religious heterogeneity through which climate change is experienced and politicized. Better knowledge about the overlaps and differences among religious traditions can inform climate policy and generate more effective coalition building across diverse interests. I close by offering a number of emerging research agendas with which religious scholars and their companions might enthusiastically engage:

- Most scholarship on Christianity and climate change has been focused on North America and Europe. Yet the most significant concentrations of Christian believers, and where growth is strongest, are in Africa, Latin America and parts of Asia. How then do African Christians, for example, bring theological reasoning to bear on the questions of poverty, ecology and technology that lie at the heart of climate change? (see Golo and Yaro, 2013, as an leading example).

- How do individual religious believers interpret doctrinal, ethical and behavioral statements on climate change issuing from faith leaders, and what range of such lay interpretations can be – and are – accommodated within religious traditions and communities? For example, empirical research should be conducted on lay readings and uses of the Pope's 2015 Encyclical.
- Given that Islam, with nearly 25% of the world's population, is the second largest religion in terms of adherents, more attention should be paid to how Muslims – in faith and in practice – either engage with or potentially engage with climate change. For example, studies should track how ordinary Muslims interpret and respond to the August 2015 Islamic Declaration on Global Climate Change.
- What are effective and ineffective means of communicating the risks of climate change, as articulated by science, with people of various faiths? A recent report from the Climate Outreach and Information Network (2015) in the UK – *Messages to mobilise people of faith* – is of interest in this regard.
- Finally, to what extent can religious framings of the causes, effects and ethics of climate change be more effective in reconciling disputed positions on climate policy than can other authoritative cultural framings? Wilkinson's 2012 study of American evangelicals could usefully be extended in other regions and to other religions.

Science is never enough to resolve problems that are cultural in origin. Neither through its promise of solid and reliable knowledge, nor through its efforts to animate social movements, can science chart a course of action in the world that will resolve political contestation. The former Chairman of the IPCC, R. K. Pachauri, was therefore profoundly wrong when he claimed in November 2014 at the launch of the Synthesis Report of the IPCC's 5th Assessment that, "All we need is the will to change, which we trust will be motivated by … an understanding of the science of climate change" (IPCC 2014). Simply understanding climate science will not provide the "will to change." On the other hand, reading climate change and accounts of human agency through the eyes of the world's religions offers fresh insights and different inspirations about what it means to be human in an age of climate change.

Acknowledgements

I would like to thank the editors of this Handbook – Mary Tucker, Willis Jenkins and John Grim – for their invitation to write this essay and Willis Jenkins for helpful comments. Wim B Drees and the journal *Zygon* are thanked for supporting my attendance at the 2014 annual meeting in San Diego of the American Academy of Religion. Ideas that have surfaced here have matured through various public and roundtable engagements with different Christian congregations and organizations in the UK and also with Robert Albro and Evan Berry of the American University, Washington.

References

Adger W.N., Dessai S., Goulden M., Hulme M., Lorenzoni I., Nelson D.R., Naess L.O., Wolf J., Wreford A. (2009) "Are there social limits to adaptation to climate change?" *Climatic Change* 93, 335–354.
Albertson D. and King C. eds. (2010) *Without nature? A new condition for theology,* Fordham University Press, New York.

Alliance of Religions and Conservation (2007) UN and ARC launch programme with faiths on climate change. Available at: www.arcworld.org/news.asp?pageID=207 (accessed 25 March 2016).

Allison E. (2015) "The spiritual significance of glaciers in an age of climate change", *WIREs Climate Change* 6:5, 493–508.

Atkinson D. (2012) Why climate change is a confessional question Available at: www.operationnoah.org/confessional-question (accessed 25 March 2016).

Barnett L. (2015) "The theology of climate change: sin as agency in the Enlightenment's Anthropocene", *Environmental History* 20, 217–237.

Bjurström A. and Polk M. (2011) "Physical and economic bias in climate change research: a scientometric study of IPCC Third Assessment Report", *Climatic Change* 108:1–2, 1–22.

Callison C. (2014) *How Climate Change Comes to Matter: The communal life of facts,* Duke University Press, Padstow, Cornwall.

Capstick S., Whitmarsh L.E., Poortinga W., Pidgeon N.F., Upham, P. (2015) "International trends in public perceptions of climate change over the past quarter century", *WIREs Climate Change* 6:1, 35–61.

Church of England (2015) Climate change: The policy of the National Investing Bodies of the Church of England Synod and the Advisory Paper of the Ethical Investment Advisory Group of the Church of England 30 April 2015. Available at: www.churchofengland.org/media/2223994/climate.change.policy.30.04.15.pdf (accessed 25 March 2016).

Climate Forum (2002) Oxford declaration on global warming. Available at: www.jri.org.uk/news/statement.htm (accessed 25 March 2016).

Clingerman F. (2015) "Roundtable on climate destabilisation and the study of religion: Theologians as interpreters – not prophets – in a changing climate", *Journal of the American Academy of Religion* 83:2, 336–355.

Climate Outreach and Information Network (2015) "Starting a new conversation on climate change with the European centre-right". Available at: http://climateoutreach.org/resources/starting-a-new-european-conversation-on-climate-change-with-the-centre-right/ (accessed 25 March 2016).

Cornwall Alliance (2015) An open letter to Pope Francis on Climate Change. Available at: www.cornwallalliance.org/anopenlettertopopefrancisonclimatechange/ (accessed 25 March 2016).

Dasgupta P. and Ramanathan V. (2014) "Pursuit of the common good" *Science* 345, 1457–1458.

de Botton A. (2013) *Religion for Atheists: A non-believers guide to the uses of religion,* Vantage, London.

Donner S.E. (2007) "Domain of the Gods: an editorial essay", *Climatic Change* 85, 231–236.

EFSRE (2015) Religion in the Anthropocene: Challenges, Idolatries, Transformations. Available at: www.hf.ntnu.no/relnateur/index.php?lenke=meetings.php (accessed 25 March 2016).

Gerten D. and Bergmann S. eds. (2012) *Religion in Environmental and Climate Change: Suffering, values, lifestyles,* Continuum Press, New York/London.

Golo B.-W.K. and Yaro J.A. (2013) Reclaiming stewardship in Ghana religion and climate change, *Nature and Culture* 8:3, 282–300.

Gifford R., Kormos C. and McIntyre A. (2015) "Behavioral dimensions of climate change: Drivers, responses, barriers, and interventions", *WIREs Climate Change* 2:6, 801–827.

Grim J. and Tucker M.E. (2014) *Ecology and Religion,* Island Press, Washington DC.

Hesed C.D.M. and Paolisso M. (2015) "Cultural knowledge and local vulnerability in African American communities", *Nature Climate Change* 5:7, 683–687.

Hulme M. (2009) *Why we Disagree about Climate Change: Understanding controversy, inaction and opportunity,* Cambridge University Press, Cambridge.

Hulme M. (2014) "Climate change and virtue: An apologetic", *Humanities* 3:3, 299–312.

Hulme M. (2015) "Better weather? The cultivation of the sky", *Cultural Anthropology* 30:2, 236–244.

Interfaith Summit on Climate Change (2014) Climate, faith and hope: Faith traditions together for a common future. Available at: http://interfaithclimate.org/the-statement/ (accessed 25 March 2016).

International Islamic Climate Change Symposium (2015) Islamic declaration on global climate change. Available at: http://islamicclimatedeclaration.org/islamic-declaration-on-global-climate-change/ (accessed 25 March 2016).

IPCC (2014) Climate change threatens irreversible and dangerous impacts, but options exist to limit its effects. Available at: www.un.org/climatechange/blog/2014/11/climate-change-threatens-irreversible-dangerous-impacts-options-exist-limit-effects/ (accessed 25 March 2016).

Jenkins W. (2005) "Assessing metaphors of agency: Intervention, perfection and care as models of environmental practice" *Environmental Ethics* 27:2, 135–154.

Jones R.P., Cox D. and Navarro-Rivera J. (2014) *Believers, Sympathizers and Skeptics: Why Americans are conflicted about climate change, environmental policy and science,* Public Religion Research Institute (PRRI), Washington, DC.

Kuruppu N. (2009) "Adapting water resources to climate change in Kiribati: The importance of cultural values and meanings", *Environmental Science and Policy* 12:7, 799–809.

Latour B. (1993) *We have never been modern* (translation by C Porter), Harvester/Wheatsheaf, New York.

Leduc T. (2010) *Climate, Culture, Change: Inuit and Western dialogues with a warming North,* University of Ottawa Press, Quebec.

Northcott, M.S. (2007) *Moral Climate: The ethics of global warming,* Dartman, Longman and Todd Ltd/Orbis., London.

Operation Noah (2012) Climate change and the purposes of God: a call to the church. Available at: http://operationnoah.org/what-we-do/ash-wednesday-declaration/ (accessed 25 March 2016).

Pope Francis (2015) *Encyclical Letter: Laudato Si' of the Holy Father Francis – On care for our common home,* Vatican Press, Rome.

Pullinger D.J. ed. (1989) *With Scorching Heat and Drought: A report on the greenhouse effect,* St. Andrew Press, Edinburgh.

Rudiak-Gould, P. (2012) "Promiscuous corroboration and climate change translation: a case study from the Marshall Islands", *Global Environmental Change* 22:1, 46–54.

Rudiak-Gould, P. (2015) "The social life of blame in the Anthropocene", *Environment and Society: Advances in Research* 6.1, 46–65.

Taylor B. (2015) "Religion to the rescue (?) in an age of climate disruption", *Journal for the Study of Religion, Nature and Culture* 9:1, 7–18.

Veldman R.G., Szasz A. and Haluza-Delay R. eds. (2013) *How the World's Religions are Responding to Climate Change: Social scientific investigations,* Routledge, Abingdon/New York.

White L. Jr. (1967) "The historical roots of our ecologic crisis", *Science* 155, 1203–1207.

Wilkinson K. K. (2010) "Climate's salvation: Why and how American evangelicals are engaging with climate change" *Environment* 52:2, 47–57.

Wilkinson K.K. (2012) *Between God and Green: How evangelicals are cultivating a middle ground on climate change,* Oxford University Press, New York.

Wolf J. and Moser S.C. (2011) "Individual understandings, perceptions and engagement with climate change: Insights from in-depth studies across the world", *WIREs Climate Change* 2:4, 547–569.

Wright N.T. (2010) *Virtue Reborn,* SPCK, London.

26

BIODIVERSITY

An inordinate fondness for living things

Thomas E. Lovejoy

In an anecdote that biologists love to recount, J.B.S. Haldane, the prominent British scientist of the first half of the twentieth century, was once asked what he could divine about the Creator from studying the Creation. His reply: He must have had "an inordinate fondness for beetles." In the late-twentieth century when G. Evelyn Hutchinson was working on his final volume of the Treatise on Limnology he said if Haldane was to be asked the question at that point he would indicate an inordinate fondness for flies (Diptera). Were he to reply today it would have to be an inordinate fondness for micro-organisms.

The foregoing reflects dramatic changes in the understanding of the diversity of life on Earth. A half century ago the "tree of life" was essentially two sturdy trunks – plants and animals – and a minor amount of unclear things down at their base. Today it is essentially a low spreading bush sporting small terminal twigs on the right: animals, plants, and fungi. The entire rest of the low bush is made up of micro-organisms, many with strange metabolisms and strange appetites – some dating from the early history of life on Earth.

While science has a pretty fair understanding of the diversity of more conspicuous organisms, such as vertebrates and flowering plants, new species are still being recognized even in such well-known groups as birds. In the Amazon, for example, 14 new species were described within the last couple of years, and new ones continue to be identified. Sometimes it is because of only recent access to remote places – such as enclaves of savannah-type vegetation – which among other things led to the discovery of the first jay to be known from the lowland Amazon. In other cases there are cryptic species: ones that look pretty much same to the eye, but have distinctive calls on either side of big rivers; later, DNA analysis often proves them to be distinct species where only one had been recognized (by science as opposed to the birds themselves) before.

But for other groups of organisms there are vast numbers of unknown species. In the Amazon River system it is estimated that only two-thirds of the fish species have been described. For invertebrates the proportion of unknown species is huge and for micro-organisms even more vast. So it becomes seriously difficult to estimate the total number of species on the planet. It must be somewhere between ten and a hundred million. This highlights not only the imperative to explore unknown life on Earth, including, for example, the biodiversity that make soils living habitats, but also to develop ways to conserve what we do not know about. An example of the latter would be the sagebrush of the American West; conserving

249

sagebrush (if thoughtfully planned) could easily conserve the unknown as well as known species of that habitat type.

The pragmatic value of biodiversity

Biodiversity has values beyond the practical, but those I will reserve to the end because it is their recognition that can help us achieve a better outcome for the marvelous diversity of life on Earth. We basically came into existence as part of ecosystems and biomes so it is not surprising that biodiversity is important and beneficial to us. In a crude sense it provides our food, clothing, shelter, and important aspects of our medicine and health – aspects of existence we partake of every day with little heed to where it comes from and how fragile the supply line might be. It is not surprising that agriculture, forestry, and fisheries are major human activities. Their total economic product may not be that large but that belies the reality that we would be bereft without them. In part, the reason for the low "value" is that we probably haven't included full cost and benefit in their market price.

The ear of corn on my plate yesterday is known scientifically as *Zea mays*, but the reality is that it and most of our other food crops are continually being upgraded and improved by new genetic stock from wild relatives. When the *New York Times* ran a front page story in the 1970s about the discovery of a new species of tropical plant, that itself was not new. What was news is that it was a wild relative of corn – the third most important grain in support of human society – but in particular that it was a perennial species (christened *Zea diploperennis*). Suddenly this discovery raised the possibility of a perennial corn agriculture, which could be of great importance to a future more sustainable form of agriculture. (Perennial mixed species agriculture is advancing at the Land Institute in Kansas.) A dividend is that it has proved resistant to some of the major diseases of its annual relative.

Biodiversity has long contributed in valuable ways to human medicine. Aspirin, the greatest-selling medicine of all time, was originally prescribed as an analgesic in ancient Greece when it was prepared as an infusion from the bark of a willow tree. The key ingredient was a molecule that Bayer figured out how to synthesize before the Second World War and was considered of such value it became part of the war settlement.

Biodiversity is an incredible treasure chest or library for medicine and the other life sciences. The diversity of organisms is constantly moving, exploring, and testing new novelties for living systems through the process of evolution. So it is continually expanding the potential to transform human existence in extraordinary ways. We might think that the blue mold that flavors Roquefort cheese is quite wonderful but a chance observation of a laboratory culture "contaminated" by a related *Penicillium* mold not only led Rutherford to penicillin but more importantly provided the entire concept of antibiotics.

In the neotropical forests where I do research, lurks a pretty nasty viper: the Bushmaster. Its venom causes the blood pressure of its normal prey (or the occasional human struck by it for defense) to crash to zero forever. Snake venom is not a good medicine because digestive juices denature it, but following the scientific puzzle so revealed, Brazilian scientists discovered the angiotensin system of regulation of blood pressure. That, in turn, led to Squibb pharmaceutical chemists devising the first of the ACE inhibitors to control hypertension. Literally hundreds of millions of people live longer, healthier and more productive lives because of the Bushmaster, and no monetary value has probably ever been estimated for this extraordinary contribution to human welfare.

A quite recent example was the discovery in 2015 of a soil fungus from Nova Scotia, which has the ability to counter the mechanism antibiotic resistant bacteria ("super bugs") used to block antibiotics. Exactly what this will lead to in the battle with superbugs is too soon to tell. Yet clearly there is a solution extant in nature and it will show the way to where we very much need to go.

In 1993 the Nobel Prize in Medicine was awarded to a laidback scientist-surfer, Kerry Mullis, for conceiving of the polymerase chain reaction (PCR). The idea behind the reaction was two steps repeated over and over again in a chain reaction: the first step using heat to cause the two strands of a chromosome to separate and then an enzyme causing each missing strand to build its missing partner. In a very short time the material of a genetic sample could be multiplied/magnified enormously.

At the time there was no known heat-resistant enzyme. So no chain reaction. Until the American Type Culture Collection in Maryland – sort of a National Zoo for microbes – was asked if they had anything from a hot environment. They did indeed: a bacterium, *Thermus aquaticus*, from a Yellowstone hot spring (now known to occur elsewhere), and it is the source of the enzyme that makes PCR work.

PCR has revolutionized diagnostic medicine (where it vastly speeds diagnosis) and forensic medicine (where tiny samples can be magnified and identified). It has empowered a huge amount of science, including the human genome project. Nobody has ever tried to put a value on the benefit, but one economist has said it has to be on the order of a trillion dollars.

Most recently I benefitted personally. Two hours out from Dulles airport I woke up to a different kind of sore throat. It didn't seem to go away so five hours later I went to a local clinic. They expected the diagnosis would be negative since I showed no overt symptoms, but they did the test: strep throat. I was correct and being treated before the onset of any symptoms.

Another way biodiversity benefits us is as environmental indicators. Each species is, of course, a unique set of solutions to a unique set of environmental and biological challenges. That means changes in biological diversity can often be the first and most-sensitive indicator of environmental change.

Limnologist Ruth Patrick analyzed the biodiversity (and particular diatoms – algae which build beautiful silica shells) of streams in the mid-Atlantic in the 1940s. She was able to demonstrate unequivocally that the number and kinds of species reflect *not only* the natural physics, geology, chemistry, and biology, *but also* the stresses put on the watershed by human activity (Patrick, 1949). This has become the basis for water quality analysis, for evaluation of the state of watersheds, including at the Environmental Protection Agency (EPA). I think of it as The Patrick Principle: that the divergence from natural biodiversity is a measure of human impact. It can apply on land as well as in the oceans.

A final way in which we benefit from biodiversity is through "ecosystem services" (Daily, 2002). This refers to the functions that biodiversity performs in the natural construct known as ecosystems. It creates a lot of often un-estimated and probably only partly appreciated value. One of the earliest examples is from middle nineteenth century Rio de Janeiro when the Emperor of Brazil ordered what may well be the earliest tropical reforestation project. The watershed of the city was deteriorating largely through inappropriate land use. That restored forest today is the Tijuca forest, the largest urban forest in the world. The Emperor would have made his decision on a practical basis without economic analysis.

In the early 1990s, New York City faced a similar problem with its watershed renowned for the taste of its water. Water quality had deteriorated with land use changes, and the EPA was about to require the city to build a water treatment plant at a cost of eight billion dollars. A better idea was brought to the head of Region 1 for the EPA and won the day: why not restore the watershed, which could be funded at 10% of the cost (with a bond issue to buy up key land rights, restore the biodiversity, and put it back to work)?

There are various critiques of the ecosystem service approach. It has been called "putting a price on nature," which some find offensive. I don't believe that is the case. It is simply bringing some of the value into decision-making. As Pavan Sukhdev (The Economics of Ecosystems and Biodiversity [TEEB]) says: not to do so is to value nature at zero. Another criticism is that it can be hard for the value to outweigh other uses: that may be true, but it adds weight to a decision to support natural systems. Some of them may be hard to calculate, e.g., how the Amazon forest makes half its rainfall; how the floodplain forest of the Amazon (itself 20% of the world's river water) provides the main nutrients for much of the fishery.

Valuing ecosystem services adds recognition to the importance of biodiversity, but it does reduce the value of an ecosystem to its lowest possible denominator. To value a forest for its carbon is like valuing a computer chip for its silicon.

Human impacts on biodiversity

No organism can exist without affecting the environment and ours is no exception. In our early history our impact was largely through hunting and gathering: harvesting animals and plants of importance to the human diet and wellbeing. While it is true that indigenous peoples tend to be more aware of their impact on their natural resources, early human populations – especially those on islands – actually overharvested numerous species to the point of extinction. In Hawaii, for example, there were multiple extinctions of species, such as rails (presumably for food), and Hawaiian honeycreepers for feathers for their cloaks. On Easter Island the local population removed all the trees.

Perhaps the most dramatic example of extinction from hunting is the passenger pigeon of North America. Its population literally existed in the billions, occurring in flocks that would darken the skies for hours. Its biological strategy was to occur without predictability and in such huge numbers that it was invulnerable to predators. That worked until there was a predator (with firearms, a telegraph, and a railroad) that could pinpoint major nesting sites and send the harvested birds by train to market at a penny a piece. The last passenger pigeon died in the Cincinnati Zoo September 1914: a female, named Martha now part of the Smithsonian collections.

Today direct harvest is in all too many instances a story of overharvest. The dire state of most global fisheries is a prominent example (Pauly, 2010). Rhinoceroses and elephants constitute a particularly alarming example because the trade for rhinoceros horn and elephant ivory is illegal and linked to international criminal organizations. In other cases "overharvest" has been deliberate and even official in terms of predator control programs, such as directed against wolves. An interesting aspect of the latter is the remarkable change in riparian vegetation in Yellowstone once wolves were reintroduced: the renewed presence of wolves dramatically reduced the pressure of elk on vegetation adjacent to watercourses and brought back characteristic vegetation and animal species and enhanced water quality.

A second major human impact – habitat modification and destruction – also has a long human history. Initially it was probably caused by use of fire to aid in hunting, but then by deliberate habitat destruction to make way for agriculture during the Neolithic revolution. That in turn made permanent human settlements possible; however, as they expanded they actually took valuable farmland out of production. Today about 40% of the land surface of the planet is under some form of cultivation and largely devoid of the biological diversity that previously existed on those lands. Significant additional amounts of land are desertified or degraded.

Approximately 47% of the original forests of the world no longer exist and the vast majority of the biological diversity of those ecosystems are gone. In my half century of working in the Amazon forest of Brazil, loss has gone from about 3% to 20 % (in an area equivalent to the 48 contiguous United States). Some other habitat types such as American prairie are down to a tiny fraction of their original extent.

Biodiversity hotspots are places in the world where habitat destruction is so advanced that a large number of species are highly vulnerable to extinction. Initially proposed by Norman Myers, and later refined with colleagues like Russell Mittermeier, they are essentially last refuges. They also tend to be areas of high human population so the pressure on the last natural habitat is very high. One way to think of them is where the conservation "fire engines" should be sent first, but addressing the conservation challenge of hotspots is by no means sufficient in itself.

Habitat fragmentation is a third major way humans are affecting biodiversity but was little appreciated as a conservation issue until the 1970s because the impacts were and are not immediately obvious. At that time the theory of island biogeography (MacArthur and Wilson, 2001) endeavored to explain the number of species on islands as an equilibrium between colonization rates and extinction rates. It became a logical next step to think that this might apply to "habitat islands," such as remaining patches of forest. The first major breakthrough involved Barro Colorado Island, a forested island (and Smithsonian field station) created by Gatun Lake (itself was created for the Panama Canal). With half a century of bird records, it was clear that a significant number of bird species had dropped out of the Barro Colorado avifauna because the island was too small.

Habitat fragmentation is now a major field within conservation biology. The forest fragments project at Manaus (Laurance et al., 2011) is the longest running of the habitat experiments started to understand the impacts of fragmentation. Of all the results so far perhaps the most important is that a 100 hectare (250 acre) forest fragment loses half of its forest interior bird species in less than 15 years. (The forest interior birds cue on shade for their habitat and do not easily or willingly leave the dark forest interior.) So fragments shed species and become simpler than they were when part of continuous habitat. There still is an enormous amount to learn about habitat fragmentation and how it works. Today 70% of remaining forest is within 1 kilometre of the edge so habitat fragmentation is playing a huge role in altering the biology of the planet (Haddad et al., 2015).

The fourth way in which humans have been affecting biological diversity is through the introduction of alien species – both deliberate and accidental. An example of the latter is the introduction of the Brown tree snake in Guam, which eliminated a number of native bird species. The chestnut blight was a European pathogen that essentially eliminated the American chestnut from the forests of the Eastern United States. A comb jellyfish from Eastern coastal waters of the Americas arrived in ballast water into the Black Sea where it destroyed a quarter of a billion dollar anchovy fishery by short-circuiting the food chain.

Today the emerald ash borer – accidentally introduced from China in wooden crating – has been causing sweeping losses of American Ash in eastern North America. Cane toads (deliberately introduced from South America) are moving in a front across northern Australia. Introduced pythons of more than one species are taking a huge toll of native bird and mammal life in the Everglades. The Snakehead, an Asian fish, is becoming quite prevalent in the Potomac River and the Chesapeake Bay. In both instances they were released into the wild with no sense of the possible impact by local citizens: in the one case by snake lovers for whom the pythons had become too large to keep at home so they simply released the snakes into the Everglades; the other by Asian-Americans who prized the fish for food and either tired of keeping them or thought it would be great to have a local supply.

A fifth way human activity is affecting biodiversity is through the synthetic chemicals we release into the environment. There are tens of thousands of such chemicals but few understood so well as DDT, the chlorinated hydrocarbon pesticide widely in use around the world in the mid-twentieth century. (Today it is banned in the United States and some other countries, but still used judiciously in some places – such as on bed nets to prevent malaria in Africa and Asia.)

DDT (Wurster 2015) while intended for insect pests is bio-accumulative, meaning that it increases in concentration as it passes up the food chain such that predators at the end of long food chains accumulate dangerous concentrations. It took a bit of scientific sleuthing as many species including the bald eagle, the osprey, and brown pelican went into dramatic declines. The Peregrine falcon, the fastest organism on the planet disappeared East of the Mississippi. Eventually pathological calcium metabolism was identified as the common cause affecting the birds' ability to lay eggs with strong shells. With elimination of DDT in the environment, populations of all those bird species rebounded dramatically.

The perils of toxic substances like mercury have been known for a long time and there are some international agreements that at least partly address this. The roster of synthetic chemicals is huge. For a few we have a fair understanding of their impact. For almost none do we understand the impact of the interactions with other compounds that do not naturally exist in nature. Rivers and streams have dilute concentrations of pharmaceuticals and cosmetic chemicals, cleaning compounds, and the like. Freshwater mussels seem to have stopped reproducing in streams not far north of New York City. Fish in the Potomac River seem to have actually changed sex in some instances. Suntan lotion has recently been identified as contributing to coral bleaching. We are living in a soup of molecules of our own creation with little sense of their implications for the health of biodiversity or ourselves.

Human activity is also distorting levels of naturally occurring compounds such as biologically active nitrogen. There is essentially twice the level of natural occurrence of nitrogen in nature – in part but not entirely from agricultural fertilizers. A lot ends up in run off creating dead zones (basically devoid of oxygen with minimal biological activity and no longer supporting important fisheries) in estuaries and coastal waters, and a notably huge one in the Gulf of Mexico. The number of dead zones globally has doubled every decade for the past four decades.

The final way in which human activity is affecting biological diversity is through climate change from rising concentrations of CO_2 in the atmosphere, which trap radiant heat. The last 10,000 years has been an unusually stable period in planet's climate. That not only nurtured the rise of human civilization but also means the planet's biodiversity and ecosystems adapted to a stable climate.

The fingerprints of climate change can be detected almost anywhere we look in nature. Biological systems are particularly sensitive to change in temperature, associated changes in moisture regimes, and in aquatic systems to changes in Ph or acidity. The oceans are already 0.1 pH units more acid (translated from the logarithmic pH scale = 30% more acid) than in pre-industrial times – which has immense implications for species dependent on the carbonate equilibrium.

Tropical coral reefs are particularly sensitive to climate change. Only short spells of warm water cause the fundamental partnership – between the coral animal and an alga – to break down. The coral animal expels the alga and the result are called bleaching events when essentially the entire living reef shuts down. Chronic bleaching can lead to reef death and this is happening more frequently every year. The Royal Society's major symposium concluded that to maintain tropical coral reefs global concentrations should stay at 350 ppm (about 50 ppm below current levels).

On land there are fingerprints of climate change almost everywhere: changes in annual cycles; changes in geographic distribution. And there are bigger changes, such as the coniferous forest mortality, because climate change has tipped the balance in favor of the native bark beetle.

More important there is a critical lesson in the history of the living planet. Early in life's history before the rise of "higher" forms of life, life was mostly aquatic and single celled. Prominent were the blue green bacteria that formed colonies that fossilize as stromatolites. They were essentially the dominant life forms and oxygenated the atmosphere that made multicellular life including ourselves possible.

One could regard that as an extraordinary ecosystem service, especially since it benefitted us. It is important, however, to remember that oxygen was a toxic waste for these organisms and by oxygenating the atmosphere they basically undercut themselves, made the planet unsuitable for themselves, and only hang on marginally today.

We are doing exactly the same thing as the stromatolites, but with CO_2. So we have a lesson to learn, which is to treat our home as a living planet and manage it (and therefore ourselves) that way.

The intrinsic value of biodiversity

There are multiple other values of the planet and its diversity of life which His Holiness Pope Francis has reminded us of in his remarkable encyclical *Laudato Si*: "each creature has its own purpose." It is, in fact, quite wondrous. As infants we are immediately attracted to other living things and recognize a commonality and kinship with them. Later we come to understand how they are part of ecosystems, and how those systems in turn provide valuable roles in the functioning of landscapes and of the planet.

If not distracted we can come to appreciate that life on Earth and living things are a continuum of change and adaptation to new possibilities. And the outcomes, however transient because of evolutionary change, are almost beyond imagining:

- Humpback whales beaming their song across entire ocean basins;
- Elephants communicating with each other at frequencies below those that humans can hear;
- Tiny migratory birds finding their way from winter habitat to the breeding grounds in different hemispheres;
- South African beetles that use the stars as navigation cues;
- Monarch butterflies doing a multi-generation migratory round trip, which in some fashion is encoded genetically; and
- Or the things that happen at the molecular level inside ourselves that keep us both functioning and – today – provide the tools in laboratories to take on hitherto intractable medical challenges.

Spending a bit of time to understand and know the wonders of nature is something that comes naturally to us. It springs from our evolutionary history and manifests itself as biophlia, an attraction and love for other living things. Edward O. Wilson likes to say "every child has a 'bug period'" and that he or she never grew out of his.

In my own case, while always enjoying the out of doors and being interested in animals, I was really awakened to the wonder of it all by my biology teacher Frank Trevor at Millbrook School in New York. Before I was 15 I understood the outline of life on Earth: what today of course we term biodiversity. And I have never been able to get enough of it ever since.

255

It is not surprising that spending time in nature provides solace and inner peace, or that we find most of Nature beautiful and pleasing to our aesthetics. That is the fly fisherman's secret: to become totally absorbed by the events and life of a stream or river. It is part of the reward of the birdwatcher. That is what drove a big increase in national park visitation after the September 11 terrorist attacks.

The way forward

There is a lot that can be done to provide a better outcome for a sustainable future that in which human aspirations are imbedded in nature. They belong more in the realm of conservation, conservation biology, and environmental science and application with a strong integration with the social sciences and social values and concerns. That is in fact what the Sustainable Development Goals recently approved by the United Nations General Assembly are about.

In the end the success of that will depend on expanding and restoring human recognition of the values for nature. Not in economic terms but what it means to be part of a wondrous living planet in which each living thing actually shares a four billion year pedigree.

We will also need to include the perspectives of the humanities – history, literature, art, philosophy, and religion – to more fully understand how people and cultures have valued biodiversity and can value nature in the future. The growing research and teaching in environmental humanities provides hope and opportunity ecologists will be able to devise more inclusive programs of protecting biodiversity and conserving ecosystems for future generations. We need to aspire to be as wise as we are clever.

References

Daily, G. 2002. *The New Economy of Nature: The Quest to Make Conservation Profitable,* Island Press.
Haddad, N.M., L.A. Brudwig, J. Clobert, K.F. Davies, A. Gonzalez, R.D. Holt, T.E. Lovejoy, J.O. Sexton, M.P. Austin, C.D. Collins, W.M. Cook, E.L. Damschen, R.M. Ewers, B.L. Foster, C.N. Jenkins, A.J. King, W.F. Laurance, D.J. Levey, C.R. Margules, B.A. Melbourne, A.O. Nicholls, J.L. Orrock, D. Song and J.R. Townshend 2015. Habitat fragmentation and its lasting impact on Earth's ecosystems. *Sci Adv*, 1:E1500052.
Laurance, W.F., J.L.C. Camargo, R.C.C. Luizão, S.G. Laurance, S.L. Pimm, E.M. Bruna, P.C. Stouffer, G.B. Williamson, J. Benitez-Malvido, H.L. Vasconcelos, K.S. Van Houtan, C.E. Zartman, S.A. Boyle, R.K. Didham, A. Andrade, and T.E. Lovejoy 2011. The fate of Amazonian forest fragments: A 32-year investigation. *Biol Conserv*, 144:56–67.
MacArthur, R.M. and E.O. Wilson, 2001. *The Theory of Island Biogeography*. Princeton University Press.
Patrick, R. 1949. A proposed biological measure of stream conditions based on a survey of Conestoga Basin, Lancaster County, Pennsylvania. *Proc Acad Nat Sci Philadelphia*, 101:277–341.
Pauly, D. 2010. *5 Easy Pieces: How fishing impacts marine ecosystems*. Island Press.
Wurster, C.R. 2015. *DDT Wars: Rescuing Our National Birds, Preventing Cancer and Creating the Environmental Defense Fund*. Oxford University Press.

27

OCEANS

Carl Safina and Patricia Paladines

Because the ocean connects all shores and gravity moves the world downhill to the sea, one cannot talk about the ocean as separate from the land and atmosphere. The ocean's woes begin on land with us, and, in addition, the carbon dioxide that is warming the air and ocean is also acidifying the seas. Because the ocean contains and reflects rapid changes, we can see how physical changes in oceans create new moral imperatives.

Scale and detail

Oceans are big. Two-thirds of Earth's surface, an estimated 99 percent or so of Earth's habitable living-space. Yet ailing. Ailing in big, oceanic ways.

But must we dwell on the negative? In the comings of the horseshoe crabs, in the migrations of birds and butterflies along the dunes of our coast, in the fish and whales that come and go from our home waters in their seasons, we can feel the driving energy inspiring all living things to strive toward survival. And so we are moved to keep it, restore it. If religions and spiritual traditions have often been about healing, let's bring religions to bear upon the healing of oceans.

Mythic and meaningful relations with oceans have been intuited by many non-Western cultures. Seri people in what is now Mexico and Indonesia's Kai Islanders believe that sea turtles are their ancestors (Safina 2006). Some Pacific Northwest natives believed salmon were people living under the sea who transformed into fish in order to feed them (Safina 1998). Other Northwesterners believe killer whales to be the ocean counterpart to Wolf, with spiritual power (Safina 2015). But for most Westerners since Genesis, everything in the sea was put there for them, for the taking.

Now, we have a new story. Is there anyone who still hasn't heard that oceans produce half the oxygen we breathe? That they absorb a third of the carbon dioxide we produce, and over 90 percent of the heat arising from our industrialized world? (Hoegh-Guldberg et al. 2015). That a large portion of the world's human population depends heavily, even exclusively, on the ocean for economic and food security? It is plain that most people have not heard. Nor have they heard that around three billion people obtain around 20 percent of their animal protein from fish (Hoegh-Guldberg et al. 2015). Or that fish is the lowest priced animal protein in many developing countries (UN DESA). Everyone must eat. Here is another point of meeting between social and ecological justice. The following are just a few factors the oceans face that should be acted upon from a social justice perspective.

Climate changes

As we've raised atmospheric carbon dioxide nearly 40 percent since pre-industrial days, the ocean has absorbed 30 percent of the CO_2 and over 90 percent of the added heat trapped by greenhouse gases. Since the time of Earth's formation over 4.6 billion years ago its climate has experienced various changes, but never at the current rate. Since the Industrial Revolution, release of carbon dioxide through the burning of fossil fuels (mainly coal, petroleum, and natural gas) has elevated the concentration of atmospheric CO_2 from 280 parts per million (ppm) in 1750 (Feely et al. 2004) to around 401 ppm today (NOAA 2015). Just as the glass ceiling on a greenhouse traps the sunlight's heat, the thickened layer of CO_2 and other "greenhouse gases" trap heat in Earth's atmosphere. One of the effects of this increased heat is the melting of Arctic land previously considered permanently frozen, known as "permafrost." Vegetation that had been trapped in ice for thousands of years now lays rotting, releasing methane gas, a more potent heat-trapping gas, which is also released during animal farming and other agricultural practices. The ocean absorbs about one-third of the CO_2 in the atmosphere. The chemical reactions that occur when CO_2 dissolves in seawater makes ocean waters acidic enough to threaten the forms of sea life that make shells (Doney et al. 2009; Hoegh-Guldberg 2015). These alterations to the atmosphere and oceans will increasingly affect all life on Earth.

Rising ocean temperatures are changing the global distribution of many fish species and other sea life (Nye et al. 2009; Cheung et al. 2010). Some models predict near-surface sea temperatures will increase one to three degrees Celsius by 2100 (Meehl et al. 2007). Geographical ranges of many warm-water fish are expanding toward the poles, while the range of cold-water dependent species is shrinking (Barange and Perry 2009). Rising ocean temperatures will negatively affect coral reefs—nursery grounds and refuge for a vast array of sea life. Within the coral polyp, most corals harbor photosynthesizing algae. The corals provide the algae with shelter; the algae provide the coral with sugars. When water temperatures increase past their comfort zone, the algae depart their coral hosts, leaving coral stark white, an effect called "bleaching." Coral can survive in a bleached state for six to eight weeks but longer periods cause the coral to starve and die. Reports of coral bleaching are new. Between 1876 and 1979 scientists recorded only three bleaching events (Sammarco and Strychar 2009). In 2002 more than half of the world's barrier reefs experienced bleaching, with most of the affected sites experiencing large-scale die-offs. (Sammarco and Strychar 2009). Mass coral deaths are biological catastrophes that also disrupt the lives of people who depend on reefs for food and economic activities, including tourism.

Rising sea levels from melting land-ice and thermal expansion of water threaten all coastal communities. Seas are rising at an unprecedented rate. The rate of rise in the last 20 years is averaging 3 millimeters per year, nearly twice the twentieth century rate. As human activity continues to release greenhouse gases, the melting of land-based ice will add to sea level rise. Predictions range from 20 to 50 centimeters by 2100 (Bindoff et al. 2007). Coastal habitats such as mangroves, marshes, and human dwellings will experience increased flooding, degradation, and even complete submersion.

Fishing

Though increased greenhouse gases are changing the oceans' heat and acidity, the main depletion of ocean life to date results from fishing. Fishing boats annually remove approximately 80 million tons of sea life, supporting over 500 million people, the majority of whom live in developing countries. Overfishing remains the most transformative activity we do to ocean life, having depleted most targeted populations to between a third and a tenth of their former abundance (Myers and Worm 2003)—thus threatening the food source of millions of people.

For the majority of people around the world, particularly in developing countries, seafood is a main source of protein and important nutrients. Fishing employs 45 million people directly and 180 million others through peripheral activities such as canning, transport, and the restaurants that serve seafood. Fisheries generate roughly US$95 billion annually. Technology facilitates removal of unprecedented quantities of aquatic life. People take over 90 million tons of marine life annually. Over 80 percent of the world's fisheries are either fully or overexploited (FAO, 2010).

Dead zones, toxicity, disease

Seagrass meadows are home to many fishes, turtles, and marine mammals. Fertilizer runoff and excess sedimentation is causing a global decline (Cullen-Unsworth et al. 2014 and Short et al. 2014). Oyster reefs have nearly disappeared due to these same stresses plus uncontrolled exploitation of oysters for food. (Ermgassen et al. 2012). The sea now has around 170 oxygen-starved areas, known as "dead zones" because most sea-life near the seafloor cannot survive there. These dead zones are found mostly on the US East Coast, Northern Europe, and Southeast Asia but they exist in many more places too. (Boesch 2008; Nellemann et al. 2008). Synthetic fertilizers have doubled the global nitrogen flow to living systems since the 1960s. The extensively documented dead zone at the mouth of the Mississippi River, first discovered in the 1970s, now covers up to 22,000 square kilometers.

Plastics in the sea

Many of us grew up searching for beach glass along the shores but during the past 50 years glass has been increasingly replaced by plastic. An estimated 5.25 trillion plastic particles weighing 268,940 tons are found in the ocean and coastal shores (Eriksen et al. 2014). This is constantly increasing. Plastic is a death threat for many marine animals as it is often mistaken for food often resulting in death.

Seafood and mercury

Today human activities account for nearly ten times the mercury that is released from natural sources (such as volcanoes). The main source of mercury is the burning of coal, which contains mercury as a natural impurity. Asia emits two-thirds of human-generated mercury; China is the world's largest mercury polluter. More than half of the mercury released in Asia falls in North America. Similarly, around two-thirds of the mercury that first gets airborne in the US lands in Europe. When airborne mercury falls to Earth, much enters water. There bacteria convert it to the compound methylmercury. In this form, mercury first enters the food web, absorbed by algae. Moving up the food pyramid it gets increasingly concentrated as it goes from algae to zooplankton, to small fish, then larger fish and whales. Methlymercury is not soluble, therefore not easily excreted. It has an affinity for adipose tissue where it lodges. The bigger and longer lived the fish, the more methylmercury its body will contain. A herring will contain less mercury than a shark. Because mercury is particularly dangerous to nervous system development in fetuses, fish consumption during human pregnancy may have serious implications.

Shipping moves hundreds of invasive species including seaweeds, shellfish, and fishes.

Protected areas can't protect from chemicals, plastic, warming, acidification, and sea level rise. We might designate boundaries around areas of exceptional ocean productivity, but plankton and fish are shifting poleward as waters warm (Nye et al. 2009; Cheung et al. 2010). Coastal policies do nothing to stop nitrate fertilizers from running off Great Plains farms into the Mississippi drainage a thousand miles from tidewater, creating a massive dead zone in the Gulf. Environmentalists seem always to be wagging their fingers, scolding people, demanding sacrifice, saying the sky is falling.

What if the sky really is falling?

World political and religious leaders cannot even agree that these are problems, never mind what the action should be. This tells us that it is not technology that holds us back, it is morality—our valuation of the world. Our inability to succeed suggests an incapacity for dealing with the global-scale problems we now create. What we learn by creating so practical a problem is that lack of restraint is wrong. Restraint, a practical imperative, becomes an ethic. What is practical becomes what is moral.

But how to practically restrain ourselves, how to heal these wounds? How to heal the overfishing that—though tamed in US waters by new federal laws mandating recovery (Safina et al. 2005)—continues raging through the world's depths and along hungry coasts? How to *deal with* the warming of the oceans, the melting of polar-regions, the intensified storms? How to stop the rising sea from taking shorelines from horseshoe crabs and sea turtles and shorebirds whose very existence will be squeezed as seas push up and humans build walls and "the beach" disappears?

How to wrap a human mind around the acidifying seas, the consequent inability of oyster larvae to form shells, the prospect that the world's great coral reefs may, in a century or so, begin to *dissolve*? The starving poor struggle for life by felling forests and scouring reefs, while the overconsuming rich commission much of that extraction. For just one example of how people kill their own golden-egg-laying geese, consider that mangrove habitats provide over 100 million people with essentials ranging from fish to wood to protection against storms, and the rate of mangrove destruction is three to five times greater than the average global forest loss (Hoegh-Guldberg et al. 2015).

What if the sky isn't falling?

Despite the constant drumbeat of environmental *gloom and doom*, we are struck by how much beauty and vitality the world still holds. On our home coast we see recovering fish populations, the now ubiquitous Ospreys and now-common Peregrine Falcons and recovering Bald Eagles that, when we were young, had been almost totally erased by DDT and other organochloride insecticides. We see it in recovering populations of whales and sea turtles. Sizeable places still brim with life, granting sanity and solace, delight and hope, and perhaps enough resilience to keep the world for those who'll follow.

Oceans connect all shores; gravity brings all things downhill to the sea. Oceans connect by receiving. Even the most oceanic of all the oceans woes—overfishing—starts and ends not in the ocean but at the cleats that tie boats to shore and the gravity-defying inland markets by which fish flow uphill; it begins in the minds of people and ends in the bellies of people.

As oceans connect, the problems lie upstream. The problems seem new but almost all stem from an old way of thinking, from an ancient valuation of the world that did not value the world much at all. Since the first human chipped the first stone tool, we have always kept the part we wanted and discarded what we didn't want.

We don't want to pay for the part we don't want, so we throw the problem "away." But "away" is us. Profits, we privatize; costs we socialize. What physics and ecology have taught us, evolution has taught us, and even the golden rule tried to teach us: there is no free lunch. People who are not even born yet, will pay for us (Safina 2009).

Other species pay most. The costs of destruction, and the risks, are all mis-valued. America once ran its economy on the energy of slaves. The war over slavery was a war over the morality of cheap energy. We lacked moral clarity about slavery. We lack moral clarity now about fossil fuels. Fossil fuels cause worldwide problems that we are demonstrably incapable of solving. The fossil fuel industries do nothing to help solve the problems; they do everything that perpetuates them.

Trajectory in our times

Conservation and the environment receded from public priorities during the decades starting in the early 1980s. Today's tortured discourse over climate is appalling compared with the broad, effectively focused consensus of environmental concern of the 1970s in the US, which brought the banning of DDT-type pesticides and the explosion of major federal environmental legislation, including the Endangered Species Act, Clean Air Act, Clean Water Act, Toxic Substances Control Act, Fisheries Conservation and Management Act, and others.

By the late 1980s as bigger and bigger cars wasted more and more of the oil we had to drill harder and harder for, concern about the environment became positively unfashionable. This was partly the fault of environmentalists. Conservation often seemed to be presented as a contest and a dichotomy: people versus nature. It seemed environmentalists were always asking people to do with less or sacrifice—for the benefit of something else. To many, conservation seemed divisive. Worse, anti-people.

Our relationship with the world (What relationship?)

We run our civilization on anachronistic valuations developed in archaic times, when no one knew the world was round and finite. Our economic system uses the accounting model developed centuries ago. The religions and the philosophies underlying Western thought: thousands of years old.

Only in the last century have we comprehended that life is evolved and related, learned of carbon and nitrogen cycles, animal population dynamics, sources and rates of water replenishment.

Our economic, religious, and ethical institutions have mechanisms for resisting change; to endure over time, they had to. They often lack mechanisms for incorporating discoveries as re-valuations. They haven't assimilated the breakthroughs: that resources are finite, and creatures fragile. They haven't adjusted to realizations about how we're pushing Earth's systems into dysfunction. (Safina 2011).

Rather, our economy sells the Ponzi scheme promising and *requiring* endless growth on a planet that does not grow. Nearly unquestioned even now is the assertion that more of us is desirable and any prospect of fewer people is a desperate economic problem. Incentives granting preeminence to short-term gain *are the main threats* to long-term gains. As we shrink forests, empty oceans, alter the atmosphere, endanger our children's times, and risk so much of the world—it all gets *added* to that grossest valuator: Gross Domestic Product. This is a system inextricably invested in destruction on a planetary scale.

Growth—making something bigger—requires putting more material through the system at faster rates. Improvement means making things better. Betterment does not require growth.

Growing can undermine improvement. Bigger communities are not always better. Insisting that *bigger is how to measure better*—motivates choices threatening the stabilities of the economies of nature *and* finance.

On climate, population, extinction—pulpits are mainly silent. Some even *welcome* the running down of the world as sign of the Second Coming; "Good News" is the acceleration of destruction. In effect, our philosophy of living, our religions, and our economic systems regard the world as given, nothing to venerate. Yes there are many exceptions and nice quotes—but look at the trends. Despite overhauled ways of knowing and revolutionized knowledge, old thinking prevails. As the word "dogma" indicates, our valuation systems doesn't say, "As we learn, so will we adjust."

Here's the real Good News: the science of ecology continually elaborates its core discovery: the connection of all. Because the greatest thing a human being can experience is a sense of connection, there is a joyful coincidence here. It means that what we need to do is what our inner soul and human spirit and all religions want: just connect.

This is no glib assertion. The bankruptcy of nature will end our quest for peace (Safina 2009). It's suicide. For our children, it is murder. We sense this dread already. Contrary to the optimism of our parents for us, their children, nowadays fewer parents sense bright, assured futures for our children.

Despite major differences in factual understandings and beliefs, science and all ethical, moral, and religious traditions converge in agreement here: the place is ours to use, not ours to lose. All wisdom traditions say we serve each other, the creation, and our children.

Hope, if not faith

Carbon from fossil fuels will be a defining issue for the rest of our lives. (Population is a more fundamental problem, but family planning and improving access to opportunity for women is easing that problem in significant parts of the world.) Carbon draws a wider net than we've ever seen. It tells us, with a clarity we have never experienced, that we are all prisoners in the net of our own ingenuity.

If population underpins everything, carbon overshadows everything. It would be difficult to imagine a broader chronic issue. (Catastrophic nuclear war would be worse, but our problem with carbon results from everyone's peaceful day-to-day living.) To run civilization we've created a productive fossil-fuel engine that just happens to be destabilizing the planet's life.

Information abounds, but information alone does not move most people because information alone does not *change* values. Information changes. Values last lifetimes. Values determine how new information will be incorporated into living. Rather than try to change values, we might be more effective working within people's existing values context. Science and conservation must be communicated not just as information, but as information delivered to people along their value-based channels of understanding. Story, personal experience, metaphor, allegory, anecdote, imagery, community, ceremony, song, theater, faith—think of it as a question of *translation*.

Sacred as ecological valuation

Scripture's scribes considered future children among the highest forms of goodness. "A good person bequeaths to their children's children" (Proverbs 13:22). The Book of Deuteronomy (20:19–20) and the Qur'an teach that in time of war, it is prohibited to destroy fruit trees; our disputes must not impoverish our children. So for economists to discount the future is to cheapen God's eternal presence, a sin against both God *and* coming generations. The future has higher value than the present, because the future is bigger, with more lives at stake.

One religious voice addresses this topic, so startling that it might become the game-changer it clearly aims to be: Francis' encyclical of June, 2015, titled *Laudato Si: On Care For Our Common Home.* Spun by the media as "about climate change," it is about climate change the way a supermarket is about pasta. Yes, there's a section featuring it, but it offers much more. Francis' encyclical is about the human relationship with nature, its current disfigurement, greed as a main source of that disfigurement, how disfigured relationships with nature generate injustice, and he concludes with the re-valuation he believes the world is in need of. Francis' strikingly bold language—e.g. "The earth, our home, is beginning to look like an immense pile of filth"—calls for, "all that is authentic rising up in stubborn resistance... for us to move forward in a bold cultural revolution." Even secular readers can resonate with this (Safina 2011).

Concerns that ocean ecologists have documented over decades, Francis *elevates to religious imperative.* He notes that oceans contain, "most of the immense variety of living creatures, many of them still unknown to us and threatened... Marine life, which feeds a great part of the world's population, is affected by uncontrolled fishing, leading to a drastic depletion... Many of the world's coral reefs are already barren or in a state of constant decline. 'Who turned the wonderworld of the seas into underwater cemeteries bereft of color and life?'" (Catholic Bishops 1988). Seeing the global connections, Francis adds, "The same mindset which stands in the way of making radical decisions to reverse the trend of global warming also stands in the way of achieving the goal of eliminating poverty." When a *pope* comes looking for "radical decisions," you know the world is different.

Clearly, science and religion need not be natural enemies. Many scientists experience wonder so profound that their feelings widely overlap those of religious people, and might, in their devotional consistency and moral imperative, be called religious in the broad sense. Likewise many religious people have scientific curiosity. "In compassion," says spiritual author Karen Armstrong, "we de-throne ourselves from the center of our world." In science, much the same. Remember Copernicus' deduction that sun and stars do not revolve around us after all; Darwin's perception of our relatedness; and Aldo Leopold's insight that our deep community is people *plus* all that supports us (Safina 2011).

Acting for the world as good

Scientific information can inform and illuminate choices, or point out likely consequences as best we understand the facts. But we will always choose based on values. Knowing that a certain decision will eliminate a species, or a course of action will destabilize the planet's climate and the sea's pH, does not tell us whether it's bad to eliminate a species or wrong to destabilize the planet. Some welcome the end of the world based on their religious beliefs. What we must do is to decide that the world is good, that survival is good. That is a religious valuation. We must elevate to the realm of sacred the work that seeks to allow life to continue. We must fuse our science-derived understandings of facts with a fundamental perception—an *apprehension*—of the world as something vast in space and deep in time, something that passes through us momentarily in the brief spark of our own lives, something that we might help steward but must never harm, something not ours—something *sacred.* Then, *only* then, might we choose well and wisely.

References

Bindoff, N.L., J. Willebrand, V. Artale, A. Cazenave, J. Gregory, S. Gulev, K. Hanawa, C. Le Quere, S. Levitus, Y. Nojiri, C.K. Shum, L.D. Talley and A. Unnikrishnan (2007). Observations: Oceanic Climate

Change and Sea Level. In: S. Solomon, D. Qin, M. Manning, Z. Chen, M. Marquis, K.B. Averyt, M. Tignor and H.L. Miller (eds), *Climate Change 2007: The Physical Science Basis. Contribution of Working Group I to the Fourth Assessment Report of the Intergovernmental Panel on Climate Change*. Cambridge University Press: Cambridge, UK and New York, NY, US.

Boesch, D.F. (2008). Global Warming And Coastal Dead Zones. *National Wetlands Newsletter* 30, 11–21.

Cheung, W.W.L., V.W.Y. Lam, J.L. Sarmiento, K. Kearney, D. Zeller and D. Pauly (2010). Large-Scale Redistribution Of Maximum Fisheries Catch Potential. In: *The Global Ocean Under Climate Change. Global Change Biology.* 16:24–35.

Cullen-Unsworth, L.C., L.M. Nordlund, J. Paddock, S. Baker, L.J. McKenzie and R.K. Unsworth (2014). Seagrass meadows globally as a coupled social-ecological system: Implications for human wellbeing. *Mar Pollut Bull* 83:387–97.

Doney, S.C., V.J. Fabry, R.A. Feely and J.A. Klypas (2009). Ocean acidification: the other CO_2 problem. *Annu Rev Mar Sci* 1:169–92.

FAO (2010). *The State Of World Fisheries And Aquaculture 2010.* Food And Agriculture Organization Of The United Nations, Rome.

Francis (2015). *Encyclical Letter Laudato Si' of the Holy Father Francis on Care for Our Common Home.*

Hoegh-Guldberg, O. (2015). *Reviving the Ocean Economy: The case for action.* WWF International, Gland, Switzerland, Geneva.

Meehl, G.A., T.F. Stocker, W.D. Collins, P. Friedlingstein, A.T Gaye, J.M. Gregory, A. Kitoh, R. Knutti, J.M. Murphy, A. Noda, S.C.B. Raper, I.G. Watterson, A.J Weaver and Z.C. Zhao (2007). Global Climate Projections. In: S. Solomon, D. Qin, M. Manning, Z. Chen, M. Marquis, K.B. Averyt, M. Tignor and H.L. Miller (eds), *Climate Change 2007: The Physical Science Basis. Contribution of Working Group I to the Fourth Assessment Report of the Intergovernmental Panel on Climate Change*. Cambridge University Press: Cambridge, UK and New York, NY, US.

Myers, R.A. and B. Worm (2003). Rapid Worldwide Depletion Of Predatory Fish Communities. *Nature.* 423:280–3.

Nellemann, C., S. Hain and J. Alder (2008). *In Dead Water – Merging Of Climate Change With Pollution, Over-Harvest, And Infestations In The World's Fishing Grounds.* United Nations Environment Programme, GRID-Arendal, Norway.

Nye, J.A., J.S. Link, J.A. Hare and W.J. Overholtz (2009). Changing spatial distribution of fish stocks in relation to climate and population size on the northeast United States continental shelf. *Mar Ecol Prog Ser* 393:111–29.

Safina, C., A.A. Rosenberg, R.A. Myers, T. Quinn and J. Collie (2005). U.S. Ocean Fish Recovery; Staying The Course." *Science* 309:707–8.

Safina, C. (2006). *Voyage of the Turtle.* New York: Henry Holt Co.

Safina, C. (2009). "Too big to fail." *Ocean Geographic.* February.

Safina, C. (2011). *The View From Lazy Point.* New York: Henry Holt Co.

Safina, C. (2015). *Beyond Words.* New York: Henry Holt Co.

Sammarco, P.W., and K.B. Strychar (2009). Effects of climate change/global warming on coral reefs: Exaptation in corals, evolution in zooxanthellae, and biogeographic shift. *Environmental Bioindicators* 4:9–45.

Short, F.T., R. Coles, M.D. Fortes, S. Victor, M. Salik, I. Isnain, J. Andrew and A. Seno (2014). Monitoring in the Western Pacific region shows evidence of seagrass decline in line with global trends. *Mar Pollut Bull* 83:408–16.

UN DESA (undated). Ocean Acidification: A Hidden Risk for Sustainable Development. *Copenhagen Policy Brief No. 1.* Online.

Zu Ermgassen, P.S.E., M.D. Spalding, B. Blake, L.D. Coen, B. Dumbauld, S. Geiger, J.H. Grabowski, R. Grizzle, M. Luckenbach, K. McGraw, W. Rodney, J.L. Ruesink, S.P. Powers and R. Brumbaugh (2012). Historical ecology with real numbers: past and present extent and biomass of an imperilled estuarine habitat. *Proc Biol Sci* 279:3393–400.

Carl Safina, an ecologist, is founder of The Safina Center at Stony Brook University, New York, United States, where he is the inaugural holder of the Endowed Chair for Nature and Humanity and steering committee co-chair of the Alan Alda Center for Communicating Science. Patricia Paladines works in the Institute for the Conservation of Tropical Environments at Stony Brook University. Their marriage survived the writing of this chapter.

28

CONSERVATION AND RESTORATION

Gretel van Wieren

This chapter tracks central debates and recent changes within discussions of conservation and restoration, with particular emphasis on critical ethical issues that face the practice of ecological restoration as we enter the climate era. It focuses on three key areas of emergent concern: meanings for restoration, novel ecosystems, and public participation. While the role of religion has not been explicitly considered in these discussions, I argue that it should be, for restoration debates often have unavoidable religious dimensions, lead into religion-like interpretation, or otherwise can be deepened by religious analysis. A final section therefore specifically treats the multiple ways in which religious and spiritual interpretations may play a role in the restoration enterprise. Even as this discussion is largely limited to a consideration of ecological restoration practice in North America, as well as to a treatment of western philosophical and Christian approaches in particular, it does not, of course, necessarily preclude its potential relevance for other contexts and traditions, though it does limit the types of sources that are drawn upon, and, in turn, the sets of the questions that these sources help to form. What this emphasizes is the extent to which the religious views, values, virtues, and norms adopted in other cultures for restoration vary, as do restoration's meanings more broadly. With this in mind, I conclude that religious communities and traditions may beneficially contribute to public dialogue about human interventions in ecosystems insofar as they are explicitly shaped by ecological beliefs and practices.

Shifting meanings for restoration

Defining restoration has never been an easy task, given that meanings are numerous and diverse, often varying dramatically from ecosystem to ecosystem and culture to culture. As a vernacular practice, perspectives on restoration shift according to the types of ecosystems (forest, grassland, wetland, river), degradations (deforestation, erosion, toxification, species loss), and repairs (bioreactivation, recontouring of land or waterways, reintroduction of native species, removal of exotics). Additionally, definitions vary depending on understandings of an ecosystem's predisturbance condition, the environmental features that are selected for regeneration, and the goals that are determined for a particular restoration project. Further complicating any definition of restoration is the fact that restoration is not a new phenomenon, broadly understood, but has a long and varied history, reaching back at least as far as there is record of people interacting intentionally to maintain their natural environments by, for example, shifting crops, fallowing land, or managing certain animal and plant species for consumptive, medicinal, or spiritual purposes (Van Wieren 2013, 35–54).

The current version of the International Society for Ecological Restoration (SER) – the largest professional society of restorationists globally – definition that most restorationists continue to rely upon reads thus: "Ecological restoration is the process of assisting the recovery of an ecosystem that has been degraded, damaged, or destroyed" (SER International Primer, 3). This barebones definition is accompanied by an expanded description of restoration in "The SER International Primer on Ecological Restoration," which includes attention to sustainable cultural practices, especially in relation to developing countries and indigenous perspectives on restoration. The Primer, for example, states: "Some ecosystems, particularly in developing countries, are still managed by traditional, sustainable cultural practices. Reciprocity exists in these cultural ecosystems between cultural activities and ecological processes, such that human actions reinforce ecosystem health and sustainability" (2). It continues by specifying the role of indigenous ecological management practices in promoting the ecological and cultural survival of indigenous peoples. Further it states that the North American emphasis on the restoration of pristine ecosystems is indefensible in large parts of the landscapes of, for instance, Africa, Asia, and Latin America where land-use must be manifestly tied to bolstering human survival.

Yet the context of climate change has pressed definitional concerns in restoration thought in tight corners. Long-standing moral concerns regarding the role of human agency in natural processes become trickier, particularly in terms of determining just how much and what types of technological interventions in earth's systems are justifiable. There are concerns about whether ecological restoration should mostly focus on the restoration of natural and social capital for human purposes, or on the regeneration of "classic" ecosystems for their own sake. The pervasiveness of invasive species in many regions has caused restorationists to debate whether they should shift attention to the emergence of hybrid and novel ecosystems. Tracking these debates, former SER president and restoration philosopher Eric Higgs has suggested that three substantial changes around how restoration is understood have occurred in the past decade. These include the ideas that multiple values and practices, social, cultural, political, economic, and aesthetic are requisite to successful restoration efforts; cultural variations should define restoration goals so that, for example, landscapes in the global South that are characterized by small-scale agriculture and subsistence farming are restored differently than wilderness areas in North America; and meanings for restoration will need to shift in the face of climate change and the proliferation of hybrid and novel ecosystems (Higgs 2012).

Despite Higg's implicit assertion that that there appears to be somewhat of a global consensus around how the restoration enterprise is understood, there remains a remarkable amount of ferment around defining its meaning. This is particularly the case in discussions about hybrid and novel ecosystems. Some philosophers such as Ronald Sandler have suggested that given the current context of climate change and adaptation the term "ecological restoration" be jettisoned in favor of the less historically oriented term "assisted recovery" (to restore, after all, implies that something from the past was lost that can be recovered) (2012, 63–80). Higgs has suggested that terms such as "regeneration" and "reconciliation" may more adequately reflect the reality of change, though still, he argues that "restoration" remains the term that best describes the present practice and also "may have the medium-term tactical advantage of forcing us to stare more closely at history" (2012, 98). William Throop argues for "healing" as the best moral, though not empirical, metaphor to guide our approach to intervening in damaged ecosystems in light of climate uncertainty (2012, 55). I have written elsewhere that "redemption" may be a good religious term, and Western cultural term more generally, to refer to restoration activities (2013, 174–9).

What each of these attempts at revision points to is the *re-storying* that is underfoot and will need to continue to be worked on in order for meanings for the human relationship to the natural world to develop in ways that make adaptive sense. This raises additional questions about the particular types of values, virtues, and norms, and the particular practices and narratives that will shape them in order to school persons to live well with their ecologies. Thus, we turn next to a second, closely related set of debates facing the restoration enterprise: that of novel ecosystems and the virtues requisite for living well within them.

Novel virtues for novel ecosystems

Perhaps the most trenchant debate that has emerged in conservation and restoration thought over the past decade has been around the matter of novel ecosystems. Novel ecosystems are "systems that differ in composition and/or function from present and past systems as a consequence of changing species distributions, environmental alteration through climate and land use change and shifting values about nature and ecosystems" (Hobbs, Higgs, and Hall 2013, 4). While there is recognition that the idea of novel ecosystems, broadly construed is not knew, given longstanding insights regarding ecological change and the existence of anthropogenic landscapes, there is also growing awareness that the rate and pervasiveness of current environmental alterations has forced new dynamics on ecosystems, so that the possibilities to return an ecosystem to a less altered state become more difficult, if impossible, without heroic measures. Some restoration scientists and practitioners have insisted that restoration theory and practice forgo consideration of novel ecosystems altogether, instead maintaining focus on the practice's defining features: attention to historical continuity, native species, and ecological integrity. Others have suggested that the reality of rapidly changing ecosystems demands consideration of the emergence of novel ecosystems, and that the restoration enterprise depends on reinventing itself in the face of such change. Still others have taken a both/and approach to the issue.

Central to this debate is the matter of historical fidelity. Popularized by Higgs in his now classic book, *Nature by Design: People, Natural Processes and Ecological Restoration*, "historical fidelity" relates to the idea central to restoration practice to be "true" and "accurate," "faithful," in a sense to the reproduction, even if not exact, of an ecosystem's predisturbance conditions (2003, 127). But what happens when it is no longer possible to determine conventional historical referents in order to "reset" and assist the recovery of an ecosystem's trajectory? Should historical considerations then be neglected in restoration efforts, and if so, can we really call what we are doing restoration?

Developing responses to such questions has provided fertile ground for philosophers working on issues of restoration and climate change. In a 2012 volume edited by Allen Thompson and Jeremy Bendik-Keymer, *Ethical Adaptation to Climate Change: Human Virtues of the Future*, Sandler argues that rapidly changing climatic conditions, which will require greater openness and accommodation in our interaction with ecosystems, weaken the justification for historical fidelity in restoration practice (72). In the same volume, Throop conversely suggests that our rapidly changing and uncertain ecological future demands greater attention to historical fidelity as a safeguard against hubris and gluttony in our manipulation of ecosystems (47–62). Higgs too favors continued attention to historical fidelity, though he revises his original conception to include seven different ways in which we might use history in our interventions (2012, 91–3). These include, for example, the ideas that historical knowledge can inform restoration efforts aimed at postcolonial reconciliation where disruption of traditional ecosystems and life-ways are being redressed, and that it can serve as "a governor on our exuberant ambitions," in the sense that researching long-term change and preexisting conditions may help us exercise prudence in determining future courses of action (93).

Taken together, a pluralistic view of history, and of restoration more generally, provides the type of diverse approaches to human intervention in ecosystems necessitated by the shifting climate context. Yet simply acknowledging the multivalent values that new applications of history may generate in understanding people's relationship to natural processes is not enough to guide responsible action in a rapidly changing and uncertain world. What we will need is for these values to be translated into particular virtues or norms fit for humans attempting to live ethically in evolving ecologies.[1]

Thompson has called such virtues, "novel virtues," for "how at least some environmental virtues of the future – the virtues of those living an ecologically sustainable form of life – may be quite different from the environmental virtues of today" (2012, 204). Fitted within a humanist view of adaption, which involves "adjusting our conception of who we are to appro-priately fit the new global context," Thompson suggests "environmental responsibility" as a cardinal virtue (7). Modeled on Iris Marion Young's social notion of collective, political responsibility, Thompson's environmental responsibility is part of, yet unique, among a larger suite of "virtues of stewardship," a "collection of character traits that are excellences of char-acter for individuals who have the role of environmental steward, which…is a responsibility shared by all humanity" (217).[2] Environmental responsibility in particular, according to Thompson, involves the moral capability to manage well "the satisfaction of a plurality of normative demands derived from our accountability for the basic conditions of life on Earth" (217–18).

Throop suggests additional virtues to guide human agency in the context of Anthropocene (2012).[3] Using healing as a primary moral metaphor for restoration activities, virtues that follow include *humility*, which involves recognition of the limitations of our ecological knowledge, especially given present uncertainties regarding the ecological future; *self-restraint*, a correlative to humility that suggests the need to quiet our own interests in order to respect and satisfy the needs of the other, and; *sensitivity*, which in ecological context implies attentiveness to the idio-syncratic particularities of ecosystems, as well as the ways in which they interrelate and overlap with other systems in the locale.[4] Acting in accordance with these three virtues will tend to promote conservative restoration goals characterized by a high degree of historical fidelity, Throop concludes. Humility, for example, suggests avoidance of approaches that claim certainty that it is possible to create some new, better ecosystem that will function with certain predictable results; self-restraint will tend to suggest approaches that are more oriented by reset-ting the autonomy of natural processes and systems prior to human disturbance, and less oriented by satisfying immediate human desires and interests. These same virtues, nevertheless, may suggest the adoption of restoration goals that favor novel forms; sensitivity, for example may point to a type of novel evolution that is less disruptive to the overall system (56–8).

The emergence of the type of moral model just noted represents among the most interest-ing edges of conservation and restoration ethics in my mind. But there are additional critical discussions that have developed over the past decade that warrant consideration in thinking ethically about human adaptation to rapidly changing ecosystems. These revolve around issues of public participation in the conservation and restoration of particular natural lands. Thus we turn in this next section to examine what has been referred to in the restoration literature as the "participation paradox" (Throop and Purdom 2006). For despite the fact that ecological restoration practice is often touted as a conservation strategy that can materially connect humans with their ecologies, this requires that people, in fact, participate in restoration efforts, something that for various reasons is not always the case.

The paradox of participation in restoration

The matter of citizen participation in ecological restoration efforts has been a topic of significant debate in restoration thought since the mid-1990s. Beginning in the early 1980s, restoration's first philosophers, Robert Elliot and Eric Katz, critiqued ecological restoration efforts for "faking nature" (1982) and as a "big lie" (1992), respectively. Early proponents of ecological restoration countered by arguing that restoration's value as a conservation strategy could be redeemed for how it had the distinctive capacity for directly involving people in caring for their local and regional ecologies and therefore beneficially contributing to a culture of nature. Philosopher Andrew Light (2005), for example, called restoration's public participatory element a form of ecological citizenship, while Higgs (2003) called it focal practice. Both ways the model of "good" ecological restoration put forward relied on public participation of local citizens, what has since been termed participatory restoration.

Despite its generally positive elements, public participation in restoration activities has nevertheless been called into question in the past decade, and along two lines. The first concern relates to wilderness restoration efforts. As Throop and Purdom have suggested, it may be that some wilderness contexts necessitate the minimization of human presence, and thus of public participation, as a way to respect wild ecosystems and promote historic preservation norms (2006). This is especially the case in areas that are defined by principles laid out by the U.S. Wilderness Act of 1964, where wilderness is understood as a place that is "untrammeled" by humans and that appear to "have been affected primarily by nature, with the imprint of man's work substantially unnoticeable" (493). This is in contradistinction to the ideal of participatory restoration, write Throop and Purdom, which involves the integration of humans in ecosystems, so making, in Higgs' terms "the connections between culture and ecology, people and place, prominent" (494). More often that not, public participation in restoration efforts requires engagement in projects for longer periods of time and in greater numbers. Were this not the case, Throop and Purdom still believe that "it is hard to see how a view of good restoration that celebrates human manipulation of an ecosystem can be reconciled with an ideal of wilderness that minimizes the 'imprint of man's work' on the landscape" (494). That is unless one takes a more permissive view of wilderness, as has Higgs, following Cronon, Callicott, and others (Callicott and Nelson 1998). This view insists that there are no natural areas, including wilderness, that are free from human manipulation, and furthermore, that the categories of culture and nature should not be divided along such strict, ontological lines. The notion that nature is "Other" is over, wrote Callicott over two decades ago (Callicott 1992). We should not conceive of conservation or restoration efforts, even in wilderness, where humans are understood as distinct from ecosystems.[5]

Still, there remains something valuable in the idea that public participation in certain wilderness restorations – and perhaps other context as well – should be minimized for nonhuman nature's sake. This has less to do with analytical philosophical arguments and more to do with the intuitive religious idea of the de-centering of the self. Closely related to the virtue of humility, which Throop and Purdom cite echoing wilderness activists, Dave Foreman and Reed Noss, restorationists themselves at times draw on the idea of a de-centered self to describe the affective feelings elicited through the experience of working directly with nonhuman natural processes and species. Restoration work can enable a de-centering of the self through the realization that humans are dependent on and interdependent with larger nature. An important part of this de-centering is the recognition that humans are limited and finite in relation to the larger community of life. Restorationists perform their work with the dual, at times tension-ridden, realization that while humans are part of and play an active role within the natural order of creation, they are not the center, or even always comfortable members, of ecological systems or processes.[6] In a double paradox of participation, then, it may be that the quality of de-centered selfhood and the closely related virtue of humility in relation to the natural world are importantly formed through direct involvement with the regeneration of wounded landscapes, though necessitate the "pulling back" of engagement in certain settings. The nub of the question, of course, is just what contexts constitute human retreat, and conversely which ones necessitate so much technical intervention that public participation is neither feasible nor desirable. The latter part of the question points to a second concern about public participation in restoration. An anecdotal example illustrates.

In 2014, I gave a public talk on ecological restoration as a form of public spiritual practice at The Ohio State University in Columbus, Ohio, where the university is engaged in a massive restoration project of the Olentangy River corridor, part of which runs through campus. The multimillion dollar project includes the removal of a dam, the rechanelization and recontouring of the river, the addition of wetlands, riffles, and pools, and the planting of native grasses, shrubs, and trees. In order to reach the venue of my presentation, I walked across the river on a bridge where several placards were posted explaining the restoration plans, which I noted in my talk. The focus of my presentation, however, was not on large-scale, hi-tech restorations such as the Olentangy project, but on community-based restorations such as those performed by the Benedictine nuns of Middleton, Wisconsin who are engaged in the restoration of prairie, oak savanna, and a glaciated lake on the over 100 acres of natural lands surrounding the monastery.[7] In the question and answer period, a member of the audience asked whether I thought the nuns could perform the type of extensive restoration effort necessitated by the Olentangy context (I half facetiously answered that nuns could do anything if they put their mind to it). Implicit in the questioner's query was a deeper worry: was participatory restoration sufficient for redressing the massive ecosystem disturbances that we face?

In some instances it may be: where volunteer-based restoration efforts are well organized and truly participatory in the sense that they involve public involvement "right from the start," as well as throughout the planning and implementation phase, they can be incredibly effective even in situations that also require hi-tech, expert inputs, which are almost always present in restoration efforts of any scale. There are multiple examples of this, including the Midewin National Tallgrass Prairie project in Joliet, Illinois, the former site of the Joliet Army Ammunition Plant which required extensive cleanup from contamination from decades of TNT manufacturing and packaging, and relies heavily on volunteers to maintain over 15,000 acres of tall grass prairie. The South African-based Working for Water (WFW) program, which since 1995 has been working to restore the Cape region's water system and provide socio-economic development for its inhabitants through the removal of invasive alien plants (IAPs), utilizes a workforce of approximately 30,000 citizen employees and volunteers.[8]

In other instances participatory restoration may not be enough to effectively deal with the extent of ecosystem and species degradations that are going to need to be reversed in the next centuries in order to stabilize the climate. Given what we now know, all tools, from participatory to hi-tech, are going to need to be on the table. What we will want to know in this scenario is how the public citizenry can become maximally involved with decision-making processes that dictate how land, including the biosphere, is used and repaired. That said, citizen involvement in political processes should not be viewed as a substitute for direct engagement with particular landed places. The former cannot address the root of the problem, which is a fundamental disconnect of industrial culture from the world of nature. This is why the argument for hands-on involvement in restoration work is so powerful and important, for there are not many activities in modern life that can connect such a broad range of people in multidimensional ways to particular landed places.

Restoration and religion

But what might be the role of religion in the conservation and restoration of ecosystems in the rapidly changing context in which we find ourselves? In earlier work, I characterized religion and spirituality as arising from the inherent human impulse to make sense of and find meaning in relation to life's big questions, including the mysteries, wonders, and beauties of the universe and of earth. I distinguished the terms by using "religion" and "religious" to refer to a systematic approach to spirituality (Benedictines as religious restorationists), and "spirituality" and "spiritual" to refer to concrete experiences or action (restoration-based spiritual experience) (2013, 17–18). There has been a tendency in ecological restoration thought to associate religion with ritual. While ritual is certainly a significant element of religious practice, it is nevertheless too narrow a focus, for there are in fact multiple ways in which religion and spirituality function in relation to restoration. Here I suggest five varieties. Each type suggests a slightly different view of religion and spirituality, and of the interface of cultural and ecological values; these are not mutually exclusive, and can function concurrently.[9]

The first and perhaps most prominent role of religious expression in ecological restoration, as noted, is the ritualization of restoration activities. Debates over the ritualization of restoration have been heated, with some restorationists arguing, for instance, that they do not need their souls saved, while others suggest that ritual can help humans productively deal with the unavoidably difficult dimensions of restoration practice (e.g., culling, ripping out, burning), and of the relationship with the natural world more generally.[10] Ritual activities may involve more traditional or secular forms. Examples include the use of indigenous spiritual practices such as the California world renewal tradition of the Klamath people in the Pacific northwest Intertribal Park restoration project, and community-oriented ceremonies such as the Bonfire and Bagpipes event in Northpark Illinois, where the annual burning of huge piles of the invasive buckthorn is accompanied by a public, community-wide evening celebration, with music and food. Ritual is understood to heighten restoration's meaning and integrate human beings in the "primordial acts of creation," as restoration ecologist, William Jordan has conceived. I have characterized this religious interpretation of restoration as "ecological symbolic action" and "public spiritual practice" (Van Wieren 2008).

Second, religious or spiritual considerations are made in setting restoration goals where sites include spaces, artifacts, or species that are considered in some sense to hold spiritual qualities for cultural and historical reasons. Representative of this idea are the restoration of sacred groves in India and of land areas sacred to indigenous peoples of North America.[11] Restoration ecologists Andre F. Clewell and James Aronson suggest this role for religion in their conception of the cultural values that restoration addresses: "We restore ecosystems to satisfy values that are shared collectively within a culture. For example, much restoration is dedicated to the recovery of impaired ecosystems in iconic places such as parks and preserves, where people gather to enjoy nature-oriented recreation and leisure, or in sacred places and sacred groves that have spiritual or religious significance" (Clewell and Aronson 109).

Closely related to the cultural significance of religion for restoration is how spiritual understandings of restoration implementation are operative in traditional ecological knowledge (TEK) systems. Robin Wall Kimmerer points out the distinctive quality of TEK in relation to modern scientific ecological knowledge systems, noting that the former is "laden with values," given how understandings of ecological relationships are "woven into and inseparable from the social and spiritual context of the culture" (Kimmerer 2000). This idea is apparent in the mission of the SER's Indigenous Peoples Restoration Network (IPRN), which states that the indigenous spiritual values of care-giving and world renewal must be central in ecological restoration. "Western science and technology…is a limited conceptual and methodological tool; the 'heads and hands' of restoration implementation. Native spirituality is the 'heart' that guides the head and hands" (Kimmerer, 3). The incorporation of indigenous spiritual and cultural values may also play a central role in postcolonial restoration efforts where indigenous peoples are reinstated as land managers of natural reserve or park areas where they were historically excluded.

Fourth, religious outlooks are used to interpret direct experiences of transformation and renewal that are yielded by hands-on restoration activities. This may involve more conventional or secular spiritual language to describe the restoration experience. Benedictine restorationists at the Holy Wisdom Monastery, for example, talk about experiencing God's presence through hands-on prairie restoration activities (Van Wieren 2013, 165). Other restorationists speak more generally about the restoration of the personal heart that occurs through restoration work. This includes feelings of interdependence and reciprocity and awe and wonder, as well as of limitation and inadequacy in terms of the human capacity to live harmoniously with the natural world.

Finally, and encompassing of the varieties already noted, religious interpretations provide narratives and stories related to the restoration process that help give meaning to the human relationship to a wounded and healing Earth. Forms of *re-storying* or *re-story-ation*, as I and others have called it, take religious shape for how they, in some sense, revolve around interpretations of the sacred, as well as present holistic visions for what it means to, say, live a faithful or virtuous life in community with others. So, for example, the Benedictine nuns of Holy Wisdom Monastery state: *We care for the land because it is good for the land. We care for the land because it is good for the Lake Mendota watershed. We care for the land because it is good for the souls of all God's people.* Secular, spiritually oriented restorationists put it differently, but here too, narrative accounts, religiously interpreted, point to understandings of a sacred Earth that warrants devotional, restorative care (Mills 1998).

While not exhaustive, these five types of religion and spirituality, taken together, broaden a view of religion in ways that heighten its significance for conceiving of, interpreting, and evaluating restoration efforts. Insofar as religious communities and traditions incorporate conservation and restoration issues in their institutional life, additional functions of religion in relation to particular ecological changes will emerge. So too as restorationists interpret their work as in some sense holding a meaning-making function, other dimensions of restoration-based spirituality will come to the fore. If faith traditions and spiritual values are to beneficially contribute to the conception of and conversation around conservation and restoration work, however, they will need to be shaped by beliefs and practices that serve to respect, care for, and give back to the natural lands in which they are necessarily embedded. Both restoration and global religious traditions face daunting challenges in terms of defining central tasks, goals, and objectives as we enter an era marked by significant uncertain and continually shifting planetary dynamics. They will need to be flexible and dynamic, bent toward the novel, while retaining continuity with the historical trajectory of life on an evolving Earth.

Notes

1   Higgs, 2012, 94.
2   Thompson draws on Young's 2004 article, "Responsibility and Global Labor Justice," *Journal of Political Philosophy* 12(4): 365–88.
3   Throop recognizes the dangers associated with the healing metaphor, including the notion that it suggests the idea that ecosystems are like organisms, and standard practices of modern medicine that emphasize a mechanistic and technological approach to patient care. Instead, he suggests that his view of healing aligns more readily with "alternative approaches to medicine and in the attitudes of many old-fashioned general practitioners" Throop 2012, 55.
4   Throop, 2012, 55–6.
5   Daniel Jenzen argues in his view of biocultural restoration, based on his decades long research in northwestern Costa Rica, that biodiversity on the landscape is greater where sustainable human practices exist. For an overview of Jenzen's work, see Woodworth, 2013, 256–86.
6   For example, restoration writer, Freeman House observes the following in relation to his efforts to restore native salmon in northwest California: "King salmon and I are together in the water…It is a large experience, and it has never failed to contain these elements, at once separate and combined, empty-minded awe; an uneasiness about my own active role both as a person and as a creature of my species; and a looming existential dread that sometimes attains the physicality of a lump in the throat" (House, 1999, 13).
7   For more on Holy Wisdom Monastery's restoration work, see: http://benedictinewomen.org/care-for-the-earth/natural-environment/ (accessed July 30, 2015).
8   For an overview of the WWF project, see Woodworth, *Our Once and Future Planet*, 49–87.
9   I am indebted to Higgs' treatment of history in restoration for helping me conceptualize religion and spirituality as a typology. See Higgs, 2012, 91–3.
10 For an excellent discussion of ritual in restoration practice, see Jordan, 2003.
11 On the former see Clewell and Aronson (2006). On the latter see Martinez (1992).

References

Callicott, J. Baird and Michael P. Nelson, eds. (1998) *The Great New Wilderness Debate*, University of Georgia Press.
Callicott, J. Baird. (1992) "Las Nature est morte, vive la nature!" *Hastings Center Report* 22:5, 16–23.
Clewell, Andre F. and James Aronson. (2006) *Ecological Restoration: Principles, Values, and Structure of an Emerging Profession*, Island Press.
Elliot, Robert. (1982) "Faking Nature", *Inquiry* 25: 129–33.
Higgs, Eric. (2012) "History, Novelty, and Virtue in Ecological Restoration", in Allen Thompson and Jeremy Bendik-Keymer, eds., *Ethical Adaptation to Climate Change: Human Virtues of the Future*, MIT Press, 81–102.
Higgs, Eric. (2003) *Nature by Design: People, Natural Process, and Ecological Restoration*, MIT Press.
Hobbs, Richard, Eric S. Higgs, and Carol M. Hall. (2013) "Introduction: Why Novel Ecosystems?", in Richard J. Hobbs, Eric S. Higgs, and Carol M. Hall, eds. *Novel Ecosystems: Intervening in the New Ecological World Order*, Wiley-Blackwell.
House, Freeman. (1999) *Totem Salmon: Life Lessons from another Species*, Beacon Press.
Jordan III, William. (2003) *The Sunflower Forest: Ecological Restoration and the New Communion with Nature*, University of California Press.
Katz, Eric. (1992) "The Big Lie: Human Restoration of Nature", *Research in Philosophy and Technology* 12: 231–43.
Kimmerer, Robin Wall. (2000) "Native Knowledge for Native Ecosystems", *Journal of Forestry* Volume 98:8, 4–9.
Light, Andrew. (2005) "Ecological Citizenship: The Democratic Promise of Restoration", in R. Platt, ed. *The Humane Metropolis: People and Nature in the Twenty-first Century City*, University of Massachusetts Press 176–89.
Martinez, Dennis Rogers. (1992) "Northwestern Coastal Forests: The Sinkyone Intertribal Park Project", *Restoration and Management Notes* 10:1, 64–9.
Mills, Stephanie. (1998) *In Service of the Wild: Restoring and Reinhabiting Damaged Land*, Beacon Press.
Sandler, Ronald. (2012) "Global Warming and Virtues of Ecological Restoration", in Allen Thompson and

Jeremy Bendik-Keymer, eds., *Ethical Adaptation to Climate Change: Human Virtues of the Future*, MIT Press, 63–80.

"The SER International Primer on Ecological Restoration," a publication of the Science and Policy Working Group of the Society for Ecological Restoration, 1st ed. Accessed 5 April 2016 (www.ser.org/resources/resources-detail-view/ser-international-primer-on-ecological-restoration)

Thompson, Allen. (2012) "The Virtue of Responsibility for the Global Climate", in Allen Thompson and Jeremy Bendik-Keymer, eds., *Ethical Adaptation to Climate Change: Human Virtues of the Future*, MIT Press, 203–22.

Thompson, Allen and Jeremy Bendik-Keymer. (2012) "Introduction: Adapting Humanity", in Allen Thompson and Jeremy Bendik-Keymer, eds., *Ethical Adaptation to Climate Change: Human Virtues of the Future*, MIT Press, 1–24.

Throop, William. (2012) "Environmental Virtues and the Aims of Restoration", in Allen Thompson and Jeremy Bendik-Keymer, eds., *Ethical Adaptation to Climate Change: Human Virtues of the Future*, MIT Press, 47–62.

Throop, William and Rebecca Purdom. (2006) "Wilderness Restoration: The Paradox of Public Participation", *Restoration Ecology* 14:4, 493–99.

Van Wieren, Gretel. (2013) *Restored to Earth: Christianity, Environmental Ethics, and Ecological Restoration*, Georgetown University Press.

Van Wieren, Gretel. (2008) "Ecological Restoration as Public Spiritual Practice", *Worldviews* 12: 237–54.

Woodworth, Paddy. (2013) *Our Once and Future Planet: Restoring the World in the Climate Change Century*, University of Chicago Press.

29

FOOD AND AGRICULTURE

A. Whitney Sanford

Problems of hunger, malnutrition, and obesity are rampant across the globe, and providing the world's population with affordable, nutritious, and safe food is becoming increasingly difficult. Food and agriculture present unique problems in the field of religion and environment because, first, everyone must eat and, second, agriculture relies on sets of relationships between humans, plants, and animals. Agriculture requires human interventions into the earth's processes in the form of planting, harvesting, and slaughtering, but we can choose how we construct those interventions. If humans are members of the biotic community, to use Aldo Leopold's words, what kind of members should we be? The world's religious and cultural traditions offer multiple frameworks to assess human roles in the biotic community. This chapter explores some factors that food and agriculture present in the field of religion and environment and will consider how the world's religious traditions offer tools—both potential and actual—to think through how we grow, produce, and distribute food while considering all members of the biotic community: human and nonhuman.

The Food and Agriculture Organization (FAO) predicts that by 2050, the world's population will reach 9.1 billion (FAO 2009, 2).[1] Most of these people will be urban and wealthy enough to demand the resource-intensive, meat-rich diets of the global North. (The terms global North and South are primarily, but not exclusively, geographic designations and indicate access to resources.) The U.N. estimates that we must increase food production by 70% to feed these people and almost double annual production of both meat and cereals. Feeding this hungry world in a just and sustainable manner is one of our greatest challenges. Ecological obstacles such as climate change, depleted fisheries, and over-salinized soils threaten current and future food production, and social challenges, such as poverty and food waste, frustrate efforts to feed hungry people.

In what follows, I discuss how religious traditions offer tools to construct frameworks that integrate equity, economic, and environmental concerns—often called the three legs of sustainability. Viewing challenges around food and agriculture through different frames and narratives provides perspectives that can help us imagine alternatives to historical and current patterns of behavior. Considering ideas and approaches from different traditions does not mean adopting these traditions blindly or without critique. Instead, asking how different traditions understand human relations with the earth and with each other might suggest new possibilities for food and agriculture.

Food and agriculture are embedded in relationships between people, animals, and the earth, and food plays a significant role in sacramental and ritual life in the world's religious traditions. While we must eat other living beings, whether plants or animals, it is important to assess the social and ecological effects of food production and consumption. Religious perspectives on greed and voluntary simplicity, for example, might help us think about limits to growth and consumption and whether our food practices are harming others. These perspectives offer more holistic perspectives on food and food production, invoking criteria such as justice and sustainability that transcend economy and profit. For example, are North American levels of food and resource consumption sustainable, and how do these levels affect other beings? The paradigm of agricultural relations leads to discussions about how we produce food, how human food production affects the biotic community, and how we might assess those effects. Writers and thinkers, such as Fred Kirschenmann and Mark Bittman, have claimed that we now have two competing food systems—a large-scale, industrial agriculture system focused on commodity crops, and a small-scale, bottom-up system focused on human-scale technologies—and that both systems claim to be able to "feed the world" (Bittman 2013; Kirschenmann and Falk 2010). Questions about agricultural technologies, such as genetically modified organisms (GMOs), reflect larger questions about how we might assess the systems according to the triple bottom line of ecology, equity, and economy. For example, who benefits from our food systems, and who—or what—loses? Considering how food production and consumption practices shape relations with the earth, animals, and other people is one means of assessing those practices.

Agricultural relations

In *Growing Stories: Religion and the Fate of Agriculture*, I argued that agriculture, and by extension, food is inherently relational (2011b). We grow food in relation with the earth and soil, and eating is rooted in our social, ethical, and religious lives. We can choose how we want to enact these relations, and we can choose our guiding metaphors—domination or cooperation. For example, Wes Jackson's ecosystem-based agriculture grows food in cooperation with—not in opposition to—natural processes, where "consulting the genius of the place," as Jackson phrases it, exemplifies a model of cooperation (2011a). Alternative agricultures, such as ecological agriculture and agroecology, highlight concepts of interdependence and cooperation in the relations between humans, soil, and plants (Altieri 1995).

Paradigms that emphasize compassion and interconnectedness offer insights into how we can—and should—produce and consume food. Viewing our food system holistically is certainly more complicated because it means understanding how food and production fit within multiple systems and relationships. To draw a parallel, the study of ecology or ecosystems is similarly complicated because we cannot simply isolate one piece of information. Scholars and practitioners, including Kirschenmann (2005) and Carolyn Merchant (1990), have critiqued the western scientific paradigm, based on a Cartesian framework that views the body and the earth as machines. "In viewing the natural world as a machine with interchangeable parts, data is extracted and isolated from its situatedness in more complicated sets of relationships" (Kirschenmann 2005, 3). For example, if the sole marker for judging corn production is the number of bushels produced per acre, its yield, then we exclude a range of other factors, including its role as fodder for animals, the impact of fertilizers on soil and water health, and the effects of monocultures on rural economic health (Sanford 2011b, 36). Kirschenmann notes that the machine metaphor forces us to overlook vital pieces of information about our world—information about relationships, interdependencies, and emergent properties—all vital, as it turns out, to economic, social, and ecological sustainability (2005).

Changing paradigms opens up new possibilities in how we understand the body, agriculture, and ecology. Consider the body as a self-regulating or homeostatic entity. This metaphor, unlike a machine, suggests that the constituent parts and processes of an organism operate interdependently; what you do to one part of the body affects the rest (Johnson 1987, 127–135). The organic metaphor raises questions about the relationships of the various processes and their interactions. Organic agriculture, restoration agriculture ("farming in nature's image"), and agroecology, for example, reflect homeostatic rather than mechanistic paradigms.

> Agroecological practices stress interactions between the various biological and non-biological components of the system. By creating a functional biodiversity, processes occur that provide ecological services such as the activation of soil organisms, the cycling of nutrients, the enhancements of beneficial insects and antagonists, and so on.
>
> *(Altieri 1995, 88)*

Viewing agriculture through language that emphasizes our interconnectedness provides the imaginative space to explore the conditions of these relationships. What are the costs and benefits of various practices and technologies? Who wins, in other words, and who loses? Focusing solely on yield, for example, does not answer the question of who benefits from those yields. In a post- North American Free Trade Agreement (NAFTA) world, Oaxacan farmers have not benefitted from higher U.S. yields in corn; instead, they have lost their livelihoods (Cummings 2002).

Alternate frames, such as relationship or compassion, are not anti-science or regressions to some mystic past, but they can help us ask different questions of our research. For example, in using the term biotic community, we must acknowledge trophic relations, such as predator–prey, that do not appear in metaphorical or theological descriptions of the biotic community, as environmental ethicist Lisa Sideris (2003, 174) argues. Nonetheless, many people around the world have been trained—and persuaded—to believe that intensive, chemical- and product-driven, industrial agriculture is the *only* way, not simply *a* way, to feed the world. Instead Wes Jackson's research on perennial polycultures that mimic the prairie ecosystem offers an alternate—and still data-driven—approach. These perennial polycultures, designed to restore health to soil ecosystems exemplify principles of interdependence and reciprocity between humans, soil, and plants. These qualities of interdependence, regeneration, and reciprocity are fundamental to alternative agricultures such as agroecology and restoration agriculture. Agroecology, for example, is good science: yields and soil health, for example, can be verified by data and are not simply a matter of belief or opinion.

Food, agriculture, and environmental degradation

The public has become increasingly aware of the environmental and social costs of large-scale industrial agriculture. Wes Jackson, Fred Kirschenmann, Tony Weis, and others have questioned whether industrial agriculture can feed the world and if so, at what cost (Jackson 2011; Kirschenmann 2010; Weiss 2007). The FAO documented the negative environmental effects of beef production in their report "Livestock's Long Shadow," and agricultural run-off has produced a dead zone in the Gulf of Mexico where large algae blooms choke off life below (Steinfeld 2006).

In a talk on resilience, Fred Kirschenmann identified four major threats to industrial agriculture: energy constraints; water shortages; climate change; and environmental degradation (Pearsall 2013). Contemporary industrial agriculture demands massive amounts of fossil fuel-based energy, including petroleum-based fertilizers, transportation, and processing. Declining availability of fossil fuels renders this form of agriculture increasingly unsustainable. Industrial agricultural practices have treated topsoil and groundwater, such as the Ogallala Aquifer, as limitless resources when, in fact, they are non-renewable in any meaningful time frame (Sanford 2011b, 8). Climate change will affect crop and livestock production as temperatures and weather patterns shift in unpredictable ways. Finally, heavy applications of pesticides and herbicides have contaminated water and soil, created pest problems, and led to soil nutrient depletion, which will produce less food. Further, the intensive production of biofuels raises questions about whether crops such as corn or sorghum should feed cars or people. In 2013, 40% of the U.S. corn crop became biofuel, 45% fed livestock, and only 15% was used for food or beverages (Conca 2014).

Environmental degradation is compounded by predictions that growing populations will demand a first-world, meat-intensive diet that will place even higher demands on nutrient-poor soils. Concerns about the ability to grow food have led to what some term "land grabs" in regions of Africa with healthy soils (Cotula et al. 2009). The UK-based International Institute for Environment and Development (IIED) "estimated that nearly 2.5 mn hectares of African farmland had been allocated to foreign-owned entities, including Korea, India, China, and nations in the Arab Gulf, between 2004 and 2009 in just five countries (Ethiopia, Ghana, Madagascar, Mali and Sudan)" (Laishley 2014). So countries with scarce arable land now control good land in regions where hunger and poverty is rampant, meaning that poor Africans will be less able to grow and purchase food as crops grown nearby are shipped to foreign lands.

The chief mantra of industrial agriculture, feed the world, is based on the assumption that the social and environmental consequences of industrial agriculture are collateral damage in a war on hunger. The premise is that industrial agriculture is inevitable and the sole means to feed the world. Critics of industrial agriculture, however, claim that alternative agricultures such as ecological agriculture and agroecology produce more food with less damage to soil and water (Altieri 1995, Kirschenmann 2010; Sanford 2011a/b), and a U.N. Special Rapporteur argued for a transition to agroecology (De Schutter 2010). Agroecology works with ecological systems to reduce the need for intensive inputs, such as pesticides and fertilizers, many of which are highly toxic to humans and animals. Several studies indicate that small, diversified farms using organic methods could significantly contribute to the world's food supply while using fewer inputs (Badgely et al. 2007; Perfecto 2010).

Others, such as Nobel Prize winner Amartya Sen, have challenged figures that call for 70% more food by 2050 and argued that waste and poor distribution rather than the gross lack of calories is responsible for hunger. The world already produces 1.5 times the amount needed to feed everyone, but the poor cannot afford to buy the food they need (Holt-Giminez et al. 2012, 595; Soil Association 2010). Further, we waste food at all stages of food production, from production to processing to storage (Stuart 2009). These arguments place many food and agriculture discussions in the realm of food policy and social ethics rather than the agricultural sciences. That is, the challenge is not so much how do we produce more food, but how do we grow food sustainably and distribute that food equitably. Rethinking agricultural relations and invoking frames of compassion and justice can help us critique those agricultural narratives that have led to environmental degradation.

Changing paradigm, changing practice

Considering agricultural practices offers an opportunity to choose how humans want to interact with the land. Religious and cultural perspectives provide models to consider our relations with other beings in a holistic manner. In one sense, a holistic approach would mean recognizing the broader effects of our food production systems on people and animals. Agrarian thinkers such as Wendell Berry, Wes Jackson, and Norman Wirzba have made us aware of the social consequences of contemporary food production, such as rural depopulation in the U.S. Midwest. Christian Theologian John Cobb suggests that we focus on "planetism" or reverence for the earth rather than economism, our current obsession with productivity and consumption (Cobb 1993).

Voices from several other traditions offer similar arguments. Winona LaDuke, an Ojibwe environmentalist, economist, and activist from the White Earth Reservation in Minnesota, argues that corporate control of food production and seed has severed the religious, nutritional, and agricultural relations associated with harvesting wild rice (2005). Vandana Shiva, an Indian scientist and environmental activist, warns us of the devastating consequences when farmers lose the cultural capital and indigenous knowledge associated with small-scale agriculture (2006). These writers claim that the industrial model of agriculture is broken and leads to increased hunger and environmental devastation. Despite representing different religious traditions, they are united in their call for holistic approaches to food production and renewed efforts to repair our agricultural relations. Focusing on agricultural relations helps us better understand the stories they tell about relations between humans and the biotic community and allows us to think critically about their story-telling as well.

Although we can easily see a number of commonalities between different religious traditions, compassion towards animals for example, there is also considerable variation in how different religious traditions (and considerable variation within traditions) conceptualize relationships between humans, the earth, and the divine. To explore this concept a bit more deeply, I will draw upon aspects of Christianity and Buddhism to imagine how these aspects might suggest different interpretations, models, and practices.

Stewardship traditions

Historically, Christians—as well as Jews and Muslims—have drawn on concepts of stewardship to determine how best humans should act upon the earth. In these traditions, a divine being who has created all beings has tasked humans with stewardship, the responsibility of caring for the earth and the beings of the earth. Theologians, scholars, and practitioners have debated how humans should fulfill this role. Christian interpretations of stewardship range from domination to forms of benevolent tyranny to gentleness (Santmire 1985).

Lynn White's article "The Historical Roots of our Ecologic Crisis," published in 1967, argued that, historically, stewardship has been interpreted, or at least enacted, as a form of domination, with little regard for the well-being of other created beings. In a subsequent article, "Christianity and The Survival of Creation," Wendell Berry excoriated his fellow Christians for abusing the earth and stated "Christian organizations, to this day, remain largely indifferent to the rape and plunder of the world and of its traditional cultures" (Berry 1993). But he also added that anti-Christian environmentalists—and some Christians—should, perhaps, actually read the Bible because it contains a multitude of passages affirming both the value of creation and our responsibility to maintain God's gifts. Today, many evangelicals have embraced Creation Care, recognizing that humans have a responsibility to care for all of God's creation, that to wantonly harm the earth is disrespectful to God, or sinful.

In his recent book *A Theology of Eating*, Norman Wirzba describes gardens as microcosms of "the complex array of relationships that join us to the soil and water and to creatures and God, relationships that have nurture and feeding at their root" (2011, 36). His extended meditation places humans, food, and gardens as "members in creation and community" and declares that to draw life from the garden, we must also serve it—and others. Service, humility, and responsibility lead to self-sacrifice, putting the needs of others ahead of our own (Wirzba 2011, 68–69). These practices offer structures to say no to food choices that unnecessarily harm animals or deprive others of food and to rethink food and agriculture on a larger scale.

Other Christian thinkers push beyond stewardship traditions, suggesting alternative ecotheological models for evaluating agricultural relations (Kearns and Keller 2007).

Buddhism

Buddhist concepts about relations between humans, divine beings, and animals differ greatly from Christian views. Buddhist practice also provides resources to help us deal with desire and the frustration of limits. While Buddhist thought and practice is extraordinarily diverse, and I am, by necessity, over-generalizing, over-arching Buddhist themes of compassion and interdependence provide different tools to rethink agricultural relations. In Buddhist thought, impermanence and change is the fundamental state of all existence, and our desire for permanence is one cause of suffering. Knowledge of this existential condition is the first step to liberation and comprises one wing of the Buddhist tradition. Compassion for ourselves and for the suffering of all beings comprises the other wing. These qualities—impermanence, interdependence, and compassion—could lead towards a new food ethic.

Compassion for all beings seems obvious. Our human capabilities, which have given us the capacity for harm and destruction, also demand that we act to relieve the suffering of others. Buddhist concepts of impermanence and interdependence also erase meaningful distinctions between beings, or myself and others. The qualities that comprise who I am at any given moment—my body, my thoughts, my personhood, for example—are transitory. David Suzuki, geneticist, environmental activist, and Buddhist, likens interdependence (or interdependent co-arising, a Buddhist philosophical term) to physical cycles of decomposition and regeneration. "I will return to nature where I came from. I will be part of the fish, the trees, the birds; that's my reincarnation" (Suzuki 2007, 198). Buddhist concepts of interdependence explain how our physical existence is intimately bound with ecological processes, and that clinging to our narrowly constructed selves causes suffering to ourselves and others. According to Thích Nhat Hạnh, recognizing our connections, and showing compassion to others is a form of self-help; we depend on healthy ecosystems just as three sticks need all three in order to stand (1999, 221–249).

Christian and Buddhist thought highlight compassion and responsibility for others, although using different philosophical and theological underpinnings. These qualities can also help us assess the effects of what we eat and how we produce food.

Two points of contention

Meat and GMOs present some of the most complicated—and most vehemently argued—issues at the intersection of religion, food, and agriculture. Representatives for the world's religious traditions offer diverse paradigms for evaluation.

Should we eat meat?

For many in the U.S., eating meat, especially beef, is part of being an American. But meat production, especially factory-farmed meat production, carries enormous environmental, health, and social costs and is responsible for unspeakable animal suffering. Thích Nhat Hạnh writes that "by eating meat we share the responsibility of climate change, the destruction of our forests, and the poisoning of our air and water. The simple act of becoming a vegetarian will make a difference in the health of our planet" (Hạnh 2008, 23). On the other hand, "of the 880 million rural poor people living on less than $1 per day, 70 percent are partially or completely dependent on livestock for their livelihoods and food security" (World Watch Institute 2011). If we choose to eat meat, changing how we eat meat, eating less meat, or switching to organic pasture-raised meats would significantly impact multiple arenas, including reducing greenhouse gas emissions, supporting pastoral livelihoods, and improving animal husbandry.

Rose Zuzworsky draws upon Christian theological resources to consider how we eat meat and asks how we might apply concepts of justice, care, and mercy to the treatment of food animals. She quotes John Cobb on the concept of justice, "it is time to call for a justice that is better expressed in the Hebrew term shalom, a right relation to the land and all that dwell therein," again recognizing that all beings are ultimately connected and interdependent (Cobb 1993, 184–185; Zuzworsky 2001, 184–187). As she notes, applying these abstract ideas to practice regarding food animals is difficult and complex. In Florida, for example, our feral hogs are destructive, invasive, and plentiful. Unlike animals in Concentrated Animal Feeding Operations (CAFOs), they roam wild until they are hunted and provide an inexpensive source of food for low-income hunters. Nonetheless, they are animals, and that fact makes hunting and eating feral hogs distasteful to many. For others who object to factory farming, eating an animal that once roamed free is preferable.

The contentious issue of meat will continue to challenge us on accepting limits and acknowledging the reality that we can't have everything we want. Practicing restraint and voluntary simplicity can move us toward re-establishing right relations with the earth and other beings—human and nonhuman.

Genetically modified organisms

Paradigms based on religious thought help us wade through the muddy and contentious debates about agricultural technologies, such as GMOs or transgenics. Battles over GMOs are waged in metaphoric languages that evoke the paradigmatic realms in which these arguments are embedded. While some proponents invoke salvific language of "feeding the world," detractors warn of "playing god" or "runaway genes"—all of which do point to very real concerns, but hyperbolic language does not lead to meaningful discussion about potential benefits or their potential social, economic, and ecological consequences (Brunk and Coward 2009). Members of the world's religious traditions have questioned the moral status of GMOs from multiple angles, including the right to choose (labelling laws), sanctity of genetic boundaries, and suitability as ritual offerings. Some of the most important questions about GMOs emerge in discussions about social justice and economics, for example, should germplasm, the building blocks of life, be owned and patented, and can poor farmers afford these expensive seeds? Religious paradigms based on compassion and justice, for instance, offer alternate perspectives on these question.

Farmers in groups such as Via Campesina, an international peasants' rights organization, as well as farmers in India I have worked with, question the benefits and consequences of GMOs, citing concerns about patents on and ownership of germplasm, loss of landraces, costs of technological packets that accompany transgenic seeds, and loss of control of their food supply. If GMO seeds are not drought-resistant, for example, and require intensive inputs, then they replicate the problems of monocultures in addition to the added new problems of ownership and patents. One persistent critique is that such seeds have not been designed to adapt to specific conditions and locations, mostly because it is too expensive for companies to do so. While these concerns about control, exploitation, and entitlement to resources affect anyone who eats, they have the greatest effect on marginalized peoples who pay the greatest social costs and reap the fewest benefits from expensive technologies.

These considerations reflect Michelle Adato and Ruth Meinzen-Dick's (Consultative Group on International Agricultural Research [CGIAR]) assertion that until recently agricultural research has focused on increasing yields of staple foods rather than alleviating poverty. In assessing agricultural technologies and practices, they advocate the "livelihood" approach that includes dimensions such as vulnerability, risk, social status, and gender that go beyond quantitative economic measures (Adato and Ruth Meinzen-Dick 2007, 20–55). The livelihood approach includes disaggregating regional and household access to food, technology, and money, and assessing which practices are encouraged. For example, do men gain sole access to new technologies and cash at the expense of women who retain—now devalued—local knowledge about species and intercropping methods?

Religious paradigms, such as compassion and justice, can help us consider the social and ecological effects of food production. Will an over-reliance on expensive agricultural technologies, such as GMOs, harm agrarian economies in the global South? Food democracy and social justice—asking who benefits, who loses—provide criteria to guide use and development of such technologies that do not either cede judgment to a narrative of inevitability and technological progress or, on the other hand, surrender to dystopic scenarios of monsters and "wild genes" that render consideration of new technologies almost impossible.

Creating new paradigms

Most religions offer models for our relationships with the divine, our relationships with each other, and our relationships with the earth. While traditions conceptualize these relations differently, most portray some variation of "right relations," that is, compassionate and sustaining relations between humans and the biotic community.

Voluntary simplicity

Changing how we eat and grow food will be difficult, and existing financial structures place significant obstacles in the path of change. In an interview with *YES! Magazine*, Fred Kirschenmann notes that existing infrastructure, such as transportation networks and corn silos, render change difficult (Pearsall 2013). My recent work with intentional communities has explored how people translate abstract values, such as nonviolence and voluntary simplicity, into agricultural and eating practices. Intentional communities, including cohousing groups, Catholic Worker Houses and Farms, and ecovillages, have created holistic frameworks to govern their material, economic, and spiritual lives, in the process experimenting with alternate forms of energy and food production, and re-introducing (and adapting) skills such as plowing with draft animals and grafting that have languished over the past fifty years. These communities are not trying to go back to the good old days—this is not nostalgia but recognition that, to go forward, we will need to adapt our lifestyles to meet existing challenges such as climate change.

Agricultural relations

If we want to feed the world in a just and sustainable manner, residents of the global North must reconsider and change what we eat and how we produce food—at the large-scale and policy levels, not just as individuals. One the other hand, residents of the global South, and impoverished peoples abroad and in the U.S., should have the means to scale up their consumption. Similarly, if we don't produce our food sustainably, the poor and marginalized will suffer further as they will have less access to arable land and water, an increasingly scarce resource. At the outset, I suggested that religious perspectives illustrate holistic paradigms to help us address the global food crisis—and our role in it. The frame of agricultural relations—or even kinship, in some cases—helps us consider the broader social and environmental effects of the food we eat and the way in which it is produced. Alternative agricultures such as organic agriculture, agroecology, and restoration agriculture, among others, invoke metaphors of mutuality and cooperation rather than domination and a war on nature. In practice, techniques such as inter-cropping or crop rotation might supplant toxic herbicides and pesticides for weed and pest control. Finally, religious perspectives offer frames and techniques to accept and embrace limits to consumption, at least in the global North.

Note

1   Some portions of this chapter were previously published in Sanford (2014).

References

Adato M. and Meinzen-Dick R. (2007) *Agricultural Research, Livelihoods, and Poverty: Studies of economic and social impacts in six countries,* Johns Hopkins University Press, Baltimore.

Altieri M.A. (1995) *Agroecology,* Westview Press, Boulder.

Badgley C., Moghtader J., Quintero E., Zakem E., Chappell M., Avilés-Vázquez K., Samulon A. and Perfecto I. (2007). "Organic agriculture and the global food supply", *Renewable Agriculture and Food Systems* 22:2, 86–108.

Berry W. (1993) "Christianity and the survival of creation", *Cross Currents* 43:2, 149–164.

Bittman M. (2013) "How to feed the world", *New York Times,* November 15, 2013. Available at: www.nytimes.com/2013/10/15/opinion/how-to-feed-the-world.html (accessed April 10, 2014).

Brunk C. and Coward H. (2009) *Acceptable Genes? Religious traditions and genetically modified foods,* SUNY Press, Albany.

Cobb J.B. Jr. (1993) "Economics for animals as well as people", in Pinches C.R. and McDaniel J.B. eds. *Good news for animals? Contemporary Christian Approaches to Animal Well-Being,* Orbis Books, Maryknoll 172–186.

Conca, J. (2014) "It's final – Corn ethanol is of no use", *Forbes.com,* 20 April 2014. Available at: www.forbes.com/sites/jamesconca/2014/04/20/its-final-corn-ethanol-is-of-no-use.html (accessed March 10, 2016).

Cotula L., Vermeulen S., Leonard R. and Keeley J. (2009) Land grab or development opportunity? Agricultural investment and international land deals in Africa, IIED/FAO/IFAD, London and Rome. Available at: ftp://ftp.fao.org/docrep/fao/011/ak241e/ak241e.pdf (accessed March 25, 2016).

Cummings, C. (2002) "Risking Corn, Risking Culture", *World Watch Magazine,* November/December, 8–19.

De Schutter O. (2010) Agroecology and the right to food, Report of the United Nations Human Rights Council to the UN General Assembly, New York, 20 December 2010. Available at: www.srfood.org/images/stories/pdf/officialreports/20110308_a-hrc-16-49_agroecology_en.pdf (accessed March 25, 2016).

Food and Agriculture Organization. (2009) How to feed the world in 2050. Available at: www.fao.org/fileadmin/templates/wsfs/docs/expert_paper/How_to_Feed_the_World_in_2050.pdf (accessed March 25, 2016).

Holt-Gimenez E., Shattuck A., Altieri, A., Herren H. and Gliessman S. (2012) "We already grow enough food for 10 billion people…and still can't end hunger", *Journal of Sustainable Agriculture* 36:6, 595–598.

Jackson W. (2011) *Consulting the Genius of the Place: An ecological approach to a new agriculture,* Counterpoint Press, Berkeley.

Johnson M. (1987) *The Body in the Mind: The bodily basis of meaning, imagination, and reason*, University of Chicago Press, Chicago.

Kearns L. and Keller C. eds. (2007) *Ecospirit: Religions and Philosophies for the Earth*, Fordham University Press, New York.

Kirschenmann F. (2005) Spirituality in agriculture, Paper prepared for The Concord School of Philosophy, Concord MA, 8 October 2005. Available at: www.leopold.iastate.edu/pubs-and-papers/2005-10-spirituality-agriculture (accessed March 25, 2016).

Kirschenmann F. L. and Falk C. (2010) *Cultivating an Ecological Conscience*, University Press of Kentucky, Lexington.

LaDuke W. (2005) *Recovering the Sacred*, South End Press, Boston.

Laishley R. (2014) Is Africa's land up for grabs? *Africa Renewal Online*, 2014 Special Edition on Agriculture. Available at: www.un.org/africarenewal/magazine/special-edition-agriculture-2014/africa%E2%80%99s-land-grabs (accessed March 25, 2016).

Merchant C. (1990) *The Death of Nature: Women, ecology, and the scientific revolution*, Harper Collins, San Francisco.

Nh t Hạnh T. (1999) *Call Me by My True Names*, Parallax Press, Berkeley.

Nh t Hạnh T. (2008) *The World We Have: A Buddhist approach to peace and ecology*, Parallax Press, Berkeley.

Pearsall P. (2013) Farmer-philosopher Fred Kirschenmann on food and the warming future *Yes Magazine*, 22 February 2013. Available at: www.yesmagazine.org/planet/farmer-philosopher-fred-kirschenmann-farms-warming-future (accessed March 25, 2016).

Perfecto I., Vandermeer J.H. and Wright A.L. (2009) *Nature's Matrix: Linking agriculture, conservation and food sovereignty*, Earthscan, London.

Sanford A.W. (2011a) "Ethics, narrative, and agriculture: Transforming agricultural practice through ecological imagination", *Journal of Agricultural and Environmental Ethics* 24:3, 293–303.

Sanford A.W. (2011b) *Growing Stories from India*, University Press of Kentucky, Lexington.

Sanford A.W. (2014) "Why we need religion to solve the world food crisis", *Zygon* 49:4, 977–991.

Santmire H.P. (1985) *The Travail of Nature*, Fortress Press, Minneapolis.

Shiva V. (2006) *Earth Democracy*, Zed Books, London.

Sideris L. (2003) *Environmental Ethics, Ecological Theology and Natural Selection*, Columbia University Press, New York.

Soil Association (2010) Telling porkies: The big fat lie about doubling food production. Available at: www.soilassociation.org/LinkClick.aspx?fileticket=qbavgJQPY%2Fc%3D&tabid=680 (accessed March 25, 2016).

Steinfeld H. et al. (2006) *Livestock's Long Shadow: Environmental issues and options*, Food and Agriculture Organization of the United Nations.

Stuart T. (2009) *Waste: Uncovering the global food scandal*, W.W. Norton & Company, New York.

Suzuki D., McConnell A. and Mason A. (2007) *The Sacred Balance*, Greystone Books, Vancouver.

Weis, A. J. (2007). *The Global Food Economy: The battle for the future of farming*, London: Zed Books.

Wirzba N. (2011) *Food and Faith*, Cambridge University Press, Cambridge.

World Watch Institute (2011) Global meat production and consumption continue to rise *World Watch Institute*, 11 October 2011. Available at: www.worldwatch.org/global-meat-production-and-consumption-continue-rise-1 (accessed March 25, 2016).

Zuzworsky R. (2001) "From the marketplace to the dinner plate: The economy, theology, and factory farming", *Journal of Business Ethics* 29:1/2, 177–188.

30

WATER

Christiana Z. Peppard

Fresh waters and their socio-natural textures are a vital part of ecological ethics in the twenty-first century. As geographer Jamie Linton puts it: "We mix language, gods, bodies, and thought with water to produce the worlds and the selves we inhabit" (2010, 3). Yet as recently as 2003, environmental philosopher Michael Nelson remarked that water existed in a "metaphysical blindspot" for ethics: it was largely out of sight but signified a crucial direction for ethical reflection (Nelson 2003). Now, scholarly spheres of vision expand across disciplines and methodologies to explore fresh water, oceans, and the brackish zones where they meet. Relentlessly dynamic and shape-shifting, the many forms and functions of water condition geomorphologies, shape ecosystems, augment or inhibit the lives of flora and fauna—including the survival of human beings and our varied societies, economies, and cultures.

Fresh water is one of the few global, earthen realities upon which all life depends. Yet it is neither uniform nor singular: it is better understood in diverse, multiple hydrologic incarnations. This chapter identifies how fresh water is emerging as a global challenge requiring interdisciplinary, constructive analyses in an era of planetary environmental degradation and pervasive economic globalization. The first section explains how religion, ecology, and water intersect. The second section identifies sites of contestation about the "global challenge" of fresh water.

Fresh water, religion, and ecology

Scholarship on religions and ecologies has set a myriad of conceptual, methodological and topical agendas (Tucker and Grim 2014; Gottlieb 2006; Jenkins 2013; Bauman 2014). But religion—like water—is a slippery substance to theorize: the singular noun contains multitudes. Some scholars, like Holmes Rolston III, have pointed out that, historically, "lofty" human goals like development, conservation, or sustainability have tended to "require" values that are "grounded in some ethical authority," which "has classically been religious" (Rolston 2006, 38). Others argue that "religion" need not be understood in a strictly creedal or institutional sense (Taylor 2009). Of course, while religious traditions can motivate ethical action or recalibrate moral norms, they are neither exhaustive nor necessarily ethically superior to other cultural movements (Cuomo 2011). This chapter presumes that religion is more than a set of creedal statements, scriptures, or concepts of transcendence: it includes individual and communal bodily practices, contextual or place-based knowledge and stories, material cultures and arti-facts, and patterns of behavior that sometimes become codified as moral systems.

The task is not merely academic. Many scholars of religion and ecology care deeply about the ethical and practical implications of their work, and many people in religious or intentional communities are at the forefront of articulating and living out ecological insights. Thus, within deceptively simply terms like "religion and ecology" are diverse ways of charting prospects for healthy twenty-first century ecological relationships—including how human beings relate to the planet's fresh water in an era of economic globalization, social inequalities, and anthro-pogenic impacts on earth systems.

Water is a globally significant substance that is essential but not uniform, universal and yet always particular (Peppard 2014). It is best understood as plural and contingent instead of singu-lar and steadfast. Still, several features are generalizable. First, water holds a central role in cosmological narratives. Second, fresh water is essential for human and ecosystem survival and flourishing. Third, water is materially, geographically, and culturally mediated. Finally, since the Industrial Revolution, human activities have dramatically affected the quality and quantity of available fresh water supply.

Cosmological narratives

Before there was life, there was water. Accounts of water's centrality in the creation of the cosmos and Earth permeate many cultures and traditions, such as the Hopi of North America, for whom water is understood to be the first existing substance (LaDuke 2005). The creation narrative in the Hebrew Bible's book of Genesis—a sacred text to Jews and Christians—describes the separation of waters before dry land appeared. Scientists concur that life began with water, and over evolutionary time the sea came to teem with living creatures—a grow-ing, differentiating, mutating abundance of multicellular organisms. Thus water is part of the narrative cosmology portrayed by Brian Swimme and Thomas Berry, who anthropomorphize plants as "[t]he first heroes to venture onto land" after eons of watery habitation (Swimme and Berry 1994, 116). Or as Paracelsus phrased it during the Renaissance: "Water was the matrix of the world and of all its creatures" (Ball 2001, ix).

The symbolic potency of water as agent of life, rebirth, and destruction is performed in many different types of religious rituals and cultural ceremonies worldwide. Mircea Eliade claimed that in religious symbolism, water "is *fons et origo*, the source of all possible existence… water symbolizes the primal substance from which all forms come and to which they will return" (Eliade 1958, 188).

Cultural–geographic specificities

Landscapes and geographic considerations figure into sacred texts and cultural stories, while religious rituals (or other traditional forms of bodily practice) can also reveal how lived reli-gions refer back to specific water sources. For example, in Christianity the rite of baptism evokes the baptism of Jesus on the banks of the Jordan River, where present-day pilgrims seek the holy associations of those waters (Havrelock 2011; Peppard 2013). At the triennial Hindu celebration of the Kumbh Mela, millions of people gather over the course of a month for bathing and purification rituals at sacred river sites.

Beyond formal religious rituals or pilgrimages, water holds cultural associations. For exam-ple, hospitality has often been linked to the provision of water to the thirsty traveler (Shiva 2002). It is also the case that many people who profess neither religious creed nor institutional affiliation have deep aesthetic or spiritual connections with waters and rituals—what some have named "deep blue" (or "dark green") religion—and evoke normative ideas from these aesthetic and experiential realities (Gerber 2003; Shaw and Francis 2008; Taylor 2009).

Embodied dependencies and outsized human impacts

Specific titrations of fresh and salt water make life possible and earth habitable for all manner of flora and fauna, including our charismatic hominid species. And while all creatures modify their contexts, industrial humanity has drastically changed the physical and social conditions under which waters flow and support life.

Until the mid-twentieth century the amount of fresh water on earth was viewed as relatively constant or at least renewable. The past few decades of hydrological and geological research have upended that assumption (Postel, Daily and Ehrlich 1996; Shiklomanov and Rodda 2003; Milly et al. 2008). Roughly three percent of all water on earth is fresh water. Of that, approximately seventy percent of the world's fresh water is encapsulated in ice, thirty percent in groundwater, and the statistically insignificant remainder is surface water (which is renewable in short time frames due to the hydrologic cycle). Over the course of the twentieth century, due in part to hydraulic technologies as well as increased demand, groundwater reserves have been depleted worldwide at rates that far exceed the rates of recharge (Famiglietti 2014; Glennon 2002).

These trends are exacerbated by climate change, since higher temperatures lead to higher concentrations of water vapor in various parts of the globe, which shift regional patterns of precipitation and soil moisture, impacting water supplies and forms of agricultural production. As individuals and societies become attuned to aspects of fresh water scarcity, practical and ethical quandaries will continue to swirl around the right courses of value and action in moral, legal, political, and economic spheres (Sandford and Phare 2011; Wescoat 2013). Thus, the next section describes seven challenges at the intersections of fresh waters, religions, and ecologies.

Global challenges: fresh water in religion and ecology

Population, consumption, and pollution

Environmentalists have long worried about human population growth that stresses or exceeds the carrying capacities of the planet's earth systems. While in an absolute sense, more people means more demand for fresh water, the identification of population numbers with consumptive use of water is analytically flat. Any discussion of population growth and water scarcity must be informed by sharp analysis of consumption patterns: industrialized nations tend to use significantly more water per capita than developing nations, for example, and by sector most of the world's water goes to agriculture (approximately 70%) while a smaller share to manufacturing/industry (22%) and the smallest proportion to domestic or survival uses (8%). Fresh water scarcity should not be parsed solely by the basic metric of population growth, since patterns of privilege and consumption are central.

Consumptive uses of water are those that do not return water to the watershed in any meaningful way. Agriculture represents a consumptive use of water, and grain production generally consumes less water than meat production. Domestic uses (such as bathing) tend to be non-consumptive. Agricultural and industrial pollution remains a particularly acute problem, since governments may or may not regulate point and non-point pollution in ways that conduce to the health of the watershed or downstream populations. Sanitation is a major concern in many areas. And cultural or religious understandings of waters can play into those dynamics, as David Haberman has shown with regard to paradoxes of purity in *River of Love in an Age of Pollution* (2006). Thus, how religious communities intersect with local and regional water consumption or pollution is a significant area for analysis by scholars and activists.

Water and conflict

Much contemporary hype about global fresh water scarcity centers on whether future wars will be fought over water. To a degree, this is a reasonable worry, because societies will go to great lengths to ensure access to fresh water, and there are cases in which water has been a factor in both inter- and intra-state armed conflict (Pacific Institute 2009, Wolf 2009). Water scarcity has been a contributing if subterranean factor in ongoing power struggles in the Middle East. Perhaps the most famous recent case of conflict over water supply occurred in Cochabamba, Bolivia, at the turn of the century when, as part of international development loan conditions, a subsidiarity of Bechtel privatized the city's water supply in ways that negatively impacted local residents and led to local and international protests (Schulz 2008). Yet water conflict negotiators emphasize that while water can exacerbate simmering tensions, it also has the capacity to bring people together because it is essential for existence (Wolf 2009). Adept scholarly navigation of religiously informed designations of water's value and cultural patterns of use will be crucial in considering these vexed contexts.

The commons, management, and governance

The idea of the commons in recent western philosophy is most famously associated with Garret Hardin's "tragedy of the commons"—the diagnosis that open-access resources in unregulated, competitive systems would be consumed to the point of destruction. This melancholy thesis has been both popular and widely criticized, and "the commons" has become a major focus of political economic theory (Ostrom 1990; Pacheco-Vega 2013). Other Western scholars are increasingly turning to notions of "public trust" in efforts to rehabilitate legal paradigms of water as a public good over and against its status as private property (Ingram and Oggins 1992; Wood 2013; Barlow 2014).

The notion of water as a "commons" or "public trust" suggests a series of moral challenges to tacit western assumptions of private property and individualized benefit. For whom are the goods of the earth—including water—intended? Who is entitled to access them, to benefit from them, and in what ways? How ought the commons to be managed, by whom, and to what ends? What are the conceptual grounds for such constructs? Increasingly, religious figures have drawn on the language of the commons to address the moral responsibilities of stewardship of vital resources such as water. The Catholic papal encyclical letter *Laudato Si'* refers numerous times to "the common home," "common good," and "global commons"—such as climate, air quality, biodiversity, and water—which must be protected and are intended for the benefit of all (Pope Francis 2015). Such notions stand over and against the idea that goods of the earth can be owned by a privileged few or exploited for the benefit of the powerful at the expense of the majority of the world's inhabitants.

Beyond dominant monotheistic religious traditions, indigenous communities and other repositories of traditional ecological knowledge and non-western metaphysical systems offer important accounts of what kind of thing water is understood to be. With the rise of global environmental justice activism, non-dominant cultures and groups have challenged the values embedded in western forms of water management (as well as the western metaphysics that may undergird those paradigms). Thus, notions of the commons, public goods, public trust, and cultural rights are ongoing sites of negotiation for activists and scholars—especially insofar as colonial powers and subsequent economic priorities have determined modes of water management, valuation, and access (Bakker 2013; Conca 2006; Groenfelt and Schmidt 2013).

Fresh water as a human right

In the 1990s, high-level global economic debates focused on whether water should be viewed as a commodity or a human right. In 2010, the UN affirmed that "access to fresh water is a fundamental human right," confirming moral claims that water justice advocates had long advocated—namely, that access to sufficient water supply is a prerequisite for the achievement of other rights and should not be contingent upon ability to pay. This ethical advance recognizes universal, essential substrates of survival, especially for underprivileged and vulnerable groups, such as women and people living in poverty (Sultana and Loftus 2015). At the same time, rights paradigms are fraught with logistical challenges. Calling water a human right does not necessarily mean water cannot be privatized, commercialized, or financialized (Bakker 2013; Schmidt and Peppard 2014).

Even as it is not yet clear *how* the right to water may be implemented, religious institutions and communities have expressed practical support and cultural, theological, ethical, or epistemic bases for the human right to water. Official documents from the Catholic Church, for example, have since 2003 argued that there is a fundamental human right to fresh water and that commodification should not trump equity: every person is entitled to access essential quantities of water because it is a gift from God, intended for all people now and in the future, and should be considered a right because it is fundamental to the achievement of all other rights (Peppard 2014, chapter 4). These notions are echoed in *Laudato Si'*: "water is a scarce and indispensible resource and a fundamental right which conditions the exercise of other human rights"; and "our world has a grave social debt towards the poor who lack access to drinking water, because they are denied the right to a life consistent with their inalienable dignity" (Pope Francis 2015). The pope also decries the "growing tendency, despite its scarcity, to privatize this resource, turning it into a commodity subject to the laws of the market."

Such pronouncements are expected to deliver moral weight and practical results because the Catholic Church is an international institution with many adherents and centralized authority structures. Many Protestant Christian organizations (including the World Council of Churches) and some schools of thought in Islam support a fundamental human entitlement to water (Gudorf 2010). Emerging consensus or critique of the human right to water in religious traditions more broadly remains to be seen.

There are important philosophical objections to the anthropocentrism of human rights. Do other animals, or ecosystems, have rights to the integrity of waters? Does water "itself," however we understand that complex concept, have a right to exist in a clean, unsullied and undammed state? It could be argued that because of the geological and evolutionary agency of water, this slippery substance itself is deserving of rights. Studies of normative views of water in comparative religious studies and cultural anthropology are beginning to emerge (Chamberlain 2008; Groenfeldt 2013). This is crucial, because fresh water is a socio-natural liquid that significantly takes the shape of its ethical container. For such reasons Sultana and Loftus, among others, have called for "the articulation of radically distinct subaltern perspectives in the democratization of the hydrosocial cycle" (2012, 13).

The environmental humanities have much to offer by problematizing received dominant traditions, giving voice to subaltern value paradigms, and probing normative commitments that are too often taken for granted in global discourses (Brown and Schmidt 2010). For example, the 2010 "Declaration of the Rights of Mother Earth," which resulted from the massive World People's Conference on Climate Change in Bolivia, deploys the language of rights but disarticulates the inherent anthropocentrism of Western paradigms while specifying how ecological destruction and economic dependency are wrought upon nations and peoples in the global South by processes of industrial and economic globalization that disproportionately benefit the global North. In this Declaration, certain kinds of entitlements are due to the Earth, "without distinction of any kind, such as may be made between organic and inorganic beings, species, origin, use to human beings, or any other status"—and the Earth (understood here as Mother) also has a right "to regenerate its biocapacity and to continue its vital cycles and processes free from human disruptions," including "the right to water as a source of life" (World People's Conference on Climate Change 2010).

Bodies of water: social impacts

Not all human beings have equal opportunity to flourish within societies, economies, and environments: it is now well established that burdens related to water and sanitation are borne disproportionately by women and girls in places where water infrastructure is unavailable. The gendered aspects of water procurement and distribution mean that as pollution and water shortages diminish available water supplies for domestic uses, many women and children (especially, but not exclusively, girls) spend their time carrying the heavy liquid rather than going to school or engaging in forms of economic activity. As Farhana Sultana and Alex Loftus have argued, "The impacts of water insecurity and injustices are clearly gendered, where women and girls in much of the global South spend countless hours fetching water for productive and reproductive needs. A gendered division of labor, as well as gendered livelihoods, wellbeing and burdens, are deeply affected by water quality, availability, provision systems and water policies" (Sultana and Loftus 2012, 8). The UN reports that girls and women are at increased risk of sexual assault when they travel long distances to procure water or find remote areas for personal hygiene (UN-Water 2006). When it comes to water, "gender intersects with other axes of social difference (such as class, race, caste, dis/ability, etc.) whereby water crises can exacerbate socially constructed differences and power relations" (Sultana and Loftus 2012, 8).

Water, sanitation, and gender are closely linked. Menstruating school-aged young women need adequate bathroom facilities, but the lack of such in many parts of the world means that adolescent women often stop attending school. Many development advocates extol how establishing water infrastructure and empowering women are positively correlated with economic development. It is also important to remember that the moral tenor of the factors surrounding gender and water ought not be reduced to the language or motivation of economic benefit: women's rights and flourishing are justice matters that deserve to be central to water ethics discourse (Brown 2010). Thus eco-feminist approaches to water and sanitation, social relations, and environmental justice present rich and urgent areas for water ethics (Gaard 2001; Neimanis 2014; Schmidt and Peppard 2014).

Water has also been used to manipulate communities towards desired political, economic, or nationalistic ends, often by withholding access to fresh water supply. The constraints put upon water by the state of Israel that impact Palestinian residents of Gaza and the West Bank represent particularly problematic forms of ethical violation that privilege some bodies over others in ways that violate the ideals inherent in the human right to water.

Finally, it is not only social bodies that depend upon sufficient quantities of fresh water: as noted previously, bodies of water extend beyond the human to include other species, ecosystem flows, and geomorphologies.

Value/price in the nexus of water scarcity and security

Water scarcity and security are now viewed as central threats to economic growth and production in the twenty-first century. Corporations and free-market advocates increasingly propose solutions that range from standard mechanisms of privatization and financialization of water resources (often linked to amorphous ideas of sustainable development) to the incorporation of water's "true" value as a form of natural capital. But in such discourses questions of cultural, social, philosophical, existence, or use values are noticeably absent, even after the designation of water as a human right. Indeed the issue of water's value often seems linked to economic paradigms that privilege price, growth, and finance and are predicated upon western legal assumptions of private property (Schmidt and Mitchell 2014). Such parameters insufficiently encapsulate the diversity of water's values in an age of ecological concern and moral multiplicities.

Scholars and practitioners of religion and ecology have a role to play in bringing alternative framings and activisms into global public discourse. Case studies can help to illuminate these issues: for example, two recent legal disputes mobilize moral claims that water is cultural inheritance and a type of commons—the Maori contestations over water in Aotearoa/New Zealand, and the Zuni wetlands settlement in Arizona (Strang 2014; Ruru 2012; Coburn, Landa, and Wagner 2014).

The question of innovation and technology

Fresh water scarcity often elicits the suggestion that technology will help to ameliorate or solve the problem. Indeed, water systems from ancient Persian *qanats* and Roman aqueducts to twentieth-century irrigation schemes demonstrate the longevities of technological and hydraulic inventiveness (Sedlak 2014).

In the past century, myriad large dams, diversions, and pumps were implemented at unprecedented scales to corral waters around the world in pursuit of higher agricultural yields and other forms of production. Such hydraulic feats continue in many places worldwide despite convincing social, environmental, and even economic contraindications. Most experts concur that a variety of technologies and innovations will be crucial for reducing demand (making systems more efficient), facilitating water recycling and reuse, or increasing supply (through desalination, for example). Yet many experts also insist that technology is not a panacea. Technologies are forms of social production. The utility of technological innovation with regard to water must be judged with reference to the frameworks within which it is formed and deployed, the contexts in which it is implemented, the question of who benefits and who bears the burdens, and the ends towards which the technology is oriented. Often, major unforeseen, unintended, or undesirable consequences can ensue—such as the displacement of environmental refugees from large dam projects, deleterious impacts on native non-human species, or environmentally toxic byproducts, to name a few. There will be no singular technological "solution" to global water challenges; rather, the challenge is to integrate technological advances within compelling ethical frameworks that protect social equity and environmental integrity in the short and long term.

Conclusion

Fresh water's multiple forms condition the possibilities of life. Religious and cultural values, practices, and socio-ethical paradigms have significant traction for how scholars of the environmental humanities parse problems, prospects, and planetary implications of fresh water as a global challenge for the twenty-first century. This chapter sought to catalyze such discussions—first, by articulating the confluence of waters, religions, and ecologies in theoretical frame; and second, by identifying seven complex, contentious, and important aspects of fresh waters in the twenty-first century.

References

Bakker K. (2013) "Constructing 'public' water: The World Bank, urban water supply, and the biopolitics of development", *Environment and Planning D: Society and Space* 31:2, 280–300.

Ball P. (2001) *Life's Matrix: A biography of water,* University of California, Berkeley.

Barlow M. (2014) *Blue Future: Protecting water for people and the planet forever,* The New Press, New York.

Bauman W. A. (2014) *Religion and Ecology: Developing a planetary ethic,* Columbia University Press, New York.

Brown P. G. and Schmidt J. J. eds. (2010) *Water Ethics: Foundational readings for students and professionals,* Island Press, Washington, D.C.

Brown R. (2010) "Unequal burden: Water privatization and women's rights in Tanzania", *Gender and Development* 18, 59–67.

Chamberlain G. (2008) *Troubled Waters: Religion, ethics, and the global water crisis,* Rowman & Littlefield, Lanham.

Coburn K. M., Landa E. R., and Wagner G. E. (2014) "Of silt and ancient voices: Water and the Zuni land and people", Buffalo, National Center for Case Study Teaching in Science (http://sciencecases.lib.buffalo.edu/cs/files/zuni.pdf)

Conca K. 2006 *Governing Water: Contentious transnational politics and global institution building,* Cambridge: MIT Press.

Cuomo C. (2011) "Climate change, vulnerability, and responsibility", *Hypatia* 26:4, 690–714.

Eliade M. (1958) *Patterns in Comparative Religion,* trans. R Sheed, Sheed and Ward, London.

Famiglietti J. (2014) "The global groundwater crisis", *Nature Climate Change* 4, 945–948.

Gaard G. (2001) "Women, water, energy: an ecofeminist approach", *Organization & Environment* 14:2, 157–172.

Glennon R. (2002) *Water Follies: Groundwater pumping and the fate of America's fresh waters,* Island Press, Washington, DC.

Gerber L. (2003) "The nature of water: Basia Irland reveals the 'is' and the 'ought'", *Ethics & the Environment* 8:1, 37–50.

Gottlieb R. (2006) *A Greener Faith: Religious environmentalism and our planet's future,* Oxford University Press, New York.

Groenfeldt D. (2013) *Water Ethics: A values approach to solving the water crisis,* Earthscan, London.

Groenfeldt D. and Schmidt J. J. (2013) "Ethics and water governance", *Ecology and Society* 18:1, 14.

Gudorf C. (2010) "Water privatization in Christianity and Islam", *Journal of the Society of Christian Ethics* 30:2, 19–38.

Haberman D. (2006) *River of Love in an Age of Pollution: The Yamuna River of Northern India,* University of California, Berkeley.

Havrelock R. (2011) *River Jordan: The mythology of a dividing line,* University of Chicago, Chicago.

Ingram H. and Oggins C. (1992) "The public trust doctrine and community values in water", *Natural Resources Journal* 32, 515–537.

Jenkins W. J. (2013) *The Future of Ethics: Sustainability, social justice, and religious creativity,* Georgetown University Press, Washington, DC.

LaDuke W. (2005) *Recovering the Sacred: The power of naming and claiming,* South End Press, Boston.

Linton J. (2010) *What is Water? The history of a modern abstraction,* University of British Columbia, Vancouver.

Milly P. C. D., Betancourt J., Falkenmark M., Hirsch R. M., Kundzewicz Z. W., Lettenmaier D. P., and Stouffer R. J. (2008) "Stationarity is dead: Whither water management?", *Science* 319, 573–574.

Nelson M. P. (2003) "Earth, air, water…ethics", *Transactions: Scholarly Journal of the Wisconsin Academy of Sciences, Arts and Letters* 90, 164–173.

Neimanis A. (2014) "Alongside the right to water, a posthumanist feminist imaginary", *Journal of Human Rights and the Environment* 5:1, 5–24.

Ostrom E. (1990) *Governing the Commons: The evolution of institutions for collective action,* Cambridge University Press, Cambridge.

Pacheco-Vega R. (2013) "On the impact of Elinor Ostrom's scholarship on commons governance in Mexico: An overview", *Policy Matters: IUCN Commission on Environmental, Economic and Social Policy* 19, 23–34.

Pacific Institute (2009) Water Conflict Chronology (www2.worldwater.org/chronology.html).

Peppard C. Z. (2014) *Just Water: Theology, ethics, and the global water crisis,* Orbis Books, Maryknoll.

Peppard C. Z. (2013) "Troubling waters: The Jordan river between religious imagination and environmental degradation", *Journal of Environmental Studies and Sciences* 3, 109–119.

Pope Francis (2015) *Laudato Si': On Care for our Common Home* Vatican City (http://w2.vatican.va/content/francesco/en/encyclicals/documents/papa-francesco_20150524_enciclica-laudato-si.html)

Postel S., Daily G. and Ehrlich P. (1996) "Human appropriation of renewable fresh water", *Science* 271, 785–788.

Rolston H. (2006) *Science and Religion: A critical survey,* 2nd Edition, Templeton Foundation Press, Philadelphia.

Ruru J. (2012) "The right to water as the right to identity: legal struggles of indigenous peoples of Aotearoa New Zealand", in Sultana F. and Loftus A. eds., *The Right to Water: Politics, governance and social struggle,* Routledge, New York 110–122.

Sandford R. W. and Phare M.S. (2011) *Ethical Water: Learning to value what matters most,* Rocky Mountain Books, Victoria, Vancouver, Calgary.

Schmidt J. J. and Peppard C. Z. (2014) "Water ethics on a human-dominated planet: Rationality, context, and values in global governance", *WIREs Water* 1, 533–547.

Schmidt J. J. and Mitchell K. R. (2014) "Property and the right to water: toward a non-liberal commons", *Review of Radical Political Economics* 46:1, 54–69.

Sedlak D. S. (2014) *Water 4.0: The past, present, and future of the world's most vital resource,* Yale University Press, New Haven.

Shaw S. and Francis A. eds. (2008) *Deep Blue: Critical reflections on nature, religion, and water,* Routledge, New York.

Shiklomanov I. and Rodda J. eds. (2003) *World Water Resources at the Beginning of the Twenty-First Century,* Cambridge University Press, Cambridge.

Shiva V. (2002) *Water Wars: Privatization, pollution, and profit,* South End, Boston.

Shulz J. (2008) "The Cochabamba water revolt and its aftermath", in Schulz J. and Crane Draper M. eds., *Dignity and Defiance: Stories from Bolivia's challenge to globalization,* University of California, Berkeley.

Strang V. (2014) "The Taniwha and the crown: Defending water rights in Aotearoa/New Zealand", *WIREs Water* 1, 121–131.

Sultana F. and Loftus A. eds. (2012) *The Right to Water: Politics, governance and social struggles,* Earthscan, London.

Sultana F. and Loftus A. (2015) "The human right to water: Critiques and conditions of possibility", *WIREs Water* 2:2, 97–105.

Swimme B. and Berry T. (1994) *The Universe Story: From the primordial flaring forth to the Ecozoic era—a celebration of the unfolding of the cosmos,* HarperOne, San Francisco.

Taylor B. (2009) *Dark Green Religion: Nature spirituality and the planetary future,* University of California Press, Berkeley.

Tucker M. E. and Grim J. (2014) *Ecology and Religion,* Island Press, Washington, D.C.

UN-Water (2006) *Gender, Water and Sanitation: A policy brief,* (www.un.org/waterforlifedecade/pdf/un_water_policy_brief_2_gender.pdf)

Wescoat J. L. (2013) "Reconstructing the duty of water: a study of emergent norms in socio-hydrology", *Hydrology and Earth System Sciences* 17, 1–10.

Wood M. C. (2013) *Nature's trust: Environmental law for a new ecological age,* Cambridge University Press, New York.

Wolf A. T. ed. (2009) *Hydropolitical Vulnerability and Resilience Among International Waters,* UNEP, New York.

World People's Conference on Climate Change and the Rights of Mother Earth (2010) "Universal Declaration of the Rights of Mother Earth", Cochabamba, Bolivia. (http://therightsofnature.org/wp-content/uploads/FINAL-UNIVERSAL-DECLARATION-OF-THE-RIGHTS-OF-MOTHER-EARTH-APRIL-22-2010.pdf)

31

ANIMALS

Paul Waldau

This chapter addresses human animals' relationship to the countless nonhuman animals with whom we share the larger Earth community. Truly a principal global challenge facing our species, this inevitable human–nonhuman intersection is seen far better today than it was only a few decades ago. Through helpful lenses afforded by interdisciplinary and multicultural approaches, humans today can identify the relevance of religious practice and scholarship to humans' inevitable connections with other-than-human animals.

This intersection, which is an essential feature of "our common home" (cf. Pope Francis 2015), faces an array of challenges generated by modern humans' denial of their own animality. Our better understanding of other animals gives us improved prospects of (1) noticing nonhuman animals' actual realities as individuals and members of their own communities, and (2) providing adequate accounts of humans' long past, complex present, and future prospects of living in a more-than-human world populated by healthy human and nonhuman animal communities.

These aims are sharpened by two crucial developments. First, our discussion about nonhuman animals has evolved greatly in the last half-century, and now features great diversity, openness, and a commitment to seeing other animals in terms of their actual realities rather than the one-dimensional caricatures that characteristically have dominated inherited beliefs and generalizations. Second, current discussions about humans' relationship to our fellow animals have made important contributions to broader intellectual currents and cultural movements by enlarging many people's notions of morality and a virtuous life.

It is important to underscore the important role that religious analyses and scholarship add to perspectives on humans' multifaceted engagement with nonhuman animals. In many circles of today's industrialized societies, the idea of ethics-driven protection of nonhuman animals is associated primarily with secular figures who have intentionally distanced themselves from religion, such as the leading animal protection philosopher Peter Singer. That animal protection has deep roots in religious traditions is easily seen in an argument that the great sages of the Axial Age—the pivotal period of global religious and philosophical ferment between 900 BCE and 200 BCE—understood the core of religion to be *respect for the sacred rights of all beings*" (Armstrong 2006, xiii–xiv).

Understanding the origins, breadth, and depth of anti-cruelty sentiments and other animal protections is but one way that the academic study of religion deepens perspectives on the human–nonhuman intersection. Another is supplied by sociological studies of how commitments to such protection continue to prevail today in some religious communities. In contemporary versions of those religious traditions that stem from Axial Age insights, the story is mixed. In medieval or modern forms of Christianity, for example, dismissals of animals outside the human species exist alongside high points of concern expressed by seminal figures such as Francis of Assisi and Albert Schweitzer. In the Axial Age tradition of noticing and taking the world beyond the species line seriously, present forms of religion and ecology advocate habitat protection, "creation care," and "stewardship," which embody ethics and practices crucial for protection of nonhuman animals and their communities.

Such features of modern religious practice have been used to support two claims regarding religion's relevance to the question of the Earth community's nonhuman citizens. First, those who wish to understand attitudes across time and place toward nonhuman animals cannot understand such issues well without awareness of the religious roots of many current attitudes or the wide-ranging phenomenon of religious communities protecting certain nonhumans through their ethical values, education, and daily practices. Concerns for other animals are, for example, a vital part of the worldview and daily life of many indigenous and small-scale society religious traditions. They are also an important feature of modern scholarship about the two largest religious traditions (Christianity and Islam)—scholars *from within these traditions* have noted the key role that animal protection continues to play in their tradition (see, for example, Hobgood-Oster 2008 and 2010; Masri 1987 and 1989; Tlili 2012). Through such diverse historical and contemporary sources, a variety of religious traditions offer remarkable awareness of the ethically charged dimensions of the human–nonhuman intersection.

A second claim about religious traditions and animals is that those who seek to understand humans' religious dimensions will not succeed *unless* they engage the long history of religions posing profound questions about humans' relationship to Earth's other-than-human living beings. Evidence for this second claim overlaps with the evidence offered in support of the first claim (cf. Waldau 2013, 174–176), but also includes the arguments cited below about cognitive development benefits that are lost when a child or adult is prevented from noticing and taking other animals seriously.

Claims about the importance of modern religious and secular circles taking seriously humans' ability to act as responsible citizens of a more-than-human community have not only affected, but also been reciprocally nurtured by, perspectives now being developed in many subfields of our sciences and humanities. The result is that the field of Religion and Ecology, the emerging discipline of environmental humanities, and the many sub-disciplines of the umbrella field known as animal studies (cf. Waldau 2013) can draw upon abundant scholarly, practical, scientific, and religious sources about humans' connections to nonhuman animals. All three fields also have an ethical cast, for each foregrounds the importance of humans meeting a challenge described succinctly in the mid-twentieth century—humans must evolve "from conqueror of the land-community to a *plain member and citizen* of it" (Leopold 1991).

Beyond human exceptionalism

Although human life interacts constantly with countless microorganisms, our native, unaided human senses cannot detect the vast majority of these "neighbors"—thus, the practical reality is that, in our daily lives, we can attend only to much larger "macro" animals (those animals visible to us in our daily lives) (Waldau 2011, 22–23). In practice, however, humans frequently attempt to ignore all but their fellow humans, often to the detriment of both the Earth community and themselves. Chief Luther Standing Bear in 1933 observed, "[T]he old Lakota

... knew that man's heart, away from nature, becomes hard; he knew that lack of respect for growing, living things soon led to lack of respect for humans too" (Standing Bear 1988). A twentieth-century naturalist added, "Whenever man forgets that man is an animal, the result is always to make him less humane" (Krutch 1949). Echoes of such wisdom undergird a medical doctor's evaluation of our present condition: "Much of the damage that we inflict on ourselves, on others, and certainly on the natural world stems from extreme adherence to the notion of human exceptionalism" (Ratey and Manning 2014, 8). Exceptionalist tendencies in humans' self-evaluation of our importance to the universe have crystalized in the last centuries into a virulent human exceptionalism, a form of human-centeredness that holds humans superior to and thus rightly entitled to privileges over all else in our more-than-human world (Waldau 2013, 144–149).

Human exceptionalism ignores two foundational facts: (1) humans are embedded in an unavoidably multispecies world, and (2) humans *need* a shared, multispecies community for a variety of scientific and humanistic reasons (Louv 2005 and 2011). Apart from the obvious benefits that responsible membership in our larger community brings to the nonhumans now under grave threat of harm from humans, benefits *inside the species line* range widely (cf. Waldau 2013, 9, 16 and 56), and include therapy for children and adults, increased ecological awareness and connectedness, and diverse opportunities for spiritual connection and integration with one's nonhuman surroundings. Of great relevance is the science-based claim that healthy cognitive development in children is retarded by removal of children from the natural world— "Yet, at the very moment that the bond is breaking between the young and the natural world, a growing body of research links our mental, physical, and spiritual health directly to our association with nature—in positive ways. ... Nature inspires creativity in a child by demanding visualization and the full use of the senses" (Louv 2005, 3 and 7). A decade earlier, Paul Shepard observed, "Children respond spontaneously to the details of nature and the names and movements of animals because animals were (and are) the path into categorical thought and, eventually, the terms of a philosophy or a cosmology" (Shepard 1996, 10). Jacques Derrida mused, "The animal looks at us, and we are naked before it. Thinking perhaps begins there" (Derrida 2002, 397). Other animals, then, because they prompt our capacity for creative imagination and enhance key reflective capabilities, nurture both experience and learning. Benefits also flow from the opening of hearts and minds to types of compassion that produce joy even as they strengthen character-creating diverse opportunities for forms of self-actualization that can occur *only* through self-transcendence (see Frankl 1992).

Scientific and empirical challenges

The respect afforded to the natural sciences in the modern world has given science-based claims about nonhuman animals great interdisciplinary power. Scientific data about animal cognition and emotion have been available since the 1970s (Griffin 1998) and today provide an important perspective because the emotional capacities of nonhuman animals were something Western societies had to *rediscover*. Many ancient peoples believed fervently that some nonhuman animals experienced emotions, and Charles Darwin also observed, "The fact that the lower animals are excited by the same emotions as ourselves is so well established, that it will not be necessary to weary the reader by many details" (Darwin 1874, 84). But gradually the Western cultural world and its scientific establishment slipped into persistent, Cartesian-inspired denials of emotional capabilities in *any* nonhuman animals. Today in many scientific research precincts, such denials have been surmounted (see, for example, Bekoff 2007; Marzluff and Angell 2012; de Waal 2013). Public perception of emotions in some nonhuman animals has also been advanced by familiarity with emotion-intensive nonhumans, such as family dogs and cats (see Bradshaw 2011 and 2013).

Nonetheless, there remain circles where doubts about other animals' emotional realities prevail, as is the case in the production-oriented academic field of animal science, which is dominated by money from industrialized agribusiness (see Waldau 2013, 68ff). In the profit-making businesses themselves, denials of food animals' individual emotional needs remain the *de facto* reality (see Scully 2002; Foer 2009).

Claims about the cognitive abilities of certain macro nonhuman animals follow a similar pattern—ancients held some other animals to be intelligent, early modern scientists and many citizens in the industrializing western world doubted such claims, and then modern science subsequently confirmed that ancient views were often far closer to the truth than were early modern doubts. Darwin himself suggested that while "[o]nly a few persons now dispute that animals possess some power of reasoning," nevertheless "many authors have insisted that man is divided by an insuperable barrier from all the lower animals in his mental faculties" (Darwin 1874, 90 and 95).

Today, despite much cutting-edge work in a wide range of sciences (cf. Griffin 1976 and 1998) dualistic schemes still dominate major modern institutions. Law and other public policy circles, all levels of education, mainline religious establishments, economics-focused scholarship, and the business establishment are bastions of unmitigated human exceptionalism. At the same time, there have been profoundly important changes in public values and social ethics in many countries, which in turn have led to great ferment in the last half-century about what future our species might choose regarding treatment of various nonhumans (see Waldau 2011 and 2013). Science-based discoveries continue to arrive regularly regarding the social and cognitive realities of many different species (e.g. de Waal 2013) and such findings consistently confirm ancient intuitions that our world is graced by multiple intelligences (Tucker 2006).

Contemporary analyses suggest, however, a number of reasons that our science traditions should be infused with humility. These include the long history of paradigm-shifting discoveries, the constant revision of one major theory after another, the problems of fraud and political correctness (Dreger 2015), the use and control of science by self-interested corporations (Oreskes and Conway 2010; Krimsky 2003), all-too-familiar patterns of resistance to change (Griffin 1998), and contemporary harms to farm animals (Pew Commission 2008). Such problems reveal that practices pursued by the scientific community have often promoted human exceptionalism and the privileges it enshrines.

To insist that humility is needed as science goes forward is in no way to deny that our sciences have made great advances in many fields. Many advocates of Religion and Ecology and the Environmental Humanities have from the beginning embraced science-based viewpoints (Tucker and Grim 2009). Other disciplines, such as Religion and Animals and the now decades-old legal education field of Animal Law, began their work more narrowly, confining themselves to the discourse of, respectively, theology and law. Today, however, these fields feature second-wave forms that are richly interdisciplinary and cross-cultural in scope (Waldau 2016).

About nonhuman animals, then, there is much that has been learned and, of course, much more to learn. The fact that any full understanding of the human–nonhuman intersection must be informed by some adequate level of science-based realism by no means implies that science alone will suffice—it will not, for informed empirical approaches to nonhuman animals are a necessary, not a sufficient, condition for making informed judgments about animals, whether human or nonhuman. As noted by a Nobel Laureate in physics,

> The image of the world around us that science provides is highly deficient. It supplies a lot of factual information, and puts all our experience in magnificently coherent order, but keeps terribly silent about everything close to our hearts, everything that really counts.
>
> *(Erwin Schrödinger, quoted in Revel and Ricard 1999, 214)*

We need to acknowledge humbly that our human languages, though truly marvelous, can bewitch our minds and thereby create false illusions. Unexamined uses of common but anti-scientific phrases such as "humans and animals," which remains the standard even in science-based education, favor the kinds of ignorance that harden into exclusions. Fortunately, it is possible today to speak in ways that do not support misleading dualisms or commit the fallacy of misplaced community by pulling humans out of the animal community (Waldau 2013, 16). To realize our full human nature and to actualize our bountiful human caring abilities, we also need ethically sensitive, religiously perceptive, and cross-culturally aware understandings to grasp the profound ways in which each human is a citizen of a variety of nested communities that, taken together, reveal how our shared Earth is a *much* more-than-human community.

The "Anthropocene" question—a segue from science to ethics

The question of nonhuman animals in our changing ecological context helps us make sense of today's debates over humanity's past actions and future choices, and influences our discussion of whether we should name a new geological epoch after ourselves. Anthropocene relies on Greek etymology—*anthropos*, "humans," + *kainos*, "new." It has been proposed in geological circles to signal that we are no longer in the Cenozoic era (*kainos*, "new" + *zōion*, "animal"), which, by scientific consensus, began about 65 million years ago when various factors, including a large comet striking the Earth, led to the extinction of dinosaurs and thereby opened the way for our kind of animal, namely, mammals. One implication of choosing this new term would be that, of all the animal species, we are the first one whose world-altering powers can be compared to the asteroid or comet from the sky responsible for the most recent epoch-ending global catastrophe.

The behaviors that have produced the changes memorialized by the proposal of the Anthropocene as a new geological era are not, of course, new behaviors. Instead, they are the latest version of a long-standing pattern of rank human-centerednesses that have licensed disregard in certain societies for the more-than-human world. Anthropocene is, in effect, a corollary of human exceptionalism, which now functions as the dominant narrative of our time. The result is that many humans have lost awareness of their own animality and of our species' membership in the larger community. But as explained below, our radically human-centered narrative promotes only *some* humans' privileges, for the harms created by modern societies, including consumption patterns, undeniably have disadvantaged many humans as well.

The Anthropocene and other animals

In part because geological uses of Anthropocene have been attracting attention, the term has already been appropriated in nongeological circles. That nongeologists have come to speak of "the Anthropocene" tells us two additional things about ourselves. First, this word has been used by some concerned citizens who want us to curtail humans' broad harming of the Earth and its communities. Speaking of the Anthropocene, then, can be seen to reflect our capacious ethical abilities.

In addition, however, the arrival of the Anthropocene has an undeniably ugly implication. Elizabeth Kolbert has shown how the apathy and self-inflicted ignorance that human exceptionalism promotes regarding lives outside our own species has led to a state of affairs she calls "the sixth extinction" (Kolbert 2014). In fact, many of the harms committed by human societies fall short of extinction but nonetheless result in "a massive diminution of the entire body corporate of animate creation … species that still survive as distinct life forms but have suffered horrendous diminishment" (Mowat 1996, 14). We should not be too eager to celebrate a pattern of extinction and other harms for our larger community that compares in any way to five previous global catastrophes.

Kolbert's concerns reveal an ethically charged backstory involved with any use of "Anthropocene," namely, the pervasive problem of humans having exceptionalized *only our own kind*. Some may assume that, on its face, this notion must be a boon for humans. But upon closer examination, there are many reasons to halt before the precipice we create by our deeply troubling tendency to name things after ourselves alone. A parallel exists in the limits of our educational field called "the humanities"—there are few places in this megafield of our "higher education" where a student can study nonhumans in robust ways unimpeded by a Protagoras-like axiom that "man is the measure of all things" (cf. Waldau 2013, 289–306). The *anthropos* undergirding Anthropocene is eerily close to the extinction-promoting assumptions of human exceptionalism. But so much contemporary science, common sense, and traditional wisdom make it *obvious* that the Earth is neither designed primarily for humans nor fully in our control. There are even hints in various narrative traditions, as in the recurring claim among North American indigenous groups that "every animal knows way more than you do" (Ratey and Manning 2014), that it might be *nonhuman* animals who fit better into this world than do humans.

Less-than-careful uses of "Anthropocene," which risk *celebrating* the series of failures that have now produced measurable global destruction of living beings, beg questions about facile, nongeological uses of the term. Consider an example that illuminates how welcoming the Anthropocene may amount to nothing less than contemporary elites justifying their own lifestyles and fascination with modern technologies and power. Diane Ackerman's (2014) *The Human Age*, Part I of which is entitled "Welcome to the Anthropocene," features an optimistic tone throughout. One reviewer challenged the book's focus on only surviving humans and not the marginalized poor who will be unable to wall off their low-lying communities from rising sea levels or deal with other destructive effects of climate disruption (Nixon 2014). De-emphasizing or, worse, ignoring altogether the problem of marginalized humans *not* surviving in the Anthropocene is deeply troubling. *From the vantage point of nonhuman animals*, however, such optimism about the benefits that privileged segments of humanity could gain in a "good" or "great" Anthropocene would likely seem mere "business as usual" in the all-too-familiar key of human exceptionalism.

Given these problems, does Anthropocene offer any solace for those who take seriously today's worldwide animal protection movement that promotes a better understanding of humans' possible roles in relation to our fellow animals? If individuals choose to deeply affirm humans' animality and our larger community by actively seeking the role of responsible, plain citizen, our species can add impetus to environmental and animal protection efforts to create a *modus vivendi* in which *whole societies*, through their public policies, laws, and social ethics, choose not their own privilege but, instead, the connections that plain citizens can enjoy in a larger, more-than-human community. This achievement would, of course, afford humans a form of life congenial to a full acknowledgement that humans are a very special kind of animal. To do so would also be consonant with the lines of evidence confirming that a multispecies world is needed for human flourishing, and would thereby promote the *full* actualization of human potential. More than two decades ago it was argued,

> Meditation on animals and our relations with them must be very nearly the oldest and most persistent form of human pensiveness; it is doubtful that we could ever really adequately know our identity as humans if we did not have other animals as a frame for our own activity and reflectivity.
>
> *(Fernandez 1995, 8)*

Making choices in favor of a multispecies community does more than recognize the ancient roots of our connection and community with other animals. Taking responsibility for an inclusive rather than an exceptionalist future not only enhances our *present* capacities to care about both human and nonhuman others. It also gives us the chance to search out the full extent of our own animal abilities and create a tradition of realistic narratives about other macro animals' realities as individuals and as members of their own communities. If we affirm ourselves in this way, we thereby attest that being an "animal" can be extraordinary. Further, we set the stage for exploring other extraordinary animals, and we create forms of education that do something other than equip our children to be ever more effective vandals of the Earth (Orr 1994, 5). As challenging as it may be to forge a future in which nonhuman neighbors are respected as fellow citizens, this projected future contrasts favorably with the bleak prospects under today's human exceptionalism where extinctions loom and disfavored nonhumans are referred to as "trash animals" (cf. Nagy et al. 2013).

Religious and educational challenges

While only some segments of contemporary religious communities today emphasize animal protection, the budding field of Religion and Animals has shown that the number of religion-affiliated people concerned for nonhumans and their communities is now rapidly increasing (see Forum on Religion and Ecology 2015; Humane Society of the United States 2015). Any suggestion, then, as to who in our wide human world now commands the long view of humans' right relation with other animals surely must mention not only the better known "world religions" along with scientific and secular animal protection circles, but also indigenous traditions as well despite the fact that our modern world continues to harm these human groups (Goldhagen 2009, 54).

The problems presented by the human exceptionalism rooted in today's educational establishment brings to mind Helvetius' suggestion that humans are not born stupid, but only ignorant—they are made stupid by education (cf. Waldau 2013, 310). Care for others within and beyond the species line will be a *sine qua non* of developing robust education capable of nurturing responsible, plain citizens of biologically diverse local communities.

Conclusion

Exploring tensions and opportunities that exist in the complex relationships between human animals and nonhuman animals helps make sense of possible human roles in today's changing ecological context. To grasp the range of these roles, *at least* the following basics should be attempted.

(1) Develop awareness of the importance of *local place*, for such consciousness enhances awareness that humans, like other animals, are healthiest when they are place-knowing

creatures. As one scholar observed, "there never was an 'is' without a 'where'" (Buell 2011). This observation has affinities with the wisdom traditions developed in many small-scale human societies (see, for example, Basso 1996; Deloria 1969).

(2) Avow that in any place we find ourselves, nearby living beings are our neighbors, including the nonhuman animals visible to us in our daily lives.

(3) Recognize that humans' *unavoidable* membership in a more-than-human world has important ramifications for who we truly are—"we cannot be truly ourselves in any adequate manner without all our companion beings throughout the earth" *precisely because* that "larger community constitutes our greater self" (Berry 2006, 5).

(4) Acknowledge that *all* of humans' different forms of social awareness, including our communal, ethical, religious, and political capabilities, are required if humans are to recognize well our ecological embeddedness.

The unassailable fact that compassionate concern for other animals can be a vibrant part of our moral universe makes it evident why today there are many different cultural and political debates over whether and how humans as a collective might, *if we choose*, again extend our sense of community to the more-than-human world. This choice is both personal and momentous, for whatever choices we make regarding our larger community will thereby make morals and shape the future of life on Earth.

References

Ackerman D. (2014) *The Human Age: The world shaped by us,* W. W. Norton, New York.

Armstrong K. (2006) *The Great Transformation,* Alfred A. Knopf, New York.

Basso K. H. (1996) *Wisdom Sits in Places: Landscape and language among the western Apache,* University of New Mexico Press, Albuquerque.

Bekoff M. (2007) *The Emotional Lives of Animals: A leading scientist explores animal joy, sorrow, and empathy, and why they matter,* New World Library, Novato, CA.

Berry T. (2006) "Loneliness and presence", in Waldau P. and Patton K. eds. *A Communion of Subjects: Animals in religion, science, and ethics,* Columbia University Press, New York.

Bradshaw J. (2011) *Dog Sense: How the new science of dog behavior can make you a better friend to your pet,* Basic Books, New York

Bradshaw J. (2013) *Cat Sense: How the new feline science can make you a better friend to your pet,* Basic Books, New York.

Brody J. (2014) "We are our bacteria", *The New York Times,* 14 July 2014 (http://well.blogs.nytimes.com/2014/07/14/we-are-our-bacteria/).

Buell L. (2011) "American literature and the American environment: There never was an 'is' without a 'where'", in Shephard J., Kosslyn S., and Hammonds E. eds., *The Harvard Sampler: Liberal education for the twenty-first century,* Harvard University Press, Cambridge 2011.

Darwin C. (1874) *The descent of Man And Selection in Relation to Sex,* 2nd edition, Thomas Y. Crowell, New York.

Deloria V. Jr. (1969) *Custer Died for Your Sins: An Indian manifesto,* Avon, New York.

Derrida, J. (2002) "The Animal That Therefore I Am (More to Follow)", *Critical Inquiry,* Volume 28, Number 2 (Winter 2002), 369–418, trans. by David Wills.

Dreger A. (2015) *Galileo's middle finger: Heretics, activists, and the search for justice in science,* Penguin, New York.

Fernandez J. (1995) "Meditations on animals: Figuring out humans", in Roberts A. ed., *Animals in African Art,* The Museum for African Art.

Foer J. (2009) *Eating Animals,* Little, Brown and Company, New York.

Foltz R. (2006) *Animals in Islamic Tradition and Muslim Cultures,* Oneworld, Oxford.

Forum on Religion and Ecology 2015 (http://fore.yale.edu/).

Frankl V. (1992) *Man's Search for Meaning: An introduction to logotherapy,* 4th ed. Beacon Press, Boston.

Goldhagen D. (2009) *Worse than War: Genocide, eliminationism, and the ongoing assault on humanity,* PublicAffairs, New York.

Griffin D. (1976) *The Question of Animal Awareness: Evolutionary continuity of mental experience,* Rockefeller University Press, New York.

Griffin D. (1998) "From cognition to consciousness", *Animal Cognition* 1:1, 3–16.

Hobgood-Oster L. (2008) *Holy Dogs and Asses: Animals in the Christian tradition,* University of Illinois Press, Urbana.

Hobgood-Oster L. (2010) *The Friends we Keep: Unleashing Christianity's compassion for animals,* Baylor University Press, Waco Texas.

Humane Society of the United States 2015 Facts & faith: Religious statements on animals (www.humane-society.org/about/departments/faith/facts/statements/).

Kolbert E. (2014) *The Sixth Extinction: An unnatural history* Henry Holt, New York.

Krimsky S. (2003) *Science in the Private Interest: Has the lure of profits corrupted biomedical research?* Rowman & Littlefield, Lanham, MD.

Krutch J. (1949) *The Twelve Seasons: A perpetual calendar for the country,* W. Sloane Associates, New York.

Leopold A. (1991) "The Land Ethic", *A Sand County Almanac, with essays on conservation from Round River,* reprint edition, Ballantine, New York.

Louv R. (2005) *Last Child in the Woods: Saving our children from nature-deficit disorder,* Algonquin, Chapel Hill, NC.

Louv R. (2011) *The Nature Principle: Human restoration and the end of nature-deficit disorder,* Algonquin, Chapel Hill, NC.

Marzluff J. and Angell T. (2012) *The Gift of the Crow: How perception, emotion, and thought allow smart birds to behave like humans,* Free Press, New York.

Masri, B. A. (1987) *Islamic Concern for Animals,* Petersfield, England: The Athene Trust.

Masri, B. A. (1989) *Animals in Islam,* Petersfield, England: The Athene Trust.

Mowat F. (1996) *Sea of Slaughter,* Chapters Publishing, Shelburne, Vermont, 14.

Nagy K., Johnson P. and Malamud R. eds. (2013) *Trash Animals: How we live with nature's filthy, feral, invasive, and unwanted species,* University of Minnesota Press, Minneapolis.

Nixon R. (2014) "Future footprints: 'The Human Age,' by Diane Ackerman", *The New York Times,* 7 September 2014 (www.nytimes.com/2014/09/07/books/review/the-human-age-by-diane-ackerman.html).

Nussbaum M. (2006) *Frontiers of Justice: Disability, nationality, species membership,* The Belknap Press of Harvard University Press, Cambridge.

Oreskes N. and Conway E. (2010) *Merchants of Doubt: How a handful of scientists obscured the truth on issues from tobacco smoke to global warming,* Bloomsbury Press, New York.

Orr D. (1994) *Earth in Mind: On education, environment, and the human prospect,* Island Press, Washington, DC.

Pew Commission on Industrial Farm Animal Production (2008) *Putting Meat on the Table: Industrial farm animal production in America* (www.ncifap.org/_images/PCIFAPFin.pdf).

Pope Francis (2015) *Encyclical Letter: Laudato Si' of the Holy Father Francis – On care for our common future,* Vatican Press, Rome.

Ratey J. and Manning R. (2014) *Go Wild: Free your body and mind from the afflictions of civilization,* Little, Brown, New York.

Revel J. and Ricard M. (1999) *The Monk and the Philosopher: A father and son discuss the meaning of life,* Schocken Books, New York.

Revkin A. (2014) "Exploring academia's role in charting paths to a 'good' Anthropocene", *The New York Times,* 16 June 2014 (http://dotearth.blogs.nytimes.com/2014/06/16/exploring-academias-role-in-charting-paths-to-a-good-anthropocene/).

Revkin A. (2015) "Varied views (dark, light, in between) of Earth's Anthropocene Age", *The New York Times,* 15 July 2015 (http://dotearth.blogs.nytimes.com/2015/07/15/varied-views-dark-light-in-between-of-earths-anthropocene-age/).

Schlosser E. (2001) *Fast Food Nation: The dark side of the all-American meal,* Houghton Mifflin, New York.

Scully M. (2002) *Dominion: The power of man, the suffering of animals, and the call to mercy,* St. Martin's Press, New York.

Shepard, Paul (1996) *The Others: How animals made us human,* Washington DC.: Island Press.

Standing Bear C. L. (1988) "Indian Wisdom", in Callicott J. B. and Nelson M. P. eds., *The Great New Wilderness Debate,* University of Georgia Press, Athens, Georgia.

Tlili S. (2012) *Animals in the Qur'an,* Cambridge University Press, New York.

Tucker M. E. (2006) "A communion of subjects and a multiplicity of intelligences", in Waldau P. and

Patton K. eds. *A Communion of Subjects: Animals in religion, science, and ethics,* Columbia University Press, New York, 645–647.

Tucker M. E. and Grim J. (2009) Overview of world religions and ecology (http://fore.yale.edu/religion/).

Van Horn G. and Aftandilian D. eds. (2015) *City Creatures: Animal encounters in the Chicago wilderness,* University of Chicago Press, Chicago.

de Waal F. (2013) *The Bonobo and the Atheist: In search of humanism among the primates* W. W. Norton, New York.

Waldau P. (2011) *Animal Rights,* Oxford University Press, New York.

Waldau P. (2013) *Animal Studies: An introduction,* Oxford University Press, New York.

Waldau P. (2016) "Second wave animal law and the arrival of animal studies", in Cao D. and White S. eds., *Animal Law and Welfare: International perspectives,* Springer, Sydney.

Whitehouse A. (2015) "Listening to birds in the Anthropocene: The anxious semiotics of sound in a human-dominated world", *Environmental Humanities* 6, 53–71.

32

POPULATION

Guigui Yao and Robert J. Wyman

No topic is more important for ecological sustainability than the expansion of human numbers. World population has been growing at a billion people every dozen years. This growth rate has been amazingly constant for the last half century (see Figure 32.1) and another billion are expected in the next dozen years (United Nations, 2011, 2013, 2015). Each new person needs somewhere to live, somewhere to grow food, somewhere to deposit their waste and pollutants, somewhere to have a job, and somewhere to get the energy and materials they need. All these somewheres are the environment. The enormous ecological footprint of a billion extra people every dozen years swamps the achievements of environmentalists (Wyman, 2013). In terms of global ecology, demography is destiny.

The good news is that the global birth rate has been falling and is now lower than it has ever been in human history (see Figure 32.2). The bad news is that, even if the fertility rate remains at its current unprecedented lows, world population will quadruple from a current seven billion to 28 billion in this century (Constant fertility variant, top line in Figure 32.1). Since the earth has entered a regime with major environmental problems as its population passed from six billion to seven billion, it is very hard to imagine that the earth's ecology could survive a further quadrupling of the population. This can be avoided only if the global fertility rate continues to fall to levels considerably below its current rate. The four lower projections in Figure 32.1 depict the future trajectory of population growth in various reduced fertility scenarios.

The future is unknowable. Global fertility may continue to fall if education and contraception become more available, if women's status rises, and if political and economic stability are achieved. However, there are any number of factors that could rapidly and drastically increase the fertility rate (Wyman, 2003). High on this list is economic collapse as happened in Asia in 1997, and again in the US mortgage debacle of 2008 whose after-effects are still destabilizing the world economy. As China changes rapidly, will its government maintain control over population growth? Wars and ethnic conflicts keep reversing decades of social progress. International support for family planning has declined; the political consensus backing such support has been broken and a large proportion of the funding for reproductive health has been devoted to combatting AIDS.

The most fragile constructs that humans make are our political and social arrangements. As population continues to grow rapidly, poor countries, where essentially all population growth

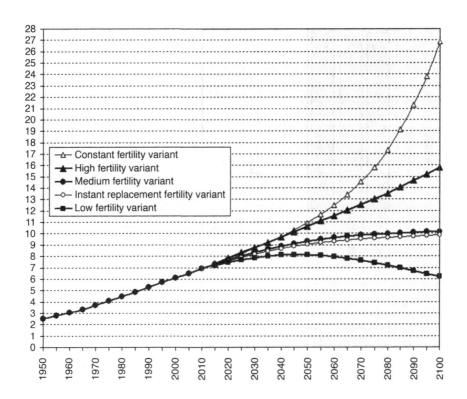

Figure 32.1 Union Nations: world population prospects: the 2010 revision

Note: Population has nearly tripled since 1950. Note the almost straight-line rise in population, the result of approximately equal numbers being added each decade. Up until the present, the growth rate has not abated; it stands at a billion extra people every dozen years. The future is unknown. Population under five scenarios is graphed here. If fertility stays as it is now (constant fertility variant, top line), population rises from a current 7 billion to 27 billion (29 billion in the UN's 2012 Revision). If fertility falls in the future, the range of possibilities is enormous, ranging from 6 to 27 billion depending on how much fertility falls.

is, may no longer be able to provide for even the most basic needs of their people—then chaos may ensue and all progress may be lost. Parts of the Middle East and Africa, with their populations exploding into violent conflict, may be in this "failed state" situation right now. Environmental degradation itself may destroy the basis of many people's lives and eliminate the conditions that have led to a decreasing birth rate. Global warming, water scarcity (Wyman, 2013), and desertification, all driven importantly by huge populations, are all increasing. Under these, and many other scenarios, the declining fertility story could turn around on a dime and population growth could start accelerating again.

We may already be seeing this effect. In its 2012 projections, the UN said it had been too optimistic in the 2010 projections (the source of Figure 32.1) and added about two billion people to the constant fertility variant for 2100. These numbers were raised again in the 2015 projections; since 2010, even the "medium" variant (the only 2015 variant published so far) for 2100 has risen by over a billion people. And these numbers do not yet reflect the global economic disruptions flowing from the US mortgage fraud crisis nor the effect of the upheavals in the Muslim world.

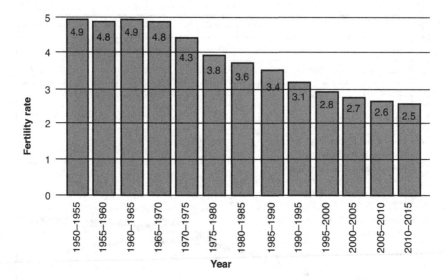

Figure 32.2 World total fertility rate 1950–2010

Note: The world birth rate has been falling and is now lower than ever before in human history. Nevertheless, at a total fertility rate of 2.5 the global population increases by 25% each generation. Population growth is almost all in the poorer countries in which childbearing begins at a young age, leading to a short generation time.

Source: United Nations (2013).

A major factor in the future of global population will be the teachings of the world's religions. Currently, pro-natal religious fundamentalism is on the rise throughout the world. In the US, the religious right has become so strong that not only abortion, but also contraception, has come under political attack. In India, the Hindu nationalist party, the BJP, is in power. Hindus worry that Muslims are out-reproducing them and Muslims retort that it is because Hindus do away with so many of their girls. The Israeli/Palestinian conflict continues unabated with both sides exhibiting extremely high birth rates in an exercise of so-called "political demography". Even Iran, whose Minister of Health received the UN's 1999 prize for developing one of the best family planning programs in the world (UN, 1999), has reversed course and eliminated its family planning program (Roudi, 2012).

In most discussions of religion and population, religious precepts are taken as givens, they are specified in the holy books, and the ways in which these precepts influence various religious traditions are discussed (Maguire 2009). This chapter includes that approach, but also the opposite, the ways in which population influences the development of religion. Population and religion is a two-way street with strong interactions both ways.

Demography influences religion

The difficulty of survival

All of the major world religions grew up in the long period when the survival of life was precarious. Demographic measures help us understand how tenuous the continuation of life was. Estimates from Roman times suggest that about 60% of girls died before reaching reproductive age (15 years old) (Livi-Bacci, 2012). Another period of death, due to the dangers of childbirth, followed. The average human survived only into his young 20s. This short lifespan did not improve much until recently; as late as the 1870s, in already quite modern Germany, almost 40% of children died by the age of 15 (Knodel, 1974). In Asia survival did not improve much until the latter part of the twentieth century. In Africa, horrific infant mortality is still common (see Figure 32.3). In modern genocides, the rate of murder of particular religious or ethnic groups can approach totality.

If all women born survived through all their reproductive years, women would have to bear an average of two children each to reproduce the population. However, in the severe regimen of death that has been the real circumstance for most of human history, only about one-third of women survive to reproduce the society. Under that circumstance, women would have to bear an average of six children just to reproduce the population. Six holds only if all women who survive engage in reproduction. In fact, there will be many who cannot or do not reproduce. Many will be infertile or too sick or malnourished or will remain unmated. The remaining women must bear many more than six children just to reproduce the population. Only those religions that demanded, and achieved, a sufficiently high birthrate have survived. The others have disappeared.

Be fruitful and multiply

The Hebrew Bible's "Be fruitful and multiply" (Gen 1:28, 9:1) epitomizes the pro-natalist stance. Should this be interpreted as a divine commandment that people have followed, or is this one interpretation of the Bible through which a society managed to survive? The traditional interpretation is that it requires unlimited reproduction. However, in the Bible, exactly the same injunction is given to the animals (Gen 1:22, 8:17, 9:1–7). Now that we are aware that human numbers are causing the extinction of many species, an anti-natal interpretation would require that human numbers be limited so that God's plan for other species is not thwarted. For creationists, each of these species is uniquely God-created, so that overpopulation is destroying God's creation.

This "ecological" interpretation is strengthened by God's command that a male and female of "every living thing" be brought aboard the ark. God establishes his covenant to maintain life, not only with Noah, but "with every living creature". In Genesis 6–9 "every" and "all" is repeated 37 times referring to the species that are covenanted with and must be saved.

The point here is not to argue Biblical exegesis, but just to show that the Bible does not require an unlimited pro-natal interpretation. The opposite is just as convincing, if not more so. Jesus, himself, believing that the end of the world was imminent (e.g., Matt 4:17), was uninterested in procreation. Matthew 19:29: *'And every one that hath forsaken ... children ... shall receive an hundredfold, and shall inherit everlasting life.'* Luke 14:26: *'if any one come to me and does not hate his father and mother and wife and children ... he cannot be my disciple'.*

Some of the early Church fathers believed that the world was already fully populated, with St. Augustine saying specifically that procreation was not "a duty of human society." (Bratton, 1992). Tertullian (~200 CE) wrote that "plague, famine, wars and earthquakes must be regarded as a blessing to civilization since they prune away the luxuriant growth of the human species". St. Thomas Aquinas agreed "that the number of children generated should not exceed the resources of the community and that this should be insured by law as needed"; without this, "the result would be poverty which would breed thievery, sedition, and chaos" (Munera, 2000). The Church fathers favored virginity over marriage and childlessness over parenthood. However, they added the non-Biblical "sex for procreation only" requirement, leaving celibacy

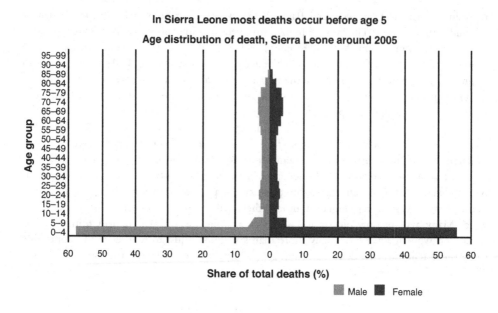

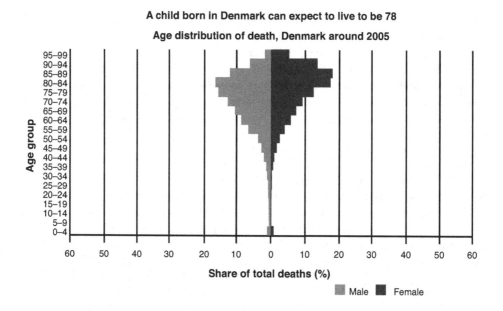

Figure 32.3 Age of death: Sierra Leone and Denmark

Note: Each horizontal bar represents an age group: its width displays the percentage of deaths that occur at that age. In Sierra Leone, more than half of all deaths occur before age 5. By contrast, Denmark has almost no infant deaths and the modal age of death is not until the 80s.

Source: World Bank 2006.

as the sole alternative to continual childbearing. (e.g. St. Jerome ~400 CE). Following the celibacy alternative, some of the most devout Christians eschew reproduction: priests, nuns, monks, and ascetics. The Shakers, like many other Christians, believed that Adam's sin was engaging in sex. Therefore, they forbade intercourse and went extinct.

Pro-natalism: no womb shall be unfilled

Abrahamic traditions have generally chosen the pro-natal interpretation. Its support for maximum childbearing is probably an important reason why those traditions have survived. The command to "be fruitful and multiply" is made specific in many particular commandments. Some of these can be understood as ensuring that "no womb goes unutilized." The Hebraic practice of the "levirate" certainly belongs in this category. If a man's brother dies, it is the religious responsibility of the surviving brother to impregnate the dead man's wife. This was considered such an important requirement that the punishment for violation is immediate death. In Gen 38:8–10, Onan's older brother dies, but "when [Onan] went in unto his brother's wife, he spilled his seed upon the ground." The passage is only three sentences long and ends abruptly with "God killed him."

A similar interpretation can be applied to the Muslim allowance of polygamy. This arose during the early wars of Muslim conquest. The Arab population was originally drawn from small groups of nomads living in harsh desert conditions that did not allow large populations. Hence, the number of Muslims was originally quite limited and the territories they set out to conquer were vast. Many males died, reducing Arab numbers even further. The influence of demography is shown in Mohammed's changing of prior practice to allow Muslim men to take up to four wives. This not only had the benefit of providing for the many war-widows, but also for allowing those same widows to continue reproducing.

Anti-natal religion

Paradoxically, anti-natalism can sometimes be required for survival. In 1624, westerners first encountered the Siraya, an aboriginal, hunter-horticulturist, head-hunting Taiwanese people. The region was sparsely populated (3.1 persons/km²) and an abundance of deer provided plenty of proteinaceous food. The Siraya engaged in fairly free sexuality, but required any woman who got pregnant, by husband or lover, before she was 35 to have an abortion (Shepherd, 1995). With good nutrition available, the women were fertile enough to reproduce the population even though starting childbirth so late and after 15 or so abortions. They considered the abortions a religious obligation and the priestesses induced the abortion by an ancient and effective massage technique still used in SE Asia (Potts, Graff, & Taing, 2007).

The Sirayan population was thriving, possibly limited only by internecine warfare (Keeley, 1996). Given the low population density, the warfare, like among many indigenous peoples, was probably *not* about competition for land or resources. Although the people of the time considered abortion part of their religion, a modern anthropological interpretation of its function was to allow the men to concentrate fully on their military training without any conflict from family obligations (Shepherd, 1995). The combination of a fairly free sexuality with the absence of both childbearing and exclusive marital fidelity is ideal for allowing young men to focus on their military responsibilities within the male comradeship group.

The Sirayan system was successful in maintaining the group when population density was low and resources were abundant. The greatest threat to their existence was violence from their neighbors. Their religion responded by evolving to favor military capability over population growth.

The message is that culture, including religion, must provide for the essential elements of survival. Members of a society are usually unaware of the survival functions of cultural practices, but often understand them to be religious requirements.

Conflicts: families, religion, and culture wars

"What the religious establishment allows and what people actually do are often two different things" (Bratton, 1992). Families struggle to limit their childbearing to their desires and their resources. This brings them into conflict with their religion. "Parents kill their infants, abandon them, neglect them in the hopes that they will die, give them into the care of wet-nurses (where they usually die)" (Mason, 1997).

Infanticide by abandonment was rife in Europe. In 1728–1757 England, >60% of 500,000 foundlings died by age two. In other European locales, death rates in foundling homes exceeded 90% (Langer, 1974). Abandonment only disappeared when contraception became a mechanism to control family size.

The conflict between religion and families was quite overt. Through the mid-twentieth century, the major Christian denominations, Catholic and Protestant alike, "viewed with alarm the growing practice of the artificial restriction of the family. They urged "all Christian people to discountenance such means as demoralizing to character and hostile of national welfare"" (Seccombe, 1993). Contraception was only fully decriminalized in the US in 1965. As the dangers of overpopulation became obvious, some Christian sources called for voluntary (and sometimes involuntary) population control (Hoff, 2012). The conflict continues in the American Pro-Life, Pro-Choice culture war.

Buddhism teaches not to take the lives of others. Nevertheless, in pre-revolution China, infanticide eliminated about 15–25% of girls (Lee and Feng, 2001). Now sex-selective abortion has replaced infanticide, resulting in approximately the same dearth of females. In Northeast Japan, between 1660 and 1870 about 40% of all pregnancies ended in abortion or infanticide (Drixler, 2013). Eventually Buddhist monks mounted a campaign successfully stopping these practices. It took 200 years. However, when abortion was legalized after WWII, under the American occupation, the abortion rate soared, cutting the birth rate in half in the whole country within eight years (Davis, 1963).

Various schools of Islamic jurisprudence cover the full range of opinions on abortion, from unconditional permission to unconditional prohibition (Shaikh in Maguire, 2003), but the lay public often considers it to be *haram* (forbidden). However, a name change may suffice. In Bangladesh "abortion," to terminate a pregnancy, is highly restricted legally, but the same procedure, when classified as "menstrual regulation," to restore a "blocked flow" of the menses, is permitted. Each name was assigned to about half of the 1.3 million abortions in 2010 (1 abortion for every 2.7 live births) (Singh, et al., 2012). In Pakistan, abortion is illegal, but, nevertheless there are about 2.25 million abortions in 2012 (one abortion for every 2.4 live births) (Sathar et al., 2014). People believe that other necessities outweigh the religious prohibition and that Allah will understand: "Then I opted for an induced abortion and prayed to God for forgiveness for the sin, which I had committed" (Tsui et al., 2011).

Religion or religiosity?

The birthrate does not correlate well with religion. Catholic France was the first country to experience a sustained fall in its birthrate, ~1800. The Protestant countries in northwest Europe followed (about 70 years later, in 1870). The southern Catholic countries lagged, but when they started controlling, in the 1970s, their fertility rates dropped to below that of the Protestant countries. Catholic Latin American countries are solidly in the middle of the pack. Their fertility started to decline earlier than it did in Asia, but never dropped as drastically as has happened in East Asia. Similarly, Muslim countries ranging from Tunisia and Indonesia, which have declined to replacement fertility, to countries like Pakistan and Mali, which have experienced little fertility decline. Muslim North Africa has seen significant fertility declines, while, in most of sub-Saharan Africa, neither Christian nor Muslim populations have experienced large fertility reductions. The Buddhist populations of East Asia started with extremely high birth rates, with the propagation of the family as the centerpiece of their religious beliefs, but then experienced extremely rapid falls to become one of the lowest birthrate regions in the world.

In the US today, religion does not dictate reproductive behavior: 99% of all women have used a contraceptive method other than "natural" family planning. This includes 98% of Catholic women, 99% of "mainline" Protestant women, and 100% of Evangelical women. The differences in current contraceptive use are also tiny, usage ranges only from 87 to 90% for the above groups (Jones and Dreweke, 2011). Only 2% of Catholic women rely on natural family planning; the same percentage applies even among Catholic women who attend church once a month or more. Even though the Catholic Church vehemently opposes abortion, the relative abortion rate for Catholic women was no different from that for all women. Catholics are almost 40% more likely to have an abortion then Protestants, whose churches hold a wide range of views (Jones et al., 2010).

While individual women may not be very obedient to the instructions of their religions, their governments may be. Governments with weak legitimacy among their populations often have to find support from a wealthy oligarchy or from powerful religious leaders. The Catholic Church, in places like Latin America and the Philippines, has played this role. In return for the priests instructing villagers in how to vote, the church can demand that contraception and abortion be made unavailable. Thus, even a very religious population can be desperate to control their fertility, but the means are not available (Hern, 1992). This situation is becoming rare in Catholic countries as they legalize contraception and abortion.

What is important is the degree of religiosity. In the US, women who attend church regularly have fewer abortions. Twenty-four percent of US women of reproductive age attend religious services weekly, but they account for only 15% of abortions; 23% of women never attend but account for 41% of abortions (Jones, et al., 2010).

In much of the world, religiosity does not work through a knowledge of the formal precepts of the religion, but rather through a "God's will" fatalism.

> Nigerian woman: "I wasn't prepared for a pregnancy. I didn't want it that early … but along the line, I got to accept it. It was a gift from God … I ran to God and he encouraged me. He gave me reasons why I should accept the pregnancy."
>
> *(Tsui et al., 2011)*

Humans are extremely susceptible to social pressure. The European Fertility Project found that socially communicated norms were crucial to Europe's fertility decline (Coale and Watkins, 1986). Lesthaegh (2015) found that fertility limitation spread rapidly from France into the culturally similar French-speaking part of Belgium. But it took 70 years for limitation to cross into the Flemish-speaking part of Belgium even though both parts were Catholic and economically matched. The drop in fertility correlated very closely with the degree of religiosity as measured by numbers voting for the Catholic political party as opposed to the Socialist party.

In Bangladesh, the degree of religiosity in a village was a more significant predictor of contraceptive use and a family's number of births than was the religiosity of the family itself (Atrash and Schellekins, 2011). Similarly, the effects of female autonomy on fertility is clearest at the community (as opposed to the individual) level (Mason and Smith, 1999).

A person's religion is only rarely based on individual choice; it is primarily inherited through family lines and maintained by social pressure. The fact that the populations adhering to the major religions occupy geographically contiguous regions tells us that the religion surrounding a person is the dominant factor.

The rate of religious conversion is miniscule compared with the effect of population growth. In the period to 2050, Pew (2015) estimates that 12.6 million people will convert to Islam. However, this is almost 100 times less than the increase due to the high Islamic birthrate. The only large conversion group is the 100 million who switch to "unaffiliated." But even this number is dwarfed by the internal population growth of the established religions.

The religions with a high birthrate maintain an expanding population of which a very large fraction are children (see Figure 32.4). In the case of Nigeria currently, and in China before the one-child policy, each year, three times as many girls enter reproductive ages as older women leave reproductive life. This large number of young people insures that their population growth will continue for decades into the future even if fertility eventually starts to fall. Currently, 34% of Muslims are younger than 15, but only 27% of Christians.

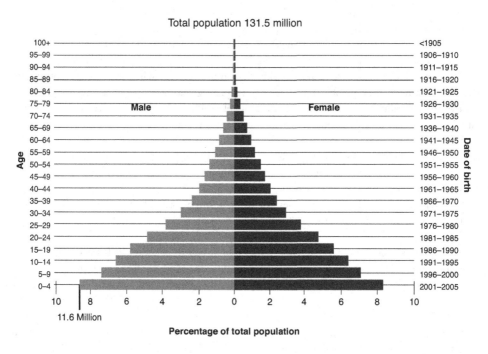

Figure 32.4 Nigeria's age structure (2005)

Note: Over 6% of the population are girls aged 10–14 who are just about to enter reproductive age, while only about 2% of the population are women aged 40 to 45 who are about to leave their reproductive ages. In short, three times as many women will begin reproduction as will end it. Thus, even if fertility per woman falls dramatically, the number of children will keep increasing.

Source: Population Action International.

Divergent fertility rates are rapidly changing the religious composition of the world (Pew Research, 2015). While there are now about 36% more Christians (now 2.17 billion) than Muslims (1.6 billion), the higher Muslim fertility rate (total fertility rate [TFR] 3.1 versus 2.7) means that by 2050 the numbers will be almost equal. Due to the global population growth, both groups will expand, the number of Christians will increase by about 35% (to 2.92 billion), while the Muslim population will increase at about double that rate (by 73% to 2.76 billion). Most of the Christian increase will occur in sub-Saharan Africa where the Christian population will double to constitute 40% of Christians in 2050. The number of Hindus will rise at about the same rate as Christians (34%). Because of the extremely low birth rate in East Asia, the number of Buddhists (TFR 1.6) will remain essentially constant. The future numerical balance of the world's religions has already been determined.

The culture wars

The last century and a half has witnessed a tremendous competition between traditional and modern ideas of sexual behavior. In traditional forms, authority, especially religious authority, determines the limits of people's sexual and reproductive behaviors. Modern ideas emphasize individual freedom in choice of sexual practices and control of childbearing.

Globally, sexual morality is a major motivating factor in the battle that rages today between various forms of autocratic and theocratic governance and modern democratic liberalism. China's one-child policy is generally abhorred in the west (Nie and Wyman, 2005). In the clash between Islam and the west, no topic is more emotional than the treatment of women. The west abhors the extreme oppression (as the west sees it) of women in some Muslim societies. Equally, many Muslims abhor the extreme freedom (as they see it) of western women. Both sides see an epidemic of violence against women in the other culture as symptoms of the failures of that culture.

The current centrality of sexual and reproductive moralities can be seen as the US culture wars are dominated by pre-marital sexuality, homosexuality, gay marriage, abortion, and sex education. In larger historical terms, at least for the west, this contest may be considered a part of the devolution of power, since the enlightenment, from authority imposed from above to individual agency, choice, and liberty.

The US is now debating where the limits to this freedom should be. Abortion and pre-marital sexuality are the flash points. National abortion rates (down 27%) and teen pregnancy (down 12%, 2010–2014) are at historic lows (CDC, 2015). The Christian Right, Pro-Life forces see this as a confirmation that their abstinence-only education programs and anti-abortion activities have been successful. Pro-Choice activists see an increase in the use of contraception as the cause of the declines and a confirmation that their family planning activities have been successful.

There is no doubt that the legislative thrust has turned pro-life; since 2011, 267 abortion restrictions have been enacted in 31 states. However, abortion rates fell in almost every state and the largest falls occurred both in states with the most aggressive anti-abortion laws and in those that have maintained unrestricted abortion access (Associated Press, 2015). This evidence suggests that legal restrictions did not cause the drop.

On the other hand, providing cost-free access to contraception has had "startlingly" positive results (Tavernise, 2015). Since 2008, Colorado has offered free intrauterine devices and long-lasting contraceptive implants to poor women and teenagers (Colorado, 2015). Over only five years (2009–13), the teen birthrate plunged by 40% and the abortion rate fell by 42%. A similar free-contraception program in Missouri found the pregnancy rate among sexually active teenagers dropped to 21% of the national rate and abortions to 23% (see Figure 32.5) (Secura et al., 2014).

Pregnancy, birth and abortion rates among girls and women 15 to 19 years of age in the CHOICE Cohort as compared with those in the US population			
Outcome	US population (all teens)	US population (sexually experienced teens)	CHOICE (Cohort)
	No. per 1000 teens		Mean No. per 1000 teens (95% CI)
Pregnancy	57.4	158.5	34.0 (25.7–44.1)
Birth	34.4	94.0	19.4 (13.3–27.4)
Abortion	14.7	41.5	9.7 (5.6–15.8)

Figure 32.5 Results of a St. Louis area program that provided free contraceptives to sexually active teenagers

Note: Pregnancy and abortion rates were drastically below national levels for sexually active teenagers and even far below rates for all teenagers, which included sexually inactive girls. CI = confidence interval.

Source: Secura et al. 2014.

In the culture wars, religious attitudes join with broader visions of how life and society should be organized. Conservatives defend the "family values" that they believe to be the mainstays of the social cohesion needed to perpetuate their vision of civilization. This includes traditional, religion-supported, gender and sexual attitudes, prescriptions, and proscriptions. Liberals emphasize the personal desires of individuals to choose gender roles, the number and spacing of their children, and a variety of sexual arrangements and practices. The passion and certitude of both sides seems to preclude a shift of religious/moral energies to a focus on the global crises of population explosion and environmental decay.

References

Amin, S., Diamond, I., and Steele, F. (1997). Contraception and Religiosity in Bangladesh. In: G. W. Jones, R. M. Douglas, J. C. Caldwell, and R. M. D'Souza (eds), *The Continuing Demographic Transition*. Oxford: Clarendon Press, pp. 268–89.

Associated Press. (2015). *Abortions Declining in Nearly all States*. Available at: http://bigstory.ap.org/article/0aae4e73500142e5b8745d681c7de270/ap-exclusive-abortions-declining-nearly-all-states (accessed 5 March 2016).

Atrash, A., and Schellekens, J. (2011). *Islam, religiosity, and Marital Fertility among Israeli Palestinians*. Available at: http://paa2011.princeton.edu/papers/110400 (accessed 5 March 2016).

Bratton, S. (1992). *Six Billion and More: Human Population Regulation and Christian Ethics*. Louisville: Westminster John Knox Press.

CDC. (2015). About Teen Pregnancy also Reproductive Health. Available at: www.cdc.gov/teenpregnancy/about/index.htm (accessed 5 March 2016).

Coale, A. J., and Watkins, S. C. (1986). *The Decline of Fertility in Europe*. Princeton University Press.

Colorado. (2015). *Reducing Unintended Pregnancy*. Available at: www.colorado.gov/pacific/cdphe/reducing-unintended-pregnancy (accessed 5 March 2016).

Drixler, F. (2013). *Mabiki*. University of California Press.

Hern, W. M. (1992). Family Planning, Amazon Style. Natural History, Dec. 1992. Available at: www.popline.org/node/328215 (accessed 25 January 2016).

Hoff, D. (2012). *The State and the Stork*. University of Chicago Press.

Jones, R. K. and Dreweke, J. (2011). *Countering Conventional Wisdom: New Evidence on Religion and Contrceptive Use*. Guttmacher Institute: New York.

Jones, R. K., Finer, L. B., and Singh, S. (2010). *Characteristics of US Abortion Patients 2008*. Guttmacher Institute: New York.

Keeley, L. H. (1996). *War Before Civilization*. Oxford University Press.

Knodel, J. E. (1974). *The Decline of Fertility in Germany, 1871–1939*. Princeton University Press. p. 152.

Langer, W. (1972). Checks on Population Growth, 1750–1850. *Scientific American, 226*, 92–9.

Lee, J., and Feng, W. (2001). *One Quarter of Humanity*. Harvard University Press.

Lesthaegh, R. J. (reissued 2015). *The Decline of Belgian Fertility, 1800–1970*. Princeton University Press.

Livi-Bacci, M. (2012). *A Concise History of World Population*. Wiley-Blackwell.

Maguire, D. C. (2009). Population, Religion, and Ecology. In: R. S. Gottlieb (ed.) *Oxford Handbook of Religion and Ecology*. Oxford University Press.

Mason, K. (1997). Explaining Fertility Transitions. *Demography, 34*, 443–54.

Mason K. O. and Smith, H. L. (1999). "Female Autonomy and Fertility in Five Asian Countries." Available at: http://swaf.pop.upenn.edu/sites/www.pop.upenn.edu/files/Auto.Fert.pdf (accessed 15 November 2015).

Munera, A. (2000) *New theology on Population, Ecology, and Overconsumption from the Catholic Perspective*. In: Coward, H. and Maguire, D. C. (eds) *Visions of a new earth. Religious perspectives on population, consumption, and ecology*. State University of New York Press.

Nie, Y., and Wyman, R. J. (2005). The One-Child Policy in Shanghai: Acceptance and Internalization. *Population and Development Review, 31*, 313–36.

Pew Research. (2015). *The future of world religions: Population growth projections, 2010–2050*. Pew Research Center: Religion and Public Life.

Potts, M., Graff, M., and Taing, J. (2007). Thousand-year-old depictions of massage abortion. *Journal of Family Planning and Reproductive Health Care, 3* (4), 233–4.

Roudi, F. (2012). *Iran is Reversing Its Population Policy*. Woodrow Wilson International Center.

Seccombe, W. (1993). *Weathering the Storm*. London: Verso.

Secura, G. M., Madden, T., McNicholas, C., Mullersman, J., Buckel, C. M., Zhao, Q., and Peipert, J. F. (2014). Provision of no-cost, long-acting contraception and teenage pregnancy. *New England Journal of Medicine, 371*, 1317–23.

Shepherd, J. R. (1995). *Marriage and Mandatory Abortion among the 17th Century Siraya*. American Anthropological Association.

Tavernise, S. (2015, July 7). Colorado Finds Startling Success in Effort to Curb Teenage Births. *New York Times*.

Tsui, A. O., Casterline, J., Singh, S., Bankole, A., Moore, A. M., Omideyi, A. K., Palomino, N., Sathar, Z., Juarez, F. and Shellenberg, K. M. (2011). Managing unplanned pregnancies in five countries. *Global Public Health, 6* (supp1), S1-24.

UN. (1999). *Iran and Vietnam Win 1999 United Nations Population Award*. Available at: http://iranian.com/News/April99/award.html (accessed 5 March 2016).

United Nations. (2011, 2013 and 2015). *World Population Propects, The 2010 & 2012 Revisions*.

Wyman, R. J. (2013). The Effects of Population on the Depletion of Fresh Water. *Population and Development Review, 39* (4), 687–704.

Wyman, R. J. (2003). The Projection Problem. *Population and Environment, 24*, 329–37.

33

CONSUMPTION

Laura M. Hartman

Arguably, the human impact on the environment may boil down to one thing: consumption. As "the physical throughput of materials and goods in human lives," consumption is the primary interface between humanity and the natural world (Hartman 2011a: 9). We eat the non-human world; we wear it, we burn it for fuel, we use it as tools and devices; when we are finished, we dump it or burn it or compost it. If we care about the ecological systems in which we live, we should seek to minimize the ill effects of our consumption, but this is easier said than done. In a global, industrialized economy, we typically do not know the effects of our choices, and sometimes we are constrained by a system not of our own making, such that we do not have a real choice in the matter of our consumption's environmental impact.

In this chapter I begin conceptually: consumption is an existential problem. Consumerism is a strong cultural force, and the way it shapes desire must be addressed. Religions typically respond by calling for individual restraint, but some thinkers call for more systemic reforms. Further study on the topic is needed.

The existential problem

Consumption may be more harmful or less harmful, but ultimately it cannot be completely benign. In this world, we eat and are eaten (even if we avoid eating animals). Through our consumption, those of us in rich countries exploit the ecosystems we pillage and we profit from low-wage workers. We benefit from harm to others; we are complicit. This may demonstrate the Buddhist insight that all life is suffering, and complicity in that suffering is nearly impossible to avoid. Christian ethicist L. Shannon Jung writes, "Complicity is a curse, a curse that we inherit by being born into a consumer society. We are caught in this web, a web not entirely of our making, but one that is part of deep cultural structures that are part of us" (26). Most religions[1] carry an ideal of peace and harmony with nature and between humans, and with this ideal comes yearning for something like sustainability: the ability to consume in a way that does not cause suffering. Religions must wrestle with the existential reality of human lives whose very existence is bought at the price of others' lives.

The most common religious response is some form of austerity, to limit the amount of harm caused by consumption. For example, Jainism responds to this problem by prescribing dietary restrictions including vegetarianism and prohibition of certain fruits and vegetables that may contain small sentient organisms (Jain 2010: 126–127). Jainism (and Hinduism as well) promotes *aparigraha*, or non-possessiveness, as a virtue, which is a way of promoting diminished consumption (*Ibid*). Mahavira, the founder of Jainism, reached a point at which he renounced clothing entirely, because of concern for the wellbeing of the cotton plants and the risk that insects might be trapped and killed in folds of cloth (Chapple 2008: 516). Living without clothing may seem extreme, but most religions offer some attempt to minimize the harm caused by our consumption.[2] Purity traditions, such as Kosher food in Judaism or Halal foods in Islam, may also limit harmful effects, since some interpret Kosher and Halal slaughter laws as more merciful to the animals.[3]

Christian Quaker abolitionist John Woolman approached the existential problem of consumption with different presuppositions than Mahavira, but similar results. He maintained that God would not have created a world that required the suffering of others in order to supply each of us with the sustenance for life. Woolman drew a line separating luxury from necessity, asserting that God surely created adequate resources to meet our basic needs, but if we transgress into luxury, then our consumption is causing undue suffering (in his case, slavery) (Hartman 2011a: 39–40). Woolman's call for simplicity has something in common with Jainism's *aparigraha*: letting go of luxury as a response to the existential problem of consumption. If we consume less, perhaps the poor and other creatures will be able to flourish.

Another religious response to this existential challenge is to find a way to direct human consumption toward holy ends. Christian poet and farmer Wendell Berry puts it well:

> To live we must daily break the body and shed the blood of creation. When we do this knowingly, lovingly, skillfully and reverently it is a sacrament. When we do it ignorantly, greedily and destructively it is a desecration. In such a desecration we condemn ourselves to spiritual and moral loneliness and others to want.
>
> *(1981: 281)*

Berry, then, recognizes the fundamental tragedy of living in a world in which life feeds upon life. As a Christian he believes the world to be fallen and imperfect, but he nevertheless affirms the potential for consumption that is harmful, yet redeemed. This sacramental view, imbuing the material world with sacred significance even when it is devoured, is present in many traditions. Judaism's laws of Kashrut also speak to this reality: "Besides being life-sustaining, satisfying, and often joyous, eating is a holy act," when performed within the limits of Kosher observance; and eco-Kashrut (eating limited by environmental concerns) is an increasingly popular way to live out this religious commitment (Brook). Native American traditions also honor non-human lives lost for the sake of human survival. The Creeks, Cherokees, and many other Native Americans, for example, only kill animals for food when truly necessary; they apologize to, and thank, the animals who are killed; and they use every bit of the animal's body, wasting nothing, so that its death has proper value (Aftandilian 2011: 200).

Consumerism

In addition to the existential challenge of living in an eat-and-be-eaten world, contemporary religions also address a unique cultural force: consumerism. Consumerism may be defined as an ideology that sees consumption as an identity-forming, meaning-making activity that rivals religious practice in its importance for humans' sense of self and community (Hartman 2011a: 6). Consumerism tends to commodify non-commercial values, turning what was priceless (say, clean clear water) into something that may be bought and sold (say, in plastic bottles). This process also tends to objectify what might otherwise be a subject and render fungible what was once one-of-a-kind (viz. the ease with which members of consumer societies switch homes, employment, religions, spouses, as if all are somehow interchangeable products rather than covenants and memberships). Consumerism's scope is so wide-ranging that it is hard for people in the industrialized world to imagine disengaging from it. This system linking the market, advertising, production, and shipping comes at an environmental cost – and arguably a spiritual cost as well.

Thinkers of practically every religious perspective decry consumerism: consumption should not function as an idol; products are no substitute for religious devotion; and true peace and happiness is to be found in a spiritual life, not in ever-increasing consumption.[4] The tendencies of commodification, objectification, and interchangeability undermine the respect for life, respect for the eternal, and respect for human beings common to most religious belief. Islam is a commonly cited example of religious resistance to consumerism, because of the ways that modern consumerism, as an agent of globalization and "westernization," has been resisted in some majority-Muslim countries (see Izberk-Bilgin). Despite this, consumerism is truly infectious, and the consumption patterns and practices among many Muslims in majority-Muslim countries still reflect the influence of consumerism (Sandicki and Jafari).

By contrast, some religious institutions embrace or benefit from consumerism. Since consumerism says that one's identity lies in what one buys, religious institutions might sell items that broadcast one's religious identity (for example, a bracelet bearing a "What Would Jesus Do?" slogan). In India, images of Hindu gods are used to advertise products; religion is leveraged to enhance consumerism.[5] Consumerism's broad scope includes religious images and even religions themselves. Indeed, as Vincent Miller has observed, adherents might view religions as so many competing products, and choose the church or other institution that suits their consumer preferences. Religions now function, for better or worse, in a type of religious "marketplace" (Miller 2005: 9–11).

Desire

Advertising, the hallmark of consumer culture, seeks to cultivate desire. Sociologist Colin Campbell asserts that Romanticism, the cultural movement originating in eighteenth- and nineteenth-century Europe, with emphases on inspiration, subjectivity, and "self-expression and fulfillment," has contributed as much to the rise of consumerism as Protestant Christianity contributed to the rise of capitalism (279).[6] Mass consumption requires never-ending desire, something that Romanticism also promotes. Drawing on this, Matthias Zick Varul has controversially suggested that Sufi Islam, with its spirituality of imagination and longing, might unwittingly fuel a rise in consumerism in some areas. Varul may not adequately appreciate that Sufi Islam, and many other religions, teach that proper human desire is directed to God or the holy rather than toward consumer products. (And even Romanticism invites people to long for what is "natural" and "pure" over what is marketed and manufactured.) Clearly, religions both affect culture and are affected by it; when it comes to encouraging consumerism, most religions do so unintentionally, if at all.

Advertising, according to Christian theologian Georg Rieger, directly cultivates unquenchable desires in order to keep the engines of production bolstering the economy (2013, 9). Such work runs contrary to religious teachings, in which God or another sacred ultimate is the proper subject of desire. Buddhism, by contrast, teaches against desire. Consumerism promotes clinging to sense-objects and clinging to the self, but these are unskillful means that lead to suffering (Hartman 2011b: 157). The alternative is mindfulness, a meditative attention to one's experiences, in order to "learn to unhook from the craving" for material possessions (Kaza 2005: 150). Buddhists seek non-attachment rather than continual desire; humans should learn "to be content with what we have," as expressed in the Buddhist concept of *santutthi,* or satisfaction (Loy 2002: 59).[7] Buddhism's effect on environmentally conscious consumption is a continuing subject of study (see Minton et al.).

Calls for restraint

The response to out-of-control desire is often a call for restraint. Christian ethicist Maria Antonaccio describes asceticism in terms that apply to many religious practices of restraint, austerity, and simplicity:

> *Askesis* [is] reorienting human desires through practices that seem to run counter to what one normally understands as "flourishing." In such practices, conventional notions of what constitutes human fulfillment are shattered by the insight that true fulfillment is of a different order than what one had thought.
>
> *(2006: 90)*

Fasting, sexual restraint, voluntary poverty, physical simplicity, dietary limits: religions employ these practices, among others, to pursue the "true fulfillment" of which Antonaccio speaks. Christopher Key Chapple discusses ascetic practice in Jainism, Buddhism, and Yoga: all of these move the practitioner toward relevant spiritual goals, and – insofar as they entail living lightly on the planet – they are also good for the environment. Indeed, fasting, Sabbath keeping, and voluntary poverty can be practices that fulfill double duty: enhancing one's spirituality while also reducing environmental burdens.

The Japanese concept of *mottainai* means both "don't be wasteful," and "what a waste," and it "expresses a feeling of regret at wasting … a resource or object" (Taylor 2011: 32). *Mottainai* is a Buddhist concept that also draws on the Shinto reverence for objects: all objects have a spirit or *kami,* and therefore should be used with care (Kawanishi). Adopted by Kenyan Nobel peace prize winner and environmental leader Wangari Maathai, *mottainai* has attained international significance as a way to understand the importance of restraint and respect in the use of all material things (Taylor 2011: 36).

Constraining one's consumption can be a mark of membership and identity for some religious groups. Whether it is fasting during Ramadan or preferring to patronize brands and establishments known to be friendly to the religious community, some Muslim groups form their identity through restraint (Karatas and Sandiki). Jewish groups engage in similar practices when they consume only kosher foods or avoid shopping on the Sabbath. Amish, Mennonite, Quaker, and other Christian churches whose group identity is tied to "simplicity" in dress and other consumption also use restraint as a hallmark of membership in the faith. Some religious renunciants – notably Roman Catholic and Buddhist – are known for their vows of poverty and simple dress. Jain and Hindu renunciants also fast, live in poverty, and avoid excess in pursuit of holiness. This is a kind of reverse-consumerism; our identity comes not from what we consume, but from what we eschew.[8]

Gandhi's Hindu perspective on ecology and economics counseled temperance and self-sacrifice to live within Earth's limits; he famously said, "Earth provides enough to satisfy every man's need, but not every man's greed" (Shinn 2000: 228). Ronald Sider, an evangelical Christian, would agree with him. In *Rich Christians in an Age of Hunger,* Sider draws on the Bible and his own experiences to argue that affluent Christians are called to live radically simple lives in order to free up as much money as possible to aid the hungry poor around the world. Sider's link between asceticism and social justice is not unique: what is good for the soul can certainly also be a vehicle for justice. But this is not necessarily the case. Renunciation may be a retreat from the world's problems rather than a way to solve them. As Willis Jenkins puts it, "Even if the wealthy discipline their desires, it will not necessarily help the poor to satisfy theirs" (2013: 263). To address the links between poverty and consumption, a more systemic approach is needed.

Systemic perspectives

Some of the most interesting recent work being done in consumption involves not individual virtue or restraint but collective transformation. This is an important approach, because purchasing eco-friendly products or cutting back on consumption is only part of the solution. Many environmental problems are more systemic in nature. Most problems that are addressed through "green consumption" are only partly solved in this way. Those with adequate money and access can buy organic food, put solar panels on their homes, and so forth. But if such measures remain supererogatory, they will only be done by a small percentage of the population – far too small to create the major transition that is necessary if humans are to live in balance with the rest of life on earth.

The following three sections describe groundbreaking work being carried out by religious thinkers addressing systemic problems relating to consumption's environmental effects.

Work and the meaning of dignified labor

Not far below the surface of the conversation about consumption is a conversation about work, money, and human dignity. If consuming certain products harms the earth with minimal benefit to humans (for example, easily broken plastic children's toys of limited educational value), perhaps those products should not be produced at all. It is worth asking which jobs are actually contributing to the world's wellbeing and which are not. The Buddhist teaching of Right Livelihood criticizes ways of earning a living that cause suffering, including "any trades that involve weapons, poisons, drugs, alcohol, or any trade in human beings or animal flesh" (King 97–98). For ecologically minded Buddhists, prohibited employment might also include work that harms the earth.

In a similar vein, Gandhi called for "a life connected to physical labor and the earth," and his own symbolic actions of spinning and weaving speak to the value of manual labor in his thought: in this context, laboring meant self-reliance and self-rule, a political and spiritual goal for Gandhi (Shinn 200: 232). Christian theologians including John Paul II and Darby Kathleen Ray have written thought-provoking meditations on work, its spiritual meaning, and its relation to consumption.

More work needs to be done on this topic, particularly work's connection to wages. Environmentally preferred consumption choices are often more expensive (compare organic produce to its conventional counterparts), so people who earn low wages rarely can afford to engage in "green" consumption. According to environmental writer Michael Pollan, low wages and low-quality food exist in a "nonvirtuous circle," one necessitating the other (2010: 2). Arguably, low-quality products justify low wages, but they both cause human and non-human suffering. Cheap products are not the solution to human poverty, nor do they solve ecological impoverishment.

Consumption's connection to work, wages, and dignity requires a more systems-oriented discussion. Religious thinkers find themselves arguing for certain policies (such as living wage laws) or collective practices (such as granting workers greater self-determination), in addition to supporting particular consumption patterns (such as fair trade products).

Time, rest, and the Sabbath

Many types of environmentally preferred consumption are not only more expensive but require more time: walking rather than driving, hanging up clothes rather than putting them in the dryer, cooking whole foods rather than buying pre-processed foods, and so forth. But the epidemic of busyness and time poverty that afflict many in the industrialized world mean that people believe they are too busy to take such measures. This means that ecologically sensitive habits require a systemic re-patterning of one's life; such new patterns, insofar as they go against the grain of mainstream consumer culture, are best inculcated in groups. Without adequate social support, an individual's pursuit of an ecologically virtuous but inconvenient habit is usually doomed to failure. How do we make the time to walk rather than drive, or to cook from scratch rather than using pre-processed foods? There are religious concepts to address this question.

Rabbi Abraham Joshua Heschel's book *The Sabbath: Its Meaning for Modern Man* is as much about economics as it is about prayer. Observing a weekly Sabbath recalibrates our relationship with time. The Sabbath is a quiet day, free from "the profanity of clattering commerce, of being yoked to toil" (1952: 13). It should be "a day of armisitice in the economic struggle with our fellow men and the forces of nature," – which indicates that Heschel and others recognize the negative environmental impact inherent in nearly all human work (1952: 28). Indeed, what if all human work ceased for one day in seven? No delivery trucks burning gas; no factories giving off waste; no computers sucking energy. Much consumption would also cease that day. The environmental impact would be significant.

Seventh-day Adventist ethicist Miroslav Kis writes that Sabbath observance allows participants to take on God's perspective, such that "the priorities of Heaven, the divine sensitivity to injustice and oppression urge them to extend the Sabbath rest to those around them," which can include laborers who are denied a day of rest and also non-human creation (1995: 99). Theologian Norman Wirzba would agree. Wirzba advocates becoming "Sabbath people, who take a humble stance within creation, who recognize creation as a gift of God given for all to enjoy" – and doing so, he argues, means that "we simply have to realign and dramatically curtail our consumption habits" (2006: 151). The practice of Sabbath-keeping can retrain desire, curtail consumption, and cultivate awareness of creation – all within a context of joy and celebration (Hartman 2011a: 130–152).

While the concept of a day of rest is not prominent in Asian religious traditions, practices of meditation, quietness, and aesthetic simplicity connect with the Biblical Sabbath concept. Arguably, living a life grounded in silence and simplicity – whether it is a day of rest or a contemplative practice – frees mental space for ecologically minded consumption (Kaza [2005]). Doing so with community support can facilitate the development of systems that support such counter-cultural practices.

The economic system as a whole

"Our economic system and our planetary system are now at war," writes activist Naomi Klein (21). She is overstating the case, but only slightly: economics as we know it is driven by consumption and geared toward never-ending growth. The earth can withstand the onslaught of our economic system – but only to a certain extent. Increasingly, we are reaching the limits of nature's resilience and recognizing the harms that our consumption-driven, growth-oriented economy is wreaking. The systems within which we exist – the economy and the earth – are on a collision course. The laws of nature are not going to change to suit our economy; we must change the laws of our economy to suit the laws of nature.

Richard Norgaard's chapter in this volume addresses economics directly. Suffice it to say here that religious thinkers who seek to address the galloping consumption that is devouring the planet often conclude that the best solution involves re-thinking the entire system of unlimited growth, unrestricted trade, and financial speculation. In this perspective, the human economy as a system interconnects with the earth economy. According to South African Theologian Puleng LenkaBula, *botho*, a Sesotho word about right relationality, and *Ubuntu*, its equivalent in Nguni, speaks to this need. It is a rich concept that promotes unity and togetherness while also honoring difference and diversity. Ultimately, *botho/Ubuntu* leads to ecological justice, because it teaches respect and restrained consumption for others' sake: a completely different economic model. There is an echo here of E. F. Schumacher's so-called "Buddhist Economics," which promotes simplicity, non-violence, and care for right livelihood above profit and competition (1975: 53–55).

Research suggests that there are fruitful points of convergence between ecological economics and Chinese traditions as well. T. N. Jenkins suggests that Confucianism's "anthropocosmic view of embedded relationships," Daoism's teachings about balance and flow, and Buddhism's teachings about restraint and care for non-human creatures, all contribute to a re-thinking of economic systems toward ecological wholeness. The Daoist concept of *wuwei* can function as a political principle, according to Liu Xiaogan, leading toward alignment with, rather than action against, natural processes. In the same vein, the move toward an "ecological civilization" in contemporary China involves re-thinking economic systems through an environmentally aware lens. Chinese "ecological civilization" draws on Daoism and Confucianism in its views of harmony and responsibility, and some would say that its economic models can function as a direct alternative to capitalism and Marxism (Gare). "Ecological civilization" may also seek to redefine happiness as "ecological happiness," happiness which is social, sustainable, and intergenerationally just: this puts it in direct conversation with the desire-cultivating dimensions of consumerism (Zhou et al. 2013). More research on Chinese traditions, consumption, and systemic reforms of the economy is needed.

The work of Lutheran Christian ethicist Cynthia Moe-Lobeda in her *Resisting Structural Evil: Love as Ecological-Economic Vocation* is worth highlighting here. Following Martin Luther and other Christian thinkers, she argues that love, not greed, should define a new sustainable economy. She stakes out an "economic moral vision… in which all people have the necessities required for a healthy life, Earth's life systems are sustained and regenerated, and none accumulate vast wealth at the cost of impoverishing others or Earth's life support systems" (246). Such a vision cannot be attained in global neoliberal capitalism as we know it, on Moe-Lobeda's diagnosis. The aim, she writes, "is to re-arrange systems of transportation, food production, housing construction, and so on, so that our daily activities are in themselves practices that contribute to healthy eco-systems and socially just human relations" (252). In other words, Moe-Lobeda joins her voice with other religious thinkers seeking out new systems. This systemic perspective means creating "the kind of society… where it is easier for people to be good," in the words of Catholic Worker co-founder Peter Maurin (Day 1997: 280).

Conclusion

Consumption questions are thorny ones. Often there are no deeply satisfactory solutions, and thoughtful consumers are left yearning for a better system. As Willis Jenkins writes, once we stop desiring what the advertisers peddle, and start desiring what religions preach, we may find that we are "learning to desire revolution" (2013: 261). Deploying concepts such as asceticism, channeling desires, prioritizing justice, and creating visions of a better system, religions are vital partners in the conversation about consumption and ecology. Until we live in that "society... where it is easier for people to be good," consumption will continue to be a vexing problem, deserving of humanity's best and most insightful solutions.

Notes

1   I am trained primarily in western religions, with an emphasis on Christianity. In this chapter I attempt to include multiple religions, but based on my own background and training there will, unfortunately, be more examples from Christianity than from other religions. A full interfaith conversation on this topic has yet to develop: further study is needed.

2   A more moderate Buddhist response acknowledges the presence of harm in consumption, but emphasizes the human ability to "choose *how much* harm we want to be responsible for" (Kaza 2010: 54).

3   This claim is controversial. See Zoethout.

4   See, notably, Loy 1997 and Sivaraksa 2000.

5   This practice began with the British and its perpetuation is a relic of colonial norms. Scholars argue that the commercial content is less impactful than the religious imagery, but this is open for debate (Inglis 1998: 69).

6   The Romantic movement itself was heavily influenced by Enlightenment individualism, as well, and Max Weber's thesis about the causal links between Protestantism and the rise of capitalism has also been much debated. For a view opposed to Weber's, see Blanchard.

7   Interestingly, teaching about *santutthi* was banned in Thailand in the 1960s, because it was clear that too much contentment would undermine the country's economic growth (Kaza 2010: 50–51).

8   There is a risk, of course, that such practice becomes fetishized to the degree that it is seen as an end in itself, rather than a means to a spiritual outcome.

References

Aftandilian D. (2011) "Toward a Native American Theology of Animals: Creek and Cherokee Perspectives", *Cross Currents* 61:2, 191–207.

Antonaccio M. (2006) "Asceticism and the Ethics of Consumption", *Journal of the Society of Christian Ethics* 26:1, 79–96.

Berry W. (1981) *The Gift of Good Land*, North Point, San Francisco.

Blanchard K. (2010) *The Protestant Ethic or the Spirit of Capitalism: Christians, Freedom, and Free Markets*, Cascade Books, Eugene, OR.

Brook D. (2009) "The Planet-Saving Mitzvah: Why Jews Should Consider Vegetarianism", *Tikkun* 24:4, 29–37.

Campbell C. (1983) "Romanticism and the Consumer Ethic: Intimations of a Weber-style Thesis", *Sociological Analysis* 44:4, 279–96.

Chapple C.K. (2008) "Asceticism and the Environment", *Cross Currents* 57:4, 514–25.

Day D. (1997 [1952]) *The Long Loneliness: The Autobiography of Dorothy Day*, Harper San Francisco, San Francisco.

Gare A. (2012) "China and the Struggle for Ecological Civilization", *Capitalism Nature Socialism* 23:4, 10–26.

Hartman L. M. (2011a) *The Christian Consumer: Living Faithfully in a Fragile World*, Oxford University Press, Oxford.

Hartman L. M. (2011b) "Economics", in Bauman W. A., Bohannon R. R., and O'Brien K. J. eds. *Grounding Religion: A Field Guide to the Study of Religion and Ecology*, Routledge, New York, 147–62.

Heschel A. J. (1952) *The Sabbath: Its Meaning for Modern Man,* Farrar, Straus and Young, New York.

Inglis S. R. (1998) "Suitable for Framing: The Work of a Modern Master", in Babb L. A. and Wadle, S. S. eds. *Media and the Transformation of Religion in South Asia,* Motilal Banarsidass, Delhi 50–75.

Izberk-Bilgin E. (2012) "Infidel Brands: Unveiling Alternative Meanings of Global Brands at the Nexus of Globalization, Consumer Culture, and Islamism", *Journal of Consumer Research* 39:4, 663–87.

Jain P. (2010) "Jainism, Dharma, and Environmental Ethics", *Union Seminary Quarterly Review* 121–135.

Jenkins T. N. (2002) "Chinese Traditional Thought and Practice: Lessons for an Ecological Economics Worldview", *Ecological Economics* 40:1, 39–52.

Jenkins W. (2013) *The Future of Ethics: Sustainability, Social Justice, and Religious Creativity* Georgetown University Press, Washington, DC.

John Paul II (1981) *Laborem Exercens.*

Jung L. S. (2012) "The Reeducation of Desire in a Consumer Culture", *Journal of the Society of Christian Ethics* 32:1, 21–38.

Karatas M. and Sandiki, Ö. (2013) "Religious Communities and the Marketplace: Learning and Performing Consumption in an Islamic Network", *Marketing Theory* 13:4, 465–484.

Kaza S. (2010) "How Much Is Enough? Buddhist Perspectives on Consumerism", in Payne R. K. ed. *How Much Is Enough? Buddhism, Consumerism, and the Environment,* Wisdom Publications, Somerville, MA 39–61.

Kaza S. (2005) "Penetrating the Tangle", in Kaza S. ed. *Hooked! Buddhist Writings on Greed, Desire, and the Urge to Consume,* Shambhala, Boston.

Kawanishi Y. (2007) "Mottainai Grandma Reminds Japan, 'Don't Waste'", National Public Radio (www.npr.org/templates/story/story.php?storyId=14054262) Accessed 11 September 2015.

King S. B. (2009) *Socially Engaged Buddhism,* University of Hawaii Press, Honolulu.

Kis M. M. (1995) "Sabbath", in Teel C. W. ed. *Remnant and Republic: Adventist Themes for Personal and Social Ethics,* Loma Linda University Center for Christian Bioethics, Loma Linda, CA.

Klein N. (2014) *This Changes Everything: Capitalism vs. the Climate,* Simon & Schuster, New York.

LenkaBula P. (2008) "Beyond Anthropocentricity – *Botho/Ubuntu* and the Quest for Economic and Ecological Justice in Africa", *Religion & Theology* 15, 375–394.

Loy D. (2002) "Pave the Planet or Wear Shoes? A Buddhist Perspective on Greed and Globalization", in Knitter P. and Muzaffar C. eds. *Subverting Greed: Religious Perspectives on the Global Economy,* Orbis Books, Maryknoll, NY.

Loy D. (1997) "The Religion of the Market", *Journal of the American Academy of Religion* 65:2, 275–290.

Liu X. (2001) "Non-Action and the Environment Today: A Conceptual and Applied Study of Laozi's Philosophy", in Girardot N. J., Miller J., and Liu X. eds. *Daoism and Ecology: Ways Within a Cosmic Landscape,* Harvard University Press, Cambridge, MA 315–40.

Miller V. J. (2005) *Consuming Religion: Christian Faith and Practice in a Consumer Culture,* Continuum, New York.

Minton E. A., Kahle L. R. and Kim C. H. (2015) "Religion and Motives for Sustainable Behaviors: A Cross-Cultural Comparison and Contrast", *Journal of Business Research* 68:9, 1937–1944.

Moe-Lobeda C. (2013) *Resisting Structural Evil: Love as Ecological-Economic Vocation,* Fortress Press, Minneapolis.

Pollan, M. (2010) "The Food Movement, Rising", *New York Review of Books,* June 10.

Ray, D. K. (2011) *Working,* Fortress Press, Minneapolis.

Rieger, G. (2013) "Christianity, Capitalism, and Desire: Can Religion Still Make a Difference?", *Union Seminary Quarterly Review* 1–13.

Sandicki Ö. and Jafari A. (2013) "Islamic Encounters in Consumption and Marketing", *Marketing Theory* 13:4, 411–420.

Schumacher E. F. (1975) *Small Is Beautiful: Economics as if People Mattered,* Harper & Row, New York.

Shinn L. (2000) "The Inner Logic of Gandhian Ecology", in Chapple C. and Tucker M. E. eds. *Hinduism and Ecology: The Intersection of Earth, Sky, and Water,* Harvard University Press, Cambridge, MA.

Sider R. J. (2005) *Rich Christians in an Age of Hunger: Moving from Affluence to Generosity,* 5th ed. W Publishing Group, Nashville.

Sivaraksa S. (2000) "The Religion of Consumerism", in Kaza S. and Kraft K. eds. *Dharma Rain: Sources of Buddhist Environmentalism,* Shambhala, Boston, 178–182.

Taylor K. (2011) "'Mottainai': A Philosophy of Waste from Japan", *Kinesis: Graduate Journal in Philosophy* 38:2, 31–41.

Varul M. Z. (2013) "The Sufi Ethics and the Spirit of Consumerism: A Preliminary Suggestion for Further

Research", *Marketing Theory* 13:4, 505–512.

Wirzba, N. (2006) *Living the Sabbath: Discovering the Rhythms of Rest and Delight* Brazos, Grand Rapids, MI.

Zhou Z., Yang R. and Daojin S. (2013) "Happiness View in Ecological Civilization Age", *Canadian Social Science* 9:6, 80–84.

Zoethout C. M. (2013) "Ritual Slaughter and the Freedom of Religion: Some Reflections on a Stunning Matter", *Human Rights Quarterly* 35:3, 651–672.

34

GENDER INJUSTICE

Heather Eaton

> Gender justice must be a central focus for religion, ecology, and environmental humanities in an Anthropocene era if people are to live with adequate life options and justice.

Humanity is gendered, and gender is one determinant of self-understanding, social interactions, and life opportunities. The most recognizable gender dichotomy is that of female and male. All societies develop distinct gender codes for women and men. However, a spectrum of gender differences is now recognised and increasingly accepted. Gender refers to a composite of one's biological sex, the behaviours, attitudes, and emotions associated with one's sex, the personal and cultural processes of gender identity formation and expression, and one's sexual orientation. Gender and sexual orientation can include considering oneself to be lesbian, gay, heterosexual, bi or trans-sexual, and/or transgender. Research in gender studies reveals that sexual orientation and gender identities are more fluid or elastic than previously understood, and exist on a continuum rather than as fixed types. Such analyses challenge prescribed roles and cultural expectations of women and men.

Gender injustices permeate every society in the world. They include the systematic oppression of women, homophobia, laws against and punishment for homosexual relationships, bisexual practices, or transgender identities, and, an overall disdain and lack of support for any type of relationship or identity that is not hetero-normative (i.e. the alignment of biological sex, sexuality, gender identity, and gender roles). The increasing awareness of gender injustices and the global push for gender justice encompass activism and research that acknowledges and appreciates the complexity of humans as gendered beings.

Gender is now a topic of study at most universities worldwide, and spans several distinct fields of inquiry, including women's or feminist studies, queer studies, and gender studies (see Hall and Jagose 2012). Much of this has roots in the women's movement. The past century saw the beginning of women moving effectively, although unevenly, into social, political, and economic spheres.[1] The impetus of the women's movement and extensive feminist analyses is threefold. The first is to acquire equity, equality, and autonomy for women. The second is to understand and to eliminate the systematic oppression of women, which occurs, with variances, in every society. To transform the ubiquitous domination of women requires in-depth analysis of how such oppression is sustained by cultural, structural, and symbolic systems. The third impetus is to stop the pervasive violence against women in all cultures. The tenacious and global oppression of women is the most widespread form of gender injustice.

Gender injustices: religion and ecology

Feminist analyses have confronted religions from every aspect and angle over the past fifty years. Texts, rituals, ethics, leadership, and institutions have responded, more or less, to these challenges. Whether theology or religious studies, Eurowestern academic programs frequently include feminist courses. Feminist studies of religion have sought to enlarge the spaces to include women, to critique and dismantle the patriarchal bases and biases, or to advocate that religions are unequivocally and irredeemably oppressive to women. LGBTQ[2] communities and analyses also interact with, and critique, religions in multiple ways. However, while these academic discourses prosper, few address ecological matters in any depth.

The field of religion and ecology developed from many disciplines, analytic trajectories, and particular concerns. The crossroads of gender, religion, and ecology also come from distinct directions. For example, as Christian liberation theologies expanded their use of (post) Marxist analyses and/or critical theories, they became astute in understanding the nature of social oppression, confronting inequality, and advocating for social justice. The scrutiny of power and privilege in relation to structures of race, ethnicity, class, gender, and culture became forceful fora for critical analyses and social change. For a time these pursuits did not see ecological topics as directly pertinent. However, today, the attention to religions and ecology is worldwide. However, for most Christian liberationist groups, ecology is relevant chiefly as ecojustice: that is, when ecological decline becomes a further weight on those already suffering from the unequal distribution of benefits and burdens.

The majority of contributions that include religion, ecology, and gender come from ecological-feminism (ecofeminism) (see Warren 2015). Ecofeminism appeared in the 1970s predominantly in North America, although in *Le Féminisme ou la Mort* (1974) French feminist Françoise d'Eaubonne called upon women to lead an ecological revolution. Theologian Rosemary Radford Ruether wrote:

> Women must see that there can be no liberation for them and no solution to the ecological crisis within a society whose fundamental model of relationships continues to be one of domination. They must unite the demands of the women's movement with those of the ecological movement to envision a radical reshaping of the basic socioeconomic relations and the underlying values of this society.
>
> *(Ruether 1975, 204)*

Numerous conferences and popular ecofeminist publications in the 1980s were forerunners to academic research particularly in philosophy, sociology, theology, and religious studies. Ecofeminism became an umbrella term for historical associations between women and nature, for resistance to domination as a mode of inter-human and human–Earth relations, and for in-depth understanding of patriarchal worldviews and social structures. Ecofeminism mirrors the plurality within feminist analyses and the diversity within environmental thought.

Those studying the roots of ecological degradation repeatedly found that the devaluing of nature and the unrestrained destruction of the natural world are deeply embedded in the worldview(s), history, and expansions of Eurowestern societies. The domination of nature had infiltrated the intellectual foundations, habits of thought, and cultural practices to the point that ecological deterioration was not discerned. Theologians probed the aspects of monotheistic traditions that devalue, or emphasize salvation from, nature, and promote otherworldly aspirations. Ecofeminists paralleled these theoretical developments by incorporating analysis of domination, especially that of women and nature.

The theoretical and religious inquiries into the Eurowestern woman/nature nexus are prolific areas of ecofeminist research. The basic realization is that the oppression of women and the natural world is built into the very mode of interpreting both (Merchant 1980). Further, Christian thought, reinforced by philosophy, has both propelled and sanctioned the *feminizing of nature* and the *naturalizing of women* and their mutual subjugation (Ruether 1992; Gebara 1999; Eaton 2005). Nothing is clear-cut about these historical processes, but overall they point to a logic of domination embedded within Eurowestern worldviews (Warren 1996 and 1997). Hierarchical dualisms and binary oppositions were exposed and critiqued: superior/inferior, male/female, heaven/Earth, culture/nature, spirit/matter, divine/demonic, order/chaos, mind/emotion, etc. These dualisms permeate the religious and philosophical history and development of Eurowestern intellectual traditions. Ecofeminists trace the entrenchment of dualistic thinking, and expose the tendencies to devalue and then oppress the inferior dyad.

Systems of domination become naturalized and customary within a worldview structured with hierarchal dualisms. Ecofeminists show how the dominations of women and nature have connections to the development of patriarchy, social hierarchies, misogyny, agricultural mastery, slavery, anthropocentrism, androcentrism, and what some call a conquest of reason over nature (Plumwood 1993 and 2001). Ecofeminists demonstrate critical links, or intersectionality, among influential and pervasive spheres of domination: gender, class, ethnicity, and nature. The relationship among hierarchies, power, agency, gender, and social change are frequent discussion topics. The acumen of ecofeminist theories is staggering, with intense debates about origins, ideological sub-structures, beliefs, practices, and identities. Theories about essentialism or gender construction are debated, as well as how domination became an established mode of social relations, which includes, but is not limited to, gender and the natural world.

The theoretical inquiries and contributions from ecofeminism pushed the field of religion and ecology to consider the ubiquitous nexus of women and nature. While ideological connections between women and nature have been found in all religions, they are diversely constructed and maintained, and they contribute to ecological ruin and the subjugation of women in different ways (Eaton and Lorentzen 2003).

Outside of academia, and beyond the boundaries of particular religions, the rise of interest in and experimentations with women's and (eco)feminist spiritualities has flourished. Ecofeminists debated the meanings of "the feminine." Most feminists work to deconstruct the binary oppositions or hierarchical dualisms within which the feminine/masculine dichotomy resides, and take an avid anti-essentialist position. Others, albeit few, support a revised interpretation of "the feminine." They consider an essentialist position to be either ontologically accurate, or strategically useful. Throughout the historical development of ecofeminist inquiries, many debates occurred surrounding the cross-cultural similarities and differences of the women/nature nexus, essentialism, to what degree religions are malleable to gender justice, and the feminist methods of privileging women's experiences.

Gender analyses, empirical data, and lived realities: gains and gaps

Feminist and gender analyses, beginning with the concern for gender injustices, have made significant contributions to mainstream studies in religion, philosophy, epistemology, history, economics, international development, psychology, critical theories, and so on. In postmodern and poststructuralist discourses, including postcolonial, eco, and literary criticism, gender explorations are original, creative, intelligent, and influential in deconstructing binary or habitual modes of thought. Feminist critiques have shaped psychoanalytic theories, including profound inquiries into embodiment, desire, identity, and multiple genres of relationality. Current deliberations in feminist materialism, posthumanist ethics, entanglement, and dynamic interactions, and other trans and postdisciplinary inquiries are both copious and innovative.[3] Ecofeminist analyses, initially prolific but faded from view for a time, are experiencing a resurgence. These academic efforts represent potent, fertile, and immense intellectual forces.

Without undervaluing their accomplishments, a large gap persists between academic feminist scholarship and the recurrent gender injustices. Feminist and gender insights tend to be found in confined cultural places, such as graduate university programs, and are often esoteric and inaccessible without advanced intellectual skills. Yet, patriarchy, while theoretically dismantled, remains operative throughout the world.

There are ongoing disconnects among (eco)feminist theoretical work, the empirical concerns of gender injustices, the ecosystem and planetary degradation, and, the complex relationships between these and gendered social lives. A brief survey of the empirical evidence of the pervasiveness of gender injustices, especially as related to women, reveals the nature of these theoretical–empirical gains and gaps.

Empirical observations: women's life struggles

There are numerous documented difficulties and injustices for women in virtually all countries.[4] The most severe is the systemic and persistent lack of basic human rights: food, water, safety, shelter, education, and health care. Gender injustices are further differentiated and exacerbated by intersectionality: structural inequalities or entrenched practices of domination reinforced through intersections of class, race, ethnicity, gender, sexuality, age, ability, and nationality. For example, an unschooled Dalit woman in rural India will experience more systemic oppression and have fewer rights and opportunities than a university-educated urban Caucasian woman in England. However, they both will face the risk of physical and sexual violence.

The United Nations deems that the gamut of violence against women is one of the most serious and pervasive human rights violation anywhere, and everywhere, in the world.[5] The World Health Organization (WHO) considers violence to be the most significant health risk to women: a global public health problem that affects between one- and two-thirds of women, depending on the culture.[6] In Haiti, India, or the Democratic Republic of Congo, violence, usually sexual, is inflicted upon over fifty percent of the women. Violence against women is difficult to transform because it is connected to the sexualizing of women's bodies and patriarchal beliefs of male rights, power, and control. The global trafficking of women and girls is the second largest international crime industry (behind illegal drugs and surpassing arms trafficking as of 2010), and is an extension of the sexual oppression of women on a global scale.

Each country and society is distinct. Prospects vary among the women of the same culture, and women have dissimilar experiences. Yet there are combinations that reinforce systemic injustices. When poverty, illiteracy, food insecurity, lack of education, and commanding religious or political patriarchal governance combine, there is considerable gender injustice. Sexual regulation of women is cross-cultural, although differentiated in matters of sexual access, prohibitions of sexual assault, legal resources, clothing options, modesty codes, and women's reproductive choices. The amount of freedom to conform or not also varies, as do the consequences. Women may comply with strict gender norms due to fears of Sati (bride burning) in India, or honour killings, or sanctioned sexual assaults used as punishment. Genital mutilation

Heather Eaton

and virginity tests occur in several countries. Dowry payments, forced marriages, child brides, early pregnancies, and son preference are not uncommon. A sexual division of labour found in most societies renders family sustenance as women's work. Women's unpaid and informal work is acknowledged to be the major portion of the global economy, but evaluated as economically unproductive and irrelevant to the gross domestic product. Many women have little access to economic resources, ownership of land, or commercial businesses.[7] All of this perpetuates the feminization of poverty globally.

The insights and activism for gender justice are gaining traction around the world, albeit most often in urban and educated contexts. Yet patriarchy functions in countless ways. The leadership and practices of corporate capitalism, political governance, military forces and policing, and legal and religious institutions are patriarchal bastions. In consumerist cultures, the beauty, cosmetic, and clothing industries persuade women to redesign themselves, while women in factories and sweatshops make the products. Objectifying and sexualizing women's bodies, from mainstream media to pornography, prostitution, and trafficking uphold patriarchy's cultural monopoly.

In highly industrialised countries or in affluent contexts, gender injustice can be subtle. Women have more life choices, educational possibilities, and are somewhat unfettered from a strict sexual division of labour. However, professions such as engineering, managerial and executive positions, military service, or political leadership remain male dominant worldwide. In general, women encounter glass ceilings, earn less than male counterparts, and may feel pressure to choose between children or professional advancement. Regardless of the country, workplace sexual harassment, and sexual assaults are widespread. Thus while there is cultural diversity, there are also cross-cultural similarities and global patterns of gender injustices.

Theorists have realized that the observable issues are only a part of well-established cultural gender asymmetries. It is not straightforward to determine the causes of systematic gender injustice. Deep-seated values are embedded in diverse social practices entangled in ideological, domestic, social, economic, political, and religious affairs that promote ideas that women are inferior to, or complementary to, heterosexual men: men being the human norm. Complex ideological substructures saturate cultural and religious symbols connected to myriad patterns of domination.

It is important to recognize that many men do not deliberately engage in the overt oppression of women. These are cultural systems that condition both women and men to accept and promote gender injustices. For example, some women internalize sexism and use patriarchal privilege to oppress other women. Many men gain little from this privilege, depending on the intersectionality of race, class, sexual orientation, and ethnicity. Furthermore, the identity and behavioural constraints for men, in strict patriarchal contexts, are often debilitating, preventing men from acknowledging vulnerabilities, or building social alliances outside of military, sports, or economic contexts. Globally, studies about men are just beginning. Meanwhile most men, and many women, know little about gender analyses.

LGBTQ research has probed multiple aspects of male, female, and LGBTQ identity formation, sexual diversity and orientation, relationships between sexuality, power and domination, and masculine cultural stereotypes, all of which confront gender injustices. Patriarchy depends on a biologically or religiously based gender dichotomy of essential differences between women and men. Yet the characteristics, sexual orientation, and public presentation allotted to females and males are not commensurate with the gender varieties of women and men. This confirms that the binary gender categories of man/masculine and woman/feminine are socially constructed ideologies. Thus as a social construct, it can be changed. Patriarchy, as an ideological construct and social order that privileges hetero-normative men, continues to reign over much of the world. It is at the heart of most, if not all, gender injustices. But it can be changed.

Feminism is a challenge and opportunity to reconsider tropes on gender. Responses to feminism range from acceptance to a backlash against women. Yet, the insistence on gender justice is reverberating globally. Little was known twenty years ago about the extent of gender injustices, signifying the lack of importance. The colossal amount of data about the oppression of women is highly significant because it reveals that such domination is becoming unacceptable. LGBTQ insights are furthering the analyses and expanding the terrain. It remains that the United Nations consistently finds that women's access to economic resources, income, and employment is uneven, and health, nutritional, and educational status, while sometimes improving, is precarious. Violence against women is pervasive. Globally and progressively women's rights are accepted into law and yet are tenuous in practice.

Gender and ecology

As the Earth's primary life-support systems of air, water, and land become unstable, people are forced into new and more burdensome relations with the natural world. Often entire communities are forced to become environmental refugees, living in camps or overcrowded urban settings. The United Nations recognizes that frequently ecological problems disproportionately affect women, and poor women are the first victims.

Yet, the gendered implications of sustainable food sources, soil erosion, potable water, deforestation, access to fuel, and reduction of common land are seldom addressed by academics working on religion, gender, and ecology. Even when they are, gaps persist between the research on the topic and transformation of the problem. Questions that need addressing are many. What sources are used to consider the intersection of religion, ecology, and gender? What questions guide the research? What is the relationship between analytic research and social change? What is the role of research on these topics if not to transform them? What institutions need to collaborate to alter the patterns and practices of domination? How can knowledge be effective to confront and amend gender injustices? What are the roles religions play in sustaining, confronting, or transforming gender injustices?

A second significant gap is between the work on gender injustices and religion, and the escalation of ecological ruin. Ecological issues are not prominent in gender and religious studies, or with those dealing gender injustices on the ground. This could be for many reasons. Gender issues can be overwhelming, and adding ecological difficulties may seem unmanageable. Given that gender matters are readily dismissed or obscured, it is arduous to continually push for gender justice. The challenges of responding to ecological degradation are massive, and also require profound and extensive social transformations. Gender is not readily added to ecological issues. It may be that the personal nature of gender issues and identities feel more immediate than climate change, for example. Or it may be that gender studies, like other disciplines, are slow to develop ecological awareness.

Gender, religions, and ecology: transformations of consciousness

Stepping back from the myriad issues and assessments the impulse for gender justice is not only a social political movement and an ethical transformation. Feminism along with gender/sexual diversity projects represent a historical revolution of consciousness: arguably the largest shift of consciousness since the Neolithic revolution. It is important not to underestimate the depth of challenge the women's movement and LGBTQ initiatives bring to human social establishments, ideological and cultural constructs, identity formation processes, and all the related and embedded symbolic facets and material manifestations.

Patriarchy has been the ruling social, ideological, and symbolic organization for at least 3,000 to 8,000 years (see Bennett 2006 and Lerner 1986), It has been, and is, the governing social form virtually everywhere. Classical religions all developed with the rise of patriarchy. Among the many theories of the origins of patriarchy, none are conclusive. There are even more theories about male subjugation of and violence towards women, and misogyny. The roots of these are buried deep in an obscured past, probably beyond the Neolithic era.

What is occurring is an awakening of consciousness regarding human social organization. Something new is emerging, sparking a historical transformation. Gender justice is about an amplified and strengthened understanding of humanity. This changes the patriarchal reference points of hierarchies, supremacies, and male dominance, bringing forth a radically new perception of human capacity, differentiation, and complexity. The subjectivity, diversity, and elegance of the human, as a species, are being expanded, enhanced, and intensified.

Religious traditions are fully implicated, as they touch many aspects of human life. The impact and importance of religious symbols, rituals, traditions, and worldviews is considerable. Given that, worldwide, women make up the majority of religious adherents, it is evident that religions are relevant and influential. From the standpoint of opposing gender injustice and promoting justice, religions are more often an obstacle than an ally. Traditional forms of religions, institutions, and religiosity tend to be part of the problem of gender injustices because they perpetuate patriarchal norms. However, feminism and gender justice are a public challenge to and confrontation with patriarchal forms of religions, and are gradually achieving change.

The field of religion and ecology has grown into a significant force in both academic study and religious practices. Transformations are occurring in most religions traditions throughout the world. Interpretations of the extent of the ecological challenges to religion vary. For some it is retrieving, re-evaluating, and reconstructing of religious traditions to find insights for ecological stewardship. For others, it is evident that religions themselves are in transition. The humanities have brought forth attentiveness to values, knowledge production, power, symbolic consciousness, and postcolonial and postmodern epistemologies, which have unsettled stable meanings of "religion." Due to the extent of the ecological crisis, as well as to knowledge of evolution, Earth sciences, physics, and cosmology, it is evident that religions are challenged to reinterpret themselves. Both epistemological revolutions as well as planetary transformations are occurring. These are compelling an immense transformation of religions. Many recognize that humanity, along with all our reflective traditions and orienting worldviews, must move towards a planetary, in addition to global, vision in order to realize the limits of the natural world.

Gender and ecology cannot simply be added to religions. Each is stirring an awakening that is new in human history. Ecological worldviews together with the movements for gender justice are two revolutions of consciousness. Together they are invoking transformations in religions. If combined, they are potent forces for an ecological civilization that includes equality and human rights.

The Anthropocene

There is increasing recognition that Earth's biosphere is entering a new era, termed as the Anthropocene: an era of anthropogenic ecological impacts on a geological scale. Various meanings to an Anthropocene epoch are being defined, debated, and contested. Here, the Anthropocene signifies that the ecological influences of *Homo sapiens* are so great that the Cenozoic era is ending, and that most planetary processes, ecosystems, evolutionary trajectories, and climate and hydrological systems are being disrupted or in decline. It signifies that we must reassess ourselves beyond ethnic communities and nation states to include a global and planetary perspective and vision. Further, from bioregional to planetary ecosystems, ecological knowledge is developing and must be embedded in a vision, social imaginary, or worldview to orient humanity in this Anthropocene era. For some, the Anthropocene is interpreted as the "rise of humanity" above the constraints of nature. Here, it means that humanity emerged from, is part of, and belongs to a biosphere of life. Only within this Earth community can humans flourish.

Earth sciences have determined that the current ecological impact is anthropogenic; however, acknowledging the validity of the Anthropocene is only one step. Thinking through the implications, grappling with choices and decisions, and planning for the consequences are quite another. This requires the humanities, including religion and the social sciences, considered as environmental humanities. The following are four areas where the environmental humanities, including religion, are important for a liveable future and decisive for gender justice.

1. Ecological issues are often analysed and described in scientific terms. Solutions are usually framed as technical interventions. The various ecological crises—climate changes, toxins, soil erosion, and environmental illnesses—are spawning technological creativity and innovation. What are often missing in ecological reporting are debates about values, attitudes, responsibilities, long- and short-term costs of choices, and preparation for ecological instability. Conversations about values, ethics, and meaning are decisive. Feminist contributions in ethics, relationality, and new materialism offer an extensive challenge to a human/nature divide. Together these disrupt anthropocentric hubris by making evident that mutual embeddedness and radical connectivity and continuity are always operative. Ecological issues and human wellbeing are reframed and intertwined. The need for science, the humanities, and feminist insights to collaborate is essential. Religious institutions are one of very few public spaces where discussions about values, ethics, and social choices are possible. Taking a leadership role in both encouraging these debates, as well as affirming the religious significance of the Earth, is a key role for religion and ecology efforts.

2. Ecological disaster data proliferates. Yet most governments seem unable or inept to address the severity of domestic or planetary ecological issues. Such a situation creates great tensions. Few citizens feel able to respond because of an insufficient scientific background, the inability to envision alternatives, or a sense of personal or political powerlessness. Apocalyptic predictions generate fear, which impairs a sense of agency. One theme within the religion and ecology work has concentrated on issues of worldviews. This is vitally important in four ways: an ecological imaginary or vision functions to promote ecological literacy; it affirms connections between a viable Earth community and religious traditions; it encourages transformations of religions for an ecological era; and it provides an alluring, rather than apocalyptic, orientation and vision. An ecological worldview is also an ethical appeal for the common good of humanity, while diminishing an anthropocentric bias and strengthening the notion of a planetary ecological community. People need a vision, and hope, in order to make difficult ethical choices.

3. A central debate is how far global, planetary, or the Anthropocene discourses erase differences, and reinscribe "the age of man." Discussions that speak of "humanity" tend to homogenize and obscure marginalised and dissimilar people and communities. In turn, this reinforces unequal power distributions. Additionally, to accentuate "diversity" (gender, multicultural) can acknowledge differences, yet leave systemic inequities, power asymmetries or structural injustices intact because there is no sharing of political power. The juxtaposition of the unity and diversity within "humanity" is challenging.

4. Nation states are politically and ideologically ill equipped to address ecological problems, often because of their global economic entanglements. Democracy is being managed as a storefront for camouflaged global forces and corporations. The rise of the Eurowestern security states are overtaking democracy, and with alliances of countries waging a 'war on terror." Sectarian violence is increasing. These are enabling a global escalation of violence and a proliferation of arms and militarized states, which exacerbate and obscure ecological problems. Further, evidence mounts that as conflicts increase, women lose rights and violence against women escalates.

The counter response is to cultivate a global ecological democracy. Terms such as inclusive democracy, global ecological citizenship, biospheric egalitarianism, and global biodemocracy are surfacing around the world. Planetary systems, such as climate change, are difficult to address not only due to jurisdiction and national sovereignty, but also because political and ecosystem boundaries are unrelated. New governing structures are required. In addition, the focus on democracy means inclusion and political participation. The decades of work in gender justice is a crucial partner in these efforts. A third component of ecological democracy is a commitment to peace, with an increasing interest in nonviolent resistance. Climate change protests, demonstrations, and political activism are forms of ecological democracy in action. These efforts are an antidote to unresponsive governments and the militarized nation state.

Religion and ecology must wade into the political fray and claim their transformative power. Religious institutions have access to myriad communities, and can encourage politically active citizens. The combined efforts of religion and ecology, gender justice, and environmental humanities possess considerable influence to enliven a politically active ecological citizenry.

The collective perception is that we live at the edge of an era, facing challenges of a type and magnitude not faced previously by human communities. There are multiple causes and uncertain solutions. Gender injustices can be transformed, as they are sufficiently understood and there are countless local, national, and global action plans. Political will is required. Ecological problems will intensify. Without serious attention to gender justice, democracy, commitments to peace, and a planetary vision, as ecological problems exacerbate so will the systems of oppression. Ecological security will be only for the privileged. Gender justice must be a central focus for religion and ecology, and for environmental humanities in the Anthropocene.

Notes

1 In this essay the terms *the women's movement* and *feminism* are somewhat interchangeable.
2 LGBTQ refers to Lesbian, Gay, Bisexual, Transgender, Queer/Questioning, Intersex, Pansexual/Polysexual, Asexual/Autoerotic/Ally.
3 See New Materialism Society (2015) Topics and bibliography (http://newmaterialismsociety.co/wordpress/topics-and-bibliography/) Accessed 3 October 2015; and New Materialism (2015) NM publications + reference bibliography (http://newmaterialism.eu/year-3-publications-4/bibliography) Accessed 3 October 2015.
4 UN Women: The United Nations Entity for Gender Equity and the Empowerment of Women. www.unwomen.org/en/what-we-do Accessed 26 March 2016.
5 United Nations Humans Rights: Office of the High Commission for Human Rights (2010) Violence against women (www.ohchr.org/EN/NewsEvents/Pages/ViolenceAgainstWomen.aspx) Accessed 3 October 2015.
6 World Health Organization (2013) Global and regional estimates of violence against women: prevalence and health effects of intimate partner violence and non-partner sexual violence www.who.int/reproductivehealth/publications/violence/9789241564625/en/ Accessed 26 March 2016.

7   UN Women (2015) Progress of the world's women 2015–2016: Transforming economies, realizing
     rights www.unwomen.org/en/digital-library/publications/2015/4/progress-of-the-worlds-women-
     2015#view) Accessed 3 October 2015.

References

Bennett J. (2006) *History Matters: Patriarchy and the challenge of feminism,* University of Pennsylvania Press,
     Philadelphia.
Eaton H. (2005) *Introducing Ecofeminist Theologies,* T&T Clark, London.
Eaton H. and Lorentzen L. A. eds. (2003) *Ecofeminism & globalization: Exploring culture, context, and religion*
     Rowman & Littlefield, Lanham.
d' Eaubonne, F. (1974) *Le féminisme ou la mort,* P. Horay, Paris.
Gebara I. (1999) *Longing for Running Water: Ecofeminism and liberation,* Fortress Press, Minneapolis.
Hall D. E. and Jagose A. eds. (2012) *The Routledge Queer Studies Reader,* Routledge, London.
Lerner G. (1986) *The Creation of Patriarchy,* Oxford University Press, Oxford.
Merchant C. (1980) *The Death of Nature: Women, ecology, and the scientific revolution,* Harper Row, San
     Francisco.
Plumwood V. (1993). *Feminism and the Mastery of Nature,* Routledge, London.
Plumwood V. (2001) *Environmental Culture: The ecological crisis of reason,* Routledge, London.
Ruether R. R. (1975) *New Woman/New Earth: Sexist ideologies and human liberation,* Seabury Press, New
     York.
Ruether R. R. (1992) *Gaia and God: An ecofeminist theology of earth healing,* Harper Collins, San Francisco.
Warren K. (1996). *Ecological Feminist Philosophies,* Indiana University Press: Indianapolis.
Warren K. ed. (1997) *Ecofeminism: Women, culture, nature,* Indiana University Press, Indianapolis.
Warren K. (2015) "Feminist Environmental Philosophy", (Summer 2015 edition) in Zalta E. N. ed.,
     Stanford Encyclopedia of Philosophy (http://plato.stanford.edu/entries/feminism-environmental/)
     Accessed 3 October 2015.

35

ENVIRONMENTAL JUSTICE

David N. Pellow and Pengfei Guo

This chapter's focus is on what we can learn from thinking through connections between main factors raised in the field of religion and ecology and the field of environmental justice (EJ) studies. A number of scholars have noted that there is not much literature linking these two fields (Immergut and Kearns 2012; see Jenkins 2013 for an insightful analysis of the limits and problems stemming from the methodological whiteness of religion and ecology), but some have begun to bridge this gap (e.g. Jenkins 2013; Ruether 2005).

This chapter reviews some of the major ideas, advances, and debates in the field of EJ studies and offers a series of perspectives on how that work might be useful for interpreting some of the most significant intersections among religion, spirituality, and EJ struggles.

Environmental justice movements

The U.S. EJ movement gained visibility and strength beginning in the late 1970s and early 1980s, as activists and movement networks confronted a range of locally unwanted toxic land uses that were disproportionately located in communities of color and working class neighborhoods across the nation. This movement fused discourses of public health, civil and human rights, anti-racism, social justice, and ecological sustainability with tactics like civil disobedience, public protests, and legal action to prevent the construction or expansion of unwanted and controversial facilities and industrial projects, such as landfills, incinerators, mines, and chemical plants. Activists also demanded that owners of existing facilities shut them down or improve their operational safety, reduce pollution levels, and provide economic benefits to local residents. This movement sought to openly integrate campaigns for justice on behalf of vulnerable human beings with the goal of ecological sustainability (see, e.g., Brisman 2008; Bullard 2000).

Scholars soon recognized that the movement had long been a global phenomenon as indigenous, people of color, and working class communities in the global South articulated their concerns over ecological and public health, cultural integrity, and sustainable economic development in the face of increasing threats from states, corporations, developers, and invading settlers. By the 1990s and 2000s, transnational networks were focused on a range of global EJ issues including mining, incineration and solid waste, agriculture, electronics, hydroelectric power, energy, free trade, climate change, and many others.

The emergence of EJ studies

The field of EJ studies developed in the U.S. during the 1970s and 1980s—and soon grew as a global field—as social scientists took note of the fact that myriad environmental hazards disproportionately affect poor communities, communities of color, and other marginalized populations (UCC 1987). Much of the scholarship in the field of EJ studies has been sparked if not inspired by the presence and visibility of the EJ movement, which has not only demanded substantive changes in corporate and government policy making and institutional practices with respect to the production, use, and distribution of industrial hazards, but has also repeatedly called on scholars to document environmental inequalities.

Much scholarship on EJ places emphasis on the documentation of disproportionate environmental and/or industrialized hazards. Hundreds of studies have documented that racial minorities, people of lower socioeconomic status, and other marginalized communities are disproportionately affected by toxic facilities, largely through their residence, but also through the location of institutions, such as schools (see, for example, Crowder and Downey 2010). More recently, important work by scholars focusing on food justice movements reveals that, as a result of a complex matrix of policy mechanisms that tend to subsidize and support the industrial agricultural system, working class communities and communities of color are often sites of hunger, malnutrition, and obesity as a result of the dominance and control of food systems by a small group of large corporations focused on making profit rather than feeding communities (Alkon and Agyeman 2011). Thus the literature on food justice is an integral component of EJ studies.

While the field of EJ studies is dominated by an emphasis on the spatial relationship between environmental threats and racial inequality (and, to a lesser extent, class), there are also important links between environmental inequalities and gender, citizenship, immigration, indigeneity, and nation. (Pellow and Park 2002; Cole and Foster 2000; Taylor 2009). Recent studies document the ways that women experience and resist discriminatory environmental policies in workplaces, residential communities, and elsewhere (Buckingham and Kulcur 2010). Immigrants in the U.S. are more likely than non-immigrants to live in residential communities with high concentrations of pollution (Hunter 2000; Mohai and Saha 2007; Bullard et al. 2007). Ethnographic studies reveal similar dynamics and demonstrate how ideologies of exclusion, privilege, and nativism support the production and maintenance of such an unequal socioecological terrain, both inside and outside the workplace and residential communities where immigrants may labor and live (Pellow and Park 2002; Park and Pellow 2011). With respect to indigenous peoples in various nations around the globe, these populations are systematically excluded from participation in environmental decision-making, frequently evicted from their lands, disproportionately exposed to pollution, and restricted from using ecological materials within their territories while states and corporations unilaterally appropriate them (Smith 2005; Agyeman et al. 2010).

Climate change is an issue sparking significant and growing attention within EJ studies, as the production of greenhouse gases and carbon emissions threatens the stability of life on this planet and also reflects deep inequalities within and across nations and regions. Several studies show that people of color, women, indigenous communities, and global South nations often bear the brunt of climate disruption in terms of ecological, economic, and health burdens—giving rise to the concept of *climate injustice* (Bullard and Wright 2012; Roberts and Parks 2007). These communities are among the first to experience the effects of climate disruption, which can include "natural" disasters, rising levels of respiratory illness and infectious disease, heat-related morbidity and mortality, and large increases in energy costs.

Debates in EJ studies

Since the field's early days, scholars have debated the nature and extent of environmental inequality across various communities, stemming from the use of different methodological approaches, levels of analysis, data sources, varying geographies, and types of hazard in question. The overwhelming majority of studies of environmental inequality conclude that there is strong evidence of the role of racial inequalities underlying this phenomenon (Mohai and Bryant 1992; Downey 2006), while some studies find evidence that other social categories like age, poverty, and class matter as much or more than race, depending on the context (Mennis and Jordan 2005). A related debate has emerged among scholars around whether EJ studies have placed too much emphasis on environmental disadvantage to the neglect of environmental privilege; i.e., the social forces that produce exclusive and relatively protected socioecological spaces for elites (Norgaard 2011; Park and Pellow 2011; Taylor 2009). Finally, some scholars have called for a move beyond the "distributive paradigm" associated with a focus on the spatial concentration of environmental hazards in vulnerable communities (Schlosberg 2007). Much of this scholarship argues that other critical matters like "recognition," participation, and capabilities are critically important yet neglected aspects of justice in EJ studies. While we agree, we would approach this subject with caution because it runs the risk of overlooking the fact that inequality is at the root of misrecognition, lack of participation, and highly uneven capabilities (the resources, opportunities, freedoms, and institutions necessary for all of society's denizens to experience full membership and inclusion) and that this is inherently a distributional question. Moreover, we would contend that, the dynamic nature of distributional politics and impacts has still yet to be fully explored. In the words of the field's founding scholar, Robert Bullard, EJ—the "principle that all people and communities are entitled to equal protection of environmental and public health laws and regulations" (Bullard 1996: 495)—is not, at root, an "environmental" issue; it is most effectively understand as a *social* problem that originates in and is reinforced by social structures and discourses. This is a core insight because EJ scholarship redefined what counts as an "environmental" issue by challenging conventional framings to be inclusive of social justice.

Environmental Justice, religion, and spirituality

A growing number of scholars are exploring the many ways that religion and spirituality play a role in shaping EJ struggles. From our review of various literatures in which religion and EJ (broadly defined) intersect, we find that there are three dynamics that are frequently at play: (1) when religious institutions, leaders, or doctrines are deployed to legitimate the commission of environmental injustices; (2) when religious or spiritual practices are harmed by the commission of environmental injustices; and (3) when religion or spirituality becomes a source of resistance and resilience against environmental injustice. In the following sections of this chapter, we explore each of these three dynamics and consider what they suggest for new directions in scholarship as well as for the future of social movements that combine religion, spirituality, ecology, and EJ.

Religion in the service of environmental injustice

Religion—particularly in the form of institutions—has all too often been used to justify or legitimate the commission of environmental injustices. The historical record is replete with instances of such practices, including, for example, the ways church leaders facilitated the conquest of Native American peoples and their lands (LaDuke 2005; Smith 2005). As indigenous scholar and activist Winona LaDuke writes, "colonialism has been extended through an expansion of centralized power by the spread of Christianity, Western science, and other forms of Western thought" (LaDuke 1994). For example, when the University of Arizona and the Vatican built the Mt. Graham International Observatory, they initiated a bitter conflict that pitted academia and the Catholic Church against the indigenous Apache nation for whom Mt. Graham is a sacred place. Despite opposition by the Apache and dozens of national and international environmental organizations, the U.S. Congress approved the construction of three telescopes at the site, including one built by the Vatican. The construction of the telescopes on Mt. Graham threatens the Apache's access to their sacred sites and destroys key elements of the mountain, while also putting at risk the endangered Mt. Graham red squirrel (LaDuke 2005). This case is one in which multiple sides of the conflict deploy religion to further their objectives, including Apache leaders, whose opposition to the Observatory project could be described as an example of religion being put to use in the service of EJ.

One of the most intractable case of geopolitics and social conflict in the Middle East, if not the world, is the Israel-Palestine struggle. While this conflict is often described in ethnic and religious terms, its environmental dimensions are far less well known but central to the tensions between these populations. The role of land in particular has always been at the center of the conflict between the state of Israel and the Palestinian Arab communities both inside and outside of Israel. In 1901, the Keren Kayemeth LeIsrael Jewish National Fund (KKL-JNF) was founded to promote land acquisition and property rights *exclusively for Jewish settlements* and later evolved to embrace the practices of afforestation, agricultural development, sustainable development, education, and tourism (jnf.org 2015). KKL-JNF has always seen its work in the European green colonial tradition of transforming an otherwise "desolate and neglected Asiatic desert of Palestine into a blooming green European terrain of forest" (Massad 2004, 61; see also Tal 2002). Unfortunately, this has resulted in the planting of forests directly on sites where Palestinian Arab villages *and mosques* once stood, thus erasing the evidence of the Arab presence in the name of environmental sustainability and Israeli Jewish sovereignty (Davis 2003). In a fascinating twist on the politics of indigeneity within and across species, the KKL-JNF planted non-indigenous conifers, pine trees (which are native to the northern hemisphere), and cypress trees on land where indigenous trees and Arabs were uprooted via the "destruction of the terraced landscape and … over 500 villages in the areas that are now supposedly 'forested'" (Masalha 2012, 61).

More traditional examples of environmental inequality abound as well in the region. For example, as a result of Israeli government practices, there is highly uneven and insufficient access to water for farming and a lack of sewage treatment among many Israeli Arab communities (Tal 2002, 332–340). There are also instances where chemical factories and heavily polluting industries that have been outlawed inside Israel simply relocate to the West Bank and other Palestinian communities (Lorber 2013).

Thus the Israel-Palestine conflict is a case where religion and EJ concerns intersect because the State of Israel was established as a Jewish state and has created a system of laws, policies, and practices that exclude many non-Jews from access to critical social services, property, land, clean air, clean water, and EJ. Moreover, the Jewish National Fund's *raison d'être* is explicitly focused on making land and non-human natures available to people of the Jewish faith. Thus religion is correlated with one's relationship to environmental threats versus environmental privileges.

Religion and spiritual practices harmed by environmental injustice

More common perhaps, are instances in which states and/or corporate actors commit acts of environmental injustice that curtail, restrict, or even make impossible the spiritual practices of affected communities. As Smith (2005) points out, this occurs frequently because colonial governments and their corporate partners often seize land and waterways that are sacred to indigenous peoples, thus disallowing access and therefore religious practices in those spaces. Smith argues that native spiritualities are "practice-centered" and "land based—they are tied to the land base from which they originate. When Native peoples fight for cultural/spiritual preservation, they are ultimately fighting for the land base which grounds their spirituality and culture" (Smith 2005, 121; also see LaDuke 1999). Unfortunately, the U.S. court system has consistently disallowed many Native peoples access to sacred sites and lands that are critical to their practice-centered spiritualities. As LaDuke, Smith, and Mark Dowie argue, the questions of spirituality and land are at the heart of the very survival of native peoples. Spirituality is also just one of many components of knowledge systems—collectively called Traditional Ecological Knowledge (TEK)—which are threatened by state and corporate development projects around the globe. Dowie explains, "'Traditional ecological knowledge,' or TEK for short, is the collective botanical, zoological, hydrological, cultural, and geographical know-how, rooted in spirit, culture, and language essential to the survival of a particular tribe or community in a particular habitat" (Dowie 2009, 108).

For example, sociologist Kari Norgaard (2011) writes about the conflict between the indigenous Karuk nation (based in California) and the U.S. government. The Karuk people rely on salmon from the Klamath river for their physical and cultural sustenance. However, the U.S. government dammed the river via the construction of the Iron Gate Dam in 1962, harming salmon runs and restricting the Karuk's access to fishing—a ritual practice and a form of nutrition considered sacred. The federal government does not recognize the Karuk's rights to the land and refuses to recognize their right to practice traditional fishing, which has resulted in the arrest and incarceration of tribal members who fish, "making it more difficult to maintain ceremonies, continue language use, maintain and strengthen cultural identity, or carry out other vital cultural practices" (Norgaard 2011, 36).

The Minnehaha Free State is another case that reveals the ways that state and corporate "development" practices threaten sacred places and spiritual practices. This was Earth First!'s first urban land occupation in the U.S. and took place in the Twin Cities of Minnesota in 1998 to 1999. The campaign was directed at the state government's plans to reroute a highway through an area that activists claimed was ecologically sensitive and sacred to Native Americans. Specifically, there were four oak trees that many Native leaders declared were a traditional site of indigenous peoples' gatherings, and nearby Cold Water Spring was an ancient water source that is regarded as sacred as well. Sharon Day, an indigenous (Ojibwe) activist, explained that the sites had important ceremonial and metaphysical significance for the native peoples, comparing them with the Vatican for Catholics or Mecca for Muslims (Day 2009). These two sites were directly in the path of the proposed reroute of Minnesota State Highway 55. A number of Earth First! and American Indian Movement (AIM) activists came together to found the Minnehaha Free State, a land occupation that delayed the highway project for nearly two years. Although the state of Minnesota officials and agencies involved in the reroute project rejected any and all claims that the land was a traditional sacred site, the activist coalition was ultimately successful at saving Cold Water Spring. This campaign involved the first collaboration between Earth First! and the AIM and remains one of the longest road occupations in U.S. history.

Religion and spirituality as a source of resistance against environmental injustice

The literature is also rich with examples that reveal how religion and spirituality have been central components of activist movements resisting environmental injustice (Binder 1999; Gedicks 2001; Kaalund 2004; McCutcheon 2011). It is well known that a number of faith-based organizations and institutions have been at the center of much of the EJ movement since its beginnings. For example, among the 500 people arrested for protesting against a toxic waste landfill arrested in Warren County, North Carolina in 1982, was the Director of the United Church of Christ's (UCC) Commission for Racial Justice (Reverend Benjamin Chavis) and the co-founder of the Southern Christian Leadership Conference (Dr. Joseph Lowery). And in 1987, the UCC published the first national study of what came to be known as environmental racism in the U.S. (United Church of Christ 1987).

A few years later, in 1991, at a gathering of several hundred activists at the First National People of Color Environmental Leadership Summit, the attendees agreed to a document known as the Principles of Environmental Justice, which began with a preamble, which reads in part: "We, the people of color gathered together at this multinational People of Color Environmental Leadership Summit, to begin to build a national and international movement of all peoples of color to fight the destruction and taking of our lands and communities, *do hereby re-establish our spiritual interdependence to the sacredness of our Mother Earth; to respect and celebrate each of our cultures, languages and beliefs about the natural world* and our roles in healing ourselves." We have added emphasis around the section above that indicates clearly the deeply spiritual nature of the EJ movement from its early days.

In the early 1990s, the struggle over the attempt of Louisiana Energy Services (LES) to build a $750 million uranium enrichment facility in a community that was 97% African American in Claiborne Parish, Louisiana, met with stiff resistance from Citizens Against Nuclear Trash (CANT). CANT drew on deep historical roots of civil rights activism, anti-racist politics, and a firm grounding in the African American church. The church was critical to the group's success and was a key site for fundraising, organizing, networking, and cultural affirmation. As Roberts and Toffolon-Weiss write, "the religious affiliation of the group provided them something perhaps even more crucial: a common language and faith that helped them reach across race, income, and political boundaries. [One activist stated] 'I'm saying this in all sincerity: God led us. We had divine guidance'" (Roberts and Toffolon-Weiss 2001, 79–80). LES was stopped in 1997 in a battle that marked the first and only time a polluter's permit was denied by a federal agency on the basis of environmental racism. And in the battle against the Shintech corporation in the same state, as Binder says, "The religious rhetoric deployed by Shintech opponents, like the religious rhetoric evident in the environmental justice movement at a national level, emphasized the moral and spiritual dimension of the dispute and affirmed the dual importance of caring for human life and the earth. [They] expressed their moral resistance to siting through prayer. They held prayer vigils to ask for 'God's protection from the Shintech menace to our air, water and lives'" (Binder 1999, 40).

The presence of religious actors and leaders in the EJ movement is ubiquitous. The EJ group Madres del Este de Los Angeles or Mothers of East LA was started by two grassroots organizers with help from a local Catholic parish priest, and they fought noxious facilities like a toxic waste incinerator, a state prison, and a fuel pipeline (Peña 2005, 202). In the same city, major leaders in the African American Church and the Catholic Archdiocese of Los Angeles were involved in historic battles in the late twentieth century to secure and save urban parks in communities of color, in the face of attempts by government and corporate actors to site garbage dumps and power plants in these spaces (Garcia and Flores 2005).

Discussion and conclusion

In this chapter we have explored three dynamics at the intersection of religion and EJ: (1) when religious institutions, leaders, or doctrines are deployed to legitimate the commission of environmental injustices; (2) when religious or spiritual practices are harmed by the commission of environmental injustices; and (3) when religion or spirituality becomes a source of resistance and resilience against environmental injustice. In this final section of the chapter, we consider how these three dynamics are interrelated and what they might suggest for the future of scholarship in religion and ecology and EJ studies.

The often contentious, and collaborative, intersections between religion and EJ issues reminds us that religion can and is frequently employed in ways that can be harmful and/or beneficial to human societies, cultures, and non-human natures. There is nothing inherently anti-ecological or environmentalist about religion and spirituality; it depends on how these traditions are articulated, framed, and deployed by institutions, leaders, adherents, and others. What remains clear is that social inequality and the way that power circulates within and across communities exert strong influences on the ways religion and spirituality operate either in the service of EJ or against it. Thus we reject any claim that a particular religious or spiritual tradition is either ideally suited or ill-suited for promoting ecological sustainability and EJ.

In our view, the field of EJ studies and its efforts to include an ever-expanding range of social categories entangled in environmental inequality (e.g., class, race, gender, sexuality, species, etc.), suggests possibilities for productive directions in scholarship linking religion and ecology. Following Ruether (2005), who argues that the field of religion and ecology would benefit from greater attention to gender, race, poverty, and social justice, we would suggest that EJ studies offers a generative example of a field that has forged these connections. EJ studies began primarily with a focus on the relationship between pollution and social class inequalities, then expanded to include racial inequalities, and more recently has grappled with the ways that citizenship, indigeneity, nation, and gender and sexuality intersect with environmental politics. On the other hand, while there is an explicit focus on religion and spirituality among scholars using the term "eco-justice" (see Hessel 2007 and Immergut and Kearns 2012), there is far less attention to religion and spirituality among scholars working in EJ studies, and we see this is an opportunity for religion and ecology scholarship to offer generative ideas and new direction for EJ studies. For example, what might EJ studies look like if we explored more seriously the role of EJ movements and the idea of EJ as *moral* forces?

References

Agyeman J., Cole P., Haluza-Delay R., and O'Riley P. (2010) *Speaking for Ourselves: Environmental justice in Canada*, University of Washington Press, Seattle.

Alkon A. H. and Agyeman J. eds. (2011) *Cultivating Food Justice: Race, class, and sustainability*, The MIT Press, Cambridge, MA.

Binder L. (1999) "Religion, race, and rights: A rhetorical overview of environmental justice disputes", *Wisconsin Environmental Law Journal* 6:1, 1–63.

Brisman A. (2008) "Crime-environment relationships and environmental justice", *Seattle Journal for Social Justice* 6:2, 727–817.

Buckingham S. and Kulcur R. (2010) "Gendered geographies of environmental justice", in Holifield R., Porter M. and Walker G. eds. *Spaces of Environmental Justice*, Wiley-Blackwell, Hoboken, 70–94.

Bullard R. D. (1996) "Symposium: the legacy of American apartheid and environmental racism", *St. John's Journal of Legal Commentary* 9:2, 445–74.

Bullard R. D. (2000) *Dumping in Dixie: Race, class and environmental quality*, Third edition, Westview Press, Boulder.

Bullard R. and Wright B. (2012) *The Wrong Complexion for Protection: How the government response to disaster endangers African American communities*, New York University Press, New York.

Bullard, R., Mohai P., Saha R., and Wright B. (2007) *Toxic Wastes and Race at Twenty, 1987–2007*, United Church of Christ, New York.

Cole L. and Foster S. (2000) *From the Ground Up: Environmental racism and the rise of the environmental justice movement*, New York University Press, New York.

Crowder K. and Downey L. (2010) "Inter-neighborhood migration, race, and environmental hazards: modeling micro-level processes of environmental inequality", *American Journal of Sociology* 115:4, 1110–49.

Davis U. (2003) *Apartheid Israel: Possibilities for the struggle within*, Zed Books, London.

Day S. (2009) Interview by author, November 6, 2009.

Dowie M. (2009) *Conservation Refugees: The hundred-year conflict between global conservation and native peoples*, The MIT Press, Cambridge, MA.

Downey L. (2006) "Environmental racial inequality in Detroit", *Social Forces* 85:2, 771–796.

Garcia R. and Flores E. (2005) "Anatomy of the urban parks movement", in Bullard R. ed., *The Quest for Environmental Justice: Human rights and the politics of pollution*, Sierra Club Books, San Francisco 145–167.

Gedicks A. (2001) *Resource Rebels: Native challenges to mining and oil corporations*, South End Press, Cambridge, MA.

Hessel D. T. (2007) Eco-justice ethics, The Forum on Religion and Ecology at Yale. (http://fore.yale.edu/disciplines/ethics/eco-justice/) Accessed 17 March 2015.

Hunter L. (2000) "The spatial association between U.S. immigrant residential concentration and environmental hazards", *International Migration Review* 34: 2, 460–88.

Immergut M. and Kearns L. (2012) "When nature is rats and roaches: Religious eco-justice activism in Newark, NJ", *Journal for the Study of Religion, Nature & Culture* 6:2, 176–195.

JNF.org (2015) Keren Kayemeth LeIsrael-Jewish National Fund website (www.jnf.org). Accessed 1 January 2015.

Jenkins W. (2013). *The Future of Ethics: Sustainability, social justice and religious creativity*, Georgetown University Press, Washington, D.C.

Kaalund V. A. (2004) "Witness to truth: Black women heeding the call for environmental justice", in Stein R. ed., *New Perspectives On Environmental Justice: Gender, sexuality, and activism* Rutgers University Press, New Brunswick 78–92.

LaDuke W. (1999) *All Our Relations: Native struggles for land and life*, South End Press, Boston.

LaDuke W. (1994) "A society based on conquest cannot be sustained", foreword in Gedicks A. *The New Resource Wars: Native and environmental struggles against multinational corporations*, Black Rose Books, New York.

LaDuke W. (2005) *Recovering the Sacred: The power of naming and claiming*, South End Press, Cambridge, MA.

Lorber B. (2013) "Keren Kayemet Le Yisrael and environmental racism in Palestine", earthfirstjournal.org. January 11

McCutcheon P. (2011) "Community food security 'for us, by us': The Nation of Islam and the Pan African Orthodox Christian Church", in Alkon A. H. and Agyeman J. eds. *Cultivating Food Justice: Race, class, and sustainability*, The MIT Press, Cambridge, MA.

Masalha N. (2012) *The Palestine Nakba: Decolonising history, narrating the subaltern, reclaiming memory*, Zed Books, New York.

Massad J. (2004) "The persistence of the Palestinian question", in Aretxaga B. ed. *Empire and Terror: Nationalism/Postnationalism in the New Millennium*, University of Nevada Press and Reno Center for Basque Studies, Reno, NV 57–70.

Mennis J. L. and Jordan L. (2005) "The distribution of environmental equity: exploring spatial nonstationarity in multivariate models of air toxic releases", *Annals of the Association American Geographers* 95:2, 249–68.

Mohai P. and Bryant B. (1992) "Environmental racism: Reviewing the evidence", in Bryant B. and Mohai P. eds. *Race and the Incidence of environmental Hazards: A time for discourse*, Westview Press, Boulder, 163–76.

Mohai P. and Saha R. (2007) "Racial inequality in the distribution of hazardous waste: A national-level reassessment", *Social Problems* 54:3, 343–70.

Norgaard K. (2011) "A continuing legacy: Institutional racism, hunger and nutritional justice on the Klamath", in Alkon A.H. and Agyeman J. eds. *Cultivating Food Justice: Race, Class, and Sustainability*, The MIT Press, Cambridge, MA, 23–46.

Park L. S.-H. and Pellow D. N. (2011) *The Slums of Aspen: The war on immigrants in America's Eden*, New York University Press, New York.

Pellow D. N. and Park L. S.-H. (2002) *The Silicon Valley of Dreams: Environmental injustice, immigrant workers, and the high-tech global economy*, New York University Press, New York.

Peña D. (2005) "Tierra y vida: Chicano environmental justice struggles in the Southwest", in Bullard R. ed. *The Quest for Environmental Justice: Human rights and the politics of pollution*, Sierra Club Books, San Francisco, 188–206.

Roberts J. T. and Parks B. (2007) *A Climate of Injustice: Global inequality, North-South politics, and climate policy*, The MIT Press, Cambridge, MA.

Ruether R. R. (2005) *Integrating Ecofeminism, Globalization, and World Religions*, Rowman & Littlefield, Lanham.

Roberts J. T. and Toffolon-Weiss, M. (2001) *Chronicles from the Environmental Justice Frontline*, Cambridge University Press, Cambridge and New York.

Schlosberg D. (2007) *Defining Environmental Justice: Theories, movements, and nature*, Oxford University Press, Oxford.

Smith A. (2005) *Conquest: Sexual violence and American Indian genocide*, South End Press, Cambridge, MA.

Tal A. (2002) *Pollution in a Promised Land: An environmental history of Israel*, University of California Press, Berkeley and Los Angeles.

Taylor D. (2009) *The Environment and the People in American Cities, 1600s–1900s: Disorder, inequality, and social change*, Duke University Press, Durham, NC.

Tucker M. E. and Grim J. eds. *Religions of the World and Ecology*, (series) Harvard University Press, Cambridge, MA.

United Church of Christ (1987) *Toxic Waste and Race*, Commission for Racial Justice, New York.

PART VII

Disciplinary intersections

Introduction: *Mary Evelyn Tucker*

This section underscores the contributions of both the humanities and the sciences (social and natural) in understanding and solving environmental issues. We are keenly aware of the rich intellectual work in environmental humanities over the last several decades, especially in the fields of history, literature, philosophy, and art. We are eager to see religious studies enter more fully into these discussions. In addition, we are highlighting key social science perspectives from environmental policy, law, and economics, as these are critical conversation partners for religious and ethical perspectives on the environment. We conclude with a broad overview of the science of ecology, which is now more fully poised to engage the human dimension of ecological issues. Indeed, one of the key issues in environmental analysis and problem solving is how to include the perspectives of both science and the humanities.

Originally environmental issues were seen almost solely through the lens of the natural sciences, especially ecology. But with the emergence of systems sciences and complexity sciences there are many efforts to understand species as part of dynamic changing ecosystems. This has reconfigured our appreciation for identifying the conditions for resilient ecosystems and the role of humans in those ecosystems. This involves a complex and still unfolding discussion in ecology and conservation biology about the role of science as using objective and empirical methods and/or becoming engaged with policy recommendations. Many ecologists recognize that some forms of engagement may be inevitable given the state of our planetary ecosystems. This is where the Earth Stewardship Initiative of the Ecological Society of America and the Millennium Alliance for Humanity and the Biosphere may be modeling new interdisciplinary ways forward, especially between the natural and the social sciences. With the expansion of the study of nature as complex adaptive systems, the social sciences are joining the conversation. Natural scientists have recognized that we need law, policy, and economics, not just to implement their findings, but as dialogue partners in their efforts. For too long we have kept these realms separate. Systems sciences and "new ecology" are efforts to draw the sciences and humanities together in ways that promote dialogue, collaboration, and resolution.

However, these interdisciplinary efforts, valuable as they are, seem to fall short of including the humanities in a clear and intentional manner. Philosophy, in terms of environmental ethics, has entered the discussions, but by and large the rest of the humanities tend to be ignored. This

345

may be due to an internal focus in the last several decades in some of the humanities on critical theory. The deconstruction of political power became an appropriate concern of theory in response to the rise of controlling ideologies in the last 100 years. However, some of these theory-driven discussions became arcane and incomprehensible to those outside the conversation. Now, however, a new appreciation for reconstruction is emerging in both the social sciences and the humanities in response to environmental problems. For example, constructive approaches to environmental law and global governance are being discussed. In addition, ecological economics is critiquing unlimited growth and identifying the problems with externalizing environmental damage. It is calling for other strategies such as new economic indicators, socially and environmentally responsible investment, and cooperative community based businesses.

Similarly, environmental humanities are developing in a variety of ways, as these articles indicate. Environmental history has grown over four decades with the contributions of a variety of figures ranging from Americanists to scholars of Asia, Africa, Latin America, and the Pacific. Environmental literature is gaining an appreciation for forms of local–global cosmopolitanism as articulated by Ursula Heise and others. Environmental artists are leading the way in awakening citizens around the world to both the beauty and destruction of nature in photography, painting, and sculpture as well as in music, theater, and dance. Philosophy has a key role in identifying ecological ethics that shape human behaviors. As this book richly illustrates the world religions, nature spiritualities, and indigenous traditions are entering the discussions with inspiring energy and creativity. Indeed, since Lynn White's critique of religions in *Science* in 1967 there have been numerous responses, especially in the area of eco-theology, for the last several decades. Now it is clear that the enduring appeal of religion for millions of people around the world make them indispensable dialogue partners as we seek solutions to our planetary challenges. And, as this section demonstrates, the interdisciplinary collaboration with natural and social scientists is already being deepened and broadened.

36

HISTORY

Donald Worster

In 1867 a bearded and gangling college dropout from the Midwest, John Muir, set off on a religious quest through the war-ravaged American South. To his friends, he explained that he intended to hike all the way to the Gulf Coast and then take passage to South America to see its magnificent rainforests. But he was also running away from the Christian beliefs that had informed his youth. The God of Christianity, he complained in his journal, was too glibly assumed to be a "civilized, law-abiding gentleman" who had created the world primarily to serve human needs, and then had more or less turned over its management to his favorite species. Muir rejected that anthropomorphism and the anthropocentrism that went along with it. Henceforth he sought religion in the outdoors, not in churches or books. The wilderness of North America, he came to believe, was suffused with divine beauty and, more than any man-made deity, was worthy of reverence and protection. "All is beauty," he wrote, "all is God."

Others of that time went through a similar shift in consciousness, and within decades they had made Muir their prophet for a new religion of nature that to varying degrees was post-Christian. They formed a movement to preserve America's remaining wilderness, in the mountains of Appalachia and the West, among forests, prairies, deserts, wetlands, and tundra, and not only in distant places but in and near cities, on the edges of agriculture, and along the banks of rivers or the ocean shores, and not only in the United States and Canada but also in other countries, from Muir's native Scotland to Australia and Borneo.

Religion, as Muir's case shows, even the religion of a single individual or a small minority, can have powerful consequences. To be sure, religion is far too large, complicated, and diverse a phenomenon to have only one kind of impact. Christianity, like the ancient and established religions of Islam, Buddhism, or Hinduism, has worn many faces toward the nonhuman realm. Add in all the other religious ideas and practices, including those that recognize only nature as their source of value, and we can grant that religion influences human behavior in ways that have shifted radically over time, always with ambiguous and ambivalent results.

Religion has been a prominent part of environmental or ecological history, which emerged in the 1970s in the United States and spread globally, but undoubtedly it deserves an even larger role. This new history aims to deepen our understanding of how humans have interacted with the natural world. With the aid of both scientists and humanists, environmental historians attempt to understand how nature has changed over time and how those changes have influenced societies—changes from fertile to impoverished soils, from forest richness to forest scarcity, from disasters of floods or droughts, from disease-causing organisms, and from climate revolutions. At the same time, this history explores the environmental impact of human economies, technologies, politics, laws, and cultures on the nonhuman world. It seeks to understand how and why humans have altered or destroyed other organisms and ecosystems, and particularly why in more recent times they have created so many environmental problems. To call this ambitious enterprise a "field" or even "subfield" of a discipline would be misleading; in truth, environmental history represents a radical shift in worldview, an expansion of the concept of history beyond humans interacting with other humans to include nature on all scales. In the deepest sense environmental historians seek to connect humans to biological and cosmological evolution and to the pulsating flow of matter and energy.

This radical turn in what we mean by history has an unmistakably materialist bias. It does not ask whether supernatural powers controlled human–nature relations or guided the direction in which the cosmos has been moving over billions of years. Environmental historians start with the fact that humans, like all forms of life, evolved out of matter. We appeared on earth as one more way of utilizing the sun's energy and, like other species, we have used that energy to replicate our own kind. That is the most basic kind of historical change, driven by strong imperatives that lie beyond conscious intention. Like all forms of matter, humans must follow the second law of thermodynamics, which says that over time the cosmic fund of energy becomes more dispersed and unable to do work. The cosmos is winding down until one day it will all go dormant. But along the way matter has "learned' to organize itself into forms and shapes that proliferate before they wear out and disappear. Living matter is superior to nonliving matter in the utilization of energy, which is why it exists, and the human organism may be the most successful mechanism for energy capture ever devised. By this point we have managed to seize much of the sun's radiation that bathes the earth and make it do work for us, at least for a while.

Every individual or society, therefore, like the structure of a molecule or the hydraulic cycle, must be understood first as a mechanism for efficient energy capture and dissipation. John Muir was such a mechanism, and so was the nation in which he grew up, with its fossil-fuel-burning, railroad-building economy. That industrial mechanism enabled him to get to California, and then to Washington, D.C., where he testified about the value of the California redwoods and the sacred beauty of Yosemite National Park. His religion was not designed to make railroads run more efficiently but to find a better fit between Americans and the natural world. By going into wild nature he hoped to find insight into the place of humanity that he could not get in cities. He described what he found there as "beauty," "love," or "God," but he also thought in terms of physics and biology, especially climate, glaciation, and evolution. And he found in the wilderness lessons about his place within the whole. Inside the cover of his first campfire journal he wrote his address this way: "John Muir, Earth-planet, Universe." It was a grandiose gesture—but also a humbling one, locating himself within a greater coherence of matter and energy.

Religion, whether Hindu, Christian, pagan, or something else, is like all human history in that it emerges from the material brain and its material surroundings. It may seem sometimes to belong to the realm of disembodied ideas and feelings— the immateriality we call culture, a complex web of perceptions, ideologies, ethics, laws, and myths that our brains have invented for practical reasons. But in fact religion has deep ecological, functional, and material connections. Whatever other functions religions may have taken on over time, they were invented in the first place to help people succeed in the physical world. Religion and culture are, from the perspective of environmental history, tools for living in the cosmos. They emerge out of specific and local natural environments—through dwelling in a desert, sitting under a tree, fishing in the sea, or hunting on a mountain trail. Religion helps people understand and adapt to those places, to become more efficacious in sustaining lives there. The fact that so many religions have existed in the past and present is due to the fact that there have been so many environments into which people have migrated. Religions, even as they take us into realms of wonder and meaning, can never neglect for long their material role—which ultimately is to help us to absorb and utilize better the energy that flows through the cosmos. Otherwise, religion would collapse and its creators might live in greater peril.

When Muir began to think of plants and animals as his "brethren" and to call for more a sensitive and restrained use of natural resources, he was pushing his own religion, based on Scottish Calvinism, in a new direction. The emergent religion of nature was a response to changing material conditions—the pioneer onslaught on America's natural resources—and it became the foundation for a set of behaviors that we call nature conservation. The goal of conservation, as it appeared in late-nineteenth century America, was to make the nation more prudent, secure, altruistic, and permanent. That conservationists like Muir managed to leave a lasting legacy, turning their ideas into laws and habits that still exist, shows that they succeeded as cultural evolutionaries—they reformed their culture and helped insure that the United States would survive. Going into the wilderness, witnessing nature free of human manipulation, was the way not only to humility but also to a more caretaking, secure existence.

So also the rise of environmental history might be called a strategy for adapting to a changing ecology—viz., the global deterioration of ecosystems—and it too aims to make human life more secure. This new history is not a religion, but it functions in many ways like one. It seeks a better understanding of the past so that we can solve problems in the present. It describes and explains our condition, points up lessons, and even promotes new values. Environmental historians, like humanists in general, do not merely carry out research and try to explain changes that have occurred in the past; they want to promote a new way of living on earth. The line between analysis and advocacy is one they cross often and unabashedly.

One of the first environmental historians to appear on the scholarly scene, as both analyst and advocate for a new paradigm, was the medievalist Lynn White, Jr. Today he is known mainly as one who blamed our environmental problems on religion, specifically on Christianity and its teachings about nature. That argument, however, came late in his career: Earlier he built a reputation by showing that the Middle Ages had been neither stagnant or wholly absorbed in prayer and piety, as popular images so often portrayed. On the contrary, White argued, that period, which lasted for a thousand years, from the fifth to the fifteenth century, was characterized by revolutionary technological innovation and was full of this-worldly concerns. "The chief glory of the later Middle Ages," he wrote in the journal *Speculum,* "was not its cathedrals or its epics or its scholasticism: It was the building for the first time in history of a complex civilization which rested not on the backs of sweating slaves or coolies but primarily on non-human power" (White, 1940). The sources of that liberating power were horses and harnesses, new-fangled plows, and water mills. Technology had improved human life dramatically. Despite being accused of technological determinism by fellow historians, White further elaborated that thesis in his book *Medieval Technology and Social Change* (1962), one of the most influential works on its subject over the past half century. Then late in his career, and rather abruptly, he turned toward environmentalism and emerged with a far darker story of technological progress.

White was one of the first historians to make the environmental turn. In 1967, only five years after Rachel Carson's *Silent Spring* exposed the risks of agricultural pesticides, he produced a blistering attack on the whole drift of modern science and technology. Like Carson, he became an environmentalist, and he turned to history to reveal and remedy what he called "our ecologic crisis." Speaking to the American Association for the Advancement of Science, he pointed an accusing finger at "the presuppositions that underlie modern technology and science." They were the same attitudes that Carson had criticized: The drive to dominate nature, arrogance toward the living world, a reckless disregard for environmental integrity and balance. White did not, however, stop with accusing the chemists. He assailed the underlying moral outlook animating all science in western societies—a worldview he traced back to the rise of monotheism, the Book of Genesis and its creation story, and the victory of Judeo-Christianity over pagan animism. Those roots went back more than two millennia (White said 1,700 years, but more recent scholarship dates the writing of *Genesis* to the sixth century BCE). That ancient religion was "the most anthropocentric" the world had ever seen, he charged, and so was the science it spawned. Christianity had set up a rigid dualism between humans and nature and taught, "it is God's will that man exploit nature for his proper ends" (White, 1967). Those religious ideas were the ultimate cause of the crisis, while science was only a proximate one.

White constructed his thesis on two premises. First, he believed that pollution, climate change, and other environmental problems of our time are due primarily to culture and, as a corollary, he maintained that every culture, "whether it is overtly religious or not, is shaped primarily by its religion." Second, he argued that it was western culture, not the cultures of Asia or Africa, which was causing environmental decline. Western culture was at its core Christian. Therefore, Christians had a terrible burden of guilt to face up to. Similar arguments had been constructed before, for example, by the British historian Arnold Toynbee, who likewise had identified cultures (or "civilizations") as the major driving forces in history, in contrast to material factors like food, population, energy, or technology. But in White's case those arguments ran into a firestorm of protest, and in some dimension that disagreement is still at the heart of environmental history. Historians are still asking whether culture, as an independent nonmaterial force, determines human/nature relations or whether culture is only a reflection of a society's technology, ecological conditions, and mode of production.

Environmental historians have not agreed perfectly over those questions. Few would grant as much power to Christianity or other religions as White did, but many would agree that it is culture that determines the moral values of a people, not their economic system. Usually, and unduly so, they would give more emphasis to the secular side of culture and particularly those values that have shaped gender, race, and class relations. It is those secular values, many would say, that must be changed. Other historians, however, would argue that moral values, ancient or modern, are mainly reflections of and adaptations to material circumstances. Whatever side they might take in this old debate, historians insist that any society is multi-causal in origin—no one factor determines its course of development. They are also likely to add that we do not need to go back 2,000 years or more to find either the cultural or the material causes of today's problems. The present may be indeed deeply conditioned by the past, but to understand modern times we need first to concentrate on more modern events and experiences.

White's critique of Christianity depended heavily on his reading of Genesis 1:28, or more accurately of the English translation known as the King James Version, completed in 1611. According to that translation, God declared in the beginning of time that "man" was made in His image and then admonished His offspring to "be fruitful, and multiply, and replenish the earth, and subdue the earth." Genesis, according to White, authorized a conquest, one that has led down the ages to more and more ecological destruction, until it has turned the planet into a toxic dump. Many Jews and Christians, as might be expected, denied this "conquest" reading of the biblical text, or tried to dismiss it as a mistranslation. The most common charge against White was that he misunderstood what was meant by man's dominion. Originally the idea carried a more benign message that humans should serve as God's stewards on earth and not see themselves as conquerors. Stewardship meant taking care of the Lord's property.

Categorizing nature as property, to be sure, might seem to allow nature no value in and of itself, but the admonition to be a good steward did not carry a license to destroy or degrade. Christianity was not intrinsically opposed to conservation.

Historians after White have asked whether that single text from the Old Testament, whatever its meaning, really matters all that much. Some very old words lifted out of context are not enough to explain all, or even most, of the world's environmental problems. Deciphering the long-term effect of words written by an obscure author during the Jews' exile in Babylonia has, consequently, not been of much interest to historians. In fact the role of Genesis did not hold even White's interest for long. Within the span of his famous essay, he shifted focus repeatedly until he ended by identifying the real culprit, not as the Book of Genesis, nor Judeo-Christianity in general, but specifically western or Latin Christianity as it emerged in the late medieval period, that is, during the thirteenth to fifteenth centuries. With that shift two millennia of history vanished suddenly from his analysis, perhaps because White realized that it would be hard to show that there had been more environmental damage under Christianity than the damage that had gone on in other parts of the world. Buddhist, Taoist, Islamist-taught peoples of Asia had caused plenty of damage too, and so also the Indigenous peoples of the Americas. Did any religion have clean hands? And were any religious doctrines or ethics really that important in the environmental history of the planet?

In a reprise of his original essay, published in 1973, White continued to place the blame on religion but only on that later period in Western Europe. "Today's ecologic situation," he wrote, "is a by-product of a forward surging technology that first emerged during the Middle Ages in the area of the Latin Church and has continued to the present." By the fourteenth century CE, God had become identified as a Great Clockmaker, whose Creation resembled a machine, not a living organism, and that mechanistic image gave new impetus to technological innovation. Such innovation White associated with environmental exploitation. By shifting the chronology forward so far, he could, if not exonerate Christianity completely, allow a Christian remedy: Moderns, he concluded, should return to an older version of their faith that was less inclined to view nature as a machine in need of tinkering. The avatar of that more nature-benevolent Christianity was the Italian friar and patron saint Francis of Assisi, who died in 1226, just before the dream of technological progress overwhelmed the Church. White advocated going back to Francis's religion with its love of all nature and its vision of "spiritual democracy of all creatures." But here again most historians have remained unconvinced; the idea of dropping back in time eight hundred years to find a nature-friendly religion for moderns has seemed neither compelling nor plausible.

Whatever connection religious ideas have made with material reality, they are not sovereign powers that can summon new eras, new technologies, or new ecologies into being by fiat. They grow out of historical circumstances and, once again, reflect the constantly shifting human imperative to capture energy and dissipate it efficiently. Their power lies to a great extent in their proven efficacy for that task. White may have been right about the sudden enthusiasm for technology that the late Middle Ages felt and expressed in religious terms. By that point trade connections were bringing to Western Europe the incredible technological legacy of Asia, especially China, including the printed book, the magnetic compass, spindles, water mills, and the arts of metallurgy. That technology seemed to open for Europeans an escape from what was becoming a limited, overexploited agro-ecological system, with declining opportunities for expansion. After the medieval clearances of Europe's great forests and the exhaustion of its eastern frontier, the newly imported technology seemed like a gift from heaven. No wonder the churches began to extol the wonders of technology and associate them with God and devotion. But, contrary to White, the great leap forward toward a more aggressive and exploitative relationship with nature did not occur in the late Middle Ages. It came at their end, when a much greater bounty than watermills or clocks fell into the laps of Europeans.

The modern world began, most historians would agree, around 1500 CE. It was then that the Europeans, and indeed everyone on the Eurasian landmass, discovered the existence of a vast western hemisphere that they had never suspected. The discovery reverberated around the old lands because it opened a largely unexploited treasure house of natural resources. Suddenly there were two whole continents and two immense oceans. In terms of natural abundance, the New World appeared like a second Earth, opening a frontier for growth and expansion that had no parallel. True, the western hemisphere had already been discovered by a few bands of wandering Asians ten thousand years earlier, but their communication with the Old World had long been lost, their population density was smaller than that of western Europe or China or India, and their level of technology was more primitive. Around them still flourished a natural environment that was rich in untapped energy and carbon, water and soil, flora and fauna. This was the windfall that Europeans now set out to appropriate.

Following the discovery came a series of profound cultural and material revolutions, closely intertwined and far exceeding in impact anything seen in medieval times. Historians of all sorts have written about those revolutions a great deal, but usually as discrete and disconnected phenomena rather than as parts of one revolutionary upheaval in human ecology. Environmental history promises, although it has not fully accomplished the task to date, to arrange those revolutions into one chain of ecological causes and consequences. The linkage included a scientific revolution that brought new knowledge of the earth and the cosmos. There was a capitalist revolution that changed the work habits, moral ideas, and institutions of society. There was an industrial revolution that generated new technology and applied it to the production of goods, and eventually shifted the economy toward the fossil fuels. There were political revolutions all over the world, beginning in the Americas, in Haiti, and in the British colonies along the Atlantic coasts of North America, and spreading to France and to other parts of Europe, Asia, and Africa. Few of the revolutionaries themselves were in a position to see how their fervor for change came from the windfall of New World resources. But historians, with the gift of hindsight, should be in a position to see the full picture, that is, if they will first grant the importance of the natural world in the making and remaking of human life.

Another revolution, as important as any of the others, came in the practices and doctrines of Christianity, although other religions around the world would eventually be transformed too. The relation of those changes to religious upheaval has not yet been fully explored by historians. Twenty-five years after Columbus's first voyage in 1492 CE, Martin Luther wrote his *Ninety Five Theses,* attacking the Catholic Church for its sale of indulgences, an event that religious historians identify as the beginning of the Protestant Reformation. How are the two men and their deeds related? Whatever internal dynamics within Europe may have led to Luther and the origins of Protestantism, the discovery of the New World was surely a precipitating event. Both Catholics and Protestants were excited, though perhaps in different ways, by the prospects of resource abundance, presented to them (they assumed) by a benevolent God, a gift that must be seized and exploited with gratitude. Spain, Portugal, France, Italy, the Netherlands, and Great Britain all went out from their homes with a sense of religious mission to claim possession of the western hemisphere; they were determined to make it their own and not leave it to the pagans to monopolize. Natural resources were the deepest attraction of the newly discovered hemisphere, but religion, as adaptive as ever, quickly took on the work of explaining, motivating, justifying, and mobilizing European societies to claim those resources. Columbus unwittingly helped set in motion a fierce competition among religionists as among nation states.

It was then, and not earlier, that the words of Genesis 1:28 acquired their modern interpretation and provided a rallying cry for ecological imperialism. Many examples of its importance could be furnished, from the mines of South America to the slave plantations of the Caribbean and American South to the deserts of Mormon Utah, all places where the so-called Biblical admonition to conquer drove people to exploit the natural world more aggressively. Older notions of natural limits, ecological vulnerability, and harmony between humans and the rest of nature were thrust aside in the scramble to possess and exploit.

An early example of the new meaning of Genesis comes from the English lawyer John Winthrop, who in 1630 led a small band of Puritans (followers of John Calvin) to found a colony at Massachusetts Bay in the New World. To justify their abandoning of Europe he penned a document titled *Reasons to Be Considered ... for the Intended Plantation in New England.* He forthrightly called on the Bible for moral support.

> The whole earth is the Lord's garden and he hath given it to the sons of men, with a general condition, Genesis 1:28, *Increase and multiply, replenish the earth and subdue it,* which was again renewed to Noah. ... Why then should we stand here striving for places of habitation ... and in the mean time suffer a whole continent, as fruitful and convenient for the use of man, to lie waste without any improvement?

Winthrop sensed that religion could offer a new justification for colonization. Literally, he meant that his band of Puritans would be right to overthrow the Indians of New England and seize the natural resources for themselves.

Thus early modern Christianity reinvented itself, and thus it became a powerful force for environmental change. That force would continue right down to the twenty-first century, before it began to lose credibility and a critical-minded child of Calvinism like John Muir would begin to ask whether it might be time to rethink that ethos of conquest. Environmental history thus locates the roots of our "ecologic crisis" not in the Middle Ages but in the dynamic forces making the modern world, which began in the aftermath of the New World's discovery. Then it was that the heady mixture of science, capitalism, industrialization, democratic consumption, and religious competition came together and, thanks to those abundant natural resources, ignited dreams of unlimited economic growth.

Five hundred years later those dreams may be coming to an end, as the western hemisphere and the rest of the world have filled up with billions of people all seeking to enjoy the plenty that has come to seem the divine birthright of humans. The material and cultural impetus given by the great windfall has spread to every part of the planet, including technologies, economic ideas and institutions, politics, and above all attitudes toward nature. Whatever their religious traditions, people everywhere have begun telling themselves that they have a mandate (secular or divine) to increase their numbers, multiply their wants, and subdue nature around them. Buddhist businessmen in Tokyo share that view with Islamic oilmen in Saudi Arabia and Protestant cattle ranchers in Wyoming. Unlike White, most historians find it difficult to point a finger of blame at Christians alone. We cannot simply blame some people for what all people want for their own.

As the human condition changes, however, and as the earth shows more damage and fewer opportunities for expansion, all religions will undoubtedly be called on to help people find a new equilibrium. The issues are different than they were in John Muir's day: Then it was the onslaught of frontier farmers and loggers against mountain, forests, and watersheds that threatened society. A century and a half later the threats are coming from climate change, biodiversity loss, failing supplies of fresh water, soil erosion and chemical degradation, and poisonous levels of phosphorous and nitrogen. The chief historical similarity lies in the ways that religion again and again serves as agent for environmental resistance or reform. More than ever, scholars of religion, environment, and history will need to work together to understand the process of adaptation and adjustment to a shrinking earth.

To say more about the content of change in the twenty-first century would take us beyond knowing about the past and into guessing about the future. The historian makes an imperfect forecaster. We can never be sure where our innate drives and cultural ideas will take us and what we will find when we get there. The only certainty is that religions of every sort will try to address the world's ecological situation and offer adaptive strategies. The environmental historian cannot say what success they will have but can offer the observation that religion has always changed in response to the human condition, including most basically the human ecological condition, and nothing is more certain than the fact that religion will change again. We cannot know in advance exactly how it will change, but we can be sure that change is on the way.

References

Albanese, C. L. (1990) *Nature Religion in America,* University of Chicago Press, Chicago.
Epstein, S. A. (2012) *The Medieval Discovery of Nature,* Cambridge University Press, New York.
Foltz, R. C. (ed.) (2002) *Worldviews, Religion, and the Environment: A global anthology,* Thompson-Wadsworth, Belmont, California.
Gottlieb, R. S. (ed.) (2006) *The Oxford Handbook of Religion and Ecology,* Oxford University Press, New York.
Grim, J. and Tucker, M. E. (2014) *Ecology and Religion,* Island Press, Washington.
Krech, S., McNeill, J. R., and Merchant, C. (eds) (2004) *Encyclopedia of World Environmental History,* Routledge, New York.
Muir, J. (1916) *A Thousand-Mile Walk to the Gulf,* Houghton Mifflin, Boston.
Stoll, M. (1997) *Protestantism, Capitalism, and Nature in America,* University of New Mexico Press, Albuquerque.
Stoll, M. (2015) *Inherit the Holy Mountain: Religion and the Rise of American Environmentalism,* Oxford University Press, New York.
Tuan, Y. (1974) "Discrepancies between environmental attitude and behavior: examples from Europe and China" in Spring, D. and E. (eds) *Ecology and Religion in History,* Harper New York, pp. 91–113.
Weiming, T. (2004) "The ecological turn in New Confucian humanism: implications for China and the world" in Weiming, T. and Tucker, M. E. *Confucian Spirituality, vol two,* Herder and Herder, New York.
White, L Jr. (1973) "Continuing the conversation," in Barbour Ian, D. (ed.) *Western Man and Environmental Ethics,* Addison-Wesley Reading, Massachusetts.
White, L. Jr. (1967) The historical roots of our ecologic crisis, *Science* 155, 1203–1207.
White, L. Jr. (1940) Technology and invention in the Middle Ages, *Speculum* 15, 141–159.
Winthrop, J. (1995) "Reasons to be considered for … the intended plantation in New England," in Heimert A and Delanco A (eds) *The Puritans in America,* Harvard University Press, Cambridge, Massachusetts.
Worster, D. (1993) "John Muir and the roots of American environmentalism," in *The Wealth of Nature,* Oxford University Press, New York.
Worster, D. (2008) *A Passion for Nature: The life of John Muir,* Oxford University Press, New York.
Worster, D. (2016) *Shrinking the Earth: The rise and decline of American abundance,* Oxford University Press, New York.

37

LITERATURE

Scott Slovic

The term "ecocriticism," or "the study of literature as if the earth mattered," was first coined in 1978 by William Rueckert in an essay titled "Literature and Ecology: An Experiment in Ecocriticism," but scholars had been studying environmental themes in literature long before Rueckert came along (Mazel 2001, 1). In his 2001 collection *A Century of Early Ecocriticism*, David Mazel traces the lineage of proto-ecocriticism in the United States from 1864 to 1964, noting such works as Norman Foerster's *Nature in American Literature: Studies in the Modern View of Nature* (1923) and Leo Marx's *The Machine in the Garden: Technology and the Pastoral Ideal in America* (1964). Rueckert's own essay, which was collected in the pivotal 1996 volume, *The Ecocriticism Reader: Landmarks in Literary Ecology*, marked a dramatically new approach to literature. Previously, scholars had traced "nature themes" in literature, treating poetry, fictional narratives, and nonfiction essays, journals, and manifestos as storehouses of cultural information, artistic documents that somehow captured traditional or contemporary "views," or "ideals" vis-à-vis the nonhuman world. Rueckert, however, understood even as he was first using the word "ecocriticism" that the field of literary studies might eventually contribute much more to ecological thought than merely tracing relevant themes in literary texts.

One of Rueckert's boldest ideas was his notion that ecological concepts are somehow embedded in the very essence of literary expression, particularly in poetry. He writes:

> A poem is stored energy, a formal turbulence, a living thing, a swirl in the flow.
> Poems are part of the energy pathways which sustain life.
> Poems are a verbal equivalent of fossil fuel (stored energy), but they are a renewable source of energy, coming, as they do, from those ever generative twin matrices, language and imagination.
>
> *(Ecocriticism Reader 108)*

Rueckert's sense of literature—and, by extension, other forms of human cultural expression—as renewable sources of energy gets right to the heart of how the arts function as media of cultural preservation, conveyance of information vital for survival, and sources of inspiration, all of which are essential to human existence. He continues his line of thought as follows:

A painting and a symphony are also stored energy. And clearly, this stored energy is not just used once, converted, and lost from the human community. It is perhaps true that the life of the human community depends upon the continuous flow of creative energy (in all its forms) from the creative imagination and intelligence, and that this flow could be considered the sun upon which life in the human community depends; but it is not true that the energy stored in a poem—*Song of Myself*—is used once, converted, and then lost from the ecosystem. It is used over and over again as a renewable resource by the same individual.

(109)

Much as Rueckert suggests here, I find myself returning repeatedly to the same artistic texts for guidance, reassurance, and motivation as I navigate the course of my own life on Earth. I suspect this is a very common tendency for readers, listeners, and viewers—we do not use up the richness of our favorite texts, but rather interpret them more deeply with each encounter. We might well have the same response to the world itself if we approached it with our senses alert and our imaginations free. I've been reading Rueckert's words for decades, and it occurs to me increasingly that his ideas about poetry and art apply equally to the physical world of nature. The first law of thermodynamics suggests that the total quantity of energy in a system always stays the same. Rueckert's reflections on the preservation of *cultural* energy seem consistent with such an idea.

Indeed, his own essay has proved to be a source of inspiration for later generations of environmental humanities scholars. In 1996, Cheryll Glotfelty and Harold Fromm used Rueckert's term in the title of *The Ecocriticism Reader*, the book that put the entire field on the scholarly map. Rueckert died in December 2006, and his obituary mentions his prominence as a scholar of Kenneth Burke, but says nothing of his role in supplying a crucial term for the burgeoning field of ecological literary studies (Blakesley 2007). In 2013, when finishing a book on the subject of literature and energy, I wrote to Rueckert's widow to let her know how important her husband had been to the entire field and also as a source of inspiration for some of my own work, specifically in the study of energy and literature, but my letter was returned unopened.

Although William Rueckert never quite knew that he had helped to open vital "energy pathways" in literary studies, encouraging multiple generations of scholars to think in increasingly nuanced ways about the power of textual studies in nudging human civilization toward appreciation of our relationship to the physical world, many of us in the field nod regularly toward his 1978 terminology as we do our work nearly four decades later. Cheryll Glotfelty offered one of the foundational definitions of ecocriticism in her introduction to *The Ecocriticism Reader*, stating that the term refers to "the study of the relationship between literature and the physical world" (xviii). It's traditional for literary critics to use literature to look at the relationship between human beings and other humans—in other words, to look at social relationships, political situations, or the internal psychological experiences of literary characters or of authors and readers. What began to happen as the field of ecocriticism developed was the expansion of the context of literary studies, growing from human/social contexts to a much broader physical, or ecological, context.

In the 1960s, '70s, and '80s, scholars who became aware of the environmental crisis, the problems with pollution, overpopulation, overconsumption of natural resources, and other issues pertaining to the physical environment began to wonder how we might use literature and other cultural texts in order to help us understand why humans behave as they do in relationship to the natural world. And thus the field of ecocriticism was born. Glotfelty's definition expresses this concern that we read literature and we read other kinds of texts in order to explore that relationship beyond merely the human social context. Environmental historian Donald Worster wrote famously in his 1993 book *The Wealth of Nature*, "We are facing a global crisis today, not because of how ecosystems function but rather because of how our ethical systems function. Getting through the crisis requires [...] understanding those ethical systems and using that understanding to reform them" (27). Of course the ethical dimension of literary experience remains fundamental, but contemporary ecocritics are also coming increasingly to understand that our apprehension and communication of environmental phenomena are fundamental to our ability to respond to various crises.

When I introduce ecocriticism, I often find it useful to offer a few small literary examples in order to show the variety of texts one might discuss. Of course, the possibilities are actually endless. But these brief poems seem appropriate to use in light of Rueckert's emphasis on the special, inspiring energy of poetry. I will discuss these poems briefly and then run through several recent approaches in ecocriticism that represent major new directions in the field.

In 1964, American poet William Stafford published "Maybe Alone on My Bike" in the *New Yorker* magazine. It's a poem about the quotidian experience of riding his bicycle home from work. Stafford was a professor at Lewis and Clark College in Portland, Oregon. This poem about riding his bike home from work expresses some rather profound thoughts that make it a meaningful ecological text and a good example of the kind of work that one could easily study ecocritically. The poem opens like this:

I listen, and the mountain lakes
hear snowflakes come on those winter wings...

And the second stanza reads as follows:

And I have thought (maybe alone
on my bike, quaintly on a cold
evening pedaling home) think!—
the splendor of our life, its current unknown
as those mountains, the scene no one sees.

(Smoke's Way 29)

This is a simple poem about the experience of riding a bicycle home from work, but in the opening stanza we see a bizarre and beautiful conflation of the human speaker, who listens to the silent snowflakes coming down through the winter air, and the nearby lakes, which are seemingly not alive, which are inorganic phenomena. Yet the mountain lakes hear the snowflakes come, just as the human speaker seeks to do. Likewise, the poet refers to owls' "radar gaze and furred ears." Owls, which are birds, are conflated with the technology, with radars, and also are conflated with mammals because their ears are called "*furred* ears," as if they had mammalian hair. So humans, lakes, owls, and radars are all brought together in this experience of tremendous sensitivity to and awareness of the physical environment. The normal feeling of ontological separation—what philosophers call "Cartesian dualism"—is erased or complicated in the opening lines of this poem.

In the second stanza, the human speaker of the poem commands himself to pay deeper attention to the world. He says to himself, "think!—the splendor of our life." So often we go through our lives not paying very close attention to the world. We skate past experience as if it were mundane, trivial, meaningless. One of the very important outgrowths of the ecocritical field is that it guides us, through cultural texts, to be more alert to the planet itself and to our own experience of the world—not to take things for granted quite so much. In a 1988 interview, the eminent poet W.S. Merwin stated: "the cause of [my] anger is, I suppose, the feeling of destruction, watching the destruction of things that I care passionately about. If we're so stupid that we choose to destroy each other and ourselves, that's bad enough; but if we destroy the whole life on the planet! [...] We can't suddenly decide years down the line that we made a mistake and put it all back. The feeling of awe—something we seem to be losing—is essential for survival" (quoted in Scigaj 187). The second stanza of Stafford's poem guides us to think about "awe," the feeling of mysterious splendor, concerning aspects of our life we might take for granted, such as the environment we pass through during our daily commute. The phrase (in parentheses) "maybe alone on my bike" is seemingly innocuous, too—like a simple bike ride home. But the strange word "maybe," in this context, suggests that the bike rider, the poet, and indeed the reader are likely not alone in our splendid lives, as we share the planet with many visible and invisible, expected and unexpected, forms of sentience.

And then, as the poem continues, we come to a parody of politicians' language—the language that goes "Oh citizens of our great amnesty." This line implies that we are being given a second chance. We are being forgiven for our inattentiveness, our lack of awareness, even our carelessness, and given a second chance to live again and more deeply pay attention to our relationships to the planet. In the final line of the poem, "I hear in the chain a chuckle I like to hear," Stafford suggests in a subtle and perhaps revolutionary way that technology may not be the main cause of the ecological damage we cause. Often environmental thinkers argue that we are too embedded in our technology. We are separated from the physical world—we don't pay enough attention to nature, and we must learn to have more direct, unmediated experience of the natural world. This is the message, for instance, of Scott Russell Sanders' "Speaking a Word for Nature" (also collected in *The Ecocriticism Reader*), where he concludes, "Thus, any writer who sees the world in ecological perspective faces a hard problem: how, despite the perfection of our technological boxes, to make us feel the ache and tug of that organic world passing through us, how to *situate* the lives of characters—and therefore of readers—in nature" (194).

But Stafford, in this poem about his bicycle, suggests that actually certain kinds of appropriate, low-impact technologies can be technologies of *contact* rather than technologies of *separation*. They can enable us to connect with the world and think more deeply about our relationship to the world. The bicycle is one such technology. When Stafford says, "I hear in the *chain* a *chuckle* I like to hear," the chuckle (emphasized in the repeated "ch" sound in the final line) implies that this is a friendly, non-problematic technology and that its use is a good thing. I would say that this poem accords very well with Cheryll Glotfelty's definition of ecocriticism because by studying this poem we might deepen our understanding of a kind of technology that does not block our engagement with the planet, but rather enhances or deepens that engagement.

Another short literary text that helps to reveal key ecocritical ideas is a poem by the Tohono O'odham poet Ofelia Zepeda called "It Is Going to Rain." Zepeda is a linguist at the University of Arizona, and she comes from the farming town of Stanfield, Arizona, near the Mexican border. The poem begins as follows:

> Someone said it is going to rain.
> I think it is not so.
> Because I have not yet felt the earth and the way it holds still in
>   anticipation...

(143–4)

The poem is only ten lines long, but it tells the story of a debate between "someone" (perhaps a weatherman who lives in a city and uses technology to determine whether or not it will rain) and a local person (perhaps a farmer and perhaps also an indigenous person, who uses his or her physical senses in order to perceive the way the environment changes when it's about to rain). The weatherman predicts that it is likely to rain, probably because he or she is relying too much upon modern science and technology. The local person argues on the basis of physical observations—the sense of the earth holding still "in anticipation," the feeling of the air, the feeling of wind that's coming, and the taste of the "sweet, wet dirt" that occurs right before it's about to rain in the desert—that it will not rain.

A key idea in the poem, I would say, is that rain is extremely important to people who rely upon it for their survival—farmers, people living in the desert. For them, whether it rains or not is a matter of fundamental survival. It's not just a matter of what clothing to wear or whether or not to bring an umbrella to work or to school on a particular day. In this poem, the speaker—the character called "I"—uses four simple sensory examples of why he or she thinks it will not rain. The number four, for native communities in North America and perhaps in other parts of the world, is a sacred number. It refers to the four directions—North, East, South, and West. By using the sacred number four for the examples of why it is likely not to rain in this case, the speaker of the poem implies the sacredness of rain and, by extension, the sacredness of water, a phenomenon essential to the survival of all living beings on the planet—certainly essential to human life. So this short poem by Ofelia Zepeda helps readers to appreciate the sacredness of water, helps us not to take water for granted. In a sense this poem echoes the idea in the Stafford poem that we should take pains not to overlook the subtle meaning in ordinary experience and ordinary phenomena. It is easy to take water for granted; likewise, it is easy to look at the surface of the simple language in Zepeda's poem and not perceive the depth of ecological sensitivity and indigenous wisdom embedded in the very structure of the text (such as the fourfold repetition of key examples). An additional dimension of this small text is its relevance to the bioregional movement, which reshapes the power dynamic in environmental decision-making to highlight the validity of local citizens and non-Western science in determining the appropriate use of resources.

When introducing an audience to ecocriticism, I like to ask, "Can you propose a literary text or some other kind of text that you believe cannot be approached ecocritically?" Once when I was doing this in Kuala Lumpur, Malaysia, a student opened her English literature textbook and pointed to Wilfred Owen's famous World War I poem called "Dulce Et Decorum Est"—meaning "sweet and decorous it is … to die for one's country." It's a poem about patriotism and how soldiers are willing to die for their country in a war, though it quickly becomes clear that Owen is harshly critiquing such blind loyalty to one's "flag." The poem seems, on the surface, to have very little to do with nature, ecology, or ecocriticism. But from my perspective, once one reads this powerful and disturbing poem, one begins to see that, on a deeper level, it's about mortality. It's about living and dying. The human characters, the soldiers, in the poem are killing each other—their mortality is graphically, vividly, brutally portrayed. The speaker of the poem challenges the reader in the concluding stanza:

> If you could hear, at every jolt, the blood
> Come gargling from the froth-corrupted lungs,
> Obscene as cancer, bitter as the cud
> Of vile, incurable sores on innocent tongues,–
> My friend, you would not tell with such high zest
> To children ardent for some desperate glory,

The old Lie: *Dulce et decorum est*
Pro patria mori.

An ecocritic might read Wilfred Owen's poem not simply as a war poem, but as a poem that defies what we call "human exceptionalism," which is another "old lie" that human beings often tell themselves. Humans are not presented here as being exceptional, or separate, from the natural order of things, from living and dying. The vivid portrayal of injured and dying human beings in this poem demonstrates that we are mortal, that we are animals.

On the other hand, the poem also explores the question of patriotism. What does it mean to be patriotic to a flag, to a country—to a human construction? Might there be other forms of patriotism that are meaningful? Environmentalists sometimes argue that we should be devoted, or loyal, not only to human society but to the physical planet that supports us or to the more specific bioregion or local place that we rely upon for our lives. I think a poem like "Dulce Et Decorum Est" raises questions of human animal existence as living and dying creatures and questions about what it means to be properly loyal. In this sense, the poem fits very well with the line of thinking displayed in the 2002 volume *Patriotism and the American Land*, in which distinguished environmental authors Barry Lopez, Richard Nelson, and Terry Tempest Williams reflect on non-militaristic, non-nationalistic modes of patriotism in the midst of post-September 11 flag-waving in the United States.

The brief literary texts and discussions I've offered above have been an attempt to show how environmental literature can work to heighten readers' sensitivity to our own lives and to the planet; to reveal power relationships between different people when it comes to making decisions about lifestyles and uses of natural resources; to explore physical and psychological attachments, or loyalties, to geography and to embodied experience; and to demonstrate the multiplicity of voices and expressive modes available to us as we reflect upon and communicate our environmental ideas. As we move into the twenty-first century, ecocriticism is now experiencing its fourth major "wave," or phase, which emphasizes, as I explained in a 2012 Editor's Note for the journal *ISLE: Interdisciplinary Studies in Literature and Environment*, "the fundamental materiality (the physicality, the consequentiality) of environmental things, places, processes, forces, and experiences. Ranging from studies of climate change literature to examinations of the substance of poetic language, there is a growing pragmatism in ecocritical practice" (619). The previous three waves of ecocriticism include the following: *Wave 1* (1980–present), emphasis on nonfiction ("nature writing"), focus on nonhuman nature and "wilderness," American and British focus, and "discursive ecofeminism"; *Wave 2* (mid-1990s–present), multiple genres, multicultural, growing attention toward local literatures around the world, environmental justice ecocriticism, urban and suburban emphasis; *Wave 3* (2000–present), global concepts of place melding with neo-bioregionalism ("eco-cosmopolitanism," "rooted cosmopolitanism," "translocality"), transnational and transcultural comparative tendency, critiques from within (relationship with theory, "literature" too limited a focus, forgotten role of feminism, lack of precise methodology), first glimmers of material ecocriticism, emergence of animal-oriented ecocriticism (evolutionary ecocriticism, animal subjectivity and agency, vegetarian ecocriticism, posthumanism).

As we continue to witness the development of the Fourth Wave, which emerged distinctively in 2008 with the publication of Stacy Alaimo and Susan Hekman's collection *Material Feminisms*, these are some of the notable recent modes and texts in the field:

• *Material ecocriticism*: Stacy Alaimo, *Bodily Natures: Science, Environment, and the Material Self* (2010); Serenella Iovino and Serpil Oppermann, eds., *Material Ecocriticism* (2014). This work is marked by the idea that story and agency are inherent in all phenemena, not only in human beings—in her book *Vibrant Matter* (2010), Jane Bennett refers to the "agentic capacity" of material phenomena (xii). Alaimo's idea of "trans-corporeality" vividly recognizes the inescapable intertwining of the human body and the more-than-human world (*Material Feminisms* 238).

- *Transnational ecocriticism*: Karen Thornber, *Ecoambiguity: Environmental Crises and East Asian Literatures* (2012); Peter I-min Huang, *Linda Hogan and Contemporary Taiwanese Writers: An Ecocritical Study of Indigeneities and Environment* (2016). Patrick D. Murphy called for a new comparative emphasis in the field in his 2000 book *Farther Afield in the Study of Nature-Oriented Literature*, but this transnational emphasis has just begun to take off in the past few years.

- *Econarratology*: Erin James, *The Storyworld Accord: Econarratology and Postcolonial Narratives* (2015); Alexa Weik von Mossner, *Cosmopolitan Minds: Literature, Emotion, and the Transnational Imagination* (2014). Recent innovations in narrative theory, such as the focus on how stories influence the cognitive processes of readers/viewers ("cognitive narratology"), have converged with the ecocritical emphasis on the role of texts in inspiring new thinking about human relations with the nonhuman environment. Econarratology also examines experimental narratives that aspire to represent the world from nonhuman perspectives (plants, stones, and nonhuman animal species).

- *Ecocritical animal studies*: Wendy Woodward, *The Animal Gaze: Animal Subjectivities in Southern African Narratives* (2008); Randy Malamud, *An Introduction to Animals and Visual Culture* (2012); Deborah Bird Rose, *Wild Dog Dreaming: Love and Extinction* (2013). What happens to the human–nonhuman relationship when other species are not merely gazed upon but are granted the agency to "look back"? Ecocritical animal studies seeks to challenge and complicate human ways of thinking about other species, often with the goal of guiding readers to recognize the injustice of our inherited attitudes, many of which are embedded in our daily habits.

- *Ecocriticism and information processing*: Rob Nixon, *Slow Violence and the Environmentalism of the Poor* (2011); Scott Slovic and Paul Slovic, eds., *Numbers and Nerves: Information, Emotion, and Meaning in a World of Data* (2015). Many of the great humanitarian and environmental crises of our time exceed our capacity to apprehend them—they happen too slowly, too far away, or on a scale that exceeds our human senses. Even when we learn about these phenomena in the form of "data," we struggle to grasp the meaning of this information. Psychology, communication studies, and other disciplines have begun to merge with traditional ecocritical analysis of texts to produce this new direction in the field focused on "information processing." This approach has particular relevance to public policy and is a strong example of "applied ecocriticism."

Two of the overarching paradigms I've emphasized above in my abbreviated ecocritical treatment of three short texts and also in the assessment of major new trends in the field are the paradigm of place and the paradigm of animality. Place could mean either a specific local place or broader conceptions, including the entire planet; ecocritics interested in place might ask, for instance, "What does it mean to be globally conscious and to understand the vast, non-local phenomena that are occurring on the planet, such as climate change and mass extinction, which are very significant in the twenty-first century?" Animality is the other key paradigm of ecocriticism—and this ecocritical perspective can be expressed by asking, "How do we more deeply understand, through our reading of cultural texts, our own existence on the planet as animals, as living creatures, and how do we use cultural texts to deepen our understanding of our relationship to other species?" I would suggest that Stafford's "Maybe Alone on My Bike" not only vividly evokes a sense of the natural environment—the place—where the speaker is riding a bike, but it also considers the relationship between the human speaker and other living and nonliving phenomenona, such as birds, lakes, street lamps, and bicycles. That first poem can be read in the context of both place and animality. The Zepeda poem is, most prominently, a reflection on living appropriately in place and appreciating the precious and necessary natural elements, such as water, that are essential for survival in a place. And the Owen poem, from which I've quoted only a small section, considers what it means to be a human animal, frail and mortal like all other animals. Within these key paradigms of place and animality, so central to the field of ecocriticism and approached in diverse ways by ecocritics all over the world, we seek greater sensitivity to information we gather about the current state of the planet that might enable us to make decisions, as individuals and as societies, consistent with the objectives of justice and sustainability.

References

Alaimo, Stacy. *Bodily Natures: Science, Environment, and the Material Self.* Bloomington: Indiana University Press, 2010. Print.

Alaimo, Stacy, and Susan Hekman, eds. *Material Feminisms.* Bloomington: Indiana University Press, 2008. Print.

Bennett, Jane. *Vibrant Matter: A Political Ecology of Things.* Durham, NC: Duke University Press, 2010. Print.

Blakesley, David. "Bill Rueckert's Obituary." 2 January 2007. Web. https://lists.purdue.edu/pipermail/kb/2007-January/002523.html. Retrieved 9 January 2016.

Foerster, Norman. *Nature in American Literature: Studies in the Modern View of Nature.* New York: Russell and Russell, 1923. Print.

Glotfelty, Cheryll. Introduction. *The Ecocriticism Reader.* Ed. Cheryll Glotfelty and Harold Fromm. Athens: University of Georgia Press, 1996. xv–xxxvii. Print.

Huang, Peter I-min. *Linda Hogan and Contemporary Taiwanese Writers.* Lanham, MD: Lexington Books, 2016. Print.

Iovino, Serenella, and Serpil Oppermann, eds. *Material Ecocriticism.* Bloomington: Indiana UP, 2014. Print.

James, Erin. *The Storyworld Accord: Econarratology and Postcolonial Narratives.* Lincoln: University of Nebraska Press, 2015. Print.

Lopez, Barry, Richard Nelson, and Terry Tempest Williams. *Patriotism and the American Land.* Great Barrington, MA: The Orion Society, 2002. Print.

Malamud, Randy. *An Introduction to Animals and Visual Culture.* New York: Palgrave Macmillan, 2012. Print.

Marx, Leo. *The Machine in the Garden: Technology and the Pastoral Ideal in America.* New York: Oxford University Press, 1964. Print.

Mazel, David, ed. *A Century of Early Ecocriticism.* Athens: University of Georgia Press, 2001. Print.

Murphy, Patrick D. *Farther Afield in the Study of Nature-Oriented Literature.* Charlottesville: University of Virginia Press, 2000. Print.

Nixon, Rob. *Slow Violence and the Environmentalism of the Poor.* Cambridge, MA: Harvard University Press, 2011. Print.

Owen, Wilfred. "Dulce Et Decorum Est." (1920.) www.english.emory.edu/LostPoets/Dulce.html. Retrieved 25 December 2013.

Rose, Deborah Bird. *Wild Dog Dreaming: Love and Extinction.* Charlottesville: University of Virginia Press, 2013. Print.

Rueckert, William. "Literature and Ecology: An Experiment in Ecocriticism." *The Ecocriticism Reader: Landmarks in Literary Ecology.* Ed. Cheryll Glotfelty and Harold Fromm. Athens: University of Georgia Press, 1996. 105–23. Print.

Sanders, Scott Russell. "Speaking a Word for Nature." *The Ecocriticism Reader: Landmarks in Literary Ecology.* Ed. Cheryll Glotfelty and Harold Fromm. Athens: University of Georgia Press, 1996. 182–95. Print.

Scigaj, Leonard. *Sustainable Poetry: Four American Ecopoets.* Lexington: UP of Kentucky, 1999. Print.

Slovic, Scott, and Paul Slovic, eds. *Numbers and Nerves: Information, Emotion, and Meaning in a World of Data.* Corvallis: Oregon State University Press, 2015. Print.

Stafford, William. "Maybe Alone on My Bike." (1964.) *Smoke's Way*. Minneapolis: Graywolf, 1983. Print.

Thornber, Karen. *Ecoambiguity: Environmental Crises and East Asian Literatures*. Ann Arbor: University of Michigan Press, 2012. Print.

Weik von Mossner, Alexa. *Cosmopolitan Minds: Literature, Emotion, and the Transnational Imagination*. Austin: University of Texas Press, 2014. Print.

Woodward, Wendy. *The Animal Gaze: Animal Subjectivities in Southern African Narratives*. Johannesburg, South Africa: Wits University Press, 2008. Print.

Worster, Donald. *The Wealth of Nature: Environmental History and the Ecological Imagination*. New York: Cambridge University Press, 1993. Print.

Zepeda, Ofelia. "It Is Going to Rain." (1997.) *What's Nature Worth?: Narrative Expressions of Environmental Values*. Ed. Terre Satterfield and Scott Slovic. Salt Lake City: University of Utah Press, 2004. 143–33. Print.

38

PHILOSOPHY

J. Baird Callicott

Origins of environmental philosophy: call and response

As a new academic field takes shape, its founders mine history for precursors and *Wilderness and the American Mind* by Roderick Nash (1967) was the mother lode. Nash's classic is much more than a history of the North American wilderness idea; it is a history of American environmental thought from the period of British colonization to the mid-1960s. Not insignificantly for this volume, Nash begins with an analysis of the wilderness idea in the Holy Bible and in the writings of the bible-besotted Puritan clerics, with which the canon of American literature begins. Among the many authors Nash reviews whole chapters are devoted to Henry David Thoreau, John Muir, and Aldo Leopold. This triumvirate thus became the giants of environmental philosophy on whose shoulders the new breed of self-identifying environmental philosophers proudly stand.

The same year that one historian provided the future field of environmental philosophy with precursors, another provided it with a raison d'être. "The Historical Roots of Our Ecologic Crisis" by Lynn White Jr. is the historical root (pun intended) not only of the new field of religion and ecology, but also of environmental philosophy—but in very different ways. For religion and ecology, White (1967) posed a challenge by tracing the environmental crisis to the doorstep of the Judeo-Christian (J-C) religious tradition: he stimulated apologists to defend the J-C tradition against his indictment; he stimulated scholars to explore the environmental implications of other faith traditions. But for environmental philosophers, White's "Historical Roots" represented both a calling and a compass.

In "Historical Roots," White's lurid and cavalier critique of biblical attitudes and values regarding the environment is intoned in the treble range of his rhetoric, but he punctuates each stanza with a bass-line drop throbbing almost sub-aurally. Back in the day, it wasn't registered consciously; the receiving organ was more the sternum than the ear; and it affected the heart more than the mind. Years after its initial impact, looking back at White's score with a more analytic eye, one can see that these drops to the low-register refrain happen five times at strategic intervals throughout the piece: (1) "What shall we do? No one yet knows. Unless we *think about fundamentals*, our specific measures may produce backlashes more serious than those they are designed to remedy"; (2) "The issue is whether a democratized world can survive its own implications. Presumably we cannot unless we *rethink our axioms*"; (3) "*What people do* about their ecology *depends on what they think* about themselves in relation to the things around them. Human ecology is deeply conditioned by *beliefs* about our nature and destiny"; (4) "*What we do* about our ecology *depends on our ideas* about the man-nature relationship"; (5) "*We must rethink* and refeel our nature and destiny" (White 1967, 1204–1207).

White insists that the axioms we have to rethink are fundamentally Christian: What about Marxism and other avowedly secular, even atheistic belief systems? Each is but "a Judeo-Christian heresy" (1205). What about the modern scientific worldview? That science was a product of Christian natural theology is a cornerstone of White's thesis: "the task and the reward of the [early Modern] scientist was 'to think God's thoughts after him.' ... It was not until the late 18th century that the hypothesis of God became unnecessary to many scientists" (1206). We philosophers trace our intellectual ancestry back to ancient Greek philosophy. So our response is: Maybe so, but the thoughts that they were thinking after God are not found in Scripture, but in the natural philosophies of Pythagoras, Empedocles, Democritus, Plato, Euclid, Hippocrates, Eudoxus, and Aristrachus.

So, to redact White: What shall we do about the environmental crisis? The first, most basic, and most important task is to think about fundamentals, to rethink our axioms. That's because what people do depends on what they think, on their ideas, on their beliefs. (That proposition is of course questionable, but it is the sort of thing that us wide-eyed philosophers were then liable to embrace uncritically.) The remedy for the environmental crisis therefore is not "applying more science and technology" because, without rethinking the assumptions underlying science and technology, they may produce new and worse environmental backlashes (White 1967, 1205). No, the remedy is more and deeper thinking and rethinking.

And which academic discipline is all about ideas, beliefs, thinking, and rethinking? Why, philosophy, of course. Thus for saving the world from environmental apocalypse, the key world savers are those philosophers heeding the environmental summons. It now seems embarrassingly megalomaniacal, but this was the early 1970s; the American cultural revolution was in full flower; a flood of pent up environmental concern had been released on the first Earth Day of April 1970; and a handful of daring young philosophers saw a light and heard a voice (White's) and realized that they had a new vocation, a new calling.

The agenda of environmental philosophy: critique and create

White (1967) also sets out a clear agenda for environmental philosophy. First comes critical thinking. White himself provides a model in his critique of Judeo-Christianity. But the Western worldview has two taproots, the other being the Greco-Roman. What do Plato and Aristotle have to say about the "man-nature relationship?" And how about the foundational moderns, Francis Bacon, René Descartes, John Locke? All of these figures and many others found themselves being unflatteringly rethought in the pages of *Inquiry*, a journal founded in 1957 by Arne Naess; *Environmental Review*, a history journal established in 1967; and *Environmental Ethics*, founded by Eugene C. Hargrove in 1979.

Second comes creative thinking. Environmental philosophers have to come up with new axioms, new fundamentals, new thoughts and ideas about the nature of human nature, the nature of Nature, and the human–Nature relationship.

The critical task is relatively easy and straightforward—Plato's metaphysics is otherworldly; Aristotle's hierarchical teleology makes everything a means to human ends; Descartes' dualism fetishizes human reason and otherizes the body and the whole physical world of which it is a part; Newton's physics is reductive and makes the physical world into a machine, which invites redesigning it to suit human purposes, etc.; etc. (Merchant 1980; Plumwood 1993). But the

365

creative task is hard and anything but straightforward. Nevertheless, White points out a couple of ways that it might be undertaken.

One possibility is to borrow from non-Western traditions of thought. At the time, thanks to the efforts of Daisetsu Teitaro Suzuki and Alan Watts, Zen Buddhism had become popular in the American counterculture. And so White (1967, 1206) proclaims that "The Beatniks, who are the basic revolutionaries of our time, show a sound instinct in their affinity for Zen Buddhism, which conceives of the man-nature relationship as very nearly the mirror image of the Christian view"—whatever that means. But he is "dubious of its viability among us." Ignoring White's doubts, a number of early papers in environmental philosophy pursued this approach. Indeed, for Westerners to convert en masse to Buddhism, Daoism, or Hinduism is not likely to happen; and even if it should, that it might improve the "man-nature relationship" in the West is also dubious. Environmental problems afflict not only the West, but all the world's places and peoples—a fact that led to some needed but largely unheeded questioning of the link between what people think and what they do (Tuan 1968). In any case, developing uniquely Buddhist, Daoist, Hindu, and Islamic environmental ethics for Buddhists, Daoists, Hindus, and Muslims is certainly a worthwhile undertaking. That indeed is precisely the work of comparative environmental philosophy, which emerged in the 1980s (Callicott and Ames 1989), as well as of religion and ecology.

A second way to come up with new axioms, new fundamentals, new thoughts and ideas about the nature of human nature, the nature of Nature, and the human–Nature relationship is to cruise Western intellectual history looking for channels of thought that are not tributary to the mainstream. This was White's own approach. Having argued that all Western thought is essentially Christian, despite appearances and claims to the contrary, he opts for "an alternative Christian view": in particular that of Saint Francis of Assisi. And sure enough, a number of early papers in environmental philosophy also pursued this approach, touting environmental salvation by Pythagoras, Leibniz, Spinoza, Whitehead, Heidegger, and, of course, Marx. But the prospect of a mass conversion of Western culture to Pythagoreanism or Process Philosophy is even less likely to happen than a mass conversion to Buddhism or Hinduism.

A third way to come up with new axioms, new fundamentals, new thoughts and ideas about the nature of human nature, the nature of Nature, and the human–Nature relationship is to explore the metaphysical foundations of post-Modern science. If Newtonian mechanics epitomizes Modern science, the special and general theories of relativity, quantum field theory, evolutionary biology, and ecology epitomize post-Modern science (Callicott 1982a; 1986). Among the precursors of environmental philosophy, this was the way Aldo Leopold did it. There is little in Leopold's literary corpus to suggest that he was intimately acquainted with the new physics and cosmology, but his slender masterpiece, *A Sand County Almanac and Sketches Here and There*, relentlessly prosecutes a single overarching theme: the exposition and promulgation of an evolutionary-ecological worldview and its axiological and normative implications (Callicott 2011).

In his typically terse and condensed style, Leopold (1949, viii) actually anticipates White's notorious critique of the J-C worldview: "Conservation is getting nowhere because it is incompatible with our Abrahamic [i.e., biblical] concept of land. We abuse land because we regard it as a commodity belonging to us." But unlike White, Leopold doesn't advocate replacing it with a Franciscan theology; rather, with an ecological worldview: "When we see land as a community to which we belong, we may begin to use it with love and respect. ...That land is a community is the basic concept of ecology." The evolutionary aspect of the evolutionary-ecological worldview is introduced later in the book: "It is a century now since Darwin gave us the first glimpse of the origin of species. We know now what was unknown to all the preceding caravan of generations: that men are only fellow-voyagers with other creatures in the odyssey of evolution. This new knowledge should have given us, by this time, a sense of kinship with fellow-creatures; a wish to live and let live; a sense of wonder over the magnitude and duration of the biotic enterprise" (Leopold 1949, 109).

Environmental ethics

The predominant modality of early environmental philosophy was ethics. Some of the early literature in environmental ethics was revolutionary only in focus and remained loyal to the anthropocentric bias of the Western tradition of moral philosophy (e.g. Shrader-Frechette 1981; Norton 1984), but the field developed largely along the lines of competing non-anthropocentric conceptions of value and moral standing.

During the same originary period, animal ethics was also developing (Singer 1975; Regan 1979, 1980). As wild animals are a prominent part of the environment, some thinkers conflate animal and environmental ethics. Environmental concerns are, however, much broader than concern for animal welfare and they are holistic (for species, for example, not for specimens) not individualistic. Animal ethics, moreover, does not discriminate between wild and domestic animals (which are often environmentally destructive) and they are more concerned with domestic than wild animals.

Acknowledging at least that animal ethics is too narrowly focused to be coextensive with environmental ethics, some environmental philosophers attempt to build on the work of animal ethicists in order to reach out and ethically touch all organisms. This endeavor is well within the constraints of mainstream twentieth-century moral philosophy, which proceeded by identifying a property shared by a class of beings that gives them intrinsic value and entitles them—and them alone—to moral consideration (Routley 1973). The classical intrinsic-value conferring and ethics-entitling property is rationality, most fully developed by Immanuel Kant in the late eighteenth century, conveniently excluding non-human beings from ethics. Kant's contemporary, Jeremy Bentham, argues that the ethics-entitling property should instead be sentience, the capacity to experience pleasure and pain. On that basis, Singer (1975) constructs his version of animal ethics, while Regan (1979, 1980) proceeds on the basis of a modified Kantian approach, arguing that because not all humans are in fact rational—thus leaving non-rational humans beyond the pale of ethics—the intrinsic-value conferring and ethics-entitling property should be being the "subject of a life" that can go better or worse from the subject's point of view, conveniently including non-human animals within the purview of ethics. Goodpaster (1978), building on Singer's animal ethic, argues that the ethics-entitling property should be "being alive," while Taylor (1981), building on Regan's animal ethic, argues that it should be being a "teleological center of life." Taylor appropriated the term "biocentrism" for this approach to environmental ethics.

Rolston (1988), in turn, builds his environmental ethic on a biocentric platform, adding a value premium for sentience and an additional value premium for rationality, thus avoiding what Naess (1973, 95) called "biocentric egalitarianism," which, counter-intuitively, would confer equal intrinsic value and moral standing on all organisms—from a fly all the way up to a human—a proposition that Taylor (1981) is prepared to accept. For Rolston, intrinsically valuable organisms represent and reproduce their species as "good kinds" (i.e., species have intrinsic value); and such good kinds are "projected" by evolutionary processes and connected to one another by ecological relationships, such that ecosystems have "systemic [i.e., intrinsic] value" (Rolston 1988).

Callicott (1982b) notes that while biocentrism is far more inclusive than animal ethics, it does not fit well with the actual environmental concerns that motivated the development of environmental ethics in the first place—the extinction of species, the destruction of biotic communities, the pollution of air and water. Following Leopold (1949), he develops a holistic "land ethic" that departs more radically than does Rolston's from the prevailing logic of mainstream twentieth-century moral philosophy. Callicott traces the conceptual foundations of the land ethic to Charles Darwin's account of the origin and evolution of ethics in *The Descent of Man* and ultimately to the moral philosophy of David Hume, an older contemporary of Kant and Bentham. Hume argues that ethics flow ultimately from the "moral sentiments," some of which, like sympathy, are directed to individuals, but others of which, like love of country, are directed to wholes. As it provides for holistic objects of the moral sentiments, Hume's moral philosophy thus underpins a holistic environmental ethic tailored to the actual environmental concerns that motivated the development of environmental ethics in the first place.

Deep ecology, ecofeminism, environmental pragmatism and pluralism, and continental environmental philosophy

As the field matured it began to self-organize along several fault lines—in environmental ethics: anthropocentrism versus non-anthropocentrism; individualism versus holism.

As a "movement," Deep Ecology is a philosophical big tent accommodating any conceptual foundations that can support its core "platform" (Drengson 1997). Naess's own "ecosophy" is avowedly not about environmental *ethics*, but about what Naess (1987) calls Self (with a capital "S") –realization. The narrow (lowercase "s") self is the atomistic sense of self, having sprung from very deep roots in Western thought: first appearing as the Pythagorean-Platonic *psyche* that lives on after the death of the body and reincarnates, becoming the immortal soul of Christianity, the *cogito* of Descartes, Freud's superego, and latterly the self of the umpteen thousand self-help books on offer at Walmart. To come into consciousness of one's "ecological self"—one's embeddedness and dependence on the surrounding life-world—is Self-realization in Naess's sense. Ethics, according to Naess (1987, 2), is about "self-sacrifice ... a treacherous basis for conservation," but when saving the environment is a matter of Self-preservation, then Self-interest alone is a powerful enough motivation. Deep Ecology a là Naess is not environmental ethics but environmental metaphysics.

Ecofeminism conjoins feminist philosophy and environmental philosophy and is no more monolithic than either of its tributary streams. Some early ecofeminists argue that women are more attuned to Nature's cycles and processes than men and thus more nurturing of Nature—this as a foundation for an environmental ethic of care (Spretnak 1986). Subsequent ecofeminists rejected such "essentialism" and pointed out that the association of women with Nature was deeply rooted in the Western imaginary, going back to ancient Greek mythology, and became explicit in the philosophies of Plato and Aristotle (Plumwood 1993). That woman–Nature association justified the "oppression" of both women and Nature along with similarly associated racial Others and other Others. Warren (2000) provides a general analysis of "oppressive conceptual frameworks" and the "logic of oppression."

Environmental Pragmatism was born of frustration with the theoretical preoccupations of environmental philosophy—debates about the intrinsic value of nature, androcentrism versus anthropocentrism, the metaphysical implications of ecology and quantum physics—to the neglect of real-world environmental issues (Norton 1991). Ironically, Environmental Pragmatism itself draws on a rich body of linguistic and epistemological theory developed by Charles Saunders Peirce, William James, and John Dewey at the turn of the twentieth century. As to linguistic theory, two metaphysical descriptions of Nature may be cognitively quite different—one organismic; the other mechanistic—but if they make no difference in our understanding of and interaction with natural phenomena, the difference represents but a choice of language (Norton 1988). Thus all the ink spilled heralding the death of reductive Newtonian mechanism and celebrating the resurrection of organicism in ecology and the new physics comes down to little more than a superficial change in lexical fashion. Moreover, the acrimonious theoretical debates in environmental philosophy are pernicious because they are divisive—pitting anthropocentrists against non-anthropocentrists, Deep Ecologists against Ecofeminists, holists against individualists (Norton 1988). Environmental philosophers should unite around common policies and regard theories as tools for policy promotion and implementation, selecting those best tailored to the environmental issue of the moment—animal ethics for this issue, land ethics for that; anthropocentrism to persuade politicians, non-anthropocentrism to mobilize hippies.

Environmental Pluralism—the embrace of multiple ethical theories—and Pragmatism are closely related. According to Stone (1988) and Varner (1991), environmental concerns have so expanded the beneficiaries of ethics—in addition to humans: animals, plants, species, biotic communities, ecosystems—that it would seem hopeless for one theory of ethics to comprehend them all. To the contrary, Callicott (1990) argues that because theories of ethics are mutually contradictory, they cannot coherently coexist in one self-consistent worldview. While *inter*personal pluralism is in the spirit of the freedom of philosophical thought—everyone is free to choose among the smorgasbord of ethical theories on offer—*intra*personal pluralism violates the fundamental philosophical commitment to the rational principle of non-contradiction. He claims that communitarian ethical theory can accommodate differing sets of duties and obligations generated by our various human, animal, and ecological community relationships (thus making for pluralism at the level of application) unified by a single theoretical framework (thus preserving self-consistency at the level of theory).

The second wave of the environmental crisis: call and response

A second wave of environmental crisis crested in the 1980s. As conceived in the 1960s, the environmental crisis consists of a worldwide aggregate of local and regional phenomena—municipal smog; oil spills here, there, and yonder; point source pollution from industrial facilities and municipalities; pesticides sprayed on crop fields; herbicides sprayed on tropical forests as a weapon of war. The second wave of the environmental crisis consists of globally scaled phenomena: the sixth mass extinction now in progress on *planet* Earth; thinning of the stratospheric ozone membrane shielding the *planet* from lethal ultraviolet radiation; and *global* climate change.

Over its first two decades, environmental philosophy was correspondingly scaled locally and regionally. And it was principally informed by ecology: a science spatially scaled to biotic communities, ecosystems, landscapes, and, at most, to biomes; and temporally scaled to the duration of ecological processes, such as succession and periodic disturbance, which are calibrated in decades and centuries. This shift in the spatial and temporal scale of environmental problems—spatially from the local and regional to the planetary and temporally from decades and centuries to centuries and millennia—requires environmental philosophy to be reworked from square one. So far, the response has been anemic and focused mostly on global climate change. In part, that focus can be justified because climate change exacerbates the extinction crisis (Urban 2015) and, as to the "hole in the ozone," more about that shortly.

Michel Serres (1990) and Dale Jamieson (1992) were the first philosophical responders to the globalization of the environmental crisis.

Serres's approach was modeled on Rousseau's social contract. He personifies Nature as retaliating against mankind for assaulting it, thus creating the need for mankind to sue for peace by way of a "natural contract." But Nature is mute. No problem, we can appoint a guardian *ad litem* to speak for it—science. Serres's solution: a global scientocracy.

Jamieson (1992, 149) argues that the mainstream, individualistic ethics paradigm *"collapses when we try to apply it to global environmental problems, such as those associated with human-induced global climate change."* Why? Because the contribution of any individual to the total amount of anthropogenic greenhouse gases emitted to the atmosphere is relatively miniscule. Thus blame for the harmful effects of climate change if assigned, *pro rata*, to individuals is correspondingly miniscule; and, more importantly, an ethical individual's *voluntary* reduction of *personal* greenhouse gas emissions will not effectively (or even measurably) reduce the total (Sinnot-Armstrong 2005). Only "mutual coercion mutually agreed upon" to echo Garrett Hardin (1968, 1247) will avert a tragedy of the global atmospheric commons—but that's not ethics, *as conventionally understood by philosophers*, that's politics; that's *collective* action.

As greenhouse gasses accumulate and tipping points are transgressed, those who will suffer most from global climate change are the yet unborn members of future generations. Parfit (1984) has shown that changes in international policy and law compelling everyone drastically to reduce their greenhouse gas emissions will also drastically change which individuals will compose future generations. With imposed lifestyle changes, different people will meet and mate, or mate at different times, and thus different people will be born. Therefore, if no such changes in policy and law are implemented, no individual living and suffering in the future can assign blame to the present generation—not even collectively, let alone individually—for harming him or her, because if such changes in policy and law were put in place now, he or she would not exist then. If climate change raises questions about obligations to future generations, Parfit's point seems to challenge the capacity of traditional ethical frameworks to answer them.

The majority of twenty-first century climate ethicists, including Jamieson (2007) himself, nevertheless continue to work within the quaint constraints of the individualistic paradigms of ethics (utilitarianism and Kantian deontology) prevailing in the previous century (Broome 2012; Gardiner 2011; Gardiner et al. 2010; Garvey 2008; Nolt 2011; Singer 2004). Nolt and Broome even go so far as to try to guilt-trip us into voluntarily reducing our personal carbon footprints by calculating the amount of harm that an individual member of the present generation inflicts on an (anonymous and indeterminate) individual member of a future generation.

According to Nolt (2011, 9), "the average American through his or her greenhouse emissions causes the suffering and/or deaths of two future people ... *over the next millennium*." By spending time, money, and effort "the average American" might cut their carbon emissions in half, causing the suffering and/or death of but one future person over the next millennium. Only by cutting one's emissions to zero can "the average American" clear their conscience of climate murder or something almost as bad! Further, it is difficult to foresee all that might happen over the next millennium. Geo-engineering might reverse global climate change. Which leads to this consideration: The same time, money, and effort that "the average American" would have to invest to cut his or her carbon emissions by half might be better spent alleviating the hunger and disease of currently existing people—and get more and more certain bang for the ethical buck, so to speak.

Addressing those "who live a normal life in a rich country," Broome (2010, 74) avers that "it can be estimated very roughly that your lifetime emissions will wipe out more than six months of healthy human life." Many people seem quite content to wipe out several years of their own healthy human life in order to enjoy tobacco, alcohol, and junk food. Will being told that they are responsible for shortening the life of some (anonymous, indeterminate) individual member of a future generation by half a year induce them to reduce their carbon footprints? As an alternative to abstemiousness, Broome (2011) offers ethical chumps a less onerous way to assuage their feelings of climate guilt: crediting themselves with carbon offsets by, say, paying to have trees planted that will draw down CO_2 from the atmosphere.

We often say, "every little bit helps." That's true in many cases, but not in tragedies of the commons, such as global climate change. Individual and voluntary reduction of personal greenhouse gas emissions will not help mitigate climate change unless practically everyone else follows suite. And that requires collective action at the level of national and international policy and law—some combination of economic incentive via taxation or cap-and-trade; subsidies for alternative energy technologies, such as solar and geothermal; and outright bans on some behaviors such as burning coal to generate electricity. Consider how the "hole in the ozone"— another of the second wave's three big issues—was solved. Was it solved by environmental ethicists convincing individuals to stop using chlorofluorocarbon (CFC) spray propellants and refrigerants, the chemical culprits? No. It was solved because the Montreal Protocol of 1987, phasing out the use of CFCs and replacing them with ozone-friendly substitutes, was signed and implemented by every member of the United Nations (Handwerk 2010).

Page (2006) and Callicott (2013) thus conclude that any environmental ethic that adequately addresses the problems constituting the second wave of the environmental crisis must abandon individualistic ethics and posit holistic agents (nation states, most obviously, and the United Nations) and patients (future generations, for example, understood collectively) as the principal players in the climate ethics arena. Voluntary individual action will not cut it (pun intended).

Handwerk (2010) notes that there are ethically relevant differences between global stratospheric ozone erosion and global climate change. The former posed an immediate and universal threat to members of the present generation in the form of skin cancers and cataracts, while the threat of climate change is mainly to members of future generations. Actually, climate change too is starting to pose immediate threats to members of the present generation, but the harms are not as universally distributed, affecting Micronesians more than Minnesotans and Bengalis more than Brits.

In response to these differences, Callicott (2013) suggests that global human civilization can serve as a surrogate for the non-existent and indeterminate referent of the phrase "future generations." Global human civilization is a presently and robustly existing collective entity—consisting of architecture, art, music, literature, science, philosophy, and government— which is intrinsically valuable. It has a duration (going back thousands of years and potentially going forward for thousands of years more) that is commensurate with the temporal scale of global climate change. The second wave of the environmental crisis hardly threatens the extinction of *Homo sapiens*, a resourceful and resilient species, but it does threaten to bring human civilization to an abrupt end in failed states populated by fanatical tribalists led by psychopathic warlords (Sygna et al. 2013).

Conclusion

The megalomaniacal self-importance to which Lynn White Jr. tempted the first generation of environmental philosophers by his bass-line refrain about the tight relationship of thinking and doing, idea and action, is, in retrospect, laughable. Sound and creative philosophical thought about the on-going and ever-worsening environmental crisis is just a piece of the puzzle now fully engaging conservation biologists, climatologists, environmental and ecological economists, political scientists, evolutionary moral psychologists, and many other kinds of natural and social scientists and humanists—but it remains a vital piece of the puzzle.

References

Broome J. (2012) *Climate Matters: Ethics in a warming world*, W. W. Norton, London.

Callicott J. B. (1982a) "Intrinsic value, quantum theory, and environmental ethics", *Environmental Ethics* 7:3, 257–275.

Callicott J. B. (1982b) "Hume's is-ought dichotomy and the relation of ecology to Leopold's land ethic", *Environmental Ethics* 4:2, 163–174.

Callicott J. B. (1986) "The metaphysical implications of ecology", *Environmental Ethics* 8:4, 301–316.

Callicott J. B. (1990) "The case against moral pluralism", *Environmental Ethics* 12:2, 99–124.

Callicott J. B. (2011) "The worldview concept and Aldo Leopold's project of worldview remediation", *Journal for the Study of Religion, Nature, and Culture* 5:4, 510–528.

Callicott J. B. (2013) *Thinking like a planet: the land ethic and the Earth ethic*, Oxford University Press, New York.

Callicott J. B. and Ames, R. T. eds. (1989) *Nature in Asian traditions of thought*, State University of New York Press, Albany.

Drengson A. (1997) "An ecophilosophy approach, the Deep Ecology movement, and diverse ecosophies", *Trumpeter* 14:3, 110–111.

Gardiner S. M. (2011) *A Perfect Moral Storm: The ethical tragedy of climate change*, Oxford University Press, New York.

Gardiner S. M., Caney, S., Jamieson, D. and Shue, H. eds (2010) *Climate Ethics: Essential readings*, Oxford University Press, New York.

Garvey J. (2008) *The Ethics of Climate Change: Right and wrong in a warming world*, Continuum, London.

Goodpaster K. E. (1978) "On being morally considerable", *Journal of Philosophy* 65:6, 308–325.

Handwerk B. (2010) "Whatever happened to the ozone hole?", *National Geographic News* (http://news.nationalgeographic.com/news/2010/05/100505-science-environment-ozone-hole-25-years/) Accessed 13 July 2015.

Hardin G. (1968) "Tragedy of the commons", *Science* 162:3859, 1242–1248.

Jamieson D. (1992) "Ethics, public policy, and global warming", *Science, Technology, and Human Values* 12:2, 139–153.

Jamieson D. (2007) "When utilitarians should be virtue theorists", *Utilitas* 19:2, 160–183.

Leopold A. (1949) *A Sand County Almanac and Sketches Here and There*, Oxford University Press, New York.

Merchant C. (1980) *The Death of Nature: Women, ecology, and the scientific revolution*, HarperCollins, San Francisco.

Naess A. (1973) "The shallow and the deep, long-range ecology movements: a summary", *Inquiry* 16:1, 95–100.

Naess A. (1987) "Self-realization: an ecological approach to being in the world", *Trumpeter* 4:3 35–42.

Nash R. (1967) *Wilderness and the American Mind*, Yale University Press, New Haven.

Nolt J. (2011) "How harmful are the average American's greenhouse emission", *Ethics, Policy, and Environment* 14:1, 3–10.

Norton B. G. (1984) "Environmental ethics and weak anthropocentrism", *Environmental Ethics* 6:2, 131–148.

Norton B. G. (1988) "The constancy of Leopold's land ethic", *Conservation Biology* 2:1, 93–102.

Norton B. G. (1991) *Toward Unity Among Environmentalists*, Oxford University Press, New York.

Page E. A. (2006) *Climate Change, Justice, and Future Generations*, Edward Elgar Ltd, Cheltenham.

Parfit D. (1984) *Reasons and Persons*, Oxford University Press, Oxford.

Plumwood V. (1993) *Feminism and the Mastery of Nature*, Routledge, London.

Regan T. (1979) "An examination and defense of one argument concerning animal rights", *Inquiry* 22, 189–219.

Regan T. (1980) "Animal rights, human wrongs", *Environmental Ethics* 2:2, 99–120.

Rolston III H. (1988) *Environmental Ethics: Duties and values in the natural world*, Temple University Press, Philadelphia.

Routley R. (1973) "Is there a need for a new, an environmental ethic?", in Bulgarian Organizing Committee ed, *Proceedings of the XVth World Congress of Philosophy*, vol. 1, Sophia Press, Varna 205–210.

Serres M. (1990) *Le contrat naturel*, Éditions François Bourin, Paris.

Shrader-Frechette K. S. (1981) *Environmental Ethics*, Boxwood Press, Pacific Grove.

Sygna L., O'Brien K. and Wolf J. (2013) *A Changing Environment for Human Security: Transformative approaches to research, policy, and action*, Routledge, New York.

Singer P. (1975) *Animal liberation: A new ethics for our treatment of animals,* The New York Review, New York.

Singer P. (2004) *One World: The ethics of globalization,* 2nd ed., Yale University Press, New Haven.

Sinnot-Armstrong W. (2005) "It's not my fault: global warming and individual moral obligations", *Advances in the Economics of Environmental Research* 5, 293–315.

Spretnak, C. (1986) *The Spiritual Dimension of Green Politics,* Bear and Co., Santa Fe.

Stone C. D. (1988) "Moral pluralism and the course of environmental ethics", *Environmental Ethics* 10:2, 139–154.

Taylor P. W. (1981) "The ethics of respect for nature", *Environmental Ethics* 3:3, 179–218.

Tuan Y.-F. (1968) "Discrepancies between environmental attitude and behaviour: examples from Europe and China", *Canadian Geographer* 12:3, 176–191.

Urban M. (2015) "Accelerating extinction risk from climate change", *Science* 348:6234, 571–573.

Varner G. E. (1991) "No holism without pluralism", *Environmental Ethics* 13:2, 175–179.

Warren K. J. (2000) *Ecofeminist Philosophy: A Western perspective on what it is and why it matters,* Rowman and Littlefield, Lanham, MD.

White Jr. L. (1967) "The historical roots of our ecologic crisis", *Science* 155:3767, 1203–1207.

39

ART

Subhankar Banerjee

Consider this simple thought experiment. You decide to spend couple of days, looking at art, in a major cosmopolitan city, New York City, for example. First, you stroll through some prominent museums of modern and contemporary art, the Museum of Modern Art and the Whitney Museum of American Art. You will likely come away with the impression that religion plays no role in modern and contemporary art. Then, you visit the Metropolitan Museum of Art. You will likely come away from there with the impression that religion played a significant role in art for centuries in different cultures of the world. You may then logically but incorrectly conclude that, modernism severed the centuries old bond between art and religion.

In his thought-provoking book, *On the Strange Place of Religion in Contemporary Art*, James Elkins writes that, "there is almost no modern religious art in museums or in books of art history" (Elkins 2004). Elkins explores the "problem of making religious art," and writes that it is "nearly impossible to mix art and religion." He concludes, "I have tried to show why committed, engaged, ambitious, informed art does not mix with dedicated, serious, thoughtful, heartfelt religion." And that, "Wherever the two meet, one wrecks the other."

This art–religion dichotomy is misleading and false. There has been some recent effort to find common ground between contemporary art and spirituality (PBS 2001). Art and ecology, on the other hand, never had a strained relationship, although serious scholarship on ecocritical art history (Braddock and Irmscher 2009; Braddock 2015), and on the relationship between contemporary art and ecology (Brown 2014; Demos 2015, 2016), only started recently. Since the turn of the century there has also been sincere efforts to explore common ground between religion and ecology. There is yet to be a systematic initiative, however, to explore bridges that may be built among all of those three branches—art, ecology, and religion—of human inquiries. This essay is a minor effort toward that goal. Instead of a broad survey, I focus on the work of a handful of visual artists whose art I have personally experienced in museums, or in films. These artists come from the East and the West, the North and the South.

Rhythms of India

In 2008, the San Diego Museum of Art, in partnership with the National Gallery of Modern Art, New Delhi, organized a traveling retrospective of the twentieth century Indian artist Nandalal Bose, *Rhythms of India: The Art of Nandalal Bose*. In a review, art critic Holland Cotter

374

wrote in *The New York Times* that, "the show delivers a significant piece of news…that modernism wasn't a purely Western product" (Cotter 2008). Nandalal Bose is widely regarded as the "father of modern art in India," because he had "revitalized his country's art from the colonial torpor into which it had fallen and ushered it into the modern era without dependence on Western models" (Chandra and Quintanilla 2008).

There was, however, exchange between artists of the Bengal school of which Nandalal Bose was a member and western artists. In 2013, the Bauhaus Dessau Foundation in Germany presented an exhibition, *The Bauhaus in Calcutta*, which reflected back on the significance of a 1922 exhibition in Calcutta, the 14th *Annual Exhibition of the Indian Society of Oriental Art*. The Calcutta exhibition paired works by Bengal school artists—Nandalal Bose, Sunayani Devi, and Abanindranath and Gaganendranath Tagore, with works by the Bauhaus artists, including Paul Klee, Johannes Itten, Lyonel Feininger, Auguste Macke, and Wassily Kandinsky. Art historian Sria Chatterjee, co-curator of the 2013 exhibition in Dessau, illuminates a particular and significant complexity in that 1922 artistic exchange. She points out that various members of the Bauhaus group looked to India for inspiration, "India for them became a locus that was perceived as the pinnacle of the spiritual ideal." Chatterjee sums up the complexity of the 1922 exchange with these words: "In an odd clash of perceptions, both the Bauhaus avant-garde artists and the Bengal school artists strove toward an integration of abstract thought and abstract form, looking to the other for inspiration but in very different ways and to very different ends," which gave rise to distinct forms of "modernisms in Germany and India" (Chatterjee 2013).

On invitation from the polymath Nobel laureate, Rabindranath Tagore, in 1919 Bose joined the faculty of the newly established Kala Bhavan, or art department. This was, at Tagore's unconventional school of education in Shantiniketan, a small town in rural Bengal, which later became the Visva-Bharati University. After moving to Shantiniketan—the rural environment, life of simplicity, and the proximity to the indigenous Santal culture—proved a fertile ground for his art. He experimented with various media and styles—from small pencil/ink/watercolor drawings on postcards to large tempera on plaster murals. He had limited financial resources and worked with locally available materials. From the twenties through the forties Bose continued to paint and draw religious themes, primarily from Hinduism and Buddhism, but Christianity and Islam were also present in his work.

A significant transition in Bose's art happened when he began to portray village life, as in his much admired painting, *New Clouds*, 1937, which depicts village women hurriedly moving through a grove of palm trees while dark monsoon clouds hang overhead, (presumably) on their way to work in the fields. In *New Clouds*, Bose harmoniously combined the social and the ecological spheres of his local environment. This so impressed Mahatma Gandhi that he considered Bose "to be the artist who best conveyed the strength, spirit, and self-sufficiency of India's rural communities" (Quintanilla 2008, 174).

During this time Bose also used his art "as a political weapon" in service of India's freedom movement (Rosenfield 2008, 23). A notable example is the famous Haripura Posters that he produced at Gandhi's behest for the 1938 Indian National Congress convention at Haripura in the state of Gujarat. The eighty-four posters, which are tempera on handmade paper paintings—depict rural life and labor (cobbler, hunter, musician, nursing mother, tailor), and the flora and fauna of the region (bird, camel, lion, fox, flower). Executed in a breezy style with vibrant earth colors, the Haripura Posters are considered by some scholars among Bose's most original work (Sihare 1983, 37). The manner in which such humble works as *Cooking*, 1937, and *Tailor*, 1937, are depicted—bring to mind the religious practice of *Karma yoga*, or the discipline of action. In such works as *Sarangiwala*, 1937, and *Camel*, 1937, Bose gave equal importance in his framing to human and the nonhuman animal, which is consistent with his search for unity among the diversity of creation, a practice he equated to "spiritual *sadhana*", or discipline (Bose 1983, 11).

Bose also started to mix the mythological with everyday themes, either explicitly or through allusion. In the stunning painting *Annapurna*, 1943, he fused religion and the everyday struggle, with a biting critique of injustice and war. Goddess Annapurna, wife of Siva, is the provider of food and nourishment. She is seated on a lotus flower with a bowl of rice on one palm and handful of rice on the other. In front of her, Siva, reduced to a skeleton, performs the dance of destruction (the *Nataraja* posture) with an empty begging bowl made from skull. The hoarding of rice from eastern India by the British for the Allied Forces in World War II led to the great Bengal famine in 1943. Three million people starved to death (Bose 2008, 109). Bose combined several contrasting techniques and themes in the painting: the graphic detail with distinct lines on the figures are placed against an atmospheric background achieved through the wash technique of earlier years, and the poised pose of Annapurna against a cool blue background is juxtaposed with the enraged pose of Siva against an ochre background aflame—alluding to the tragedy of hunger, starvation, and death in a land of abundance—due to war. *Annapurna* is an exemplary work of political ecology.

Figure 39.1 Annapurna, 1943 by Nandalal Bose. Wash and tempera on paper
Source: Courtesy of National Gallery of Modern Art, New Delhi, India.

Nandalal Bose left an indelible mark in the history of art by bringing together eco-spiritual and ethico-political sensibilities in his prolific oeuvre. In the last decade of his life, Bose strove to find the divine in the increasingly minimalist depictions of nature. He wrote:

> I try to see the Rhythm of Life in every form common and uncommon. ... previously I sought for divinity only in the image of the gods and goddesses, now I try to find it in 'sky, water and mountains'.
>
> *(Bose 1983, 13)*

The famous religious painting, *Sati*, 1907, Bose created while still a student at the Government School of Art in Calcutta, and the minimalist spiritual drawing, *The Waves*, 1963, he made three years before his death—could be considered the bookends of a devotional career, in which the "practice of art is as good as a lifelong worship" (Bose 1956).

Nearly three decades after Nandalal Bose made his last drawing, Indian artist Subodh Gupta started to exhibit his work. An eco-spiritual link can be established between the works of these two artists, as both successfully elevate humble everyday themes to the realm of the spiritual. Bose depicted everyday life (*Cooking*, 1937) and labor (*Tailor*, 1937), while Gupta depicts objects we use (*Two Cows*, 2003–8) and consume (*Atta*, 2010)—for survival. An artist does not have to be religious to have religion play a significant part in his/her art. This is true of Gupta and his art. "No," was the answer he gave when art critic Aveek Sen asked if he wouldn't call himself religious. But then when Sen asked about the "knowledge" that the artist is exploring in his art, Gupta responded: "Well, it is *Life*—life related to religion" (Gupta and Sen 2014, 66). Art does not get created in a vacuum but instead within a particular culture, a particular environment (Braddock 2009, 26).

Finding God in deep forest

In early March 2014, I saw three separate overlapping exhibitions of twentieth-century Canadian artist Emily Carr's paintings at the Vancouver Art Gallery: *Emily Carr: Deep Forest*, *Scorned: Emily Carr*, and *Emily Carr in Haida Gwaii*. After a dormant period that lasted fifteen years, Emily Carr returned to painting with renewed focus in 1928 and created some of the most memorable forest paintings during the ensuing decade. Today, she is regarded as one of the most significant modern Canadian artists of the twentieth century. Carr worked with a limited color palette (green, brown, grey, blue), and often with inexpensive materials due to financial hardship. Her best forest paintings suspend between admiration and apprehension. "Moss and ferns, and leaves and twigs, light and air, depth and colour chatting, dancing a mad joy dance," Carr wrote in her journal (Carr 1966, 193). That is only half the story. Her forest is not inviting; it is ominous, as she herself wrote about the forests she was painting, "nobody goes there" (Carr 1966, 207). Indeed, who would want to go into the *Forest, British Columbia*, 1931–32, and then walk through the thick understory of the *Wood Interior*, 1932–35, and at the end of the day try to walk past the *Old Tree at Dusk*, 1932?

That Carr's forest is impenetrable to humans does in no way indicate that she had believed in or advocated for nature that does not include humans, a dominant theme in the American land conservation movement at the time, which advocated for *wilderness* that did not permit human habitation. During this same phase of her career Carr also painted *British Columbia Indian Village*, 1930, and *Indian Church*, 1929, where forest and human habitation and culture seem to coexist with harmony. Her impenetrable forest could be read instead as a statement of revolt against how many people in British Columbia perceived the old-growth forests at the time—simply as lumber. Carr was likely preserving the forests around her home by creating an idealized forest in paintings—a *Deep Forest* (Carr 2014).

Lawren Harris, an influential member of the Group of Seven in Canada, had introduced Carr to theosophy, and the two began an intense correspondence about art and spirituality that lasted between 1928 and 1934. "To both of us religion and art are one," Carr later commented. In *Deep Forest*, Carr successfully fused art, ecology, and religion.

In October 1933, Carr went to Chicago. During the trip she experienced two contradictory pictures of the United States: the dust bowl landscape of ruin passing by her train window, and the aggressive optimism of the world's fair, A Century of Progress, in Chicago. The trip sparked a significant transformation in Carr's spirituality, in her art, and the ecology in her art. She finally snapped "the theosophy bond" and returned to Christianity, and eventually found her God in pantheism (Shadbolt 1979).

She was no longer interested in an idealized forest. Two opposing forces began to appear regularly in her paintings: stumps that draw attention to industrial logging, as in *Stumps and Sky*, 1934, and young trees that draw attention to rebirth of forests, as in *Reforestation*, 1936. *Scorned as Timber, Beloved of the Sky*, 1935, is an exemplary work of political ecology in which the spiritual (a whirling sky) meets the ecological reality of industrial logging (sparseness of trees and stumps in the foreground). While she mourned the loss of the old growth forests near her home, she was also hopeful that the forests would regenerate:

> There is nothing so strong as growing. … Man can pattern it and change its variety and shape, but leave it for even a short time and off it goes back to its own, swamping and swallowing man's puny intentions. No killing or stamping down can destroy it. Life is in the soil.
>
> *(Carr 1966, 301)*

Toward the end of her life Carr returned to painting in the same style as her earlier *Deep Forest* works and painted three wonderful canvasses, *Cedar*, *Quiet*, and *Cedar Sanctuary*, in 1942, the last time she was able to go into the woods. She passed away three years later. After all, Carr found her God, not inside a church, or up in the sky, but in the forest: "He was like a great breathing among the trees…and filled all the universe" (Carr 1966, 329).

More than half a century after Carr made her last painting, German photographer Thomas Struth presented his forest series, *Paradise* (Struth 2002). Struth said in an interview that his "approach to the jungle pictures might be said to be new" (Struth 2002a). It would be helpful, however, to think about Struth's *Paradise* in relation to Carr's *Deep Forest*. There are some differences. The photographs in *Paradise* have a nearly homogenous universalist aesthetic, perhaps an attempt to find unity among the world's forests, while paintings in *Deep Forest* have varied aesthetic, to depict complexities of local conditions. Struth went around the world to photograph forests, while Carr painted the forests near her home. Struth's photographs are devoid of politics, while many of Carr's paintings address the political realities of her time. These differences aside, there seems to be significant common ground between *Deep Forest* and *Paradise*. Struth said that he avoided isolating "single forms" and presented the "allover" aspect of ecology. Carr did the same and that is why both *Deep Forest* and *Paradise* seem to be impenetrable. And not unlike Carr, who said about the sky, "I'm glad I know now that that is not where we have to climb to find Heaven" (Hunter 2006, 211), Struth attempts to show that paradise is not up there in the sky but here on earth among the dense foliage that "could be understood as membranes for meditation" (Struth 2002a).

Eco-spirituality in postwar Japanese cinema

If you look closely, any significant environmental issue of our time would seem Rashomon-like—meaning conflicting points of view co-exist. The phrase is based on Japanese director Akira Kurosawa's 1950 film, *Rashomon*, a masterpiece of world cinema.

The plot of *Rashomon* is comprehensible. Briefly stated—a samurai has been murdered and his wife has been raped in a forest. Four people have been summoned at the court for questioning: the bandit (presumed convict), the wife (who was raped), the husband (the dead samurai speaking through a medium), and the woodcutter (a witness). They provide four conflicting versions of the incident. We hear the testimonies through the stories that the woodcutter and the priest are telling a commoner while rain keeps pouring outside. What emerges is the dark side of the self-serving and self-aggrandizing human. Those two attributes that *Rashomon* brings to the fore have long been central concerns of religion.

The actions (rape, murder, police chase) take place inside forests, while the testimonies and storytelling take place in the tribunal courtyard and under the gate. Kurosawa envisioned that the "people going astray in the thicket of their hearts would wander into a wider wilderness" (Kurosawa 1983, 182). It would seem that Kurosawa is using "wilderness" as a place of exile. This interpretation does not hold, however. When Kurosawa writes about the mountain forest surrounding Nara, one of the two forests depicted in *Rashomon*, he speaks with affection. Kurosawa wrote:

> In those days the virgin forest around Nara harbored great numbers of massive cryptomerias and Japanese cypresses, and vines of lush ivy twined from tree to tree like pythons. It had the air of the deepest mountains and hidden glens. … Once a black shadow suddenly darted in front of me: a deer from the Nara park that had returned to the wild. Looking up, I saw a pack of monkeys in the big trees above my head.
> *(Kurosawa 1983, 184)*

Why would a Tokyo dweller express such deep sentiments about wild nature? Reflecting back on his childhood days, he wrote, "At that stage of my life I didn't understand very much about people, but I did understand descriptions of nature…I was influenced by them" (Kurosawa 1983, 46). Why did he then insist that, "people going astray in the thicket of their hearts" wander through and commit criminal activities in the wilderness, a place he appreciated so much? Kurosawa's wilderness could be considered a curious mixture of "a realm over which God's sway did not extend" (Coetzee 2001), and where it is possible for criminal activities to take place, and at the same time, it is a spiritual place that can heal the troubled soul of the human animal. The origin of the latter interpretation can be traced back to his childhood when he had experienced nature's power to cure his physical weakness while exploring the forests around the Toyokawa Village with a "mountain samurai's existence" (Kurosawa 1983, 65). Not unlike *Rashomon*, Kurosawa's wilderness holds contradictions. As the storytelling in *Rashomon* progresses, the priest begins to lose faith in humanity. But in the end, one single generous act by the woodcutter restores that faith. The optimism in Kurosawa is all the more striking given that his film career began in the midst of a war, in which he and his family had suffered greatly, including near starvation (Kurosawa 1983, 140–1).

Kurosawa was acutely aware of the "absence of spiritual progress" in postwar Japan and the fact that industrialization had turned the beautiful country "into a smog-ridden, suppurating wasteland" (Richie 1998, 199). In his 1975 film, *Dersu Uzala*, he critiques industrialization and celebrates a spiritual way of living through the character of the Siberian hunter Dersu Uzala.

Rashomon and *Dersu Uzala* are the two eco-spiritual pillars in the long and prolific career of one of the most admired directors in the history of cinema.

Critics have pointed out one unpraiseworthy attribute in *Rashomon* and *Dersu Uzala*— woman is depicted as a negative force. Another Japanese director, Hayao Miyazaki, whose career began as Kurosawa's came to an end, corrected the gender imbalance by going in the opposite direction. In Miyazaki's films the protagonist is always a young girl. More importantly, likely no filmmaker has thought for as long as and as imaginatively about ecology and spirituality in cinema as the grandmaster of animation, Hayao Miyazaki. *My Neighbor Totoro*, 1988, and *Spirited Away*, 2002, could be considered the eco-spiritual pillars of his prolific career. In *My Neighbor Totoro*, Miyazaki attempts to mend the tattered bond between human and nature. In one scene, standing in front of Totoro's tree before giving the "forest spirits a proper greeting," the father tells his two children, Satsuki and Mei, that the magnificent tree "has been around since long ago, back in the time when trees and people used to be friends."

In *Spirited Away*, on the other hand, Miyazaki creates a fantastical world, and addresses, among other topics, mass consumption and pollution. The spiritual in *Spirited Away* is primarily achieved through negation—the loss of spiritual life in postwar Japan that gave rise to material desires. Mass consumption is the leading cause of the rapid destruction of the ecological fabric on this Earth, including climate change, ocean acidification, and the Sixth Extinction. The theme of mass consumption runs throughout *Spirited Away*: Chihiro's parents ate too much and turned into pigs; the witch Yubaba's spoiled baby lives inside a pile of stuff; the river spirit regurgitates massive amounts (presumably) thrown into the river by humans; and No-Face over-consumes food, including three bathhouse workers. The young girl Chihiro is immune from the lure of mass consumption and greed. She heals No-Face and Yubaba's baby by leading them from inside the bathhouse to the outside. Miyazaki places particular emphasis on *being-outside* to curb material desires.

The Earth is faster now

Some of the key attributes of eco-spiritual art that I have mapped in this text are that:

- It can instill deep appreciation for the natural world, critique its destruction, and point toward renewal, which can prove to be a powerful tool to advance progressive environmental reforms (Dunaway 2005).
- It can address not only the ecological but also the social, economic, political, and the ethical.
- It can not only address the external, like the environment that surrounds us, but also the internal, like human nature.
- An artist does not have to be religious for religion to pervade his/her art.
- The modern artists I mention worked with meager financial means, whereas, the contemporary artists are working in an environment of general affluence, but at the same time the general affluence led to mass consumption, which then became a subject for contemporary artists.
- The arrival of modernism did not sever the centuries old bond between art and religion; and that ecology has often been the glue in that bond during the past one hundred years.

What good is eco-spiritual art for? Without overstating the usefulness of art in solving our rapidly advancing ecological crisis several points are evident. My encounter with the pacing of art, ecology, and spirituality happened in India when I was rather young. I did not grasp its meaning at the time, however. It was literary art, not visual—Rabindranath Tagore's book of poems, *Gitanjali*. He opens one of the poems with these words:

> The morning sea of silence broke into ripples of bird songs;
> and the flowers were all merry by the roadside; and the
> wealth of gold was scattered through the rift of the clouds
> while we busily went on our way and paid no heed.

The poem continues its dual branches of speeding up ("We quickened our pace more and more" and "My companions…hurried on") and slowing down ("I surrendered my mind without struggle to the maze of shadows and songs"), and ends on a devotional note. The human animal never did slow down and instead continued to quicken the pace since Tagore penned that poem more than a century ago.

"The Earth is faster now," Mabel Toolie told her nephew Caleb Lumen Pungowiyi, a Saint Lawrence Island Yupik Eskimo from Savoonga, Alaska, who passionately advocated for the preservation of Arctic cultures (Krupnik and Jolly 2002). Pungowiyi said that his aunt "was saying that because the weather patterns are [changing] so fast now" that traditional indigenous knowledge is not being able to keep pace with it to predict the weather as they used to be able to do. Reality outpaced the predictions from all the scientific models in case of the rapid melting of the Arctic sea ice. We need to slow ourselves down first if we are to slow down the Earth. Eco-spiritual art can act as "membranes for meditation" and shine a spotlight on the path toward such a slowing down.

Eco-spiritual art can also play a significant role in the research and establishment of moral ecology as a discipline of academic inquiry. Moral ecology is simply ecology in which justice for the nonhuman biotic life and for the indigenous and other marginalized and/or poor human communities around the world play a central role. One significant example of moral ecology is Pope Francis' encyclical, *Laudato Si: On Care for Our Common Home* (2015). How does art play a role in moral ecology? Nandalal Bose's *Annapurna* and Emily Carr's *Scorned as Timber, Beloved of the Sky*, are not only works of political ecology but also of moral ecology. Moreover, the cover art of this volume is a work of moral ecology. It is a photograph I made in Siberia on a cold (minus 65 degrees F without windchill) mid-November afternoon in 2007. It shows Even reindeer herder Matvey Nikolayev on a reindeer gathering the herd. The picture is responding to a tragic episode in Siberia's history. The story goes that Siberian shamans used to be able to fly around on reindeers to help manage the health and integrity of the herd for the community. During the twentieth century the Soviet communists in their attempt to wipe out shamanism from Siberia had gathered all the shamans they could find and put them on helicopters and threw them out of open doors saying something like this: You know how to fly, so you will survive. But the state attempt to sever the millennia-old spiritual bond between indigenous peoples and the reindeer failed. The Even reindeer herders have succeeded in "Outliving the Empire" (Vitebsky 2005) and are themselves, examples of moral ecology.

References

Avery, K. J., Harvey, E. J., Kelly, F. and Applegate, H. 2003. *Hudson River School Visions: The Landscapes of Sanford R. Gifford,* Metropolitan Museum of Art, New York.

Bose, N. 1956. *On Art* Kalakshetra Publications, Madras; translated from Bengali by Surendranath Tagore and Kanai Samanta.

Bose, N. 1983. The Discipline of Art in *Nandalal Bose (1882–1966) Centenary Exhibition,* National Gallery of Modern Art, New Delhi.

Bose, S. 2008. "Universalist Aspirations in a "National" Art: Asia in Nandalal Bose's Imagination", in Quintanilla S R *Rhythms of India*, San Diego.

Braddock, A. and Irmscher, C. eds 2009. *A Keener Perception: Ecocritical Studies in American Art History* University of Alabama Press, Tuscaloosa.

Braddock, A. C. 2009. "Ecocritical Art History", in *American Art*, Summer 2009.

Braddock, A. C. 2015. "From Nature to Ecology: The Emergence of Ecocritical Art History", in Davis J, Greenhill, J. A., and LaFountain, J. D. ed. *A Companion to American Art* Wiley-Blackwell, Hoboken.

Brown, A. 2014. *Art and Ecology Now* Thames & Hudson, London.

Carr, E. 1966. *Hundreds and Thousands: The Journals of Emily Carr,* Clarke, Irwin & Company Limited, Toronto.

Carr, E. 2014. *Emily Carr: Deep Forest,* Vancouver Art Gallery exhibition, Vancouver, 21 December 2013–9 March 2014.

Chandra, P. and Quintanilla, S. R. 2008. "Nandalal Bose and the History of Indian Art", in Quintanilla S R *Rhythms of India*, San Diego.

Chatterjee, S. 2013. "Writing a Transcultural Modern: Calcutta, 1922", in Regina Bittner, Kathrin Rhomberg eds. *The Bauhaus in Calcutta: An Encounter of the Cosmopolitan Avant-Garde* Hatje Cantz, Ostfildern.

Coetzee, J. M. 2001. "The Picturesque, the Sublime, and the South African Landscape", in *David Goldblatt: Fifty One Years* Museu d'Art Contemporani de Barcelona, Barcelona.

Cotter, H. 2008. Indian Modernism via an Eclectic and Elusive Artist *The New York Times,* 20 August.

Demos, T. J. 2015. "Rights of Nature: The Art and Politics of Earth Jurisprudence", in *Rights of Nature: Art and Ecology in the Americas,* Nottingham Contemporary, Nottingham.

Demos, T. J. 2016. *Decolonizing Nature: Contemporary Art and the Politics of Ecology,* Sternberg Press, Berlin.

Dunaway, F. 2005. *Natural Visions: The Power of Images in American Environmental Reform,* The University of Chicago Press, Chicago.

Elkins, J. 2004. *On the Strange Place of Religion in Contemporary Art,* Routledge, New York

Gupta, S. and Sen, A. 2014. "Conversation: Subodh Gupta & Aveek Sen", in *Subodh Gupta: Everything Is Inside,* Penguin Studio, New Delhi.

Hunter, A. 2006. "Emily Carr: Clear Cut", in *Emily Carr: New Perspectives on a Canadian Icon,* Douglas & McIntyre, Vancouver.

Krupnik, I. and Jolly, D. eds 2002. *The Earth is Faster Now: Indigenous Observations of Arctic Environmental Change,* Arctic Research Consortium of the United States, Fairbanks

Kurosawa, A. 1983. *Something Like an Autobiography,* Vintage Books, New York.

PBS. 2001. Art in the Twenty-First Century, Season 1: Spirituality

Quintanilla, S. R. 2008. *Rhythms of India: The Art of Nandalal Bose,* San Diego Museum of Art, San Diego.

Richie, D. 1998. *The Films of Akira Kurosawa,* University of California Press, Berkeley.

Rosenfield, J. M. 2008. "Introduction", in Quintanilla, S. R. *Rhythms of India*, San Diego.

Shadbolt, D. 1979. *The Art of Emily Carr,* Douglas & McIntyre, Vancouver.

Sihare, L. P. 1983. Nandalal Bose: His Aesthetic Percepts and Styles, A Few Problems in *Nandalal Bose (1882–1966) Centenary Exhibition,* National Gallery of Modern Art, New Delhi.

Struth, T. 2002. *Thomas Struth: New Pictures from Paradise,* Schirmer/Mosel, Munich.

Struth, T. 2002. "Interview—Thomas Struth: Talks about his Paradise series", in *Artforum,* May 2002.

Vitebsky, P. 2005. *The Reindeer People: Living With Animals and Spirits in Siberia,* Houghton Mifflin Company, New York.

40

POLICY

Maria Ivanova

In September 2015, Pope Francis addressed governments as they gathered at the seventieth anniversary of the United Nations (UN) to adopt a new action agenda, *Transforming Our World: the 2030 Agenda for Sustainable Development*, and a new set of global Sustainable Development Goals. He spoke at the UN General Assembly delivering a compelling call for transformation of our laws, values, and norms so that we ensure respect for the environment, elimination of exclusion, enhanced responsibility, and integral human development. He emphasized that "in all religions, the environment is a fundamental good," that human beings are part of the environment and can only survive and develop if the environmental conditions are favorable, and that all living creatures have intrinsic value in their interdependence with other creatures (Francis I 2015). The impassioned call to care for our planet shaped the tone at the governmental summit. Leader after leader repeated Pope Francis's call to tackle the environmental crisis and act to stem climate change. Religion and ecology and environmental policy and governance thus came together more clearly than ever as Pope Francis emphasized that "government leaders must do everything possible to ensure that all can have the minimum spiritual and material means needed to live in dignity," including housing, employment, food and water, spiritual freedom, and education (Francis I 2015).

In this chapter, I seek to illustrate how the field of environmental policy and governance intersects with the field of religion and ecology, and explore how the two fields could inform each other. I enter this space through the nexus of individuals and institutions, including their responsibilities for providing the minimum means to live in dignity, and through the role of education and scholarship. As Pope Francis has recently tackled this intersection directly in his Encyclical *Laudato Si* of June 2015, and in his address at the seventieth UN General Assembly, I draw heavily on these recent documents. There is a broad body of thought on religion and ecology that is pertinent to policy and governance that I am not able to engage within the limits of this chapter.[1]

Environmental policy and governance deals with the sources and structures of power as they relate to ecological and social realities. It analyzes institutions at the national and international level, actors such as states, businesses, or non-governmental organizations (NGOs), and tools such as cost–benefit analysis, risk assessments, and indicators. It overlooks the role of individuals as agents. Religion and ecology, on the other hand, often focuses on moral conditions and imperatives. The individual is a prime focus of inquiry as the field deals with the human/nature relationship and human responsibility toward the environment. It is concerned with spirituality, ethics, and morality. Both fields grapple with causes and symptoms of the ecological crisis and with the role of norms in finding solutions to pressing global problems. For policy and governance, the norms are embodied in institutions and individuals have little to no space in the theoretical discourse. For religion and ecology, individuals are a core driver in the creation and propagation of norms, even when that happens through communities, traditions, or religion-related institutions. The role of individuals in institutions as founders and leaders is an important intersection where the two fields could enrich each other's concepts, methods, and conclusions.

Education provides another entry point into the confluence of the two fields. Education is the basis for implementation of an ambitious policy agenda and for reclaiming the environment, Pope Francis noted in his UN speech. What kind of education do we provide in the field of environmental policy and governance and in religion and ecology? Where do these fields intersect in the classroom and what are differences? I bring to the discussion an analysis of the concepts covered in courses in the two fields across universities around the world in the hope of highlighting common ground and ways for moving forward that would bring together analytical and ethical approaches to studying individuals and institutions as agents for change.

The chapter begins with a brief history of global environmental governance highlighting the original integrated vision the founders of the environmental institutions had. It then offers an analysis of the education universities provide in environmental policy and governance and in religion and ecology, respectively, because it is in the classroom that we shape the conscience of many who would one day be charged with the conduct of international affairs. The chapter concludes with a vision for engaged scholarship that would shape both individual values and institutional trajectories and with reflections on the skills we seek to instill in our students as they learn to be scholars and global citizens.

Environmental policy and governance: the beginnings

The first clear articulation of environmental problems dates back to the 1960s when air and water pollution attracted strong public attention in Europe and North America and the public put strong political pressure for immediate policy action. As Bill Ruckelshaus, the first Administrator of the US Environmental Protection Agency noted,

> We had rivers that caught on fire. We had the desire of the people living in Denver to see the mountains again and people in Los Angeles to see one another. We had smell, touch and feel kind of pollution problems that have now been dealt with.
>
> *(Ivanova and Ageyo 2009)*

There was a rise in public demonstrations and publications about the impact of environmental degradation such as Rachel Carson's *Silent Spring* (1962) (about the impacts of pesticides on the environment and health) and Jean Dorst's *Before Nature Dies* (1972). Environmental institutions sprung up at the national level in the US and Europe, governments created laws and regulations, and a number of environmental NGOs emerged as important actors.

It was clear, however, that the problems did not stop at national borders and that collective action was required at the international level. At the initiative of Sweden, governments agreed to hold the first major international environmental conference in 1972. One hundred and thirty-one nation states gathered in Stockholm in 1972 for the UN Conference on the Human Environment, widely known as the Stockholm Conference, under the slogan "One Earth." The main purpose was to set the stage for environmental issues to command attention at the highest levels of international politics and motivate political commitments and practical actions

> to protect and improve the human environment and to remedy and prevent its impairment, by means of international cooperation, bearing in mind the particular importance of enabling the developing countries to forestall the occurrence of such problems.
>
> *(United Nations 1972)*

The environmental agenda was thus crafted from the outset with the full recognition that environmental protection and international cooperation were interconnected in a moral imperative.

The integrated dimensions of the natural and social systems were at the core of the preparations for the 1972 Stockholm Conference. The conference team, led by Maurice Strong, recognized early on that in the period of post-colonialism, developing countries might be opposed to what could be perceived as a Northern anti-pollution agenda. A Canadian industrialist and businessman with an avid interest for international affairs, development, and all UN matters, Strong was appointed Secretary-General of the Stockholm Conference in 1970 because of his skills as a coordinator, collaborator, and convener. One of the key achievements of Strong and his team was the commitment to active participation they elicited from developing countries. It took great energy to convince developing countries that this was not a "green imperialism conference." Strong flew from capital to capital to meet personally with presidents and prime ministers recognizing their concerns, acknowledging the social dimensions of the ecological crisis, and articulating the interconnectedness between society, economy, and the environment. He emphasized that environmental issues could adversely impact economic development through lowering groundwater levels, soil erosion, increasing desertification, depleted fisheries, and other similar problems, and that the poor and underprivileged would be most affected by such degradation.

However, the tension between environment and economic growth plagued the negotiations throughout the conference because developing countries saw industrialization and development, a coveted economic trajectory, as the cause of environmental degradation (Ivanova 2007, 344). Furthermore, they viewed the implementation of environmental protection measures as a costly effort. Or as the Brazilian delegate Bernardo de Azevedo Brito put it: "I do not believe we are prepared to become new Robinson Crusoes… Each country must be free to evolve its own development plans, to exploit its own resources and to define its own environmental standards" (United Nations General Assembly 1972). A military dictatorship at the time, Brazil pushed for unbridled economic growth where "the gains [were] loudly proclaimed and the costs swept under the rug of censorship" (Schneider 1991, 265). Since Brazil emerged as the informal leader of the Group of 77, the political negotiating block in the UN comprising all developing countries, the harsh ideological dynamic often resulted in negotiation stalemate. Yet other developing country leaders emphasized the integral nature of the environmental agenda. The speech by India's Prime Minister Indira Gandhi at the Stockholm Conference is illustrative. While she is mostly quoted for singling out poverty as the cause for environmental degradation in the developing world, she also emphasized that "the environmental crisis which is confronting the world will profoundly alter the future destiny of our planet. No one among us, whatever our status, strength or circumstance, can remain unaffected" (Gandhi 1972).

The Stockholm Conference produced two important institutional outcomes. First, states agreed on the Stockholm Declaration, which sets forth twenty-six principles stressing the impact of anthropogenic activities on the environment, and addresses issues such as nature conservation and wildlife protection, non-renewable resources, toxic substances, marine pollution, and population growth. Principle 21, on the responsibility of a state to not cause damage to the environment of others, has become part of international law, as upheld by the International Court of Justice (Bodansky 2010, 28). Second, governments created the UN Environment Programme (UNEP) as the "anchor institution for the global environment" (Ivanova 2005) to review the state of the environment, coordinate environmental issues across the UN system, catalyze action across actors, and build capacity in countries as needed.

It also produced broader societal impact as it opened up a new narrative of what today Pope Francis calls integral ecology, "one which clearly respects its human and social dimensions" (Francis I 2015, 137). The discussions about the problems, the solutions, and the institutions focused on the need for an integrated approach that considers the interactions between human and natural systems and the variations in the socioeconomic conditions. As Maurice Strong noted during the 2009 Global Environmental Governance Forum in Glion, Switzerland,

> Right from the beginning, it was recognized that the worst thing you could do is sectoralize the environment; because inherently the decisions and the actions that affect the environment are taken largely through the economy. They have social impacts, they have economic impacts, and they have environmental impacts".
>
> *(Ivanova and Ageyo 2009)*

Over time, the number of environmental institutions, laws, regulations, instruments, and policies has grown, yet the problems continue to grow largely unabated. Many scholars assume that the root cause of global environmental problems is the inability or unwillingness of states to prioritize the necessary actions themselves or to delegate appropriate authority to the international institutions they have created. The key premises for this explanation derive from the assumptions that "governments are naturally and properly reluctant to empower an international organization to make, or even to significantly constrain, their national environmental and economic policy choices" (Gaines 2003, 359) and that failure to accomplish anything of significance in terms of governance reform arises out of a lack of political will among states (Moe 1989). Based on this premise, the core thesis and dominant narrative is that international environmental institutions are deficient by design because nation-states have no interest in creating powerful international bodies that might jeopardize national self-interest and would therefore deliberately design such agencies to be weak.

Archival materials of the intergovernmental discussions and the accounts of the founding architects, however, reveal a surprising and compelling story about institutional design for the environment. Governments created a new international environmental body with a clear vision for an agile and able intergovernmental agency that would bring coherence, competence, and connectivity in an institutional landscape of independent agencies with existing priorities. Contrary to conventional wisdom, UNEP was not created to be weak. Its mandate was ambitious, its institutional form flexible, and its funding mechanism solid and envisioned to grow over time. In reality, however, UNEP has been "under-funded, over-loaded and remote... relatively obsolete, [and] eclipsed in resources and prestige by other international institutions that have taken on new environmental responsibilities" (Haas 2004, 2). It is therefore important to distinguish cause and effect in international policy and governance. The tendency in the literature is to assume that the cause of dysfunction in international institutions is the unwillingness of states to create these institutions in a way that would make them "strong." This highlights one of the core approaches in the policy and governance field—taking the state as a unitary actor. There is no historical evidence that governments purposefully created a weak international environmental institution. Indeed, the creation of an international environmental organization in the early 1970s was nothing short of a feat, and required a great deal of leadership, vision, and political mobilization.

In essence, UNEP was the creation of individuals with clear priorities and sharp instincts. They knew which functions they wanted the new institution to perform and devised a mechanism appropriate to the task despite opposition from some governments and many of the UN agencies. John W. McDonald was director of economic and social affairs at the Bureau of International Organization Affairs at the US State Department and is one of UNEP's creators. As many other state officials, his name is not to be found in historical documents. He served on the US delegation preparing for the conference and negotiating in Stockholm. McDonald had done this before. He had been instrumental in the creation of several UN offices, including the UN Fund for Population Activities, UN Volunteers, and the post of UN disaster relief coordinator and recognized the need for a central structure for all environmental efforts. "That became my mantra," said McDonald, "you had to have a new agency to actually make this happen" (McDonald 2009).

The report of the Committee on International Environmental Programs of the National Academy of Sciences, commissioned by the US State Department in preparation for the Stockholm Conference, arrived at the same conclusion:

> We recommend the establishment of a unit in the United Nations system to provide central leadership, to assure a comprehensive and integrated overview of environmental problems, and to develop stronger linkages among environmental institutions and the constituencies they serve.
>
> *(Environmental Studies Board 1972)*

Ultimately, the result of the Stockholm conference was the sum of the actions of individuals in the conference team and in governments. Their names do not appear on the documents but their impact as individuals and as a collective shaped the international institutions we currently rely on for addressing global environmental problems.

During the past half century, it has become increasingly clear that dealing effectively with global environmental change does indeed require a functional global environmental governance system to tackle the economic and political causes of the problems. Indeed, where international institutions have functioned well, problems are being resolved. Pope Francis noted this dynamic in his Encyclical, highlighting

> the Basel Convention on hazardous wastes, with its system of reporting, standards and controls, ... the binding Convention on international trade in endangered species of wild fauna and flora, which includes on-site visits for verifying effective compliance. Thanks to the Vienna Convention for the protection of the ozone layer and its implementation through the Montreal Protocol and amendments, the problem of the layer's thinning seems to have entered a phase of resolution.
>
> *(Francis I 2015, 168)*

Similarly, progress cannot be made on a number of issues—from biodiversity to climate change—when negotiations stall because of narrow economic self-interests, short-term thinking, and disregard for the common good. Ultimately, the failure of the economic and governance systems to incorporate environmental concerns reflects a failure of values.

Recalibrating our moral and ethical values will be necessary and a new ethic of global citizenship is essential for effective resolution of the twin problems of environmental deterioration and social exclusion.

Educating for the agreement between humanity and the environment

"We are faced with an educational challenge," Pope Francis writes in his Encyclical on care for our common home.

> The existence of laws and regulations is insufficient in the long run to curb bad conduct, even when effective means of enforcement are present. If the laws are to bring about significant, long-lasting effects, the majority of the members of society must be adequately motivated to accept them, and personally transformed to respond. Only by cultivating sound virtues will people be able to make a selfless ecological commitment.
>
> *(Francis 1 2015, 211)*

It is through education that such values are generated and cultivated. Traditionally, environmental education has focused on scientific knowledge, awareness raising, and understanding of cause and effect relationships in the natural environment. In the contemporary classroom, however, students seek a source of existing information, the creation of new knowledge, and empowerment to realize their ambitions.

Academia's primary focus has traditionally been on *what is and why*, i.e. on the state of the world and the reasons behind it. In an environmental classroom, however—whether focused on governance or on religion—students come with aspirations to instigate change in the world. We are therefore compelled to seriously explore *what ought to be* and envision solutions drawing from the causal mechanisms in the disciplines we cover. Yet the ultimate challenge in the classroom is to take the third step and reflect on *how to get there*, a question that demands serious and humble understanding of causal mechanisms in natural, human, and spiritual systems. And knowledge of such pathways often comes from beyond academia making engagement with policymakers or religious leaders a valuable learning tool. Trying to understand the intersection between the two fields of environmental policy and governance and religion and ecology, we need to enter the respective classrooms and examine the approaches educators use.

To get a glimpse into these classrooms, I have reviewed thirty-eight courses in environmental policy and governance at thirty-three universities in seventeen countries (plus two global online courses offered by the United Nations Institute for Training and Research [UNITAR]) and twenty-seven courses in religion and ecology at twenty-two universities in five countries. Though perfunctory and limited by the availability of syllabi online, this approach reveals the substantive and procedural aspects of teaching environmental policy and governance and religion and ecology, and offers insights into the tools faculty use to help students understand the world, envision change, and enact their ambitions. Courses on global environmental governance are taught in departments of political science, international relations, public policy, environmental studies, and law and are often cross-listed attracting students from all disciplines. Courses in religion and ecology are taught in departments of religious studies, theology, anthropology, sociology, and environmental studies and interdisciplinary studies. The overlap between the two fields, therefore, occurs only in environmental studies departments. Most notably, at the initiative of two of the co-editors of this book, Mary Evelyn Tucker and John Grim, Yale University used this overlap to develop a joint Masters degree program in religion and ecology, the first of its kind in North America. The Yale School of Forestry & Environmental Studies and Yale Divinity School offer the degree, which is strongly supported by co-appointed faculty and by the Forum on Religion and Ecology at Yale.

Substantively, the two fields explore related issues but from different perspectives. Learning about the global environment whether through a policy or a religious lens is ultimately driven by the ambition to address problems in the environmental domain. The common denominator among courses in the two fields is indeed the acknowledgement that humanity is facing an urgent ecological crisis. The perspectives on the origins and consequences of this crisis, however, might be different since the two fields see the relationship between humans and nature differently. Environmental justice is clearly a common core concern in both fields and is also cross-referenced with other areas, such as law. It offers a perspective on the tension between anthropocentrism and ecocentrism and between the utilitarian and intrinsic value of nature. Sustainability also appears as common concern but the concept might carry a different meaning in each field. For instance, from a religion and ecology standpoint, a sustainable existence might be rooted into ethical principles independent of laws and regulations.

Environmental policy and governance courses deal explicitly with institutions at the national and global level as key mechanisms for governance. The emphasis ranges from institutions as sets of rules, norms, and decision-making procedures to international organizations as the physical embodiments of the institutions. The history of the international regimes of institutions and organizations around the various issue areas—climate, biodiversity, chemicals, desertification, etc.—features prominently in many global environmental governance courses. Religion and ecology courses focus on the relationship between humans and the environment. This often includes an examination of the self and of community engagement. Religion and ecology courses appear to be more introspective, focusing more on how humans or religions see their relationship with nature, which moral or ethical standards should be upheld, and what the relationship between spirituality and sustainability should be. These types of reflections are usually missing in environmental governance and policy courses, which focus on possible modes of action to address the environmental issue, from policy to law to activism.

Environmental governance and policy courses are largely anthropocentric and one-sided, mostly concerned with the effects humans have on the environment. The web of life approach, which considers the interdependence of psychological, biological, physical, social, and cultural phenomena—is not nearly as prominent in these courses as it is in religion and ecology courses where interrelationships are emphasized. Religion and ecology courses are also, for the most part, diverse in terms of the schools of thought they cover, addressing Western, Asian, and native/indigenous traditions and civilizations, including longitudinal studies. Environmental policy and governance courses are largely Western-centric, focusing on and emphasizing democratic regimes and traditional (in the Western sense) concepts such as hierarchy, power, human and ecological security, and governance.

The approach to understanding and addressing the ecological crisis also differs in the two fields. The role of scientific knowledge as a basis for decision-making and the science-policy interface is a major tenet in the policy and governance field. Religion and ecology does not seem to put as much emphasis on the scientific knowledge about environmental issues as environmental policy and governance courses do. Although science is part of the curriculum of some courses (or at least the scientific understanding of why a natural resource is being depleted, for example), the emphasis is not so much on how to communicate science better (as in the science–policy interface studies in the policy courses), but on how to communicate and translate the environmental crisis into scriptures and religious traditions and vice-versa. In other words, the process through which religions understand the environmental crisis is different from the social and natural sciences.

Ultimately, leadership will be the vital factor for success in addressing the contemporary ecological challenges and will have to be present at all levels, from community organizers to religious leaders to national and international executives and policymakers. Academic institutions can help cultivate, motivate, and connect champions for integral ecology when they overcome disciplinary blinders and foster intellectual curiosity, academic audacity, and vivid imagination. As Pope Francis emphasized,

> [e]nvironmental education should facilitate making the leap towards the transcendent which gives ecological ethics its deepest meaning. It needs educators capable of developing an ethics of ecology, and helping people, through effective pedagogy, to grow in solidarity, responsibility and compassionate care.
>
> *(Francis I 2015, 210)*

Through teaching inside and outside the classroom and engagement of students in research work, educators can create a stimulating intellectual climate that enables students to transcend disciplinary divides and empowers them to engage in solving global problems as scholars and professionals. Our classrooms are already global: the people we educate come from all over the globe, are likely to work all over the globe, and are already connected to communities all over the globe. The philosophy of our classrooms today is therefore likely to be propagated across the world in the immediate and the very near future, one person at a time. To begin building integral human development, we will need bold, unflinching leadership to shatter stereotypes, create a climate of cooperation, and devise an analytically sound and morally grounded agenda for action. Such leadership would draw on the talent, enthusiasm, and energy of the new generations of environmental scholars and activists who are now entering our classrooms.

Moving forward

In 1916, the American philosopher, psychologist, and educational reformer John Dewey wrote that "[t]he most significant question which can be asked about any situation or experience proposed to induce learning is what quality of problem it solves" (Dewey 1916, 182). Environmental policy and governance and religion and ecology as fields attempt to inform and inspire to solve the global ecological crisis. Engaging educators from the two fields and enticing greater communication among them and their students might help us transcend the limitations and boundaries of our disciplines in a new way and create the foundation for education grounded in knowledge, inspiration, and responsibility. At their intersection, the two fields might come together around a set of common goals. Through research and education, they could help reframe the environment–economy dichotomy, cultivate shared values and philosophies, and animate communication.

Reframe the environment–economy dichotomy. Redefining the connection between the environment and the economy in a new paradigm for human progress is essential. Sustainable development, the paradigm for understanding the relationship between economic growth, social welfare, and environmental protection, has largely failed to reform economic decision-making in the way originally intended. We need a new vision of an integral human development focused less on short-term rewards and externalized risk and more on long-term values of sustainability and social justice. To this end, the two fields could spearhead the development of new knowledge, a new narrative, and new policies.

Cultivate shared values and philosophies. Ultimately, the failure of the economic system to incorporate environmental concerns reflects a failure of values. Reorienting our moral and ethical values will be a necessary condition for change in our behavior toward the environment. A new ethic of global citizenship is also essential for effective, legitimate, and equitable global environmental governance. While many academics are wary of the proposition that they shape shared values, the ideas and knowledge we generate cannot but influence the ethics of new generations of scholars and practitioners. As Abraham Lincoln once noted, "The philosophy of the classroom of one generation is the philosophy of the government of the next generation."

Animate communication. International institutional structures, tangible backyard environmental problems, and spiritual fulfillment are disconnected issues for most people. Academics from the two fields of environmental policy and governance and religion and ecology could establish this connection and facilitate communication to generate strong grassroots demand and support for improved environmental decision-making at all levels. They could conceptualize and create an integral ecology communication hub, which harnesses existing information tools and resources through new communication technologies. They can reframe the narrative, in institutional accounts and in scholarly and published accounts, to place the intellectual and scientific focus of international environmental institutions in the context of a more engaging and gripping story illustrating the interconnections between humans and nature, between planet and prosperity, and between governance and survival.

Importantly, we have to recognize the sources of such knowledge. The role of individuals in institutions is often neglected in high-level academic analyses of systemic forces. Yet it is individuals who imagine and instigate institutional change. And it is academia that has the capacity to honor such contributions.

Note

1 See for example, "The Assisi Declarations: Messages on Humanity and Nature from Buddhism, Christianity, Hinduism, Islam & Judaism." – www.arcworld.org/downloads/THE%20ASSISI%20 DECLARATIONS.pdf

References

Bodansky D. (2010) *The Art and Craft of International Environmental Law,* Harvard University Press, Cambridge.

Carson R. (1962) *Silent Spring,* Houghton Mifflin, New York.

Dewey J. (1916) *Democracy and Education,* Macmillan, New York.

Dorst J. (1972) *Before Nature Dies,* Houghton Mifflin, New York.

Environmental Studies Board (1972) Institutional arrangements for international environmental cooperation: A report to the Department of State by the Committee for International Environmental Programs, National Academy of Sciences, Washington DC.

Francis I (2015) Encyclical letter on care for our common home (http://w2.vatican.va/content/ francesco/en/encyclicals/documents/papafrancesco_20150524_enciclicalaudatosi.html) Accessed 9 October 2015.

Gaines S. E. (2003) "The problem of enforcing environmental norms in the WTO and what to do about it", *Hastings International and Comparative Law Review* 26:3, 321–85.

Gandhi I. (1972) Life is one and the world is one: Prime Minister Indira Gandhi speaks to the plenary at the Stockholm conference on the human environment, 14 June 1972, United Nations Conference on Human Environment, Stockholm.

Haas P. M. (2004) "Addressing the global governance deficit", *Global Environmental Politics* 4:4, 1–15.

Ivanova M. and Ageyo J. (2009) *Quest for Symphony,* [Motion picture]. United States.

Ivanova M. (2005) "Can the anchor hold? Rethinking the United Nations Environment Programme for the 21st Century" *Yale School of Forestry and Environmental Studies Publication Series Report* 7, New Haven.

Ivanova M. (2007) "Designing the United Nations Environment Programme: A story of compromise and confrontation", *International Environmental Agreements: Politics, Law and Economics* 7:4, 337–61.

Ivanova M. (2015) "Teaching global environmental governance", in Pattberg P. and Zelli F. eds., *Edward Elgar Encyclopedia of Global Environmental Politics and Governance,* Edward Elgar Publishers, Cheltenham 161–74.

McDonald J. W. (2009) Reflecting on the past, moving into the future, Presentation at Global Environmental Governance Forum, Glion, Switzerland, 28 June 2009.

Moe T. (1989) "The politics of bureaucratic structure" in Chubb J. E. and Peterson P. E. eds., *Can the government govern?* Brookings Institution, Washington DC 451–94.

Schneider R. (1991) *Order and Progress: A political history of Brazil,* Westview Press, Boulder.

Sengupta S. and Yardley J. (2015) "Pope Francis addresses UN, calling for peace and environmental justice" *New York Times* 25 September 2015 (www.nytimes.com/2015/09/26/world/europe/pope-francis-united-nations.html?_r=0).

United Nations (1972) Declaration of the United Nations Conference on the Human Environment (Stockholm Declaration), 16 September 1972, United Nations Environmental Program, Stockholm.

United Nations General Assembly (1972) 27th session: Summary record of the 1466th meeting. Second Committee, Official Record. A/C.2/27/SR.1466. New York.

41

LAW

Religious influences on environmental law

John Copeland Nagle

The history of environmental law

The law has long governed human activities that affect the natural environment, even though the term "environmental law" is of much more recent vintage. The common law of property and torts provides the foundation for the evolution of the law, first in England and then in the United States. The common law first emerged during the twelfth century as judges selected by Henry II followed each other's decisions to create a unified common law throughout England, instead of relying on different local codes. The premise of the common law is that judges are charged with identifying what the law is, which places great weight on following the precedents of cases decided by earlier courts. The common law developed rules for property and torts that still shape the law today, albeit supplemented by an increasing number of statutory and regulatory enactments.

Property law governs what you can do with your land. The common law of property is famously complicated, as any twenty-first century law student laments. Most of the complicated features, though, decide how ownership interests are shared over multiple generations and between multiple parties. The law governing what one may do with one's land is more straightforward. Generally, title to property affords a landowner the authority to do anything on the land that does not constitute a nuisance or waste. Nuisances include both public nuisances (which harm an interest enjoyed by the public at large) and private nuisances (which substantially interfere with another individual's use and enjoyment of their own land). A nuisance is actually a tort – the common law category for injuries to someone else – but they are a type of tort that is attached to the nature of property. Waste is a narrow legal concept that limits the ability of a landowner to use their land in certain ways that impair the value of the land for those who already hold an ownership interest in that land that will become possessory in the future.

These common law rules are centuries old, but they had relatively little application to what we would now describe as environmental issues until the nineteenth century. The reason, quite simply, is that there were very few disputes involving the condition of the land or "environmental harms" until the rapid industrial and agricultural development that began during the nineteenth century. At first, the law encouraged such development. A number of property law doctrines operated to promote the active use of land rather than letting it remain in its natural

393

condition. At the same time, throughout the nineteenth century the federal government sought to dispose the land that it acquired through purchases, treaties, and wars. The Homestead Act is the most famous example of that phenomenon. Signed into law by President Lincoln amidst the Civil War, the Homestead Act gave 160 acres of land to anyone who worked to cultivate it. Such laws, especially including the similar award of land to railroad companies that built transcontinental railroads during the second half of the nineteenth century, demonstrated the federal government's resolve to divest itself of land in favor of private owners.

All of that began to change as twin concerns about pollution and conservation arose by the end of the nineteenth century. Water pollution threatened health, particularly as rapidly growing cities dumped their sewage into the same bodies of water that they used for drinking. Air pollution was first conceived as a "smoke problem" that threatened aesthetics. Nuisance law was the first response to both kinds of pollution problems. But the difficulty in identifying culpable parties when thousands of businesses and individuals combined to contribute to pollution prompted a different strategy. State and municipal governments soon enacted ordinances that regulated polluting activity, though such laws remained novel and localized.

Concern about pollution of the urban environment was accompanied by calls to conserve the wild environment. These calls were voiced by John Muir, who championed the beauty of the Yosemite Valley in California; William Hornaday, who fought to protect wildlife from his position as head of the Bronx Zoo; Gifford Pinchot, who encouraged the newly established United States Forest Service to conserve large western forests; and Theodore Roosevelt, who used his position as President to fashion a host of new environmental protections. The federal government began to create national parks, national forests, and national wildlife refuges as areas subject to special environmental management. President Ulysses Grant approved the creation of the world's first national park in Yellowstone in 1872, though the decision was made easier by the consensus that the land was not suited for any other purposes. More controversially, President Roosevelt asserted the authority to establish the first national wildlife refuge to protect bird populations from being exterminated in Florida. The United States joined with Great Britain, Canada, Japan, and Russia to negotiate the first international treaty to protect wildlife, namely the fur seals that were being slaughtered in the Arctic. But conservationists lost some battles, too. Most famously, the federal government agreed to allow the growing city of San Francisco to ensure its water supply by building a dam across the Hetch Hetchy Valley, notwithstanding the pleas of John Muir and others that Hetch Hetchy was just a beautiful as the rest of Yosemite National Park.

This burst of energy at the beginning of the twentieth century gave way to more gradual development of the law during the first half of the twentieth century. The federal government stopped trying to dispose of its lands, and instead continued to establish new national parks, including parks in the eastern United States that facilitated more visitation. Pollution remained a problem, but most people considered it a problem well within the competence of local authorities. As recently as 1960, the United States Supreme Court said that pollution was a local problem, not a national one. The federal government did not really act against pollution until the middle of the twentieth century, and then it merely sought to assist the state efforts through subsidies and research. And engineering handbooks taught that "the solution to pollution is dilution."

That changed suddenly during the 1960s. Popular concern about pollution, overpopulation, loss of natural areas, and the disappearance of wildlife became a leading social and political issue. Congress responded enacting a series of environmental laws to address each problem. President Nixon signed the National Environmental Policy Act (NEPA) into law during the first week of 1970, and President Carter approved the Comprehensive Environmental Restoration,

Cleanup and Liability Act (CERCLA, also known as the Superfund law) in December 1980. Between those bookended dates, the 1970s also saw the enactment of the Clean Air Act (CAA), the Clean Water Act (CWA), the Endangered Species Act (ESA), the Safe Drinking Water Act (SDWA), the Federal Land Management Policy Act (FLPMA), the Resource Conservation and Recovery Act (RCRA), the Federal Insecticide, Fungicide, and Rodenticide Act (FIFRA), and numerous other environmental laws.

The federal environmental law canon emerged without significant influence from religious teaching. By contrast, churches and religious arguments played a prominent role in the federal civil rights legislation that was enacted just a few years before the environmental laws of the 1970s. The absence of religious influences environmental debates of the 1960s and 1970s is especially striking given the role that such influences played before and after that. Historians are beginning to recover the story of how religious ideas influenced environmental policies throughout the nineteenth century and well into the twentieth century. For example, Mark Stoll argues that "[a] high proportion of leading figures in environmental history had religious childhoods." More specifically, "[e]specially before the 1960s, a very large majority of the figures of the standard histories of environmentalism grew up in just two denominations, Congregationalism and Presbyterianism, both in the Calvinist tradition" (Stoll 2015, 2). Stoll describes the parallel Christian and environmental thinking of a diverse collection of luminaries including the theologian Jonathan Edwards, landscape artists Thomas Cole and Frederick Edwin Church, Forest Service head Gifford Pinchot, and Park Service founder (and Cotton Mather descendant) Stephen Mather. His claim is that Reformed Protestant ideas guided their environmental thinking whether or not that remained observant believers themselves.

A second recent book, Evan Berry's (2015) *The Religious Roots of American Environmentalism*, agrees that modern American environmental thought is deeply shaped by its relationship with Christian theological tradition. Berry traces Christian thought about nature back to the Middle Ages, and in doing so he counters Lynn White's thesis that Christian teaching was uniformly hostile to the world around us. Berry follows the trail in thinking to the United States in the early twentieth century, and specifically to the Mountaineers Club based in Seattle. That organization "played an instrumental role in the development of Mount Rainier National Park and the establishment of both North Cascades and Olympic National Parks." In doing so, "its leaders framed their purposes in religious terms, not because such terms were merely convenient or persuasive but because their project grew from fertile religious soil and always bore traces of its origins" (Berry 2015, 86). Berry sifts through the hymnals used by the club during its expeditions to show how devoutly they conceived their relationship to God and all He had created.

Religious thinking about the environment waned during the twentieth century until it played an insignificant role during the debates of the 1960s and 1970s. Then it re-emerged. There is now an unprecedented abundance of writing on environmental topics across the board from a wide range of religious perspectives. Much of this writing offers valuable insights into how we think about the natural world. But little of the writing engages the additional, and tricky, question of how to integrate those insights into the law. A more robust theory of jurisprudence, describing the appropriate role of law in society, is needed to complement the growing body of moral, ethical, and religious insights generated by today's host of writers, thinkers, and activists.

The standard compilations of environmental law contain dozens of statutes and cover thousands of pages. And that is only United States federal environmental statutes. It does not include federal regulations, state and local laws, the laws of other countries, and international treaties. The sheer abundance of environmental law makes a comprehensive survey impossible. But it is possible to understand how the law operates by looking at a few of the most important environmental statutes in the United States today. Five will be summarized here: the Organic Act governing national park management, the Wilderness Act, the CAA and the CWA, and the ESA.

National parks

The creation of Yellowstone National Park in 1872, along with the reservation of Yosemite Valley from development in 1864, marked the first significant legal actions to conserve the natural heritage of the United States. Religious ideas of natural beauty informed the growing appreciation of the scenic landscapes throughout the United States. Their manifestations included the paintings of Thomas Cole and the writings of John Muir, both of whom celebrated the work of God in nature.

For their first 44 years, the national parks were managed as separate entities by different agencies. Then Congress enacted the Organic Act in 1916, which created the National Park Service (NPS) and instructed the new agency how to manage the national parks. The heart of the Organic Act is its statement that the purpose of national parks "is to conserve the scenery and the natural and historic objects and the wildlife therein and to provide for the enjoyment of the same in such manner and by such means as will leave them unimpaired for the enjoyment of future generations." The dual goals of conservation and enjoyment often support the same management actions. Often, one can enjoy a national park while conserving it at the same time. Hiking, nature photography, and wildlife observation are among the many activities that are consistent with both enjoying a park and conserving it. Conversely, other activities threaten both the enjoyment and the conservation of a national park. Dams have played a particularly prominent role in national park disputes. Mining, logging, climate change, and constructing residential subdivisions are additional examples of actions that interfere with both the enjoyment of a park and its conservation. Those are easy cases under the Organic Act.

But there are many times when the goals of enjoyment and conservation conflict. Snowmobiles provide a memorable opportunity to enjoy winter in Yellowstone National Park, but they can threaten the conservation of wildlife, air quality, and natural soundscapes. Scenic flights provide an unparalleled view of the Grand Canyon, but they interfere with the national park's natural quiet and with the visual experience of people enjoying the scene from the ground. Cell phone towers enable visitors to communicate with friends or with park rangers in the event of an emergency, but they can obstruct the natural scenic view and interfere with the wilderness experience. Roads provide the primary means of access for nearly all visitors to nearly all national parks, but the same roads can be devastating to a park's environmental qualities.

The Organic Act does not adjudicate such conflicts. To be sure, scholars and advocates have gleaned opposing preferences for enjoyment or for conservation from the Organic Act's language and from the purpose of the national parks. The most recent version of the NPS management policies, for example, states a preference for conservation (National Park Service 2006, 11). But no court has overturned a NPS decision to favor enjoyment instead of conservation – or vice versa – because it conflicted with the Organic Act.

That would seem to leave enjoyment and conservation on an equal playing field subject to the discretion of the NPS. But laws that protect certain features of the environment push national park management toward preservation. The very characteristics of a national park – rare wildlife, wilderness areas, clean air and water, abundant wetlands, historic structures, free-flowing rivers – subject park management to the additional requirements of the ESA, the

Wilderness Act, the CAA, the CWA, the National Historic Preservation Act (NHPA), and the Wild & Scenic Rivers Act (WSRA), among many other federal environmental statutes. The NPS, therefore, must manage national parks consistent with these other conservation commands. And the courts have overturned NPS management decisions that would have authorized opportunities to enjoy national parks because those decisions violated these other federal environmental statutes.

But the tilt toward conservation accomplished by these environmental statutes sometimes faces a statutory push back in the direction of enjoyment. Congress mandates specific management policies for individual parks in two different ways. Many acts establishing a new national park contain provisions directing the NPS to permit or prohibit certain activities. Congress also legislates in response to particular NPS actions to require a contrary management policy in a specific national park. Such specific statutory commands typically favor greater opportunities to enjoy a national park, though Congress occasionally calls for greater conservation than the NPS planned to provide. Similarly, informal congressional oversight of the NPS often encourages certain activities to be allowed in a national park, and the NPS often heeds those suggestions even though they are not legally binding.

Wilderness areas

Wilderness areas are the apogee of land conservation. The Wilderness Act defines "wilderness, in contrast with those areas where man and his own works dominate the landscape," as "an area where the earth and its community of life are untrammeled by man, where man himself is a visitor who does not remain." The stated purpose of conserving such lands is to protect their "ecological, geological, or other features of scientific, educational, scenic, or historical value."

The spiritual values of wilderness are left unstated in the act, but they featured prominently during the decade of congressional debate preceding the enactment of the law in 1964 and in subsequent debates over the establishment of specific wilderness areas. Sigurd Olson, the director of the NPS, told the 1965 Wilderness Conference that "the spiritual values of wilderness" are "the real reason for all the practical things we must do to save wilderness." And while approving new wilderness areas in 1984, President Reagan remarked that "as Americans wander through these forests, climb these mountains, they will sense the love and majesty of the Creator of all of that" (Reagan 1984; cf. Cannon and Riehl 2004, 232–48). The proponents of wilderness preservation identified four specific spiritual values associated with wilderness: it leaves land the way it was created by God, it is a place of encountering God, it provides spiritual renewal, and it offers escape.

The Wilderness Act now governs the use of more than one hundred million of acres of land owned by the federal government throughout the United States. Congressional legislation is required before the act applies, no matter how wild or untrammeled the land is in fact. Once it applies, the Wilderness Act enjoys the reputation of being the most stringent law governing the use of the natural environment. Motorized vehicles, structures, and commercial enterprises are excluded from wilderness areas. The courts consistently read the act strictly to prohibit questionable activities. But there are two ways in which activities that are inconsistent with wilderness are nonetheless allowed in wilderness areas. Congress may draw the official boundaries of a wilderness area to exclude places that are wilderness in fact but where Congress wants to allow activities that are inconsistent with the wilderness, or Congress may allow those activities even though they occur within the designated wilderness areas.

The National Environmental Policy Act

"Hundreds, perhaps thousands," of ministers preached about environmental issues on the Sunday before the first day Earth Day in April 1970 (Rome 2013, 175). A few months before, NEPA proclaimed a "congressional declaration of national environmental policy":

> The Congress, recognizing the profound impact of man's activity on the interrelations of all components of the natural environment, particularly the profound influences of population growth, high-density urbanization, industrial expansion, resource exploitation, and new and expanding technological advances and recognizing further the critical importance of restoring and maintaining environmental quality to the overall welfare and development of man, declares that it is the continuing policy of the Federal Government, in cooperation with State and local governments, and other concerned public and private organizations, to use all practicable means and measures, including financial and technical assistance, in a manner calculated to foster and promote the general welfare, to create and maintain conditions under which man and nature can exist in productive harmony, and fulfill the social, economic, and other requirements of present and future generations of Americans.

NEPA states a broad environmental policy, but the actual legal tools that NEPA provided to achieve those goals are more modest. The most important feature of NEPA is the requirement that federal agencies prepare an "environmental impact statement" before they pursue a project that could have a substantial impact on the environment. The resulting Environmental Impact Statements (EIS) have produced a wealth of information about the environmental impacts of planned projects. Most recently, the EIS for the planned Keystone XL pipeline confirmed a number of issues regarding water quality and climate change that helped to persuade President Obama to block the project from going forward. NEPA, however, does not demand any substantive result; it does not impose any constraints on a project, no matter how environmentally harmful it may be. The signature value of NEPA is providing information, but the leaves it to others to decide what to do with that information.

Federal pollution laws

For centuries, the most common connotation of "pollution" involved spiritual harms. Today's familiar environmental understanding of pollution has yielded laws that rely on federal regulation to achieve their purposes. That regulation takes a variety of different forms. The CAA, for example, uses a variety of tools to achieve different purposes. It requires the United States Environmental Protection Agency (EPA) to set standards for the amount of the most common pollutants that human health can tolerate, and then it requires state environmental officials to develop policies to meet those standards. It requires federal approval of gasoline and additives that may affect air quality. It imposes especially stringent federal limits on "hazardous air pollutants." It conditions federal highway funding on state development of highway and transportation plans that minimize air pollution. It employs a market-based, cap-and-trade system for large coal-fired electric utility plants whose emissions could result in acid rain.

The CWA is similar to the CAA in many ways, but the CWA contains its own unique legal tools. Rather than trying to achieve an overall goal of environmental quality, the CWA prescribes the technology that each industry must use to reduce its water pollution. The CWA also charges states with identifying the desired uses of their lakes and rivers, which yield state water quality standards that vary for each body of water. The CWA treats municipal sewage treatment plants both as the recipients of pollution from countless individual sources and as polluters who discharge treated water into rivers and lakes and therefore are subject to federal regulation. The CWA relies on vast federal subsidies to help local governments pay for costly new sewage treatment facilities. The CWA supports federal regulation of land use decisions to the extent that a developer wants to fill in a wetland, which can only be done with a federal permit that often conditions the development on minimizing the environmental harm and mitigating the harm by protecting other wetlands.

The Endangered Species Act

The ESA is the most revered and reviled of federal environmental laws. Its champions praise it for saving the bald eagle from extinction, for blocking many misconceived development projects, and for providing a tool to protect ecosystems ranging from the southern California coast to the majestic forests of the Pacific northwest. Its detractors accuse it of sacrificing timber jobs for obscure owls, nearly completed dams for tiny fish, and small farmers for unknown rodents. The basis for these claims lies in the unparalleled stringency of the ESA's provisions. Most other environmental statutes contain numerous opportunities for environmental interests to be balanced against other human needs. The ESA, by contrast, has long been viewed as requiring efforts "to halt and reverse the trend toward species extinction, *whatever the cost* " (Tennessee Valley Authority v. Hill 1978, emphasis added).

It was not until 1966 that Congress enacted the first federal statute aimed at saving vanishing wildlife and plants. The Endangered Species Preservation Act, Pub. L. No. 89–669, 80 Stat. 926 (1966), directed the Secretary of the Interior to use existing land acquisition authorities to purchase the habitat of native fish and wildlife that were threatened with extinction, and it instructed the Secretary to "encourage other Federal agencies to utilize, where practicable, their authorities" to further the preservation effort. Congress expanded the effort three years later with the Endangered Species Conservation Act of 1969, Pub. L. No. 91–135, 83 Stat. 275 (1969), which authorized the creation of a list of species "threatened with worldwide extinction" and prohibited the importation of most such species into the United States. Almost immediately, the 1969 law was criticized for not going far enough. The law did not prohibit the hunting or collecting or killing of a listed species, it did not regulate conduct that destroyed the habitat of a species, and it did not offer any protection at all to plants. Throughout 1972 and 1973, Congress considered a range of proposed bills that would provide much more powerful protections for any wildlife or plant species that was facing extinction. But the debate in Congress always referred to bald eagles, grizzly bears, whooping cranes, alligators, whales, and other prominent species now described as "charismatic megafauna." Few members of Congress wanted to be seen as opposed to such popular animals, and few were. The Senate approved its bill 92–0, and after several minor changes, the final bill passed the House 355–4. So on December 28, 1973, President Nixon signed the Endangered Species Act —what we have known since as the ESA. As Nixon explained, "Nothing is more priceless and more worthy of preservation than the rich array of animal life with which our country has been blessed" (Nixon 1973).

The first reported case under the new law involved a water dispute between cattle ranchers in Nevada and the endangered Devil's Hole pupfish. Shortly thereafter, though, another case emerged that has colored the perception of the ESA ever since. Much to the chagrin of the United States Fish & Wildlife Service (FWS) and many of the members of Congress who had just voted for the ESA, the listing of the snail darter as endangered resulted in a Supreme Court decision holding that the nearly completed Tellico Dam could not be finished because the resulting reservoir would wipe out the fish. That decision caused Congress to amend the statute, albeit in a relatively modest fashion. Several other amendments occurred in 1982, but since then the law has remained virtually unchanged.

The congressional failure to amend the law is not for want of trying. During the early 1990s, environmentalists pressed to expand the coverage of the ESA to include whole ecosystems that were imperiled by human developments or other causes. Conversely, the ESA was blamed for causing economic dislocation throughout the Pacific northwest as a result of the listing of the northern spotted owl. The first sustained effort to reform—or gut, depending on your perspective—the ESA occurred in 1994. Speaker of the House Newt Gingrich established an ESA task force that held hearings across the country in areas that had chafed under the restrictions of the law. Landowners and developers told horror stories of widows losing their life's savings when the presence of an endangered songbird prevented them from building on their land and of farmers facing federal prosecutions for attempting to prevent fires in a manner that harmed endangered kangaroo rats. Several bills were introduced to amend the ESA by requiring more rigorous scientific evidence before a species could be listed, helping private landowners who confront a listed species on their property, and speeding recovery efforts so that a species could be delisted. The bills stalled in the face of a certain presidential veto and pressure from environmentalists, religious leaders, moderate eastern politicians, and others who were intent on saving rare wildlife. The Secretary of the Interior at the time, Bruce Babbitt, wrote that "religious values remain at the heart of the Endangered Species Act" (Babbitt 1996). The process has repeated itself since then, with the same result.

State and local environmental law

A similar collection of legal tools characterizes the other leading federal environmental statutes. State and local governments have become more active in pursuing environmental goals, too, and there are a small but growing number of regional entities charged with managing specific rivers, watersheds, or ecosystems that cross state lines. Meanwhile, little has changed in the federal environmental statutes since they were first enacted by Congress during the 1970s. Congress has been stalemated as proponents of greater, lesser, or just different regulation have failed to achieve a consensus in favor of their goals but have been able to block the pursuit of others. Instead, change has come from the administrative agencies charged with implementing the laws, and from the courts that interpret them and adjudicate cases involving them. The result has been significant progress in achieving the goals set forth in the environmental statutes enacted by Congress during the 1970s. But lots of problems still remain, and new ones have emerged, such as climate change and invasive species.

References

Babbitt B. (1996) "Between the Flood and the Rainbow: Our Covenant to Protect the Whole of Creation", *Animal Law Review* 2:1, 8.

Berry E. (2015) *The Religious Roots of American Environmentalism,* University of California Press, Oakland.

Cannon J. and Riehl J. (2004) "Presidential greenspeak: How presidents talk about the environment and what it means", *Stanford Environmental Law Journal* 23, 195–272.

The National Environmental Policy Act (1970) codified at 42 U.S. Code § 4331.

National Park Service Organic Act (1916) codified at 16 U.S.C. § 1.

National Park Service (2006) *Management Policies 2006,* U.S. Department of Interior and National Park Service, Washington D.C.

Nixon R. (1973) Statement on signing the Endangered Species Act of 1973, San Clemente, CA 28 December 1973, Online by Peters G. and Woolley J. T., The American Presidency Project (www.presidency.ucsb.edu/ws/?pid=4090).

Reagan R. (1984) Remarks on signing four bills designating wilderness areas, June 19, 1984 *The Public Papers of President Ronald W. Reagan* Ronald Reagan Presidential Library (www.reagan.utexas.edu/archives/speeches/1984/61984d.htm).

Rome A. (2013) *The Genius of Earth Day: How a 1970 teach-in unexpectedly made the first green generation*, Hill and Wang, New York.

Stoll M. R. (2015) *Inherit the Holy Mountain: Religion and the rise of American environmentalism*, Oxford University Press, New York.

Tennessee Valley Authority v. Hill (1978) 437 U.S. 153, 184.

42

ECONOMICS

Economism and ecological crisis

Richard B. Norgaard, Jessica J. Goddard and Jalel Sager

Few people pray for abundant harvests of grain in the fall and many lambs and good pastures in the spring. Few people awake to the rising sun or the crow of a rooster. Rather, our cosmos is now the economy. We awake to a radio announcer speaking solemnly of stock markets rising or falling in financial capitals to our East, work in hierarchical corporate and governmental organizations while praising the freedom of markets, and are dependent for our food and clothing on others working at considerable distances in a great global economic machine. Church steeples that once reached for the heavens now cower below towering buildings bearing corporate names. City lights and polluted air curtain us from the awe-inspiring magnificence of the starry heavens; few are even aware of the phases of the moon. Our cosmos is now human-centered; and we manage our friendships and connections to reality on backlit screens.

Our global economy requires something like a global faith. This may be seen as a shared ordering principle, akin to the "organic solidarity" of industrial societies proposed by Durkheim (2014 [1893]). Consider what would happen if people began to lose their faith in markets. Imagine, for example, that a significant number of people, fearing that supermarkets would soon not have food, rushed to the store to stock up as much as they could and then, quitting their jobs, went home to convert their backyards into vegetable gardens. Others would become worried and rush the supermarkets for the remaining stocks of food, setting out to grow their own food as well, though many only have flowerpots. Eventually the economy would collapse and real starvation would occur, in a chaotic breakdown of the system. Taking this thought experiment to its logical end demonstrates the role of economic faith in shaping the seemingly natural rhythms of our life. We call this faith economism.

Economism as public and personal religion

Economism has become, in a sense, the planetary religion: an implicit, pervasive, personal and public consciousness that supports the new world of markets. Laborers, white-collar technocrats, entrepreneurs, capitalists, financiers, inventors, managers, academics, and "people of the cloth" around the world work together, 7.3 billion globally, through shared beliefs politically and culturally sustained. The writer Karl Polanyi richly described how markets transformed many aspects of society (1957 [1944]), but he would be amazed to see the world today.

Economism is a system of shared understandings that keep us working, consuming, and investing in the economy, maintaining it so that it maintains us. We all hold these understandings that are no more than beliefs to some degree. Just as importantly, academic economic thought and public policy derived from the writings of economists helped shape an economic reality over the last two centuries that reinforces those same beliefs, partially validating them.

Today's economistic beliefs, as well as the deified markets that constitute their mental landscapes, are confronted by increasingly stark social and environmental contradictions. Once-abstract environmental limits now are becoming manifest in the form of intensified heat waves, droughts, wild fires, storms, floods, and rapidly melting glaciers and ice caps. At the same time, inequalities, vanishing public budgets, and corrupted, ineffective governments are stressing our social fabric. Many people simply deny or ignore the warnings from scientists and social activists. At the apex of society, most academic economists continue to model, and make prescriptions for, the economy in the same way they did before the financial crisis; most politicians in the US continue to privilege the needs of oil companies over climate policy. This creates cognitive dissonance for the broader public. Economism is a sustaining blast in the midst of this dissonance and is thus the focus of this chapter. We examine the evolution of economism into the dominant ordering principle for our social and environmental world and how it has supplanted many of the moral and religious traditions of the past. We defend and develop the implications of these claims through a selective synthesis of historical evidence.

Religion and economy as ordering systems

Religion and economy are often thought of as wholly separate realms. Yet they have been intricately entwined in virtually every society through history. Societies maintain and reproduce themselves through an economy, a system of institutional arrangements and beliefs. Only in very recent history have markets—in their idealized form the animating deities of economism—played a central role where their growth has been necessary. Human maintenance and reproduction entail interacting with the environment to direct natural energy and material flows, yet those flows for most of human history remained largely local, developing into globe-encircling rivers only after the beginning of the fossil-powered industrial age. With the coming of the latter age, rapid expansion of markets, greater complexity and wider participation, helped foster a new "ethos of exchange."

Karl Marx (1954 [1887]), among others, explored this territory in the nineteenth century:

> Labour is, in the first place, a process in…in which man of his own accord starts, regulates, and controls the material re-actions between himself and Nature…By thus acting on the external world and changing it, he at the same time changes his own nature.
>
> *(Capital, vol 1, page 175)*

"Man of his own accord" constructs the image of humans laboring voluntarily—until we ask: "what is this accord?" The "accord" amounts to coordination by norms of behavior. Religion and culture have historically sustained the material maintenance and reproduction of society by constructing norms. Mainstream religious norms encourage the repression of immediate gratification, stress duty to family, and encourage care for others. "Good" morals in many cases are expeditious economic norms (Weber 1963 [1922]).

What we today call economies, a term which must contain all systems of production in human history—create the spaces in which religion operates and provide resources that support religion. Beginning some 15,000 years ago, the increased economic productivity of farming facilitated a slow but inexorable increase in human population. The increase in productivity also supported the evolution of more complex societies with specialized roles for different individuals. It also brought new difficulties, and new norms, for keeping all working together and distributing their collective bounty (Tainter 1988). Over time, interrelated mechanisms evolved to keep individuals working to maintain societies (Ross 1907). At the same time, the increased productivity of agriculture facilitated the construction of larger, more ornate places of worship, religious specialization, and hierarchy—and increasingly elaborate and comprehensive teachings. Economies and religion intertwined in new ways, together constructing new forms of authority and power, while defining new classes through culture and economic position.

Science and economics challenge religious authority

In the West during the fourteenth century, scholars began to challenge religious teachings about the natural order, most strikingly about the cosmos, and the relationship of sun, moon, stars, earth—and, implicitly or not, heaven. This was the beginning of conflicts in "natural" authority, between science and religion, still being played out to this day. Technological advancements, partly spurred by science, introduced better boats and uses of gunpowder that soon entangled Christianity with exploration, conquest, and a colonial economic order around the world.

During the seventeenth and eighteenth centuries, secular scholars begin to challenge the moral authority of Christianity itself. Portions of the landed classes supported this rebel scholarship, often to contest religious–political relationships, or resolve situations such as the Thirty Year's War (1618–48), in which Catholics and Protestants and affiliated economic interests clashed (Toulmin 1990). The resulting "moral philosophy" developed lines of thought that evolved into today's social sciences. The social sciences seek to explain the nature of people and society, to justify and bound authority, and to account for the nature of value—all without reference to God or religious texts.

Rationalizing social organization and explaining value from the ground up was not easy. The arguments that eventually became dominant portrayed society as built up from individuals, assumed they had given tastes, argued that they had basic rights, and explained that they were coordinated through a mutually advantageous social contract that included a right to land for subsistence needs. This usufruct form of private property steadily morphed into the idea that people had absolute rights to private property, a right to their own separate pieces of the natural order to do with it as they saw fit. The concept of a social contract also focused on legitimate authority, leaving the issues of correct relationships among people and with nature to religion. The success of these lines of thought with respect to authority laid the foundations for both political revolutions and much of modern economic theory.

Economics was the first and is still the dominant offshoot from moral philosophy. Adam Smith, trained in social philosophy, wrote *The Theory of Moral Sentiments* (1759) before turning to economics. Smith's *The Wealth of Nations* (1776) initiated a train of reasoning that provided new, plausible, and implementable roles for markets and government. Smith's work also addressed the nature of social relations, of value, and even of ultimate ends. In short, economists began developing a complete package of explanations in realms heretofore thought of as belonging to religion. With the increasing acceptance of economic rationality during the nineteenth century, religious teachings were challenged and religious institutions reinvented themselves to work with the times.

The economic boom was not simply a matter of facilitating the role of markets. The boom was fueled by the simultaneous introduction of coal and, soon after, petroleum-based technologies. Economic activity per capita increased about 20-fold during the nineteenth and twentieth centuries. This, combined with the population growth it supported, entailed a stupendous increase in energy and material flows from, and back into, nature. The combustion of fossil fuels effectively reversed millennia of hydrocarbon accumulation, releasing back into the atmosphere the carbon that life had sequestered over millennia. This dramatic transformation of our relationship to nature was not regulated by new religious norms to protect people and their Earthly home. Such norms never had a chance to evolve given the rapid pace of change. Rather, the whole process was regulated by the norms of the market, by new ethical stances that privileged today's production, even greed, unbalanced by the needs of future generations.

The world's religions have wrestled with, accommodated, and even, to some extent, encouraged these changes. Religions periodically rebalance in the face of rapidly changing economic organization, material conditions, and distributions of wealth and power. Across religions, the adaptations have by no means been the same. In the eighteenth and nineteenth centuries, a few sects, such as the Shakers (Quakers), the Harmony Society, the Mennonites, and the Amish settled in rural America to escape having to compromise religious norms with the economic transformation underway in Europe. They established communal-property economies, Christian communisms before Marx lashed the term to atheism, where care for others was central (Hillquit 1910). The Harmony Society named their community in Pennsylvania "Economy".

Religious freedom played a central role for many who immigrated to North America, and new settlers frequently related their religious beliefs to economic opportunity, moral progress, and a destiny they could make manifest. This combination proved especially toxic in America's westward expansion, a process of conquering wild nature and too often slaughtering "primitive" Native Americans. At the same time, the prevalence of religiosity meant early American economists were less prone to avoid the contradictions between religion and economics. They were often brought up in the church, some even trained in theology. Their worldview combined religious righteousness and market mechanics. A good economy helped the poor. All deserved land and a fair chance. Real work was not only virtuous in itself but contributed to the success of everyone. Yet they also clearly recognized the corrupting power of Mammon.

In 1825, Reverend John MacVickar, professor of moral philosophy and political economy at Columbia College, included these words in a portrayal of how markets work in the conclusions to his *Outline for Political Economy*:

> This picture, however, presupposes virtue in the people. Political Economy is a science which guards against involuntary not voluntary error. It enters into harmonious alliance with religion, but cannot supply its place. It must find public men true to their trust, otherwise it renders them more ingenious in their abuse of power.
>
> *(187)*

MacVickar was trained in theology as well as economics, so it is not surprising that he dwelled on the relationship between the two. Many other early economists shared his training. A century and a half later, however, the mental separation between economics and traditional religious values was effectively complete. In the process, economics had become its own religion, and men were indeed often using it to be more ingenious in their abuse of power.

The rapid economic changes attending industrialization in the 1880s generated complementary responses from both religious and economic leaders. Progressive religious leaders during this era developed a new "Social Gospel" (Evans 2013) that was shared by many economists and social theorists (Ross 1907). Richard T. Ely (1854–1943), the founder of the

American Economics Association, provides a prominent example of this early mix of economics and religion. He firmly believed that economics could only work successfully in tandem
with ethics (Furner 1975):

> Christianity moderates desires, sets a higher aim than wealth before people, but digni-
> fies the man who gains his bread by honest toil, and enjoins diligence and an
> improvement of all talents committed to us. It teaches us to love our fellows, and this
> has encouraged enlightenment of the masses, and enlightenment increases prosperity.
>
> *(Ely 1889a, 133)*

> Luxury is materialistic and selfish: it retards the mental and spiritual development of
> a people, and tends to impoverish a nation. Luxury breeds luxury as sin begets sin.
>
> *(Ely 1889b, 37)*

At the end of the nineteenth century, however, younger scholars, trained in universities rather
than theology schools, sought to distance themselves from moral questions and become more
scientific, objective, and "professional" (Furner 1975). Meanwhile great luxury and enormous
fortunes had been shaped in the industrializing US. As society changed and became more
complex, "progressive" religions accepted both the findings of science and the views of
economic experts, while also adapting to more materialistic lifestyles and new wealth and
power. The encroachment of science and economics upon the territory of religion was
becoming more serious.

Max Weber's hypothesis that Protestantism facilitated capitalism suggests that religion can
also affect economics (Weber 1905). The facilitation of economics through religion becomes
especially true during the twentieth century in America, as developed in the next section. But
we are also at a point in our selective history where we need to note that Western religions
might have tried harder to assert the values of love, care, and sharing. As history unfolded, the
compromises of Catholicism and Protestantism with economic power found them opposing
the immediate violence of socialist revolution as they backed into the moral morass of not
having opposed or effectively resisted the collapse of the capitalisms of Germany, Italy, and
Spain into fascism (Drucker 1939).

The creed of the American century

The Great Depression raised serious doubts about the directions of capitalism, especially its
stress on individualism and private property. While economic experts had assured the public
that free markets naturally and quickly return to equilibrium with full employment after a
disturbance, evidence to the contrary accumulated with each year. Not until John Maynard
Keynes's great work (1936) did the economics profession begin to move away from this position. In his promotion of new government-led programs to get the economy moving,
President Franklin D. Roosevelt preached a "social gospel" much like that of nineteenth-
century religious leaders and economists. He stressed obligation to others and to the good of
society as a whole in order to defend higher taxes, public investments in the economy, and
helping the poor (Kruse 2015). Given the ecological disaster of the Dust Bowl, related to ill-
informed economic expansion, this social gospel also included respect and care for land.

Counter movements developed almost immediately. The most prominent involves Frank
Knight, one of the founders of the Chicago School of Economics, who wrote extensively
during the Depression on how economics must be practiced in the context of ethical

principles. Knight even coauthored a book with a theologian (Knight and Merriam 1945). From this broad perspective, Knight argued that at a fundamental level economics had to be a religion itself, the basic tenets of which must be hidden from all but a few:

> The point is that the "principles" by which a society or a group lives in tolerable harmony are essentially religious. The essential nature of a religious principle is that not merely is it immoral to oppose it, but to ask what it is, is morally identical with denial and attack.
>
> There must be ultimates, and they must be religious, in economics as anywhere else, if one has anything to say touching conduct or social policy in a practical way. Man is a believing animal and to few, if any, is it given to criticize the foundations of belief "intelligently."
>
> To inquire into the ultimates behind accepted group values is obscene and sacrilegious: objective inquiry is an attempt to uncover the nakedness of man, his soul as well as his body, his deeds, his culture, and his very gods.
>
> *(Knight 1932, 448–9)*

> Certainly the large general [economics] courses should be prevented from raising any question about objectivity, but should assume the objectivity of the slogans they inculcate, as a sacred feature of the system.
>
> *(Knight 1932, 455)*

While few economists have read these words, the profession followed the path Knight espoused here while ignoring his other writings on the importance of ethics to economics. In our own lectures, students often gasp with the sudden realization that these passages, portrayed so clearly on a PowerPoint slide, help explain why economics is taught so uncritically. Here we see one plausible explanation of how economics took on religious authority. Keeping the "ultimates" of economics—such as the assumptions about property being divisible, about free competition, and fully informed actors—hidden has facilitated the rise of economic expertise in governance, yet seriously inhibited and distorted the development of an honest academic economics. It has also arguably helped move economics into a substitute, rather than complementary, role vis-à-vis religion by generating taboos and dogma in place of a scientific paradigm that can be questioned and changed. These effects together hinder the mass academic and public consciousness necessary to raise the ecological and social crises as urgent moral issues, above the religion of economics. They in fact support the dismissal of these crises as unavoidable effects arising from the unimpeachable expansion and rule of the market.

The threat to capitalists posed by President Roosevelt's New Deal policies led a combination of market-fundamentalist religious leaders, individual capitalists, and major corporations to initiate a religious counter-insurgency in the early 1950s: Libertarian Christianity. They argued that liberty and the freedom to choose were God's will and best maintained by free markets. The Reverend Billy Graham rose to fame in this movement, preaching to large gatherings around the country and hobnobbing with legislators and presidents. Libertarian Christianity led to prayer breakfasts in the nation's capitol, the expression "Freedom under God" in our political discourse, and the phrase "In God We Trust" on our currency, the very medium of market exchange (Kruse 2015). The movement related to, and complemented, the work of Knight's Chicago School on the steady expansion of the market's purview. Here economism perhaps begins to acquire its modern hue, partly through the fusion of these two "religions."

Complementing this Christian movement, Ayn Rand, an economic refugee of the Soviet

Union, wrote fervently during mid-twentieth century on the virtues of private property, individualism, freedom of choice, and the efficiency of markets. To her, these are sacred elements of the modern world. Governments meddled with these market fundamentals and were clearly the devil. Further, both caring and collective action were sinful (Burns 2009). Ayn Rand titled one of her books *The Virtue of Selfishness*. An avowed atheist, she disguised how sacredly she held her beliefs behind the scientific sounding term "objectivism" (Rand 1966). Meanwhile, somewhat ironically, it was the strong atheism of Soviet-style socialism that helped unite the forces of traditional and market religion against it.

Complementing the rise of neoliberalism in the latter twentieth century, "prosperity gospel" attracted the next generation of Christian believers (Bowler 2013). Evangelizing that God's blessings arrive through market success has been especially effective among the poor (Harrison 2005) and in developing countries. Economism sometimes works alongside and through religion in overt syncretism, resembling in some ways other historical blends of religion, such as Afro-Christian.

Government, the very mechanism by which democratic peoples lay down the principles that guide the market in more appropriate, caring directions, became a target for the new economic priesthood. Amidst Cold War political denunciations of Soviet-style socialism, the Chicago School of Economics aggressively promoted free markets and denigrated government. Milton Friedman was the public voice equating capitalism with freedom (Friedman 1962), but many Chicago faculty bent theory to this end. Chicago School market philosophy captured much of the economics profession over the following quarter century, in part because pursuing the mathematization of economics, as other schools stressed, left a moral void that Chicago could fill, however awkwardly (Jones 2012, Burgin 2012). Dominant neoliberal economists support the deification of markets, a few such as Friedman have served as high priests of economism, and many today are unwittingly immersed in it, which helps distort moral issues underlying the economy we have created, the resulting human condition, and state of the planet.

Replacing the market god in the twenty-first century

If governments—our social contract with society—exist to act in the public good, they must also set the rules of the market for the public good and enforce them. The vast inequalities and racism manifest in the US today, for example, reflect a flawed social contract explained partly because economism says there is no need to have and use government and partly because government is made dysfunctional by the great infusion of private money into politics. In the postwar Keynesian "golden age," governments did take some steps toward transfers of wealth and care, keeping the wealth in society—and thus the power—relatively balanced. Economic history and economism reveals that this has been chipped away through a 60-year battle against the redistributionist state, which has produced an increasingly anti-democratic, corporate oligarchy.

Economism has become powerful because it provides a coherent, complete set of beliefs. Currently, economism allows individuals and communities in the US to:

a. Explain the emergence and nature of our modern human cosmos;
b. Explain and rationalize one's place in the cosmos as well as the increasing inequality around us;
c. Justify the economization of more and more realms of life to make the world better and better;

d. Rationalize the way in which we enter into employment contracts and interact through markets with each other and nature;

e. Rationalize how individual greed is good, in opposition to long-standing, religiously rooted moral teachings stressing care for others;

f. Rationalize personal transcendence through consumption—the meaning of life is to own and consume more than thy neighbor; and

g. Rationalize the growth of GDP as progress, human destiny, and social transcendence.

These beliefs coordinate how we work and live together, individually, yet at the same time collectively, maintaining and reproducing ourselves. They also provide both individual and social reasons to do so. It is in this sense that economism can be seen as an important strain of religion. It invokes moral authority in political discourse; it masks, counterbalances, sometimes even opposes the natural authority of science; it is partially institutionalized in prosperity gospel; and it poses new challenges to traditional religions. Economism serves effectively as a stand-alone religion for many people who settle for the dubious freedoms of the market and consumer choice, as well as the promises of a more wondrous material future. Economism hasn't inspired a Bach to compose a powerful toccata and fugue for organ, let alone a St. Matthew Passion, but it supports an advertising industry perhaps as creative, seemingly as sincere, and certainly as productive in generating material that reifies its idols.

The destruction of the environment hurts the poor first; it is the poor who are least able to bear the burden. The poverty that is driving millions of people to seek asylum in Europe has economic and environmental roots that intertwine. Pope Francis wrote an unprecedented encyclical calling upon the moral and ecological crises born of unfettered capitalism wherein he writes: "Inequity affects not only individuals but entire countries; it compels us to consider an ethics of international relations. A true 'ecological debt' exists, particularly between the global north and south..." (Pope Francis 2015 para 51).

The Pope's call to moralize our economic policies and environmental stewardship calls for a major religious shift, but it was not drafted from scratch. Despite economism's pervasive hold on dominant discourse and a governance system that legitimizes unlimited prosperity for the 1%, there are bottom-up social movements, scholars in marginalized disciplines, and marginalized scholars in dominant disciplines, critiquing and seeking an end to the exploitation of people and planet alike.

Alternative imaginable futures of sustainability, caring for land, and redistributive economic policies are demanding more consideration. Incorporating virtues of care and stewardship in today's public economics comes up against economism, which has powerfully inscribed mathematical formalism as a prerequisite to policy choices. The fact that caring is a process whose outcomes cannot be limited to a monetary value such as GDP further challenge its incorporation in today's thinking. This is why alternatives require a transformation of values on a societal scale (Kellert and Speth 2009). We need to read and cross-reference existing economics with anthropologists, historians, natural scientists, political scientists, economic geographers, social theorists, theologians, and beyond. Plurality in approaches to sustainability opens the avenues for communication and hence understanding for what requires caring, how care might be provided, and how it is received.

Academic economists arguing for "Sustainable De-Growth" and seeking to change the modern economic curriculum ("Re-Thinking Economics") are responding to the fact that public economism and the academic economics that support it are not suited to the moral challenges of sustainability. Academic economic studies of our current ecological crisis often appeal to the wider moral realm, but fail to meaningfully incorporate it. Their absolution comes through economism, the same economism explaining and justifying the public consumption of cheap fossil fuels that perpetuate ecological degradation. This is important because it means that experts define both problem and solution in ways that close off opportunities for moral communication and leadership.

Similar to the overshadowed existence of alternative ideas in academic economics, individual communities rejecting the axioms of Economic Man lifestyles exist (Alperovitz 2013)—e.g., co-housing, co-operative businesses, service-based consumption—and their demands for free services, their productivity through collaboration, suggest social choices that erode the totalizing economism of market capitalism. These examples are not free of "ism"-qualities and market beliefs, but they are different. They reflect realities where people are looking for another God other than market or Church. We realize the damage being done to the social and ecological world, and we can't afford to ignore it anymore—in religions implicit or explicit. In June 2015, Pope Francis made this destruction explicit within one major religion while speaking to all. Islamic leaders followed suit with a strong statement in August 2015 that also implicated our economic beliefs (Islamic Declaration on Climate Change 2015). It is now necessary to do similarly in our common global religion of economism.

The emergence of economism, and its mutualism with the neoliberal economy and religion, needs to take its place among the explanations of how humanity has come upon several forms of social and ecological crises. To find a path away from the brink, up the slippery slope, and into a safe haven, we need to deal with how economism came to be so powerful, to understand what it has wrought. We then need to drastically transform our collective beliefs about what it means to be human and to live a good life. Only then may we reconfigure the economy accordingly so that human life, and other species too, can live on.

References

Alperovitz, G. 2013. *What Then Must We Do?: Straight talk about the next American Revolution*, Chelsea Green, White River Junction.

Bowler, K. 2013. *Blessed: A History of the American Prosperity Gospel*, Oxford University Press, Oxford.

Burgin, A. 2012. *The Great Persuasion: Reinventing Free Markets since the Depression*, Harvard University Press, Cambridge.

Burns, J. 2009. *Goddess of the Right: Ayn Rand and the American Right*, Oxford University Press, New York.

Drucker, P. F. 1939 (1995). *The End of Economic Man: The Origins of Totalitarianism*, Transaction, New Brunswick.

Durkheim, E. 2014 (1893). *The Division of Labor in Society*, Simon and Schuster, New York.

Ely, R. T. 1889a. *An Introduction to Political Economy*, Chautaugua, New York.

Ely, R. T. 1889b. *Social Aspects of Christianity*, T Y Crowell, New York.

Evans, C. H. 2013. "Social Christianities and Social Gospel" in Evans C H ed *Histories of American Christianity: An Introduction*, Baylor University Press, Waco, Texas.

Friedman, M. 1962. *Capitalism and Freedom*, University of Chicago Press, Chicago.

Furner, M. O. 1975. *Advocacy and Objectivity: A Crisis in the Professionalization of American Social Science 1865–1905*, University Press of Kentucky, Lexington.

Harrison, M. F. 2005. *Righteous Riches: The Word of Faith Movement in Contemporary African American Religion*, Oxford University Press, Oxford.

Hillquit, M. 1910. *History of Socialism in the United States*, 2nd ed Funk & Wagnallis, New York.

Islamic Declaration on Climate Change 2015. http://islamicclimatedeclaration.org/islamic-declaration-on-global-climate-change/ (checked August 31, 2015).

Jones, D. S. 2012. *Masters of the Universe: Hayek, Friedman, and the Birth of Neoliberal Politics*, Princeton University Press, Princeton.

Kellert, S. and Speth, G. 2009. *The Coming Transformation: Values to Sustain Human and Natural Communities*, Yale School of Forestry and Environmental Studies, New Haven.

Knight, F. H. 1932. The Newer Economics and the Control of Economic Activity, *Journal of Political Economy* 40, no. 4: 433–76.

Knight, F. H. and Merriam, T. W. 1945. *The Economic Order and Religion,* Harper, New York.

Kruse, K. M. 2015. *One Nation Under God: How Corporate America Invented Christian America,* Basic Books, New York.

MacVickar, J. 1825. *Outlines of Political Economy,* Wilder and Campbell, New York.

Marx, K. 1954 (1887). *Capital,* v1 Progress Publishers, Moscow.

Polanyi, K. 1957 (1944). *The Great Transformation: the political and economic origins of our time,* Beacon Press, Boston.

Pope Francis 2015. *Laudato Si: On Care for our Common Home* Encyclical Letter http://w2.vatican.va/content/francesco/en/encyclicals/documents/papa-francesco_20150524_enciclica-laudato-si.html (checked August 31, 2015).

Rand, A. 1966. *Capitalism: The Unknown Ideal,* New American Library, New York.

Ross, E. A. 1907 (1973). *Sin and Society: An Analysis of Latter-Day Inequity,* Harper Torch Books, New York.

Tainter, J. A. 1988. *The Collapse of Complex Societies,* Cambridge University Press, Cambridge.

Toulmin, S. 1990. *Cosmopolis: The Hidden Agenda of Modernity,* Free Press, New York.

Weber, M. 2002 (1905). *The Protestant Ethic and the Spirit of Capitalism,* Trans. by Baehr P Wells, G C Penguin, New York.

Weber, M. 1963 (1922). *The Sociology of Religion,* trans. by E Fischoff, Beacon, Boston.

43

ECOLOGY

Nalini M. Nadkarni

The discipline of ecology poses an array of questions that include the most ancient, the most modern, the most theoretical, and the most relevant to understand the living things and their roles on our planet. The ways of ecological knowing have been useful to document the distribution and abundance of living things, and how people can use Earth's resources in ways that leave the environment healthy for future generations. The term "ecology", derived from the Greek οἶκος, meaning household, family, or home. It is the scientific study of living things and their homes – including humans – and their past, present, and future environments, including the ways they respond to their surrounds, interactions among species, and the processing of energy and materials in ecosystems (Cary Institute for Ecosystem Studies 2016).

Many of these questions are of interest to those who study religion, though the vocabulary and ways of knowing differ. As a guide to the topics and approaches to understanding that ecologists use, I present a brief history of the discipline; outline the major questions and themes ecologists explore through their approaches of observation, experimentation, and modeling; and the ways they collaborate with other scientists and other disciplines and ways of knowing, with a focus on religion.

History

Ecology has a complex history, due in large part to its interdisciplinary nature. It is rooted in the long tradition of natural history, which describes and documents the diverse of nature to identify, classify, understand biota, their habitats, and distributions. The earliest work in ecology started with taxonomist Carl Linnaeus' research principles on the economy of nature that matured in the early eighteenth century (Blunt 2001). Linnaeus first proposed *the balance of nature*, which proposes that that nature is usually in a stable equilibrium, i.e., a small change in some particular parameter will be corrected by some other event that will bring the system back to its original "point of balance" with the rest of the system, an outlook that remains in the way many humans view nature today. This work appears in the work of Charles Darwin in *The Origin of Species* where he adopted the usage of Linnaeus' phrase on the economy or polity of nature (Stauffer 1957). Biogeographer Alexander von Humbolt was also foundational, recognizing ecological gradients, which provided a precursor to our modern understanding of direct relationships of species with differing environments (Rosenzweig 2003). In the early twentieth century, ecology developed an analytical type of natural history, exemplified by the German biologist Ernst Haeckel, who coined the term "ecology" in his book *Generelle Morphologie der Organismen* (Haeckel 1866).

Prior to *The Origin of Species,* there was little understanding of evolution, the dynamic and reciprocal relations between organisms, their adaptations, and their modifications to the environment, i.e., the ability of species to respond to changes over thousands and millions of years. This scientific paradigm changed the way that researchers approached the ecological sciences (Acot 1997). This dynamic framework also caused a major shift in describing the distribution of biota. In 1905, the American ecologist, Frederic Clements championed the idea of plant communities as a superorganism, proposing that ecosystems progress through regular and pre-determined stages of development that are analogous to developmental stages of an organism, which maintains the integrity of the whole (Clements 1916). Henry Gleason, who documented that ecological communities develop from unique and coincidental association of individual organisms, challenged this paradigm (Gleason 1939). Around the same time, Charles Elton pioneered the concept of food chains, defining ecological relations using concepts of food-chains, food-cycles, food-size, and food-webs, and described numerical relations among different functional groups and their relative abundance (Elton 1927).

The field of ecology has grown enormously, and numerous histories are available (Worster 1994; Egerton 1977; Golley 1993). The Ecological Society of America (ESA) has published a fascinating timeline of ecologists (http://esa.org/history/biographies/biographies-of-ecologists/) and a 100-year "timeline" of ecology and the ESA www.tiki-toki.com/timeline/entry/113161/Ecological-Society-of-America-through-the-years/

The current field of ecology

Although these roots of ecology reach back for centuries, the modern synthesis of ecology is young. The discipline attracted formal attention at the end of the nineteenth century and entering public awareness during the 1960s environmental movement. Below are five of the current areas of ecological study.

Complexity, scope, and levels of organization in ecology

Complexity of space

The field of ecology has grown to encompass a broad scope in the realms of intellectual and social considerations. Major fields that ecologists study include: paleoecology through present-day phenomena; evolutionary, population, physiological, community, and ecosystem ecology, as well as biogeochemistry. Major approaches include descriptive, comparative, experimental, mathematical, statistical, and interdisciplinary approaches. Ecologists' questions include a wide range of spatial scales – from microscopic scales of individual ions of nutrients that move within food-webs to continental and biosphere-level scales of climate patterns that affect species migrations. To structure the study of ecology into a manageable framework of understanding, ecologists organize the world as a nested hierarchy of organization, ranging in scale from genes, to cells, to tissues, to organs, to organisms, to species, to populations, to communities, to ecosystems, and up to the level of the biosphere (Wiens 1989). For example, a single tree provides multiple spatial arenas: generations of an aphid population exist on a single leaf. Inside each of those aphids exist diverse communities of bacteria. Tree growth is, in turn, related to local site variables, such as soil type, moisture content, slope of the land, and forest canopy closure. However, more complex global factors, such as climate, must be considered for the classification and understanding of processes leading to larger patterns spanning across a forested landscape. New tools to literally "see" and make images from outer space – the field of remote imagery – has increased the breadth of ecologists' spatial bounds, as they can now gain access to data on both the distributions of organisms and the environmental characteristic that stretch across continents.

Complexity of time

Ecological investigations also require the documentation and integration of phenomena at a broad set of temporal scales, from the lifetime of a bacteria that carries out leaf decomposition to the millennial cycles of forest development. Many temporal phenomena that concern ecology – such as hurricanes and droughts – are unpredictable, which makes their study logistically difficult, and requires long-term time-scaled. Mark Harmon, researcher at Oregon State University, has initiated the 200-year study of large log decomposition in coastal Oregon forests, a time frame difficult for funding agencies, but necessary to document these processes (Harmon 1992).

Complexity of composition

Biodiversity describes all varieties of life from genes to ecosystems and is a result of the complex interplay among ecological, evolutionary, and geographical processes that influence which species live in an ecosystem and where they live within it. Many ecologists are transfixed on the many ways to measure, and represent biodiversity, which includes species, ecosystem, and genetic diversity. The promise of studying biodiversity is that understanding the ecological relationships between species and their environment will help explain and predict the presence or absence of that species at different places and times. Migration of a bird species, for example, is dependent on conditions of vegetation along its traditional flyway, the presence of other birds where it nests, and its own life history. An understanding of biodiversity has practical applications for ecosystem-based conservation planners as they make ecologically responsible decisions in management recommendations to consultant firms, governments, and industry.

Dynamics

Although some ecosystems seem static relative to the time frames of humans, all of Earth's biomes are in a constant state of flux, with energy, materials, and organisms moving within and between them. Characteristics of biota and ecosystems that are resilient (tending to return to their original state) are of great interest from both theoretical and applied standpoints. Enormous steps in the understanding of forest dynamics occurred when ecologists became able to record and analyze large data sets for such studies as Stephen Hubbell's measurements of many thousands of trees, saplings, and seedlings in the 50-hectare plots in a Panamanian rainforest (Hubbell 2001). New technology such as LIDAR (Light Detection And Ranging) has allowed forest ecologists to map elevations of entire forest, from canopy to ground, creating three-dimensional maps with remarkable precision (Simard et al. 2011). Electronic sensors are being used to monitor the "vital signs" of ecosystems such as temperature, relative humidity, and sunlight input (Holden et al. 2013)

Complexity of human effects

The study of ecology encompasses the wide range in the degree of impact of humans on ecosystems, from the nearly pristine stretches of tundra in Alaska to urban row housing and pocket parks of Baltimore. Until the 1980s, most ecologists located their field sites in the most remote and pristine areas, and learnt about ecosystem composition, structure, and function as far removed from human effects to operate in the most "natural" way. However, as the effects of humans have become more pervasive and extensive, many ecologists responded with being more inclusive of studying sites and questions where humans have clear impact. This was most apparent when the Long-term Ecological Research program, a National Science Foundation (NSF)-funded program created in 1980 to provide a backbone of long-term sites for major biomes, chose to include two urban Long-term Ecological Research sites, within the cities of Baltimore and Phoenix (1997). These two sites have fostered a tremendous amount of transformative research, demonstrating that excellent ecology can be addressed in human-dominated landscapes.

Emerging themes in ecology

The "PPP Paradigm"

One of the primary paradigms of ecology is to document the *patterns* that are manifested by organisms with respect to these characteristics, e.g., establishing that a gradient of water exists in soils, and then documenting the distribution along that gradient. Ecologists then try to figure out the *processes* that create those patterns, which can involve the physiology of the plant or animal. This frequently requires the ecologist to carry out an experiment, such as transplanting, fertilizing, or watering to determine the mechanisms that produce those patterns. With knowledge of pattern and process, ecologists can make *predictions*, such as the future distributions of a species if the area becomes drier or wetter.

Legacy structures

A key revelation in both basic and applied ecology has been the importance of "legacy structures", living or dead entities that exist in an older system that are retained in a altered system following disturbance, and which carry out valuable functions. For example, the long-time practice of clear-cutting entire forests has persisted in forest harvests despite negative consequences on soil, wildlife, and reforestation. Recently, ecological research in the Pacific Northwest has guided forest managers to leave a number of "legacy" trees – living, standing dead, and as fallen logs – which improve wildlife habitat, protect soils, and provide natural seed (Franklin 2000), and these practices are being replicated in other biomes.

Keystone species

Robert Paine first described a keystone species as an organism that has a disproportionately large effect on its environment, often playing a critical role in providing food for other species. For example, Paine (1966) documented that the predatory sea starfish is removed from the ecosystem, the mussel population explodes uncontrollably, driving out most other species, while the urchin population annihilates coral reefs. John Terborgh identified fig trees as keystone species because they provide fruit for arboreal birds and mammals when there is little other food available (Diaz-Martin 2014). This concept has implications for conservation, as keystone species may be small in numbers but important for ecosystem health.

Humans and ecology

Ecological information can help humans improve the environment, manage natural resources, and protect human health. The following examples illustrate ways that ecological knowledge has positively influenced human lives. In 2014, a group of international scientists centered at Stanford University drew up and signed a message on *Scientific Consensus on Maintaining Humanity's Life Support Systems in the 21st Century: Information for Policy Makers* (Barnosky et al. 2014). Their main message is that science has unequivocally demonstrated that human impacts concern: a) climate *disruption* (more, faster climate change than since humans first became a species; b) species extinction (not since the dinosaurs became extinct have so many species died out so fast, both on land and in the ocean); c) wholesale loss of diverse ecosystems (humans have transformed more than 40% of Earth's ice-free land, and no place on land or in the sea is free of the direct or indirect influences of humans; d) pollution (environmental contaminants in the air, water, and soil are at record high levels that harm humans and wildlife); and e) human population growth and consumption patterns (seven billion people alive today will grow to 9.5 billion by 2050, and pressures of heavy material consumption among the middle class and wealthy will intensify. They concluded that by the time today's Millennials reach middle age it is extremely likely that Earth's life-support systems will be irretrievably damaged by the magnitude, global extent, and combination of these human-caused environmental stressors unless humans take concrete, immediate actions to ensure a sustainable future. Understanding the theory, patterns, and processes of the past and present distribution and abundance of plants and animals and their relationships to their environments will be a necessary part of predicting future distributions, and mitigating the negative consequences of human activities.

Pollution from laundry detergents and fertilizers: In the 1960s, ecological research identified two of the major causes of poor water quality in lakes and streams – phosphorous and nitrogen – which were found in large amounts in laundry detergents and fertilizers. A prime example was the work headed by ecologists Robert Edmondson, who studied the basic chemistry of Lake Washington (Edmondson and Lehman 1981). With this information, citizens took steps to help restore their communities' lakes and streams – many of which are once again popular for fishing and swimming.

Health and biomedical materials: Ecologists have discovered that many plants and animals produce "secondary chemical compounds" that protect them from predators and diseases, such as tannins, pigments, and distasteful compounds against herbivory. Some of these same chemicals have been synthesized by scientists or harvested from the organism and used to treat human diseases. For example, the Pacific Yew tree (*Taxus brevifolia*) produces taxol, a substance now used in cancer treatments. Now synthesized in the laboratory, this chemical has been called "one of the most important new anti-cancer drugs in the past 50 years" (Rowinsky et al. 1992).

Ecosystem services: Ecologists have discovered that nature can be valued not only because of its species diversity or resilience to disturbance, but also because species and ecosystems provide economic values. For example, marshes filter toxins and other impurities from water, reducing water treatment costs for human communities downstream. The shade that urban trees cast on the homes and business structures in cities can reduce costs of air-conditioning. A burgeoning area of ecology is now documenting such "ecosystem services", which can serve to bolster conservation efforts. However, because it is difficult or sometimes impossible to assess the aesthetic, spiritual, or ethical values of a. organism or ecosystem, this approach can overlook valid values when making decisions about the natural world (Daily 1997).

Collaborative fields

The discipline of ecology is a powerful approach to understand the world, but it is only one way of knowing. Our knowledge and communication of the scientific aspects of ecology has not yet been successful in providing societally acceptable solutions to the environmental challenges that humans face. Many ecologists now recognize that they must go beyond the study and documentation of patterns, processes, and predictions of biota and their environment so that they can better communicate and apply what they discover. In many cases, ecologists cannot do this within their own disciplines, but must connect with the approaches, tools, and capacities of other fields.

There are many more ways to collaborate and interweave ecology with different disciplines than ever before. At its core, ecology is a scientific discipline that has as its academic cousins the fields that are concerned with the physical environment (atmospheric sciences, soil science, hydrology, microclimatology). However, in the past two decades, ecology has become more aligned with human activities, requiring collaborations with social scientists, especially dealing in urban venues (anthropologists, sociologists, those concerned with social and environmental justice). Third, many questions are now being posed that involve community stakeholders such as policy-makers and recreational land users.

Ecology and religion

As an ecologist who is dedicated to conserve the biodiversity and functions of ecosystems, I have made efforts to create my own professional and personal bridges between ecology and religion. I was raised in a family of mixed religious traditions (Hindu and Orthodox Jewish parents), and learned in my childhood that those of different faiths can live in harmony, finding commonality across widlly different beliefs and traditions. This background has guided my efforts to interweave my knowledge of forest ecology with religious thinking and commmunities. Over the past decade, I have drawn upon the religious texts of the major world religions to learn how trees and forests are described and valued by their readers. I learned that in religions, trees are viewed as important for their practical, spiritual, and aesthetic value. I delivered sermons on "trees and spirituality" in a wide range of churches, synagogues, and temples, using the pulpit and other faith-based institutional communication systems to promote an understanding and sense of stewardship for nature. I have also created tree maps of churchyards with both biological and religious information to emphasize that nature around places of workshop are as sacred as what is held inside the church (Nadkarni 2004). These activities have led to my participation in other religious venues such as the World Parliament of Religions (2015), where interactions revealed areas of common ground, common interest, and mutual listening.

Other efforts on the part of faith-based groups and ecologists have been creating pathways for discussion and understanding in the form of joint meetings, symposia, and other interdisciplinary scholarly works. From the standpoint of the science of ecology, the valuable connections to religion have only recently been explored, but are now gaining increasing attention, even in the most academic of settings. From the standpoint of both religion and ecology, this has been the work of the Yale Forum on Religion and Ecology since its founding in 1998 (http://fore.yale.edu/). Dr. Gregory Hitzhusen (Ohio State University), one of the most prominent ecologists who is active in this area, has called upon biologists and representatives of religions to recognize the many points of connection that exist between ecology and religion (Hitzhusen 2006, Hitzhusen and Tucker 2013). Ecologists are becoming involved with faith-based groups: a) providing scientifically sound information to religious groups (e.g., Interfaith Power and Light) who disseminate information to their own congregations; and b) contributing knowledge and physical help to ongoing efforts for direct conservation efforts such as tree-plantings and raising funds to mitigate negative impacts of human activities on species preservation and land conservation. In the 2014 annual meeting of the Ecological Society of America (over 8000 professional and academic ecologists in attendance), ecologists organized five sessions on this topic, which highlighted the work that ecologists are carrying out in universities, community settings, and places of worship across the country. In the 2015 ESA meeting, this increased to an entire day of panels, papers, and symposia on this topic.

E.O. Wilson's book "The Creation" (Wilson 2006) is an epiphany of common ground. Whether we believe that the diverse organisms on Earth were created by God or through the processes of evolution, the conclusion we must draw is that biodiversity requires our awareness and care. Therefore, religious congregations and ecologists should be working side by side as they reforest hillsides, protect wetlands, or lobby for land protection. With this outlook, the threads of religion and of ecology become inextricably intertwined, making the strong fabric so needed in the world today.

Acknowledgments

This chapter drew from material presented in "What is Ecology? from the European Organic Congress (http://environment-ecology.com/). Activities described were funded by grants from the NSF (DEB-1141833 and EHR 1514494).

References

Acot, P. 1997. The Lamarckian cradle of scientific ecology. *Acta Biotheoretica* 45:185–193.
Barnosky, A., Brown, J.H., Daily, G.C., Dirzo, R., Ehrlich, A.H., Ehrlich, P.R., Eronen, J.T., Fortelius, M., Hadly, E.A., Leopold, E.B., Mooney, H.A., Myers, J.P., Naylor, R.L., Palumbi, S., Stenseth, N.C. and Wake, M.H. 2014. Introducing the Scientific Consensus on Maintaining Humanity's Life Support Systems in the 21st Century: Information for Policy Makers. *The Anthropocene Review* 1:78–109.
Blunt, W. 2001. *Linnaeus: The compleat naturalist*. London, UK: Frances Lincoln.
Cary Institute for Ecosystem Studies. 2016. Available at: caryinstitute.org/discover-ecology/definition-ecology (accessed 29 March 2016).
Clements, F. 1916. *Plant Succession; An analysis of the development of vegetation*. Carnegie Institution of Washington.
Daily, G. 1997. Nature's Services: Societal Dependence on Natural Ecosystems. Washington, DC: Island Press.
Diaz-Martin, Z., Swamy, V., Terborgh, J., Alvarez-Loayza, P. and Cornejo, F. 2014. Identifying keystone plant resources in an Amazonian forest using a long-term fruit-fall record. *J Trop Ecol* 30:291–301.
Edmondson, W. and Lehman, J. 1981. The effect of changes in the nutrient income on the condition of Lake Washington. *Limnol Oceanogr* 26:1–29.
Egerton, F.N. (ed.). 1977. *History of American Ecology*. New York: Arno Press.
Elton, C.S. 1927. *Animal Ecology*. London, UK: Sidgwick and Jackson.
Franklin, J.F. 2000. Threads of continuity. *Conserv Biol* 1:8–16.
Gleason, H.A. 1939. The individualistic concept of the plant association. *The American Midland Naturalist* 21:92–110.
Golley, F. 1993. *A History of the Ecosystem Concept: More than the sum of the parts*. Yale University Press, New Haven.

Haeckel, E.H.P.A. 1866. Generelle Morphologie der Organismen. Allgemeine Grundzüge der organischen Formen-Wissenschaft, mechanische Begründet durch die von Charles Darwin reformirte Descendenz-Theorie. Berlin, Germany: Georg Reimer.

Harmon, M.E. 1992. Long-term experiments on log decomposition at the H. J. Andrews Experimental Forest, Oregon. USDA Forest Serv. Gen. Tech. Rep. PNW-GTR-280. 28 p.

Hitzhusen, G.E. 2006. Religion and environmental education: Building on common ground. *Can J of Env Education* 11:9–25.

Hitzhusen, G.E., and Tucker, M.E. 2013. The potential of religion for Earth Stewardship. *Front Ecol Environ* 11:368–76.

Holden, Z., Klene, A., Keefec, R.F. and Moisend, G. 2013. Design and evaluation of an inexpensive radiation shield for monitoring surface air temperatures. *Agri For Meteor* 180:281–6.

Hubbell, S. 2001. *Stephen P. Hubbell: The Unified Neutral Theory of Biodiversity and Biogeography*. Princeton University Press.

Nadkarni, N. 2004. Not preaching to the choir: communicating the importance of forest conservation to nontraditional audiences. *Conserv Biol* 18:602–6.

Paine, R.T. 1966. Food web complexity and species diversity. *The American Naturalist* 100:65–75.

Rosenzweig, M. 2003. Reconciliation ecology and the future of species diversity. *Oryx* 37:194–205.

Rowinsky, E., Onetto, N., Canetta, R.M and Arbuck, S.G. 1992. Taxol: the first of the taxanes, an important new class of antitumor agents. *Semin Oncol* 19:646–62.

Simard, M., Pinto, N., Fisher, J. B. and Baccini, A. 2011. Mapping forest canopy height globally with spaceborne lidar, *J Geophys Res* 116:G04021.

Stauffer, R.C. 1957. Haeckel, Darwin, and ecology. *Quart Rev Biol* 32:138–44.

Wiens, J. 1989. Spatial Scaling in Ecology. *Funct Ecol* 3:385–97.

Wilson, E.O. 2006. *The Creation: an appeal to save life on Earth*. New York: W.W. Norton & Company.

Worster, D. 1994. *Nature's Economy: A History of Ecological Ideas*. Cambridge, UK: Cambridge University Press.

INDEX

420

constructive theology 73–4
consumerism 317–18
consumption 4, 287, 316; calls for restraint
319–20; consumerism and 317–18; desire and
318–19; dignified labour 320–1; economic
system as a whole 321–2; endless rising levels
of 92; excessive 90; existential problem
316–17; luxury and 317; mass 380; minimal
35; systemic perspectives 320–2; time, rest and
the Sabbath 321
continuing revelation 99
contraception 104n11, 306, 310, 313; cost-free
access 313; in the United States of America
(USA) 311
Cooking (Bose) 375
Cook, Katsi 146
Copernicus 263
coral bleaching 258
coral reefs 254, 263
corn 161
corn agriculture 250
Cornwall Alliance 239
cosmology(ies) 242; Maori 130
cosmos 286, 348
cosmovision(s) 24; definition 107, 108; Pacific
130; and the Sun 160; *see also* indigenous
cosmovisions
Cotter, Holland 374–5
Cox, Paul 99
creation 13; human role 98; Maori philosophy
130–1; Mormon doctrines 97, 98–9, 99; *see
also* Anishinaabeg
Creation Myths of the World (Leeming) 121
creative thinking 365–6
critical thinking 365
Cry of the Earth, Cry of the Poor (Boff) 195–6
cultivation 55
cultural energy 356
cultural knowledge 124
cultural revolutionaries 349
culture: religion and 350; Western 350
culture wars 313–14

Daedalus 11
Dahlan-Taylor, Magfirah 83
Dahl, Arthur 94
Dalai Lama 43, 241
Dao 185
Daoism: mountains and 184–5; rivers and 185;
wuwei 322
Darlington, Sue 48
Darwin, Charles 263, 296, 297, 368, 412
Davies, Lincoln 104n12
Day, Sharon 340
DDT 254
dead zones 259, 277
d'Eaubonne, Françoise 327

de Botton, Alain 245
decentred selfhood 269–70
deep ecology 368
Deep Ecology (Grim) 110
deeper moralization 126
Deep Forest (Carr) 378
democracy: global ecological 334
deontology 52
dependent origination, doctrine of 49
Derrida, Jacques 67–8, 296
Dersu Uzala (Kurosawa) 379–80
Desai, Narayan 176
Descartes, René 365
Descent of Man, The (Darwin) 368
desertification 305
desire 318–19
developing countries: environmental protection
measures 385
Dev, Guru Arjun 178–9
Devi, Sunayani 375
Dewey, John 368, 390
dharma megha samadhi 174
Dharma Teachers International Collaborative
and Climate Change 43
diagnostic medicine 251
diaspora 64; *see also* African diaspora
dignified labour 320–1
disease 259
distributional politics 338
diversity 333
divine command 62
divine covenants 62
divinity 37, 38
division of labour, sexual 330
domination 74; systems of 327–8
Dorst, Jean 384
'double,' the 123
Dowie, Mark 340
Down to Earth 36
dream of progress 4
droughts 110
Druids 214
dualism 4–5
dualistic thinking 328
Duchrow, Ulrich 116
Dujiangyan irrigation system 185
Dulce Et Decorum Est (poem) 359–60
Dungy, Camille T. 202
Duran, Eduardo 142, 145
Durkheim, Emile 402
Dust Bowl 406

Earth Charter 111
Earthdiver(s) 141, 145
Earth First! 223, 224, 225; Twin Cities of
Minnesota campaign 340
earth justice 204

earthkeeping 75–6
Earth Liberation Front (ELF) 224
Earth Ministry 11
Earth Stewardship Initiative 345
earth, the: treatment by humans according to
  Judaism 61–2
East Malaysia 126
ecclesiology 74
Eckel, David 45
Eco Ashram 177
eco-Buddhists 43–4
ecocritical animal studies 361
ecocriticism 60, 355–7, 359; information
  processing and 361; waves of 360–1
Ecocriticism Reader, The 355, 356
Ecodefense: A Field Guide to Monkeywrenching
  (Foreman and Haywood) 223–4
ecofeminism 74–5, 327, 328, 329, 368
eco-halal 83
eco-Judaism 64–6
eco-justice 342
eco-Kosher 65
ecological agriculture 278
Ecological Buddhism 43
ecological democracy 334
ecological economics 346
Ecological Economy of Tai Tokerau Forestry 136
ecological elementals 122
ecological happiness 322
ecological justice 203
ecological knowledge 30, 124–7
ecological restoration *see* restoration
Ecological Society of America 345, 418; Earth
  Stewardship Initiative 6
ecologies of religion 29
ecology 28–31, 216, 412; collaborative fields
  417; complexity of composition 414;
  complexity of human effects 415; complexity
  of space 413–14; complexity of time 414;
  dynamics 414; history 412–13; humans and
  416–17; keystone species 415–16; legacy
  structures 415; PPP paradigm 415
Ecology Age 222
Ecology and Justice Series 10
ecology and religion *see* religion and ecology
Econarratology 361
economic growth 261–2
economic justice 74
Economic Man 410
economic reform: Bahá'í teachings 92
economics 402; economism 402–3, 408–9,
  409–10; ethical principles 406–7; Great
  Depression 406; Libertarian Christianity
  movement 407; religion and economy as
  ordering systems 403–4; replacing the market
  god 408–10; science and, challenging
  religious authority 404–6

economism 402–3, 408–9, 409–10
Economy of Mana 135–6
eco-pesantrens 84
eco-phenomenology 67
ecopsychology 221
ecosophy 368
eco-spiritual art 380–1
ecosystems: degradation 3, 35; novel 267–9
ecosystem services 416–17; and biodiversity
  251–2
ecotheology 71–2; fragmentation 73
eco-theophany 123
ecotones 229
ecowomanism 201, 203–5
Eden Village 65
Edmondson, Robert 416
education 384; agreement between humanity
  and the environment 388–90
Edwards, Jonathan 395
egoism 54
Elat Chayyim Center 65
Elder Nash, Marcus B. 97, 104n11
elephants 255
Eliade, Mircea 121, 123, 286
Elkins, James 374
Elliott, Robert 269
Ellis, Fiona 230
Elton, Charles 413
Ely, Richard, T. 405–6
emerald ash borers 253
empathy 54, 56, 57
Encyclopaedia of Religion and Nature (Taylor) 10,
  221
Endangered Species Act (ESA) 395, 399–400
Endangered Species Conservation Act (1969)
  399
Endangered Species Preservation Act 399
enduring mystery, nature 233–6
energy 348; cheap subsidies 89; renewable and
  sustainable 89; stored 356
enhancing life 74
environmental activism 28–9, 39; Jewish 64–5,
  66–8; Latin American anthropocentrism 195
environmental degradation 278, 305, 385
environmental education 388–90
environmental ethics 52, 367–8; and Buddhism
  47–8
environmental hazards 337
environmental history 347–54
environmental humanities 8–9, 295, 333, 345;
  development of 346; Judaism 66–8; religious
  studies in 22–4
Environmental Humanities 23
environmental imagination 126
Environmental Impact Statements (EIS) 398
environmental justice (EJ) 202, 336, 342, 389;
  climate change 337; debates in EJ studies 338;

Printed in the United States
by Baker & Taylor Publisher Services